AF412622

PROCEEDINGS OF SYMPOSIA
IN PURE MATHEMATICS
Volume 39, Part 1

THE MATHEMATICAL HERITAGE
of
HENRI POINCARÉ

AMERICAN MATHEMATICAL SOCIETY
PROVIDENCE, RHODE ISLAND

PROCEEDINGS OF SYMPOSIA IN PURE MATHEMATICS
OF THE AMERICAN MATHEMATICAL SOCIETY
VOLUME 39

PROCEEDINGS OF THE SYMPOSIUM
ON THE MATHEMATICAL HERITAGE OF HENRI POINCARÉ
HELD AT INDIANA UNIVERSITY
BLOOMINGTON, INDIANA
APRIL 7–10, 1980

EDITED BY
FELIX E. BROWDER

Prepared by the American Mathematical Society
with partial support from National Science Foundation grant MCS 79-22916

1980 *Mathematics Subject Classification.* Primary 01-XX, 14-XX, 22-XX, 30-XX, 32-XX, 34-XX, 35-XX, 47-XX, 53-XX, 55-XX, 57-XX, 58-XX, 70-XX, 76-XX, 83-XX.

Library of Congress Cataloging in Publication Data
Main entry under title:

The Mathematical Heritage of Henri Poincaré.

    (Proceedings of symposia in pure mathematics; v. 39, pt. 1–)
    Bibliography: p.
    1. Mathematics–Congresses. 2. Poincaré, Henri, 1854–1912–Congresses.
I. Browder, Felix E. II. Series: Proceedings of symposia in pure mathematics;
v. 39, pt. 1, etc.
QA1.M4266   1983   510   83-2774
ISBN 0-8218-1442-7 (set)   ISBN 0-8218-1449-4 (part 2)
ISBN 0-8218-1448-6 (part 1)   ISSN 0082-0717

Table of Contents

PART 1

PART 2

Section 5. Topological methods in nonlinear problems

Section 6. Mechanics and dynamical systems

Section 7. Ergodic theory and recurrence

Section 8. Historical material

Introduction

On April 7–10, 1980, the American Mathematical Society sponsored a week-long Symposium on the Mathematical Heritage of Henri Poincaré, held at Indiana University, Bloomington, Indiana. This volume presents the written versions of all but three of the invited talks presented at this Symposium (those by W. Browder, A. Jaffe, and J. Mather were not written up for publication). In addition, it contains two papers by invited speakers who were not able to attend, S. S. Chern and L. Nirenberg. The Organizing Committee for the Symposium consisted of F. Browder (Chairman), W. Browder, P. Griffiths, J. Moser, S. Smale, and R. O. Wells.

The casual reader may ask: What is the mathematical heritage of Henri Poincaré? How can it be described or delimited? In a certain sense, the essays presented here provide the best answer. To introduce them, let us try to answer the question in a summary form. During the period of his mathematical activity (which as the attached Bibliography of Poincaré's works indicates very sharply, was intense to a remarkable degree), Poincaré worked on a wide variety of mathematical topics stemming both from pure mathematics and from its applications. A central feature of his work was the close relation between his massive involvement in the research activity of his time in celestial mechanics and all the different varieties of physics, both theoretical and experimental, and the very deep and original insights that Poincaré developed in areas today classified under core mathematics. Poincaré made contributions of the most fundamental kind to the study of Riemann surfaces and of discontinuous groups, algebraic geometry, analytic functions of several complex variables, and non-Euclidean geometry. He was for all practical purposes the founder of many major fields of contemporary mathematics, including dynamical systems, algebraic topology, differential topology, ergodic theory, and the study of nonlinear problems using the ideas of topology. As a recent history of functional analysis by Dieudonné testifies, he can also be considered as a major seminal figure in that field as well as in the study of the general theory of partial differential equations.

As Poincaré himself described it, he was a 'pragmatist' in mathematics, both in his practice and in his theoretical self-conception. In the middle of the twentieth century, his pragmatist attitudes toward mathematical practice often were unfashionable in an environment where mathematical abstraction and an emphasis on formal elaboration of mathematical doctrine were a central concern. In recent decades, the tide has turned decisively toward mathematical creativity, as opposed to an emphasis upon rigorization and formalization. Today, even the classical figures of the old Bourbaki claim Poincaré (along with Elie Cartan) as their major precursor. (See the article by Dieudonné, *The work of Bourbaki during the last thirty years* [Notices Amer. Math. Soc. **29** (1982), 618–623].) Poincaré's concept of mathematics stresses intuition (geometric and analytical), creativity, and a strong emphasis upon a major relation of mathematics with the natural sciences.

The contents of this volume speak to this heritage. We regret very much the lack of a contribution by Jurgen Moser, who along with V. I. Arnold, represents in the sharpest and highest form the heritage of Poincaré in the direction of celestial mechanics, a field in which many of Poincaré's most original mathematical inventions were rooted. There are other gaps that one might have wished to fill (asymptotic methods in applied mathematics, or bifurcation theory, for example). One could well produce another volume to supplement the present one, with much more attention to the impact of Poincaré's works and ideas on the development of theoretical physics, or the impact of his views and writings on the foundational controversies of the early part of the twentieth century. In any case, we have before us a very substantial (if not complete) development of some of the most important aspects of the Poincaré tradition as described above, in some of the most active and vital areas of contemporary mathematical research.

Let me close with a remark that needs to be made publicly with respect to the appropriateness of this entreprise as an activity of the American Mathematical Society. If one traces the influence of Poincaré through the major mathematical figures of the early and mid-twentieth century, it is through American mathematicians as well as French that this influence flows, through G. D. Birkhoff, Solomon Lefschetz, and Marston Morse. This continuing tradition represents one of the major strands of American as well as world mathematics, and it is as a testimony to this tradition as an opening to the future creativity of mathematics that this volume is dedicated.

Felix E. Browder

Summary Chronology of the Life of Henri Poincaré

Born: 29 April 1854, in Nancy, France.

Educated in Nancy: (His teacher in Speciale, Elliot à Liard wrote in 1872 to a friend, "J'ai dans ma classe à Nancy, un monstre de mathématiques, c'est Henri Poincaré".)

First mathematical paper: in Nouvelles Annales des Mathématiques, 1873.

Entered: École Polytechnique, Paris, 1873.
Entered: École des Mines, Paris, 1875.
Doctorat d'État: 1879.
Appointed: Maitre des Conferences d'Analyse in Paris, 1881.
        Maitre des Conferences, Mathematical physics, 1885.
        Chaire de Physique mathématique et Calcul des probabilités at the University of Paris, 1886.
        Chaire d'Astronomie mathématique et Mechanique Celeste, in Paris, 1896.
Elected: Membre de la Section de Géometrie de l'Academie des Sciences, 1887. President de l'Academie, 1906.
Elected: to l'Academie Francaise, 1908.

Died: in Paris, July 17, 1912.

Poincaré

Section 1
GEOMETRY

Proceedings of Symposia in Pure Mathematics
Volume **39** (1983), Part 1

Web Geometry

SHIING-SHEN CHERN[1]

Introduction. Poincaré published two papers on surfaces of translation [10, 11].[2] They were among his lesser known papers. In the following pages I wish to show that the subject he touched is an exciting one and deserves further investigation.

1. Lie's theorem on surfaces of double translation and its developments. A surface M of translation in R^3 is defined by the parametric equations

$$(1) \qquad x^\lambda = f^\lambda(u) + g^\lambda(v), \qquad 1 \leqslant \lambda \leqslant 3,$$

where x^λ are the coordinates in R^3 and f^λ, g^λ are arbitrary smooth functions. It is immediately seen that the tangent lines to the u-curves (respectively the v curves) are independent of v (resp. u) and define a curve C_u (resp. C_v) in the plane at infinity.

M is called a *surface of double translation* if it is a surface of translation in a second way, i.e., given also by the equations

$$(2) \qquad x^\lambda = h^\lambda(s) + k^\lambda(t), \qquad 1 \leqslant \lambda \leqslant 3,$$

such that exactly two of the equations

$$(3) \qquad f^\lambda(u) + g^\lambda(v) - h^\lambda(s) - k^\lambda(t) = 0, \qquad 1 \leqslant \lambda \leqslant 3,$$

are independent. In 1882 Sophus Lie proved the remarkable theorem [7]:

If M is a surface of double translation in R^3, the four curves C_u, C_v, C_s, C_t in the plane at infinity defined by the tangent lines to the four families of parametric curves belong to the same algebraic curve of degree four.

The theorem means that the solutions of the functional equations (3) on a surface arise from an algebraic structure. Lie's proof makes use of the integrability conditions of over-determined systems of partial differential equations. In fact, from (1) we have

$$(4) \qquad \partial^2 x^\lambda / \partial u \partial v = 0,$$

which means that the parametric curves form a conjugate net, i.e., their tangent directions separate harmonically the asymptotic directions at each point. If the surface M is given in the nonparametric form

$$(5) \qquad z = z(x, y),$$

<hr>

Reprinted from Bulletin Amer. Math. Soc. (N.S.) 6 (1982), 1–8.

1980 *Mathematics Subject Classification.* Primary 53A60, 14D25.

[1]Work done under partial support of NSF grant MC577-23579.

[2]Numbers in brackets refer to the Bibliography at the end of the paper.

this condition is expressed by an equation

$$(6) \qquad R(p, q)r + Q(p, q)s + T(p, q)t = 0$$

where

$$(7) \qquad p = z_x, \quad q = z_y, \quad r = z_{xx}, \quad s = z_{xy}, \quad t = z_{yy}.$$

A surface M of double translation satisfies, besides (6), another equation

$$(6') \qquad R'(p, q)r + Q'(p, q)s + T'(p, q)t = 0.$$

An investigation of the integrability conditions of the system of equations (6), (6') involves long and tedious calculations. In particular, the fourth-order partial derivatives of z come into play. The work was a true tour de force, but Lie reached his goal.

Poincaré was quick to recognize the importance of Lie's work, and to observe its relation with the theory of abelian functions. In [10, 11] he gave two proofs of Lie's theorem based on abelian functions and algebraic geometry rather than on partial differential equations.[3] Although the proofs are perhaps not complete, Poincaré introduced fresh ideas and new viewpoints. As a consequence of Poincaré's work it follows that a surface of double translation can be defined by equating a theta function to zero. As a result the surface

$$(8) \qquad x_1 x_2 x_3 = a_1 x_1 + a_2 x_2 + a_3 x_3,$$

where the a's are constants, is a surface of double translation. The best proof of Lie's theorem was given by Darboux, using the theory of residues [5].

It should be remarked that Lie started his program on surfaces of translation through his work on minimal surfaces. It was known to Monge that an analytic minimal surface is a surface of translation (1) with parametric curves which are minimal or isotropic curves.

Lie proceeded to study the high-dimensional case. A translation manifold in R^{n+1} is the hypersurface defined by the parametric equations

$$(9) \qquad x^\lambda = \sum_{1 \leqslant \alpha \leqslant n} f^\lambda(u_\alpha), \qquad 1 \leqslant \lambda \leqslant n + 1,$$

where x^λ are the coordinates in R^{n+1} and the f's are smooth functions in the respective variables. Lie tried to determine all hypersurfaces of double translation and settled the case $n = 3$ in a long paper [8]. In the same paper he promised to return to the general case. He had several posthumous papers on the subject, without bringing the problem to a satisfactory conclusion [9]. The high-dimensional case was also considered by Poincaré. It was W. Wirtinger who in 1938 completely solved the problem, using Chow coordinates for projective varieties [14]. We will show below that web geometry offers a broader setting where the subject can be integrated.

---

[3]Lie was unhappy with Poincaré's intrusion: He said: "Unfortunately the distinguished author (i.e., Poincaré), whose achievements in other fields were recognized by nobody more than myself, failed to understand my investigations. I can only say that his works on translation surfaces and translation manifolds deal with results which are entirely special cases of my general theorems." (Ges. Abh, Bd II, Teil II, 527)

The conclusion that the functional equations imply an algebraic structure is a powerful one. This was utilized by B. St. Donat in 1975 to give a new proof of Torelli's theorem that a compact Riemann surface is determined up to isomorphism by its period matrix (or more exactly, by its polarized Jacobian variety) [12].

2. Web geometry of Blaschke-Bol [2]. Web geometry had its debut in 1926–27 on the beaches of Italy when W. Blaschke and G. Thomsen realized that the configuration of three foliations of the plane by curves, has local invariants. The distinguished geometric figure in this case is the hexagon (Figure 1). When all such hexagons are closed, for any point O and any neighboring point P on the first curve through O, the web is called *hexagonal*. Thomsen proved that a hexagonal web is locally homeomorphic to three families of parallel lines.

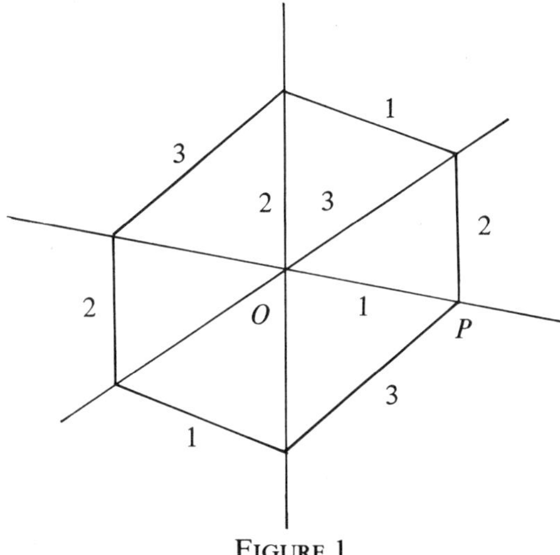

FIGURE 1

The relation of this subject to algebraic geometry is immediate. Before Blaschke and Thomsen started their work on web geometry, Graf and Sauer proved in 1924 a theorem which, in web geometry language, can be stated as follows: *If the curves of a hexagonal web are straight lines, they are the tangent lines of an algebraic curve of class three.*

Generally, a d-web in the plane is defined by d foliations in a neighborhood of the plane by curves such that through every point the d leaves have mutually distinct tangents. If x, y are the coordinates in the plane, a d-web can be defined by

$$(10) \qquad u_i(x, y) = \text{const}, \qquad 1 \leqslant i \leqslant d.$$

We shall assume that the functions $u_i(x, y)$ are smooth and satisfying the condition grad $u_i \neq 0$.

An equation of the form

$$(11) \qquad \sum_{1 \leqslant i \leqslant d} f_i(u_i) \, du_i = 0$$

is called an *abelian equation*. The maximum number of linearly independent abelian equations, is called the rank of the web, to be denoted by π. It can be proved that

$$(12) \qquad \pi \leqslant \tfrac{1}{2}(d - 1)(d - 2).$$

We see easily that for $d = 3$, the rank one webs are exactly the hexagonal webs. For $d = 4$, we have $\pi \leqslant 3$, and the 4-webs of rank 3 satisfy 3 linearly independent abelian equations

$$(13) \qquad \sum f_i^\lambda(u_i)\, du_i = 0, \qquad 1 \leqslant \lambda \leqslant 3.$$

By setting

(14)

$$x^\lambda = \int f_1^\lambda(u_1)\, du_1 + \int f_2^\lambda(u_2) = -\int f_3^\lambda(u_3)\, du_3 - \int f_4^\lambda(u_4)\, du_4, \qquad 1 \leqslant \lambda \leqslant 3,$$

we get in R^3 a surface of double translation. Lie's theorem can be interpreted as saying that a 4-web of rank 3, is locally equivalent to one consisting of straight lines; the latter must then be the tangent lines of a curve of class 4.

The deep question is whether any d-web of maximum rank $(d - 1)(d - 2)/2$ is locally equivalent to one all whose leaves are straight lines. Bol gave an example showing that this is not always the case for a 5-web of rank 6. Bol's example consists of four pencils of lines, no three of whose vertices are collinear, with the fifth family consisting of the conics through the four vertices. Of the six abelian equations one involves the transcendental function which is Euler's dilogarithm. The latter has played a role in several recent mathematical investigations, such as the volume of a simplex in an odd-dimensional noneuclidean space, the Pontrjagin number of a combinatorial 4-dimensional manifold and, more generally, in algebraic K-theory. A primary reason must be the fact that it satisfies an abelian equation. For a recent work cf. [**15**].

3. High-dimensional webs. A d-web of dimension $N - k$ in R^N consists of d foliations of a neighborhood U of R^N by submanifolds of dimension $N - k$; k is called the codimension of the web. As an example consider an algebraic variety V of dimension k and degree d in a projective space P^m of dimension m. A linear subspace P^{m-k} of dimension $m - k$ meets V in d points through each of which pass $\infty^{k(m-k)} P^{m-k}$'s. This gives in the Grassmann manifold $G(m - k, m)$ of all P^{m-k}'s in P^m d foliations of dimension $k(m - k)$. Since

$$\dim G(m - k, m) = k(m - k + 1),$$

the variety V gives rise to d foliations of codimension k in $G(m - k, m)$, which is locally R^{kn}, $n = m - k + 1$. In order to keep algebraic geometry in sight, we will consider on the basis of this example, only webs of codimension k in R^N, with $N = kn$. Even so the subject is a wide generalization of the geometry of projective algebraic varieties. Just as intrinsic algebraic varieties are generalized to Kähler manifolds and complex manifolds, such a generalization to web geometry seems justifiable.

The tangent spaces to the d leaves through a point $x \in R^{kn}$ give d linear subspaces of codimension k in the tangent space T_x to R^{kn} at x or, what is the same, d linear subspaces Ω_i of dimension k in the cotangent space T_x^*. We will suppose that they are in general position. For $k = 1$ the meaning of this is clear: no $kn = n$ of the d lines in T_x^* lie in a hyperplane of T_x^*. For $k > 1$ the right notion has to be introduced. We gave one in [4], again based on the example of projective varieties.

Analytically suppose the ith foliation of the web be defined by the equations

(15)

$$u_{i1}(x) = \text{const}, \qquad , \ldots , \qquad u_{ik}(x) = \text{const}, \qquad 1 \leqslant i \leqslant d, \quad x \in R^{kn}.$$

We will suppose the u's to be smooth functions satisfying

(16)
$$\Omega_i(x) = du_{i1} \wedge \cdots \wedge du_{ik} \neq 0.$$

For a fixed i the functions $u_{i1}, \ldots, u_{ik}$ are defined up to a diffeomorphism and $\Omega_i(x)$ is defined up to a factor. The notation is so chosen that it defines the linear space $\Omega_i \subset T_x^*$ introduced above; we will call it the ith web normal.

An *abelian equation* is an equation of the form

(17)
$$\sum_{1 \leqslant i \leqslant d} f_i(u_{i1}, \ldots, u_{ik})\Omega_i = 0.$$

The maximum number of the linearly independent abelian equations is called the rank of the web. In [4] we proved that the rank has an upper bound $\pi(d, n, k)$, which depends only on d, n, k. This bound is sharp. In particular, for $k = 1$ we have

(18)
$$\pi(d, n) \underset{\text{def}}{=} \pi(d, n, 1) = (1/2(n - 1))\{(d - 1)(d - n) + s(n - s - 1)\}$$

where s is defined by

(19)
$$s \equiv -d + 1 \mod n - 1, \qquad 0 \leqslant s \leqslant n - 2.$$

This number $\pi(d, n)$ has an important meaning in algebraic geometry. In fact, Castelnuovo proved that $\pi(d, n)$ is the maximum genus of an algebraic curve of degree d in P^n, which does not belong to a hyperplane P^{n-1}. By taking such a curve C^* in the dual space P^{*n}, we have through each point of P^n d hyperplanes belonging to C^* and it follows from Abel's theorem that the d-web so constructed is of rank $\pi(d, n)$. In relating web geometry to algebraic geometry it should be remarked that our web geometry is over real numbers while the corresponding notions of algebraic geometry refer to the complex field. The transition is not immediate, but it is possible owing mainly to the fact that we are dealing with local properties in web geometry.

It is clear that an important problem is to determine the d-webs of codimension 1 and maximum rank $\pi(d, n)$ in R^n. If the leaves are all hyperplanes, the answer is given by the following converse to Abel's [6] theorem (generalization of theorems of Graf-Sauer and Lie):

Consider a d-web of codimension 1 *in a neighborhood of* R^n *whose leaves are hyperplanes such that an abelian equation*

$$\text{(20)} \qquad \sum_{1 \leqslant i \leqslant d} f_i(u_i) \, du_i = 0$$

holds, with $f_i(u_i) \neq 0$. *Then the leaves belong to an algebraic curve of degree d in the dual projective space.*

For this theorem it is sufficient to have one abelian equation. The crucial question is thus the linearization problem: Is a d-web of codimension one and rank $\pi(d, n)$ in R^n linearizable, i.e., is it locally equivalent to one having hyperplanes as leaves? Bol's example in §2 shows that the answer is no for $n = 2, d = 5$.

Consider next the case $n \geqslant 3$. For $n + 1 \leqslant d \leqslant 2n - 1$ we have $\pi(d, n) = d - n$, and there are simple examples of nonlinearizable d-webs of rank $d - n$, which depend on arbitrary functions. Lie's case of hypersurfaces of double translation corresponds to $d = 2n$. In this case we have $\pi(2n, n) = n + 1$. For a $2n$-web of rank $n + 1$ the abelian equations

$$\text{(21)} \qquad \sum f_i^\lambda(u_i) \, du_i = 0, \qquad 1 \leqslant \lambda \leqslant \pi = n + 1,$$

can be written

(22)

$$\sum \int_{1 \leqslant i \leqslant n} f_i^\lambda(u_i) \, du_i = -\sum \int_{n+1 \leqslant i \leqslant 2n} f_i^\lambda(u_i) \, du_i, \qquad 1 \leqslant \lambda \leqslant n + 1.$$

As a generalization of (14) these common expressions can be regarded as the coordinates in R^{n+1} of a hypersurface of double translation.

An essential step in the proof of the Lie-Wirtinger theorem on hypersurfaces of double translation is to show that a $2n$-web in R^n of codimension one and rank $n + 1$ is linearizable, from which the theorem follows from the converse to Abel's theorem. In this particular case the linearization follows by a simple argument, using an idea of Poincaré. It goes as follows: For $x \in U \subset R^n$ let $Z_i(x)$ be the point in a projective space of dimension $\pi - 1$, whose homogeneous coordinates are $[f_i^1(u_i), \ldots, f_i^\pi(u_i)]$. The mapping which sends $x \in U \subset R^n$ to the space $\{Z_1, \ldots, Z_d\} \subset P^{\pi-1}$ spanned by the Z_i's is called *Poincaré's mapping*. If $u_1, \ldots, u_n$ are regarded as a local coordinate system in U, we have, from (21),

$$f_1^\lambda(u_1) + \sum_{n+1 \leqslant i \leqslant d} f_i^\lambda(u_i) \frac{\partial u_i}{\partial u_1} = 0, \qquad 1 \leqslant \lambda \leqslant \pi.$$

It follows that Z_1 is a linear combination of $Z_{n+1}, \ldots, Z_d$. Since a similar equation holds for any Z_i (instead of Z_1), we have

$$\text{(23)} \qquad \{Z_1, \ldots, Z_d\} = d - n - 1.$$

In the case $d = 2n, \pi = n + 1$, the Poincaré mapping maps the points of U into hyperplanes of P^n, such that the ith leaf goes to the point Z_i. The Poincaré mapping followed by duality in P^n maps the points of U into the points of the dual space P^{*n} such that all the leaves of the web go to hyperplanes. This proves the linearization, and the Lie-Wirtinger theorem follows.

For $n = 3$ Bol proved the remarkable theorem: *In R^3 a d-web of codimension 1 and rank $\pi(d, 3)$, $d \geqslant 6$, is linearizable.*

Griffiths and I tried to generalize Bol's theorem to R^n [3]. So far we have only succeeded to prove the theorem under the additional condition of normality. The exact statement is: *In R^n a normal d-web of codimension 1 and rank $\pi(d, n)$, $d \geqslant 2n$, is linearizable.* For the definition of normality we refer to [3].

In recent years works on web geometry have also been done by M. A. Akivis and V. Goldberg in the Soviet Union. It is not clear to me how much these works are related to those of Poincaré. The reader is referred to [1] for further information.

Tschebotarow generalized Lie's idea to the study of surfaces which admit imprimitive systems relative to an arbitrary Lie group (instead of the group of translations) [13]. To my knowledge this generalization has not been further pursued.

4. Unsolved problems. In the following I wish to list a few most immediate unsolved problems:

1. Determine all d-webs of curves in the plane having maximum rank $\frac{1}{2}(d - 1)(d - 2)$, $d \geqslant 5$.

2. Is the above linearization theorem in R^n, $n \geqslant 4$, true without normality?

3 (Griffiths'). The hexagonal condition is meaningful for a 3-web of dimension k in R^{2k}. The construction described in the beginning of §3, with $m = k + 1$, so that $n = 2$, defines from a cubic hypersurface in P^{k+1} a 3-web of dimension k in R^{2k}. It can be shown that such a web is hexagonal. Is the converse true, i.e., does every hexagonal 3-web of dimension k in R^{2k} arise from such a construction (up to a local diffeomorphism)?

ADDED DURING PROOF. I am indebted to V. Goldberg that the answer to this question is "no".

4. Lie's treatment makes heavy use of overdetermined systems of partial differential equations. Give a proof of the Lie-Wirtinger theorem by PDE.

REFERENCES

1. M. A. Akivis, *Webs and almost Grassmann structures*, Soviet Math. Dokl. **21** (1980), no. 3. Further references to works by V. Goldberg and other Soviet mathematicians on the subject can be found in this paper.

2. W. Blaschke and G. Bol, *Geometrie der Gewebe*, Springer, Berlin, 1938.

3. S. S. Chern and P. A. Griffiths, *Abel's theorem and webs*, Jahresberichte der deut. Math. Ver. **80** (1978), 13–110; also, Corrections and Addenda, same Journal, **83** (1981), 78–83.

4. ______, *An inequality for the rank of a web and webs of maximum rank*, Ann. Scuola Norm. Pisa, Serie IV, **5** (1978), 539–557.

5. G. Darboux, *Théorie des surfaces*, t.1, (1914), 151–161.

6. P. A. Griffiths, *Variations on a theorem of Abel*, Invent. Math. **35** (1976), 321–390.

7. S. Lie, *Bestimmung aller Flächen, die in mehrfacher Weise durch Translationsbewegung einer Kurve erzeugt werden*, Arch. für Math. Bd. 7, Heft 2 (1882), 155–176; also Ges. Abhandlungen, Bd. 1, Abt 1, 450–467.

8. ______, *Das Abelsche Theorem und die Translationsmannigfaltigkeiten*, Leipziger Berichte 1897, pp. 181–248; also Ges. Abhandlungen, Bd. II, Teil II, paper XIV, pp. 580–639.

9. ______, Ges. Abhandlungen, Bd. 7.

10. H. Poincaré, *Fonctions abeliennes*, J. Math. Pures Appl., Serie 5, **t.1**, (1895), 219–314; also Oeuvres **t.IV**, pp. 384–472, in particular p. 430.

11. ______, *Sur les surfaces de translation et les fonctions abeliennes*, Bull. Soc. Math. France **29** (1901), 61–86; also Oeuvres **t.VI**, pp. 13–37.

12. B. Saint-Donat, *Variétés de translation et théorème de Torelli*, Comptes Rendus Paris **280** (1975), 1611–1612.

13. N. Tschebotarow, *Über Flächen welche Imprimitivitätssysteme in Bezug auf eine gegebene Kontinuirliche Transformationsgruppe enthalten*, Recueil Math. (Sbornik) **34** (1927), 149–204.

14. W. Wirtinger, *Lies Translationsmannigfaltigkeiten und Abelsche Integrale*, Monatshefte Math. Phys. **46** (1938), 384–431.

15. David B.. Damiano, *Webs, abelian equations, and characteristic classes*, Thesis, Brown University, 1980.

16. John B. Little, *Translation manifolds and the converse of Abel's theorem*, Thesis, Yale University, 1980.

DEPARTMENT OF MATHEMATICS, UNIVERSITY OF CALIFORNIA, BERKELEY, CALIFORNIA 94720

Proceedings of Symposia in Pure Mathematics
Volume 39 (1983), Part 1

Problems on Abelian Functions
at the Time of Poincaré and Some at Present[1]

JUN-ICHI IGUSA

This is an expanded version of our symposium lecture; it consists of two parts. In the first part we have tried to explain the problems on abelian functions at the time of Poincaré with a brief follow-up; in the second part we have explained, among others, a problem of Riemann and Weil on Jacobi's formula as one of the problems on abelian functions at present.

1. Abelian functions by Poincaré.

1-1. If the variable x and a general solution y of a linear differential equation with polynomial coefficients are algebraically dependent, the periods of abelian integrals of the first kind associated with the curve $f(x, y) = 0$ satisfy certain relations. In his earliest works on abelian functions Poincaré examined such relations in some special cases. He also used a similar relation in a joint paper with Picard of 1883 on a "theorem of Riemann". Poincaré later developed a general theory of reducible integrals. This theory played some role in almost all of his works on abelian functions. We shall start by recalling the theorem of Riemann:

There are three related theorems concerning a complex torus. If f is a meromorphic function on $\mathbf{C}^g$, an element a of $\mathbf{C}^g$ such that $f(z + a) = f(z)$ for a variable z in $\mathbf{C}^g$ is called a period of f; the set of all periods of f forms a closed subgroup of $\mathbf{C}^g$, called the period group of f. Let Λ denote a lattice in $\mathbf{C}^g$, i.e., a discrete subgroup of $\mathbf{C}^g$ with compact quotient; then a meromorphic function f on $\mathbf{C}^g$ whose period group contains Λ, i.e., a meromorphic function on the complex torus $T = \mathbf{C}^g/\Lambda$ considered as a function on $\mathbf{C}^g$, is called an *abelian function* relative to Λ and a holomorphic function Θ on $\mathbf{C}^g$ with the property

$$\Theta(z + a) = \mathbf{e}(L_a(z))\Theta(z)$$

for every a in Λ, in which $\mathbf{e}(t)$ stands for $\exp(2\pi \sqrt{-1}\, t)$ and $L_a(z)$ is an affine linear function of z depending on a, is called a *theta function* also relative to Λ.

Finally a complex g-by-$2g$ matrix ω is called a *Riemann matrix* of degree g if there exists a skew-symmetric integral matrix C of degree $2g$, called a *principal matrix*, such that

$$\omega C'\omega = 0, \qquad (1/2\sqrt{-1}\,)\omega C'\bar{\omega} > 0;$$

Reprinted from Bulletin Amer. Math. Soc. (N.S.) 6 (1982), 161–174.

1980 *Mathematics Subject Classification.* Primary 14K20; Secondary 14K25.
[1] This work was partially supported by the National Science Foundation.

the second condition means that the hermitian matrix $(1/2\sqrt{-1}\,)\omega C'\bar{\omega}$ is positive-definite. Such a matrix with

$$C_0 = \begin{pmatrix} 0 & 1_g \\ -1_g & 0 \end{pmatrix}$$

as a principal matrix appeared in Riemann's paper [**28**] of 1857 as a period matrix of abelian integrals of the first kind associated with an algebraic curve of genus g. At any rate with this terminology the three theorems, called the *Riemann-Weierstrass theorems*, can be stated as follows:

(RW-1) *Any $g + 1$ abelian functions are algebraically dependent*;

(RW-2) *every abelian function f can be expressed as a quotient of two theta functions*;

(RW-3) *if there exists an abelian function relative to Λ with a discrete period group, any g-by-2g matrix ω whose 2g columns generate Λ is a Riemann matrix.*

We recall that a divisor of a complex manifold is an integral linear combination of its subvarieties of codimension 1; it is called positive if the coefficients are nonnegative. The zeros and poles, counted with their multiplicities, of a meromorphic function $f \neq 0$ on the manifold are positive divisors; and their difference is called the divisor of f. The divisors of an abelian function $f \neq 0$ and a theta function $\Theta \neq 0$ may be considered as divisors of T. With this terminology (RW-2) can be replaced by the following theorem:

(RW-2$^\sharp$) *Every positive divisor of T is the divisor of a theta function $\Theta \neq 0$.*

As a consequence, if $f \neq 0$ in (RW-2), the two theta functions can be chosen so that their divisors have no component in common. On the other hand the converse of (RW-3) is true and straightforward; in a paper of 1921 Lefschetz [**16**] proved the following theorem:

If ω is a Riemann matrix of degree g and Λ is the subgroup of $\mathbf{C}^g$ generated by its 2g columns, necessarily a lattice in $\mathbf{C}^g$, there exists a finite set of theta functions giving rise to a biholomorphic map of $T = \mathbf{C}^g/\Lambda$ to a subvariety A of a complex projective space.

And in 1949 Chow [**5**] proved the following theorem:

Every closed subvariety of a complex projective space is an algebraic variety and a meromorphic map between two such varieties is a rational map.

In particular the subvariety A in the Lefschetz theorem is a smooth algebraic variety unique up to an isomorphism and it is called the *abelian variety* associated with ω. If ω is a special Riemann matrix coming from an algebraic curve, the abelian variety A is called the *jacobian variety* of the curve.

1-2. We shall now go back to 1883: it was known that Riemann had mentioned (RW-3) in a slightly different form to Hermite at the time of his visit to Paris in 1860. Hermite regarded the period relation in the general case as extremely remarkable and included it without proof in his note [10] to the sixth edition of Lacroix's text book, vol. 2, which appeared two years later. On the other hand, in 1869 Weierstrass published a short paper [36], in which (under the assumption in (RW-3)) he made the following statement: there exist $g + 1$ algebraically dependent abelian functions by which every abelian function, all relative to Λ, can be rationally expressed. Actually he made a stronger statement and he later outlined a proof of the above statement in a letter to Borchardt of 1879. We might mention that Weierstrass was not sure about (RW-2) in 1869 and became sure about it only in 1879; we might further mention that his detailed but not complete treatment [37] was published in 1903 after his death. Therefore although Riemann's paper of 1857 certainly was a classic in 1883, the theory of general abelian functions was still up in the air at that time.

The short paper [20] by Poincaré and Picard was written under the above circumstances. In that paper they announced an outline of a proof of (RW-3), depending on (RW-1) by Weierstrass, and stated (RW-2) as its immediate consequence; there we see the fruitful idea to investigate T via Riemann's theory applied to a suitable 1-dimensional subvariety of T. Later in 1891 Appell [2] gave a satisfactory proof of (RW-2) for $g = 2$ by using Poincaré's theorem of 1883 in the Acta Mathematica to the following effect: every meromorphic function on $\mathbf{C}^2$ can be expressed as a quotient of two holomorphic functions on $\mathbf{C}^2$. This theorem was generalized by Cousin in his thesis of 1895 and Appell's proof became valid for all g.

In 1897 Poincaré announced outlines of proofs of (RW-1) and (RW-2) in [24]; a detailed proof for (RW-2) appeared the following year in [25] and one for (RW-1) appeared five years later in [26]. The above second proof of (RW-2) by Poincaré was influenced by Appell; he simply modified the proof of his theorem in the Acta paper of 1883, on which Appell's proof depended, so that for an abelian function the two entire functions became theta functions. This proof was reproduced by Weil [40] in a Bourbaki seminar of 1949; Weil's proof was indeed a new proof in as much as the classical potential theory was eliminated. As for (RW-1) Poincaré tried to show that the "image" of T by $g + 1$ abelian functions was algebraic; his idea to use the degrees of subvarieties of a complex projective space was interesting but the proof was not quite satisfactory. It was developed on a sound basis in the already mentioned paper by Chow of 1949. And in 1955 Siegel [35] gave a complete and yet elementary proof to the following general theorem:

Any $g + 1$ meromorphic functions on a g-dimensional compact complex manifold are algebraically dependent; if there exist g algebraically independent meromorphic functions, there exists another meromorphic function so that every meromorphic function on the manifold can be rationally expressed by those $g + 1$ functions.

We must also mention Serre's influential works on this and other related theorems; as an example we refer to his Bourbaki talk [32] of 1954, which preceded his FAC and GAGA, for an elegant proof of Chow's theorem.

1-3. We shall pass to reducible integrals: as we have recalled, Poincaré first worked on special examples; then he learned general theorems of Weierstrass through a paper by Kowalevski [14] of 1884. And he wrote [21] in the same year to supply proofs to those Weierstrass theorems and also to generalize them; two years later he wrote another paper [22] on reducible integrals with supplements and applications. In the following we shall review Poincaré's main theorem in its original form; but first we shall recall some definitions.

We say that two Riemann matrices ω_1, ω_2 of the same degree g are *equivalent* if there exists a pair (λ, L) of a complex invertible matrix λ of degree g and an integral matrix L of degree $2g$ satisfying $\lambda\omega_1 = \omega_2^l L$. If we pass to the corresponding abelian varieties A_1, A_2, this means the existence of a surjective homomorphism from A_1 to A_2 necessarily with a finite kernel. At any rate we have an equivalence relation in the set of all Riemann matrices of degree g.

We recall that every nonsingular skew-symmetric integral matrix C of degree $2g$ can be written as $C = {}^l L C_0 L$ for some integral matrix L of degree $2g$; in particular if we define the Pfaffian $\mathrm{Pf}(C)$ of C as $|\det(L)|$, we get $\det(C) = \mathrm{Pf}(C)^2$. If ω is a Riemann matrix with C as a principal matrix, necessarily C is nonsingular and in the above notation ω is equivalent to $\omega^l L$; the point is that $\omega^l L$ has C_0 as a principal matrix. Therefore every equivalence class of Riemann matrices contains one with C_0 as a principal matrix.

If a Riemann matrix ω has C_0 as a principal matrix, its right square submatrix λ is invertible and if we write ω as $\omega = \lambda(\tau 1_g)$, we will have

$$^l\tau = \tau, \qquad \mathrm{Im}(\tau) > 0;$$

the second condition means that the imaginary part of τ is positive-definite. The set $\mathbf{S}_g$ of all such τ's forms an open convex cone in the vector space of symmetric complex matrices of degree g; and in honor of his fundamental works, especially [33, 34], it is called the *Siegel upper-half space* of degree g. If M is a real matrix of degree $2g$ satisfying ${}^l M C_0 M = C_0$ and if we write

$$M = \begin{pmatrix} a & b \\ c & d \end{pmatrix}$$

with submatrices a, b, c, d of degree g, then

$$M \cdot \tau = (a\tau + b)(c\tau + d)^{-1}$$

defines a holomorphic action of the real symplectic group $Sp_{2g}(\mathbf{R})$ on $\mathbf{S}_g$. Poincaré's formulation of his "complete reducibility theorem" involves the action of $Sp_{2g}(\mathbf{Q})$ on $\mathbf{S}_g$ and it is as follows:

(P-1) *Let ω denote a Riemann matrix of degree g with C_0 as a principal matrix such that for some complex g'-by-$2g'$ matrix ω' and for some integral $2g$-by-$2g'$ matrix R, where $1 \leqslant g' < g$, the upper g'-by-$2g$ submatrix of ω can be written as $\omega'\,{}^l R$; then there exist an element M of $Sp_{2g}(\mathbf{Q})$ and a point τ' of $\mathbf{S}_{g'}$ such that $(\tau' 1_{g'})$ is equivalent to the necessarily Riemann matrix ω' and*

$$M \cdot \tau = \begin{pmatrix} \tau' & 0 \\ 0 & \tau'' \end{pmatrix},$$

in which $\omega = \lambda(\tau 1_g)$.

It follows from (P-1) that the equivalence class of an arbitrary Riemann matrix contains one of the form

$$\begin{bmatrix} \omega' & & & \\ & \omega'' & & \\ & & \ddots & \\ & & & \omega^{(r)} \end{bmatrix},$$

in which ω', ω'', $\ldots$, $\omega^{(r)}$ are irreducible or pure Riemann matrices. However Poincaré was not aware of the uniqueness of the classes of those component matrices; this was later observed by Scorza; cf. Albert [1].

An outline of Poincaré's proof of (P-1) in the current terminology is as follows: we regard elements of $\mathbf{Q}^{2g}$ as column vectors and convert $\mathbf{Q}^{2g}$ into a symplectic space over $\mathbf{Q}$ via the skew-symmetric bilinear form $B(x, y) = {}^{t}xC_0y$; then the columns of R span a $2g'$-dimensional nondegenerate, hence symplectic, subspace. Therefore we can find a symplectic basis $x_1, \ldots, x_g$, $y_1, \ldots, y_g$ for $\mathbf{Q}^{2g}$ such that $x_1, \ldots, x_{g'}, y_1, \ldots, y_{g'}$ form a symplectic basis for that subspace. If we define M as the inverse of the square matrix of degree $2g$ with $x_1, \ldots, x_g,\ y_1, \ldots, y_g$ as its columns, then M has the required property.

Actually Poincaré considered $\mathbf{Z}^{2g}$ as a symplectic module over $\mathbf{Z}$ via the same $B(x, y)$ and tried to find a suitable $\mathbf{Z}$-basis for the submodule generated by the columns of R; and, e.g., in the simplest case where $g' = 1$ he showed the existence of M in $Sp_{2g}(\mathbf{Z})$ such that the first row of $M \cdot \tau$ took the form

$$\left(\tau' \, k^{-1}\!\mu \, 0 \ldots 0\right),$$

in which $k = \mathrm{Pf}({}^{t}RC_0R)$ and $0 \leqslant \mu < k$. This is equivalent to a theorem of Weierstrass; cf. [14, p. 400].

At any rate, as an immediate consequence of (P-I), Poincaré observed that the $Sp_{2g}(\mathbf{Q})$-orbit of $\sqrt{-1}\,1_g$ was dense in $\mathbf{S}_g$; cf. [22, p. 340]. He regarded this fact as a key to solve various problems: "C'est là une circonstance qui donnera, je n'en doute pas, la clef de bien des problèmes".

1-4. We shall explain how Poincaré applied the above observation to a problem on theta functions; we shall first recall a theorem of Frobenius: let Θ denote a theta function relative to a lattice Λ in $\mathbf{C}^g$ and denote the linear part $L_a(z) - L_a(0)$ of $L_a(z)$ by $\lambda_a(z)$; then the difference $B(a, b) = \lambda_a(b) - \lambda_b(a)$ is an integer for every a, b in Λ. Therefore if we choose a $\mathbf{Z}$-basis $a_1, \ldots, a_{2g}$ for Λ, we get a skew-symmetric integral matrix E of degree $2g$ with $B(a_i, a_j)$ as its (i, j)th coefficient. And if E is nonsingular, then $\omega = (a_1 \ldots a_{2g})$ becomes a Riemann matrix with E^{-1}, multiplied by a suitable positive integer, as a principal matrix. Furthermore the set of all theta functions relative to Λ with the same $\mathbf{e}(L_a(z))$ forms a vector space over $\mathbf{C}$ and, according to Frobenius [8], its dimension is equal to $\mathrm{Pf}(E)$.

Poincaré consistently restricted his attention to the case where $E = -kC_0$ for some positive integer k; in that case the dimension is k^g. A theta function with $E = -kC_0$ is called a theta function of *order* k; for such theta functions we may assume that Λ is the lattice generated by the $2g$ columns of $(\tau 1_g)$, in which τ is in $\mathbf{S}_g$. In a paper [19] of 1883 and later in [22] already mentioned, Poincaré proved the following theorems:

(P-2) *Let* $\Theta_1, \ldots, \Theta_g$ *denote g theta functions of orders* $k_1, \ldots, k_g$; *then the number N of common zeros, each counted with its multiplicity, of*

$$\Theta_1(z - b_1), \ldots, \Theta_g(z - b_g)$$

in $T = \mathbf{C}^g / \Lambda$ *is equal to* $k_1 \ldots k_g g!$ *provided that it is finite;*

(P-3) *let* $\zeta_1, \ldots, \zeta_N$ *denote representatives in* $\mathbf{C}^g$ *of the common zeros; then*

$$\sum_{i=1}^{N} \zeta_i \equiv k_1 \ldots k_g (g - 1)! \sum_{j=1}^{g} b_j + c \quad \mod \Lambda,$$

in which c is independent of $b_1, \ldots, b_g$.

The above theorems were later generalized by Wirtinger [41]. Although they are now well understood, Poincaré's proofs are still interesting: he observed in (P-2) that the number N was a continuous function of τ and the parameters in $L_a(z)$; hence N was a constant. In order to determine this constant he specialized τ to a diagonal matrix and thus reduced the general case to the elliptic case where $g = 1$.

The proof of (P-3) is more involved: by using a generalization of Abel's theorem he showed that the sum depended continuously and only on τ and the parameters in $L_a(z)$ once $b_1, \ldots, b_g$ were fixed. By using the transformation theory he then showed that the formula to be proved was invariant under the action of $Sp_{2g}(\mathbf{Q})$. On the other hand the $Sp_{2g}(\mathbf{Q})$-orbit of $\sqrt{-1}\,1_g$ was already shown to be dense in $\mathbf{S}_g$. Therefore it was enough to verify the formula again in the case where τ was a diagonal matrix.

1-5. We have reviewed Poincaré's idea to use $(\mathbf{S}_1)^g = \mathbf{S}_1 \times \cdots \times \mathbf{S}_1$ embedded in $\mathbf{S}_g$; he also used an infinitesimal neighborhood of $(\mathbf{S}_1)^g$ in his paper [23] of 1895. In that paper he examined the condition for τ to be a "jacobian point", i.e., a point of $\mathbf{S}_g$ coming from an algebraic curve of genus g. This problem was treated by Schottky [31] seven years earlier in the first nontrivial case where $g = 4$. Poincaré's method is entirely different from Schottky's and it is as follows:

First of all any theta function of order 1 can be converted into

$$\theta(z) = \sum_{p \in \mathbf{Z}^g} \mathbf{e}\left(\left(\frac{1}{2}\right)^t p\tau p + {}^t pz\right);$$

the conversion consists of an affine linear transformation in $\mathbf{C}^g$ and of a multiplication by a "trivial theta functions", i.e., the exponential of a polynomial of degree 2 in the coefficients of z. We shall denote by τ_{ij} and z_i the (i, j)th coefficient and the ith coefficient of τ and z, respectively. If the off-diagonal or lateral coefficients τ_{ij}, $i \neq j$, of τ are all zero, we will have

$$\theta(z) = \theta_1(z_1) \ldots \theta_g(z_g),$$

in which

$$\theta_i(u) = \sum_{p=-\infty}^{\infty} \mathbf{e}\left(\left(\frac{1}{2}\right)\tau_{ii}p^2 + pu\right)$$

for $1 \leqslant i \leqslant g$.

A hypersurface in $\mathbf{C}^g$ is called a *translation surface* if at every point of the hypersurface its neighborhood admits a parametric representation of the form

$$z = \sum_{i=1}^{g-1} \phi_i(t_i),$$

in which $\phi_1, \ldots, \phi_{g-1}$ are holomorphic functions of one variable. If τ is a jacobian point, the hypersurface in $\mathbf{C}^g$ defined by $\theta(z) = 0$ is a translation surface; this follows from a theorem of Riemann. Poincaré took a point τ of $\mathbf{S}_g$ with very small lateral coefficients and examined the condition that $\theta(z) = 0$ defined a translation surface in a very small neighborhood of $\zeta = (\zeta_1, \ldots, \zeta_g)$ where $\theta_i(\zeta_i) = 0$ for $1 \leqslant i \leqslant g$. And in the case where $g = 4$ he proved the following theorem:

(P-4) *Put* $a_{ij} = 2\pi\sqrt{-1}\ \tau_{ij}$, $r_1 = a_{12}a_{13}a_{42}a_{43}$, $r_2 = a_{13}a_{14}a_{23}a_{24}$, $r_3 = a_{14}a_{12}a_{34}a_{32}$, *and*

$$R = r_1^2 + r_2^2 + r_3^2 - 2(r_2r_3 + r_3r_1 + r_1r_2);$$

then $R(\tau)$ *becomes a higher order infinitesimal if* τ *is a jacobian point with very small lateral coefficients.*

Actually Poincaré wrote down his condition in an irrational form, but the two conditions are equivalent. Schottky, on the other hand, found a modular form J of weight 8 for the full modular group $Sp_8(\mathbf{Z})$ such that $J(\tau) = 0$ at every jacobian point τ. If we expand J at a variable point τ in which the lateral coefficients are very small, we get

$$J = 2^{-16} \cdot \prod_{i=1}^{4} (\theta_{00}\theta_{01}\theta_{10})_i(0)^8 \cdot R + \cdots;$$

the $\theta_{ij}(u)$ for $i, j = 0, 1$ are the standard elliptic theta functions and the subscript i on the right-hand side indicates that the modulus is τ_{ii}. The relation of the conditions found by Poincaré and Schottky was discussed by Rauch [27] with some comments about their proofs.

1-6. Poincaré's works on abelian functions proper ended by his paper of 1902. However a consequence of his later papers of 1910—1911 on normal functions has to be mentioned. As we have recalled, if V is an algebraic curve of genus g, the complex torus associated with its period matrix is biholomorphic to an abelian variety J, called the jacobian variety of V. It follows from Abel's theorem that J can be obtained also as follows: we take the group of divisors of V each with the property that the sum of its coefficients is 0. This group contains the group of divisors of rational functions on V, different from the constant 0, and the quotient group with a natural structure of an algebraic variety becomes an abelian variety isomorphic to J.

If we start from an algebraic surface as in Poincaré or more generally from any smooth algebraic variety V, then the period matrix of simple integrals of the first kind on V can be obtained by reducing the period matrix associated with a general curve on V and, in particular, it is a Riemann matrix; thus we get an abelian variety A which generalizes J. Another aspect of J can also be generalized. We take the group of divisors of V each homologous to 0 as an integral cycle. This group contains the group of divisors of rational functions on V, different from the constant 0, and the quotient group again becomes an abelian variety, say P. Poincaré's theory of normal functions implies the existence of a surjective homomorphism from A to P with a finite kernel, i.e., the Riemann matrices of A and P are equivalent. This was really an important achievement in the theory of algebraic varieties at that time. In the general case, however, the abelian varieties A and P, called the *Albanese* and the *Picard varieties* of V, respectively, are not isomorphic; and the exact relation between A and P had been left in the shadows until our thesis [11] of 1952.

We shall conclude the first part with the following comments: first of all in his paper of 1938 Weil [38] introduced "hyperabelian functions"; this theory has not yet been fully explored. On the other hand if we pass to abelian varieties, the theory of abelian functions loses its transcendental character. And in fact a purely algebraic theory of abelian functions over an algebraically closed field of arbitrary characteristic was created by Weil [39] in 1948. Also a purely algebraic theory of theta functions was developed during 1966 —1967 by Mumford [17]. We might mention that their works contain generalizations and new results even in the complex case. Finally we mention that Mumford's lecture note [18] and Freitag's report [7] will give a good coverage of the results and problems up until 1976.

2. A problem of Riemann and Weil.

2-1. We shall start by recalling some definitions: let Θ denote a theta function of order 1; then, after a linear transformation in $\mathbf{C}^g$ and a multiplication by a trivial theta function, it becomes

$$\theta_m(\tau, z) = \sum_{p \in \mathbf{Z}^g} e\left(\left(\frac{1}{2}\right)^t \left(p + \frac{1}{2}m'\right)\tau\left(p + \frac{1}{2}m'\right) + {}^t\left(p + \frac{1}{2}m'\right)\left(z + \frac{1}{2}m''\right)\right),$$

in which m is the column vector with m', m'' in $\mathbf{R}^g$ as its first and the second entry vectors. If $m = 0$, we get back to the theta function $\theta(z)$ in 1-5. We shall assume that m has integral coefficients; then we have

$$\theta_m(\tau, -z) = e(m)\theta_m(\tau, z), \qquad e(m) = (-1)^{{}^t m' m''}.$$

Therefore $\theta_m(\tau, z)$ is even or odd according as the "characteristic" m is even or odd in the sense that $e(m) = \pm 1$. In view of $\theta_{m+2n}(\tau, z) = \pm \theta_m(\tau, z)$ for every n in $\mathbf{Z}^{2g}$ we may assume, if necessary, that the coefficients of m are 0, 1. The number of theta functions then becomes 2^{2g}; of these $2^{g-1}(2^g + 1)$ are even and the remaining $2^{g-1}(2^g - 1)$ are odd. If m is even, we put $\theta_m(\tau) = \theta_m(\tau, 0)$; the holomorphic function θ_m on $\mathbf{S}_g$ so defined is called a *Thetanullwert*. We shall denote by $\mathbf{C}[\theta]$ the ring generated over $\mathbf{C}$ by the Thetanullwerte.

We shall change our notation in Part 1 and denote an element of $Sp_{2g}(\mathbf{R})$ by σ instead of M; however we shall denote its submatrices by a, b, c, d as before. If l is a positive integer, we define a normal subgroup $\Gamma_g(l)$ of $Sp_{2g}(\mathbf{Z})$ by the condition $\sigma \equiv 1_{2g} \bmod l$. We further define a subgroup $\Gamma_g(l, 2l)$ of $\Gamma_g(l)$ by the additional condition

$$\operatorname{diag}(a\,'b) \equiv \operatorname{diag}(c\,'d) \equiv 0 \quad \bmod 2l;$$

we have denoted by "diag" the conversion of a square matrix into a column vector by its diagonal coefficients. If l is even, then $\Gamma_g(l, 2l)$ is normal in $\Gamma_g(1) = Sp_{2g}(\mathbf{Z})$.

The correspondence $(\sigma, \tau) \to \det(c\tau + d)$ defines a holomorphic automorphy factor of $Sp_{2g}(\mathbf{R})$; and if σ is in $\Gamma_g(4)$, then

$$\det(c\tau + d)^{1/2} = \theta(\sigma \cdot \tau)\theta(\tau)^{-1}$$

defines a holomorphic automorphy factor of $\Gamma_g(4)$. If we have a power $\rho(\sigma, \tau)$ of such an automorphy factor of a subgroup Γ of $Sp_{2g}(\mathbf{R})$, we can let Γ act on the vector space of all holomorphic functions on $\mathbf{S}_g$ as

$$(\sigma \cdot f)(\tau) = \rho(\sigma^{-1}, \tau)^{-1} f(\sigma^{-1} \cdot \tau).$$

If Γ is a modular group, i.e., a subgroup of $\Gamma_g(1)$ containing $\Gamma_g(l)$ for some l, any Γ-invariant function, which satisfies a regularity condition at "cusps" in the case $g = 1$, is called a *modular form* relative to Γ. If $\rho(\sigma, \tau)$ is the kth power of $\det(c\tau + d)$ for some nonnegative integer k, we say that the modular form f has *weight* k; and if $\rho(\sigma, \tau)$ is the kth power of $\det(c\tau + d)^{1/2}$ for some Γ contained in $\Gamma_g(4)$, we say that f has *degree* k. For instance every Thetanullwert is a modular form of degree 1 relative to $\Gamma_g(4, 8)$. We shall denote by $A(\Gamma)$ the ring consisting of finite sums of modular forms relative to Γ; then $A(\Gamma)$ forms a graded integral domain over $\mathbf{C}$ and we have the following theorem:

The graded ring $A(\Gamma_g(4, 8))$ is precisely the integral closure of $\mathbf{C}[\theta]$ within its field of fractions.

This theorem was proved in **[12]** by using the theory of compactifications by Baily **[3]** and Cartan **[4]**. The two rings in the theorem are the same for $g \leqslant 2$ but are different for $g \geqslant 3$; and yet the morphism

$$\operatorname{proj}\big(A(\Gamma_g(4, 8))\big) \to \operatorname{proj}(\mathbf{C}[\theta])$$

—the notation will be explained in a moment—is always bijective.

2-2. We shall recall the *main theorem* in the theory of compactifications: if $A = A_0 + A_1 + A_2 + \cdots$ is a finitely generated graded integral domain over $A_0 = \mathbf{C}$, the set of homogeneous maximal ideals of A other than $A_1 + A_2 + \cdots$ becomes a projective algebraic variety $\operatorname{proj}(A)$. The graded ring $A(\Gamma)$ for any modular group Γ of degree g is such a ring and it is integrally closed. Therefore $\operatorname{proj}(A(\Gamma))$ is defined and it is a normal variety. Furthermore the quotient variety $\Gamma \backslash \mathbf{S}_g$ is biholomorphic to a Zariski open subset of $\operatorname{proj}(A(\Gamma))$ such that the complement is a finite union of $\operatorname{proj}(A(\Gamma'))$ for some modular groups Γ' of degrees less than g. Since $\operatorname{proj}(A(\Gamma))$ for $\Gamma = \Gamma_g(1)$ was first introduced by Satake **[30]** as a Hausdorff space, it is sometimes called the *Satake compactification* of $\Gamma \setminus \mathbf{S}_g$. At any rate the main theorem and other standardized theorems imply that for $g \geqslant 2$ every meromorphic function on $\Gamma \setminus \mathbf{S}_g$ can be expressed as a quotient of two modular forms and in particular every holomorphic function on $\Gamma \setminus \mathbf{S}_g$ is a constant.

We shall explain three problems; the first one is called the *Schottky problem*:

Find an explicit description of a finite basis for the homogeneous ideal of $A(\Gamma_g(1))$ generated by those modular forms which vanish at every jacobian point.

Since the ideal is $\{0\}$ for $g \leqslant 3$, the first nontrivial case is $g = 4$; even in that case the problem is not quite settled because no proof has yet been published for the fact that the Schottky invariant J generates the ideal. The second problem is related to Poincaré [21, 22] and also to Kowalevski [14].

Find an explicit description of a finite basis for the homogeneous ideal of $A(\Gamma_g(1))$ generated by those modular forms which vanish at every reducible point.

The second problem seems much easier than the first problem; and yet answers have been given only for $g \leqslant 3$. The third problem is a refinement of the main theorem and it is as follows:

Denote by $A_{\mathbf{Z}}(\Gamma_g(1))$ the ring consisting of finite sums of modular forms relative to $\Gamma_g(1)$ with integral Fourier coefficients: show that $A_{\mathbf{Z}}(\Gamma_g(1))$ is finitely generated over $\mathbf{Z}$.

This problem has to be completed as follows: show also that $\text{proj}(A_{\mathbf{Z}}(\Gamma_g(1)))$ gives the "scheme of moduli" of all "principally polarized abelian varieties" of dimensions at most equal to g. If the problem is settled in the so-completed form, then that will imply the irreducibility of the variety of moduli over an algebraically closed field of arbitrary characteristic. At any rate the third problem has been settled only for $g \leqslant 2$; in those two cases the structure of $A_{\mathbf{Z}}(\Gamma_g(1))$ is also known.

2-3. We shall now explain a problem of Riemann and Weil; first we shall introduce a notation. Let $m_1, \ldots, m_g$ denote g odd characteristics and put

$$D(M)(\tau) = \pi^{-g}\big(\partial(\theta_{m_1}, \ldots, \theta_{m_g})/\partial(z_1, \ldots, z_g)\big)(\tau, 0),$$

in which M stands for the $2g$-by-g matrix $(m_1 \ldots m_g)$; this is the *Nullwert of the jacobian* of $\theta_{m_1}(\tau, z), \ldots, \theta_{m_g}(\tau, z)$ with the normalizing factor π^{-g} and it is a holomorphic function on $\mathbf{S}_g$. In a letter dated April 14, 1976 Weil proposed the following problem:

(W-1) *Is $D(M)$ always a polynomial in the Thetanullwerte with integral rational coefficients?*

At that time he needed an affirmative answer to the following weaker problem:

(W-2) *Can $D(M)$ be expressed as a quotient of two such polynomials?*

In the same letter he referred to Frobenius [9] and a quotation by M. Noether from Riemann's manuscripts [29, p. 66] for the background of his problems. Noether found in "Nr. 25, Bogen 6" expressions for $D(M)$ as "Summen von Produkten" of $g + 2$ Thetanullwerte for $g = 3, 4, \ldots, 7$; this was quoted later in Krazer and Wirtinger [15, p. 772]. Frobenius, on the other hand, gave a survey of earlier results, unaware of Riemann's manuscripts, and proved such formulas for $g \leqslant 4$ with a final detail for $g = 4$ left as a conjecture; this was nineteen years after Riemann's death. Actually for $g = 3$, 4 Frobenius proved the formulas under certain conditions on $m_1, \ldots, m_g$.

Our first result on the above problems was that (W-2) had an affirmative answer; an outline of the proof is as follows: if we denote by $\mathbf{Q}[\theta]$ the ring generated over $\mathbf{Q}$ by the Thetanullwerte, then $\mathbf{Q}[\theta]$ is a $\mathbf{Q}$-structure on $\mathbf{C}[\theta]$. Let $\mathbf{Q}[\theta]^*$ denote the integral closure of $\mathbf{Q}[\theta]$; then a theorem in algebraic geometry tells us that the $\mathbf{C}$-span of $\mathbf{Q}[\theta]^*$ is the integral closure of $\mathbf{C}[\theta]$, which by our theorem in 2-1 coincides with $A(\Gamma_g(4, 8))$; and $D(M)$ is a homogeneous element of $A(\Gamma_g(4, 8))$ of degree $g + 2$. Finally $D(M)$ is in $\mathbf{Q}[\theta]^*$ because its Fourier coefficients are all integers.

2-4. We shall next explain our theorem in [13]; we shall use the following terminology and notation: a sequence of $2g + 2$ characteristics $m_1, \ldots, m_g, n_1, \ldots, n_{g+2}$ is called a *special fundamental system* if $m_1, \ldots, m_g$ are odd, $n_1, \ldots, n_{g+2}$ are even, and for every a, b, c in the sequence

$$e(a)e(b)e(c)e(a + b + c) = -1$$

holds. We put $M = (m_1 \ldots m_g)$ as before and $N = (n_1 \ldots n_{g+2})$; we also put

$$P(N) = \theta_{n_1} \ldots \theta_{n_{g+2}}.$$

Finally for any integral domain R and any positive integer r we shall denote by $\mathbf{O}_r(R)$ the orthogonal group of $x_1^2 + \cdots + x_r^2$ over R and by G_r, Π_r the subgroups of $\mathbf{O}_r(\mathbf{F}_2)$ obtained from $\mathbf{O}_r(\mathbf{Z}_2)$, $\mathbf{O}_r(\mathbf{Z})$, respectively, under the reduction mod 2; we have denoted by $\mathbf{Z}_2$ the ring of 2-adic integers and by $\mathbf{F}_2$ the field with two elements. Then our theorem can be stated as follows:

(J-1) *If $D(M)$ is different from the constant 0 and is a polynomial in the Thetanullwerte, then for some N the $2g + 2$ columns of (MN) form a special fundamental system and the formula*

$$D(M) = \sum_{V \in G_{g+2}/\Pi_{g+2}} \pm P(NV)$$

holds; the expression on the right-hand side is unique.

Since the formula is unique, all known formulas have to be that formula; and by Riemann it has to be correct up to $g = 7$. We tried, therefore, to remove the ugly "if" in our theorem; but we were able to prove the formula only up to $g = 5$ by the method of Frobenius with a consequence of the theory of compactifications to give a finishing touch.

In a paper of 1979, which we completely overlooked, Fay [6] also proved the formula in (J-1) up to $g = 5$. Furthermore by using some properties of theta functions with a hyperelliptic modulus he discovered the following remarkable fact:

(F) *The formula in (J-1) does not hold for $g = 6$.*

Fay's finding was a relative impossibility. However if we put (J-1) and (F) together, we get the following absolute impossibility: for $g = 6$ no $D(M)$ other than the constant 0 is a polynomial in the Thetanullwerte. Therefore (W-1) has a negative answer and the quotations of Riemann by Noether and by Krazer and Wirtinger are all wrong! We have tried to remedy this unfortunate situation and found a right formulation of (J-1); and it is as follows:

(J-1$^\sharp$) *Let X and Y denote the* C-*spans of all $D(M)$'s and $P(N)$'s, respectively; then the dimension of $X \cap Y$ is either 0 or equal to* $\mathrm{card}(Sp_{2g}(\mathbf{F}_2))/\mathrm{card}(G_g \times G_{g+2})$. *Furthermore in the second case it has a* C-*basis consisting of elements of the form*

$$\sum_{U \in G_g/\Pi_g} \pm D(MU) = \sum_{V \in G_{g+2}/\Pi_{g+2}} \pm P(NV),$$

in which the expressions on both sides are unique up to a simultaneous sign change.

This theorem can be proved in the same way as (*J*-1); cf. [**13**, pp. 440–441]. The only difference is that we incorporate the results in the Appendix of that paper where we observed the "symmetry" in M and N of such formulas. At any rate (J-1$^\sharp$) implies (F): if $g = 6$ or more generally if $g \geqslant 6$, then $[G_g : \Pi_g] \geqslant 2$, hence by the uniqueness no $D(M)$ can be written as $\Sigma \pm P(NV)$.

2-5. By what we have explained we have $\dim(X \cap Y) \geqslant 1$ for $g \leqslant 5$ while $\dim(X \cap Y)$ is either 0 or

$$2^{18}3^45 \cdot 7 \cdot 11 \cdot 13 \cdot 17 \cdot 31 = 56,006,655,344,640$$

for $g = 6$; our *conjecture* is that $\dim(X \cap Y) \geqslant 1$ for all g. We have examined this conjecture from a representation-theoretic viewpoint by using Mackey's theorem. We can easily see that if the conjecture is true for a certain $g \geqslant 2$, it is true also for $g - 1$. Therefore we may assume that g is even; then we can let $\Gamma_g(1)$ act on X and Y via the $(\frac{1}{2}g + 1)$th power of $\det(c\tau + d)$ as an automorphy factor. We see that all $D(M)$'s and $P(N)$'s are eigenvectors for the representations of $\Gamma_g(2)$ in X and Y, hence they give rise to its characters, say ψ_M and ψ_N. The point is that we have $\psi_M = \psi_N$ if and only if the $2g + 2$ columns of (MN) form a special fundamental system. Furthermore if we denote such a character by ψ, all solutions of $\psi_{M'} = \psi$ can be written as $M' \equiv MU$ mod 2 with U mod 2 in G_g; similarly all solutions of $\psi_{N'} = \psi$ can be written as $N' \equiv NV$ mod 2 with V mod 2 in G_{g+2}. Finally if we denote by Γ the subgroup of $\Gamma_g(1)$ consisting of all σ in $\Gamma_g(1)$ satisfying

$$\psi(\sigma\sigma_0\sigma^{-1}) = \psi(\sigma_0)$$

for every σ_0 in $\Gamma_g(2)$, then $\Gamma/\Gamma_g(2)$ is isomorphic to $G_g \times G_{g+2}$; and we get the following theorem:

(J-2) *There exists a degree* 1 *character* χ *of* Γ *which extends* ψ; *the representations of* $\Gamma_g(1)$ *in X and Y share one and only one irreducible representation* ρ *with multiplicity one and* ρ *is the representation of* $\Gamma_g(1)$ *induced* *by χ. Furthermore the $[\Gamma_g(1): \Gamma]$ elements which appear on the left- and right-hand sides of the formula in $(J-1^\#)$, respectively, form $\mathbf{C}$-bases for the unique subspaces of X and Y where the representation ρ takes place.*

Our conjecture is, therefore, that the two representation spaces are the same, i.e., ρ appears only once in the representation of $\Gamma_g(1)$ in $X + Y$. This space is contained in the space, say $\mathbf{Z}$, of all modular forms of weight $\frac{1}{2}g + 1$ relative to $\Gamma_g(4, 8)$ and the conjecture will follow if ρ appears at most once, hence exactly once, in the representation of $\Gamma_g(1)$ in $\mathbf{Z}$.

In rounding off this article we shall explain an approach suggested to us by Mumford; as we understand it, the idea is as follows: in $D(M)$ and $P(N)$ we take the coefficients of M and N arbitrarily from $\mathbf{Z}[\frac{1}{2}]$; if we can find a conjectural formula involving such $D(M)$'s and $P(N)$'s which is preserved under the action of $Sp_{2g}(\mathbf{Z}[\frac{1}{2}])$, then the problem will be reduced to the elliptic case in view of the fact that the $Sp_{2g}(\mathbf{Z}[\frac{1}{2}])$-orbit of $\sqrt{-1}\, 1_g$ is dense in $\mathbf{S}_g$. We might mention that this is one of Poincaré's ideas which we have recalled in Part 1. It is very likely that this idea, rediscovered by Mumford, will eventually settle the problem.

References

1. A. A. Albert, *A note on the Poincaré theorem on impure Riemann matrices*, Ann. of Math. (2) **36** (1935), 151–156.

2. P. Appell, *Sur les fonctions périodiques de deux variables*, J. Math. **7** (1891), 157–219.

3. W. L. Baily, *Satake's compactification of V_n*, Amer. J. Math. **80** (1958), 348–364.

4. H. Cartan, *Fonctions automorphes*, Séminaire E.N.S. (1957/58).

5. W.-L. Chow, *On compact complex analytic varieties*, Amer. J. Math. **71** (1949), 893–914.

6. J. Fay, *On the Riemann-Jacobi formula*, Göttingen Nachrichten (1979), 61–73.

7. E. Freitag, *Siegelsche Modulfunktionen*, Jahresber. Deutsch. Math.-Verein. **79** (1977), 79–86.

8. F. G. Frobenius, *Über die Grundlagen der Theorie der Jacobischen Funktionen*, Crelles J. **97** (1884), 16–48 & 188–223; Werke II, 172–240.

9. ______, *Über die constanten Factoren der Thetareihen*, Crelles J. **98** (1885), 244–263; Werke II, 241–260.

10. C. Hermite, *Note sur la théorie des fonctions elliptiques* (1862); Oeurvres II, 125–238.

11. J. Igusa, *On the Picard varieties attached to algebraic varieties*, Amer. J. Math. **74** (1952), 1–22.

12. ______, *On the graded ring of theta constants*, Amer. J. Math. **86** (1964), 219–246.

13. ______, *On Jacobi's derivative formula and its generalizations*, Amer. J. Math. **102** (1980), 409–446.

14. S. Kowalevski, *Über die Reduktion einer bestimmten Klasse Abel'-scher Integrale 3ten Ranges auf elliptische Integrale*, Acta Math. **4** (1884), 393–414.

15. A. Krazer and W. Wirtinger, *Abelsche Funktionen und allgemeine Thetafunktionen*, Enzykl. Math. Wiss. II B 7 (1921), 604–873.

16. S. Lefschetz, *On certain numerical invariants of algebraic varieties with application to abelian varieties*, Trans. Amer. Math. Soc. **22** (1921), 327–482.

17. D. Mumford, *On the equations defining abelian varieties*. I-III, Invent. Math. **1** (1966), 287–354; **3** (1967), 75–135; 215–244.

18. ______, *Curves and their Jacobians*, Univ. Mich. Press, 1975.

19. H. Poincaré, *Sur les fonctions Θ*, Bull. Soc. Math. France **11** (1883), 129–134; Oeuvres IV, 302–306.

20. H. Poincaré and E. Picard, *Sur un théorème de Riemann relatif aux fonctions de n variables indépendentes admettant 2n systèmes de périodes*, C. R. **97** (1883), 1284–1287; Oeuvres IV, 307–310.

21. H. Poincaré, *Sur la réduction des intégrales abéliennes*, Bull. Soc. Math. France **12** (1884), 124–143; Oeuvres III, 333–351.

22. ______, *Sur les fonctions abéliennes*, Amer. Math. **8** (1886), 289–342; Oeuvres IV, 318–378.

23. ______, *Remarques diverses sur les fonctions abéliennes*, J. Math. **1** (1895), 219–314; Oeuvres IV, 384–468.

24. ______, *Sur les fonctions abéliennes*, C. R. **124** (1897), 1407–1411; Oeuvres IV, 469–472.

25. ______, *Sur les propriétés du potentiel et sur les fonctions abéliennes*, Acta Math. **22** (1898), 89–178; Oeuvres IV, 162–243.

26. ______, *Sur les fonctions abéliennes*, Acta Math. **26** (1902), 43–98; Oeuvres IV, 473–526.

27. H. E. Rauch, *Schottky implies Poincaré*, Advances in the Theory of Riemann Surfaces, Ann. of Math. Studies, no. 66, Princeton Univ. Press, Princeton, N. J., 1971, pp. 355–364.

28. B. Riemann, *Theorie der Abel'schen Functionen*, Crelles J. **54** (1857), 115–155; Werke, pp. 88–142.

29. ______, *Werke; Nachträge* by M. Noether and W. Wirtinger, 1902.

30. I. Satake, *On the compactification of the Siegel space*, J. Indian Math. Soc. **20** (1956), 259–281.

31. F. Schottky, *Zur Theorie der abelschen Functionen von vier Variabeln*, Crelles J. **102** (1888), 304–352.

32. J.-P. Serre, *Faisceaux analytiques*, Sém. Bourbaki **95** (1954).

33. C. L. Siegel, *Über die analytische Theorie der quadratischen Formen*, Ann. of Math. (2) **36** (1935), 527–606; Werke I, 326–405.

34. ______, *Einführung in die Theorie der Modulfunktionen n-ten Grades*, Math. Ann. **116** (1939), 617–657; Werke II, 97–137.

35. ______, *Meromorphe Funktionen auf kompakten analytischen Mannigfaltigkeiten*, Göttingen Nachrichten (1955), 71–77; Werke III, 216–222.

36. K. Weierstrass, *Über die allgemeinsten eindeutigen und 2n-fach periodischen Functionen von n Veränderlichen*, Monatsbericht (1869), 854–857; Werke II, 45–48.

37. ______, *Allgemeine Untersuchungen über 2n-fach periodische Functionen von n Veränderlichen*; Werke III, 53–114.

38. A. Weil, *Généralisation des fonctions abéliennes*, J. Math. **17** (1938), 47–87.

39. ______, *Variétés abéliennes et courbes algébriques*, Act. Sci. Ind. **1064** (1948).

40. ______, *Théorèmes fondamentaux de la théorie des fonctions thêta*, Sém. Bourbaki **16** (1949).

41. W. Wirtinger, *Zur Theorie der 2n-fach periodischen Funktionen*, Monatsh. Math. **7** (1896), 1–25.

DEPARTMENT OF MATHEMATICS, JOHNS HOPKINS UNIVERSITY, BALTIMORE, MARYLAND 21218

Proceedings of Symposia in Pure Mathematics
Volume **39** (1983), Part 1

Hyperbolic Geometry: The First 150 Years

JOHN MILNOR

This will be a description of a few highlights in the early history of non-euclidean geometry, and a few miscellaneous recent developments. An Appendix describes some explicit formulas concerning volume in hyperbolic 3-space.

The mathematical literature on non-euclidean geometry begins in 1829 with publications by N. Lobachevsky in an obscure Russian journal. The infant subject grew very rapidly. Lobachevsky was a fanatically hard worker, who progressed quickly from student to professor to rector at his university of Kazan, on the Volga.

Already in 1829, Lobachevsky showed that there is a natural unit of distance in non-euclidean geometry, which can be characterized as follows. In the right triangle of Figure 1 with fixed edge a, as the opposite vertex A moves infinitely far away, the angle θ will increase to a limit θ_0 which is assumed to be strictly less than $\pi/2$. He showed that

$$a = -\log \tan(\theta_0/2)$$

if the unit of distance is suitably chosen. In particular,

$$a \approx (\pi/2) - \theta_0$$

if a is very small. (In the interpretation introduced by Beltrami forty years later, this unit of distance is chosen so that curvature $\equiv -1$.)

FIGURE 1. A right triangle in hyperbolic space

By early 1830, Lobachevsky was testing his "imaginary geometry" as a possible model for the real world. If the universe is non-euclidean in Lobachevsky's sense, then he showed that our solar system must be extremely small, in terms of this natural unit of distance. More precisely, taking the vertex A in Figure 1 to be the star Sirius and taking the edge a to be a suitably chosen radius of the Earth's orbit, he used the (unfortunately incorrect) estimate

$$\pi - 2\theta \cong 1.24 \text{ seconds of arc} \cong 6 \times 10^{-6} \text{ radians}$$

Reprinted from Bulletin Amer. Math. Soc. (N.S.) 6 (1982), 9–24.

1980 *Mathematics Subject Classification.* Primary 01A55, 01A60, 51M10; Secondary 57R15, 20H10.

for the parallax of Sirius to conclude that the diameter $2a$ of Earth's orbit is less than 6×10^{-6}. (The correct parallax of $0.37''$ for Sirius would have given a sharper estimate. In fact, the first reliable measurements of parallax were made by Bessel several years later, in 1838.)

By late 1830, he was working out the volumes of pyramids, and other polyhedra in 3-dimensional non-euclidean space. Such computations are not easy, and can be quite interesting. (Compare the Appendix.)

A few years later, in 1832, J. Bolyai, a flamboyant officer in the Austro-Hungarian army, published an independent account of non-euclidean geometry. However, perhaps because he was discouraged by Gauss, he did not pursue the subject as far as Lobachevsky.

Although Lobachevsky and Bolyai were the first with the courage to publish, it should be noted that Gauss himself had been privately working on similar ideas for many years. In a letter to Taurinus in 1824 he wrote (roughly translated):[1]

"The assumption that the sum of the three angles [of a triangle] is smaller than 180° leads to a geometry which is quite different from our (euclidean) geometry, but which is in itself completely consistent. I have satisfactorily constructed this geometry for myself so that I can solve every problem, except for the determination of one constant, which cannot be ascertained a priori. The larger one chooses this constant, the closer one approximates euclidean geometry. . . . If non-euclidean geometry were the true geometry, and if this constant were comparable to distances which we can measure on earth or in the heavens, then it could be determined a posteriori. Hence I have some-times in jest expressed the wish that euclidean geometry is not true. For then we would have an absolute a priori unit of measurement."

His words to Bessel, in 1830, have an even more modern ring:

" . . . we must admit with humility that, while number is purely a product of our mind, space has a reality outside of our mind, so that we cannot completely prescribe its laws a priori."

For the first forty years or so of its history, the field of non-euclidean geometry existed in a kind of limbo, divorced from the rest of mathematics, and without any firm foundation. However, Gauss' theory of curved surfaces [1827], and Riemann's theory of higher dimensional curved manifolds [1868] provided a way of integrating non-euclidean geometry into more respectable branches of mathematics. In fact Riemann briefly described the theory of manifolds of constant curvature α, citing the metric

$$ds = \sqrt{dx_1^2 + \cdots + dx_n^2} \left/ \left(1 + \frac{\alpha}{4}\left(x_1^2 + \cdots + x_n^2\right)\right)\right.$$

as an example.

The turning point came in 1868, with the publication of two papers by E. Beltrami. In the first, Beltrami showed that two-dimensional non-euclidean geometry is nothing more nor less than the study of suitable surfaces of

---

[1]Gauss [1900, pp. 187, 201]. For other early workers in non-euclidean geometry, see for example Coxeter [1942].

constant negative curvature. He introduced the term *pseudosphere* of *radius R* for a surface of curvature $-1/R^2$. (In practice, he used the term "pseudosphere" only for complete, simply connected surfaces.) In this first paper, he was baffled by the 3-dimensional case. However, after encountering Riemann's inaugural address, which had been delivered in 1854 but published only after his death, in 1868, Beltrami published a second paper, on n-dimensional pseudospherical geometry. He started with what I will call the *hemisphere model*. Points of the n-dimensional non-euclidean geometry are identified with interior points of the hemisphere

$$y = \sqrt{a^2 - x_1^2 - \cdots - x_n^2}, \qquad y \geqslant 0,$$

in $(n + 1)$-dimensional euclidean space, provided with the Riemannian metric

$$ds = R\sqrt{dx_1^2 + \cdots + dx_n^2 + dy^2} \,/y.$$

He noted that this model is simply connected, and that the "limit space", consisting of boundary points with $y = 0$, is infinitely far from the interior in this metric. We will refer to such boundary points as *points at infinity*.

Projecting orthogonally down to the disk

$$x_1^2 + \cdots + x_n^2 \leqslant a^2,$$

he showed that each pseudospherical geodesic maps precisely to a euclidean straight line segment. Thus he obtained the *projective disk model*, later popularized by Klein.

On the other hand, projecting the hemisphere stereographically onto a disk, he obtained the *conformal disk model*, with the metric

$$ds = \sqrt{d\xi_1^2 + \cdots + d\xi_n^2} \,/ \left(1 - (\xi_1^2 + \cdots + \xi_n^2)/4R^2\right)$$

which had been noted already by Riemann.

Finally, performing an inversion in a boundary point of this disk, he obtained the *half-space model*, with coordinates $x_1, \ldots, x_{n-1}$ and $y \geqslant 0$, and with metric $R(dx_1^2 + \cdots + dx_{n-1}^2 + dy^2)^{1/2}/y$. He pointed out that this half-space metric had been used already by Liouville in the 2-dimensional case.

Klein [**1871**] reinterpreted Beltrami's projective disk model in terms of projective geometry. Following Cayley [**1859**], he took as his starting point the expression

$$\frac{1}{2}\log\frac{|q - a|\,|b - p|}{|p - a|\,|b - q|}$$

for the non-euclidean distance between two points p, q, as illustrated in Figure 2. (The factor $1/2$ is inserted so that curvature will be -1.) Here $|q - a|$ denotes the euclidean distance from a to q. In this paper he introduced the term *hyperbolic geometry* for the non-euclidean geometry of Lobachevsky and Bolyai.

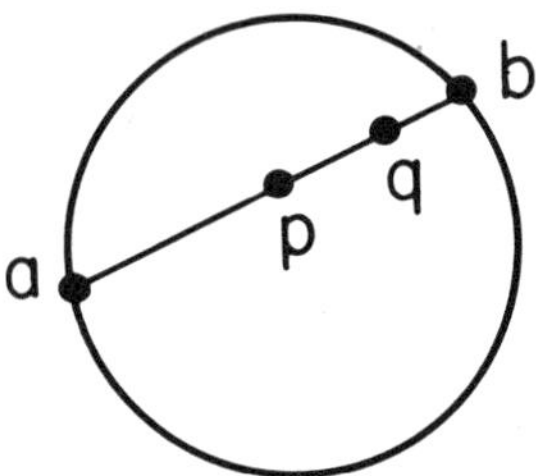

FIGURE 2. The Cayley formula for noneuclidean distance,
in the projective disk model

Poincaré [**1882**] re-introduced the Liouville-Beltrami upper half-plane model; and used it to identify the group of orientation preserving isometries of the hyperbolic plane with the group, now usually called $PSL_2\mathbf{R}$, consisting of all fractional linear transformations

$$z \mapsto (az + b)/(cz + d)$$

with real coefficients and with determinant $+1$. As a foundation for the study of automorphic functions, he emphasized the importance of finding discrete groups of hyperbolic isometries. He noted (§XIII) that examples of such discrete groups had been described already by Schwarz [**1873**], and others.

In [**1883**], Poincaré showed that the analogous group of orientation preserving isometries of 3-dimensional hyperbolic space can be identified with the group $PSL_2\mathbf{C}$ consisting of all fractional linear transformations of the plane of points at infinity, with complex coefficients. Again he studied the problem of finding discrete subgroups.

Picard [**1884**] described one particularly interesting example. The ring $\mathbf{Z}[i] \subset \mathbf{C}$ of Gaussian integers gives rise to a discrete group $PSL_2\mathbf{Z}[i] \subset PSL_2\mathbf{C}$, consisting of fractional linear transformations with Gaussian integer coefficients. A fundamental domain for this group, acting on hyperbolic 3-space, is noncompact, but has finite volume.

Bianchi [**1891**] described the analogous construction using the integers of an arbitrary imaginary quadratic number field. The precise volume of a corresponding fundamental domain was computed much later by Humbert [**1919**], in infinitely many cases. (See the Appendix.)

Inspired by examples due to Clifford, Klein [**1890**] posed the following problem:

"... to classify all connectivities which can possibly arise among closed manifolds of constant curvature."

It must be noted in this context that the concept of a (global) manifold was never defined in any satisfactory way during the nineteenth century. For smooth manifolds in euclidean space, the definition given by Betti [**1871**] works only if the normal bundle is trivial. Poincaré [**1895**] used a similar definition and also suggested extending a coordinate patch by analytic continuation. Hadamard [**1898**] clarified the subject by giving a lucid description of surfaces in 3-space in terms of overlapping coordinate patches. Weyl

[**1913**] paved the way for our modern concept of smooth manifold by defining an abstract surface in terms of overlapping coordinate patches. In particular, he gave a clear discussion of the universal covering surface. The generalization to n dimensions was now straightforward. (For a particularly nice presentation, see Cartan [**1928**].) During the same span of time, the study of combinatorial manifolds progressed from the rather vague definitions of Dyck [**1890**] and Poincaré to the precise definition of Brouwer [**1912**]. (See also Tietze [**1908**], Veblen and Alexander [**1913**].) The related concept of topological space is due particularly to Hausdorff [**1914**].

Let us return to the study of manifolds of constant curvature. Killing [**1891**] related Klein's classification problem to the study of discrete groups of isometries, as follows. If E is either n-dimensional hyperbolic space, n-dimensional euclidean space, or the n-dimensional sphere, and if Γ is a discrete group of isometries which acts freely on E, then he showed that the quotient E/Γ is a manifold of constant curvature. He called such a quotient a *Clifford-Klein space form.*

Much later, Hopf [**1925**] clarified this work by introducing a definition of *completeness*, showing that Killing's Clifford-Klein space forms are precisely the *complete Riemannian manifolds of constant curvature.* The concept of completeness was further developed by Cartan [**1928**] and Hopf-Rinow [**1931**].

By definition, a *hyperbolic manifold* is a Clifford-Klein manifold with curvature equal to -1. In the years following Killing's paper, a number of examples of hyperbolic manifolds were described. Compare Borel [**1963**]. The study of such manifolds is currently an extremely active area of research, particularly due to the work of Thurston [**1978, 1981**].

Here let me describe two particularly simple examples. The first is due to Gieseking [**1912**]. Start with a regular hyperbolic 3-simplex Δ having all four vertices on the sphere of points at infinity. Each of the six dihedral angles of Δ is then equal to $\pi/3$. (Compare the Appendix.) Now identify two faces of Δ by means of a rotation of $2\pi/3$ about a common vertex. Similarly, identify the opposite two faces by rotating about a common vertex. Then all six edges will be identified. Since the six dihedral angles add up to 2π, they fit smoothly together so that we obtain a non-singular hyperbolic manifold, even in a neighborhood of the common edge. This Gieseking manifold is nonorientable. It is noncompact, but is complete with finite volume.

The second example is due to Riley [**1975**]. Consider the figure eight knot K, as shown in Figure 3. The fundamental group Π of the complement $S^3 - K$ is generated by two loops α, β, as illustrated, which are subject to a single defining relation $(\alpha\beta^{-1}\alpha^{-1}\beta)\alpha = \beta(\alpha\beta^{-1}\alpha^{-1}\beta)$. Suppose that we look for a faithful representation

$$\Pi \to PSL_2\mathbf{C},$$

and require that the image must be a discrete subgroup. The generators α, $\beta \in \Pi$ will correspond to unknown matrices A, B, which must satisfy a single matrix equation.

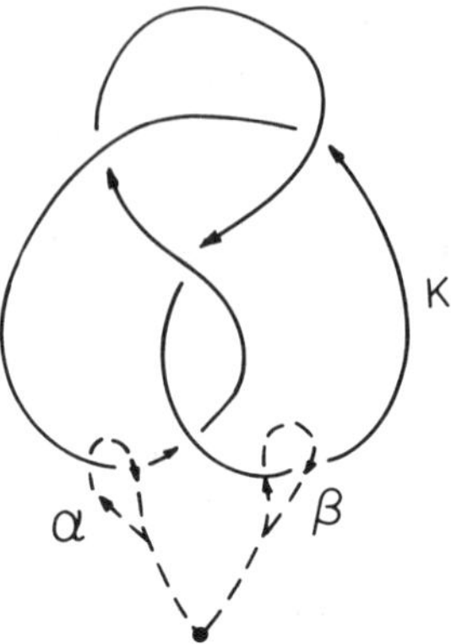

FIGURE 3. The figure eight knot

The possible choices for A and B can be drastically narrowed as follows. Note that every nontrivial element of $PSL_2\mathbf{C}$ is contained in a unique maximal abelian subgroup which is conjugate either to the group consisting of all diagonal matrices $\begin{bmatrix} \lambda & 0 \\ 0 & \lambda^{-1} \end{bmatrix}$ or to the group of all unipotent matrices of the form $\begin{bmatrix} 1 & \lambda \\ 0 & 1 \end{bmatrix}$, according as the given element does or does not have two linearly independent eigenvectors. Now the generator α of Π is contained in a free abelian group $\mathbf{Z} \oplus \mathbf{Z} \subset \Pi$, corresponding to the fundamental group of the boundary of a tubular neighborhood of K. It is easy to check that $\mathbf{Z} \oplus \mathbf{Z}$ cannot be embedded as a discrete subgroup of the group of diagonal matrices, which is isomorphic to the multiplicative group $\mathbf{C}^{\cdot}$. Hence α, and similarly β, must map to unipotent elements of $PSL_2\mathbf{C}$, with eigenvalue 1. Using a basis for $\mathbf{C}^2$ consisting of an eigenvector for α and an eigenvector for β, we may assume that

$$\alpha \mapsto A = \begin{bmatrix} 1 & 1 \\ 0 & 1 \end{bmatrix}, \qquad \beta \mapsto B = \begin{bmatrix} 1 & 0 \\ -\omega & 1 \end{bmatrix}$$

where ω is some unknown complex constant. Now computation shows that these two matrices satisfy the required relation if and only if $1 + \omega + \omega^2 = 0$, so that $\omega = (-1 \pm \sqrt{-3})/2$. *Thus the requirement that the "peripheral subgroup"* $\mathbf{Z} \oplus \mathbf{Z} \subset \Pi$ *maps isomorphically to a discrete subgroup of* $PSL_2\mathbf{C}$ *leads to a homomorphism*

$$h: \Pi \to PSL_2\mathbf{C}$$

which is uniquely determined, up to inner automorphism and complex conjugation.

Since the image $h(\Pi)$ is contained in the Bianchi group $PSL_2\mathbf{Z}[\omega]$, it is certainly discrete in $PSL_2\mathbf{C}$. In fact Riley proves that h maps Π isomorphically onto a subgroup $h(\Pi) \subset PSL_2\mathbf{Z}[\omega]$ of index twelve. Using the Haken-Waldhausen theory of 3-manifolds, he then proves that *the complement* $S^3 - K$ *is actually homeomorphic to the hyperbolic manifold* $H^3/h(\Pi)$.

In an interesting complement to these two examples, Thurston [**1978**] shows by direct geometric construction that $S^3 - K$ is homeomorphic to the 2-fold orientable covering of the Gieseking manifold. *In particular, the hyperbolic* *manifold $S^3 - K$ has a kind of triangulation into two regular ideal hyperbolic 3-simplexes, with their boundaries pasted together by suitable isometries.*

The uniqueness of this Riley representation $\Pi \to PSL_2\mathbf{C}$ illustrates the following fundamental result.

RIGIDITY THEOREM. *If two hyperbolic manifolds of finite volume, with dimension $n \geqslant 3$, have isomorphic fundamental groups, then they must necessarily be isometric to each other.*

In the compact case, this was proved by Mostow [**1971, 1973**]. (See also Margulis [**1970**].) For the noncompact case, it was proved by Prasad [**1973**], making essential use of Mostow's work.

It follows that geometric invariants such as volume, the lengths of closed geodesics, and the eigenvalues of the Laplacian operator, are also topological invariants. For an example to show that these particular geometric invariants are not sufficient to determine the topology of a hyperbolic 3-manifold, see Vignéras [**1980**].

In contrast to this Mostow-Prasad rigidity theorem, Thurston has proved a fundamental result which can be stated in a preliminary form as follows.

NONRIGIDITY THEOREM. *Suppose that $M = H^3/\Gamma$ is an orientable hyperbolic 3-manifold which is noncompact, but has finite volume. Then there exists an infinite sequence of hyperbolic manifolds*

$$M_j = H^3/h_j(\Gamma)$$

which have strictly smaller volume, and which approximate the original manifold M as $j \to \infty$ in the sense that the homomorphisms

$$h_j\colon \Gamma \to PSL_2\mathbf{C}$$

tend to the inclusion $\Gamma \subset PSL_2\mathbf{C}$ in the topology of pointwise convergence. Furthermore, the volumes of the M_j tend to the volume of M.

Of course, according to the Rigidity Theorem, these approximating manifolds M_j cannot be homeomorphic to M, or to each other. In particular, the representations h_j cannot be faithful. Their behavior can be described more precisely as follows.

To fix our ideas, suppose that the noncompact manifold M has just one end. Then a neighborhood of infinity in M is smoothly covered by a unique field of half-infinite geodesics, which converge exponentially towards each other as we go out to infinity. Any orthogonal trajectory to this field of geodesics is a flat torus. The fundamental group of such a torus-near-infinity is called a *peripheral subgroup* $\mathbf{Z} \oplus \mathbf{Z}$, embedded in the fundamental group Γ of M.

As in the discussion of Riley's example, the original representation $\Gamma \subset PSL_2\mathbf{C}$ must map this peripheral subgroup $\mathbf{Z} \oplus \mathbf{Z}$ to a unipotent subgroup, having just one eigenvector with a double eigenvalue of 1. If we perturb this representation slightly within the space of all homomorphisms $\Gamma \to PSL_2\mathbf{C}$, then the single eigenvector will split into two nearby but linearly independent eigenvectors. Hence the subgroup $\mathbf{Z} \oplus \mathbf{Z}$ will map into some subgroup of

$PSL_2\mathbf{C}$ which is isomorphic to the multiplicative group $\mathbf{C}^{\cdot}$. Such a representation of $\mathbf{Z} \oplus \mathbf{Z}$ cannot be faithful, with discretely embedded image. For most choices of perturbation, the image of $\mathbf{Z} \oplus \mathbf{Z}$ will simply be a nondiscrete subgroup of $\mathbf{C}^{\cdot}$. But for certain carefully chosen perturbations $h = h_{pq}$, the images of the two generators x, y of $\mathbf{Z} \oplus \mathbf{Z}$ will be subject to a new relation of the form $h(x^p y^q) = I$; and these images will generate a discretely embedded free cyclic subgroup of $\mathbf{C}^{\cdot}$. In fact Thurston shows that such a perturbation exists whenever p, q are relatively prime integers with $|p| + |q|$ sufficiently large. Furthermore, the image $h_{pq}(\Gamma)$ is always a discrete subgroup of $PSL_2\mathbf{C}$; and h_{pq} tends to the inclusion homomorphism as $|p| + |q| \to \infty$.

Topologically, the resulting quotient manifold $M_{pq} = H^3/h_{pq}(\Gamma)$ can be obtained from M by cutting off a neighborhood of infinity bounded by a flat torus $S^1 \times S^1$ and pasting in its place a solid torus $S^1 \times D^2$ which is attached in such a way as to introduce the required relation $x^p y^q = 1$ into the fundamental group. Thurston calls this operation a *Dehn surgery*.

Thus, if M has only one end, the approximating manifolds M_{pq} will always be compact. If M has several ends, then one or more of them can be capped off by analogous Dehn surgeries. In particular, M can always be approximated arbitrarily closely (in a suitable sense) by *compact* hyperbolic manifolds.

Here is a concluding example to test the reader's powers of visualization. Wielenberg [**1978**] has shown that the complement of the *Whitehead link W* of Figure 4 can be given the structure of a hyperbolic manifold H^3/Γ. In fact Γ can be taken as a subgroup of index twelve in the Picard group $PSL_2\mathbf{Z}[i]$. As noted by Whitehead [**1937**], the complement $S^3 - W$ is homeomorphic to the complement of the twisted version on the right. (Proof. Cut along a disk spanning C, rotate through any integral multiple of 2π, and then paste back together.) Now by filling in a solid torus neighborhood of C we obtain the manifold $S^3 - K$ from the hyperbolic manifold $S^3 - (K \cup C) \cong S^3 - W$. According to Thurston's theorem, this Dehn surgery must yield a new hyperbolic manifold, in all but a finite number of cases.

The knots K obtained in this way are called *Whitehead doubled knots*. Thus Thurston shows that *the complement of a Whitehead double has a hyperbolic structure whenever the number of twists is sufficiently large.* Furthermore, the volumes of these manifolds tend from below to a finite limit. Note that the figure eight knot occurs as a special case.

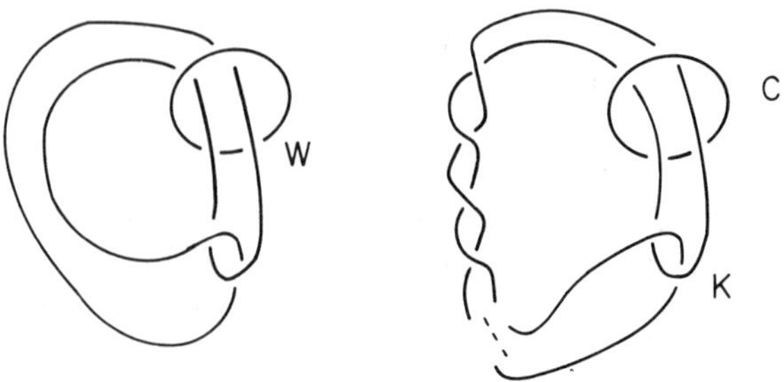

FIGURE 4. The Whitehead Link and a twisted version of it

Appendix. Volume in hyperbolic 3-space. Lobachevsky's formulas for the volumes of certain polyhedra in hyperbolic space can best be expressed in terms of the function

$$\text{л}(\theta) = -\int_0^\theta \log|2\,\sin u|\,du.$$

The Russian л is in honor of Lobachevsky, although he himself never used precisely this expression. Compare Clausen [**1832**], Coxeter [**1935**]. A graph is sketched in Figure 5. The basic properties can be listed as follows.

LEMMA 1. *This function* л(θ) *is odd, periodic of period* π, *and satisfies the identity*

$$\text{л}(n\theta) = n\sum_{k\bmod n}\text{л}(\theta + k\pi/n)$$

for every integer n.

Evidently, л(θ) is a continuous function, smooth except at the zeros of $\sin\theta$, taking its maximum at $\theta = \pi/6$.

PROOF OF LEMMA 1. Start with the trigonometric identity

$$2\,\sin n\theta = \prod_{k=0}^{n-1} 2\,\sin(\theta + k\pi/n)$$

which can be verified, for small positive values of θ, by substituting $z = e^{-2i\theta}$ in the equation

$$|z^n - 1| = \prod_{k\bmod n}|z - e^{2\pi ik/n}|.$$

The general case follows by analytic continuation.

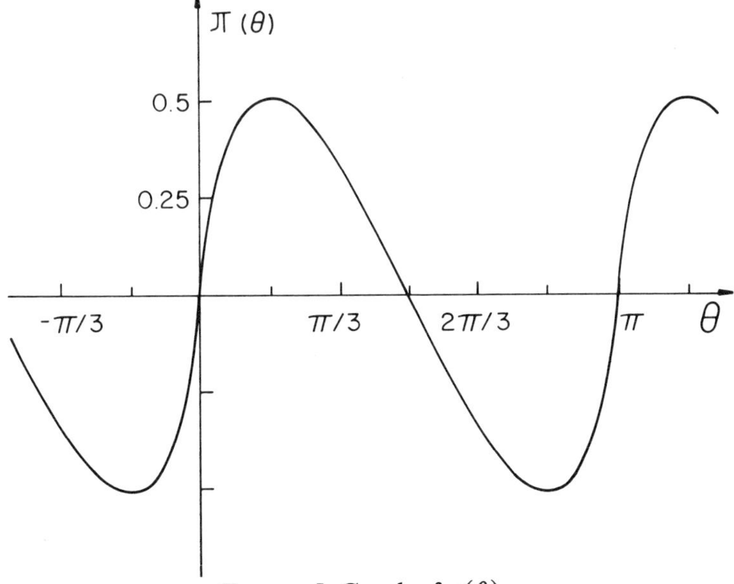

FIGURE 5. Graph of л(θ)

Integrating the negative log absolute value of both sides we obtain

$$(1) \qquad \text{л}(n\theta)/n = \sum_{k=0}^{n-1} \text{л}(\theta + \pi k/n) + \text{constant}.$$

For example, if $n = 2$, then

$$\text{л}(2\theta)/2 = \text{л}(\theta) + \text{л}(\theta + \pi/2) + \text{constant}.$$

Substituting $\theta = \pi/2$ or $\theta = 0$ in this last equation and subtracting, we see that $\text{л}(\pi) - \text{л}(0) = 0$. Since the derivative $\text{л}'(\theta) = -\log|2 \sin \theta|$ is periodic of period π, this proves that л itself is also periodic of period π.

Finally, to prove that the constant in equation (1) is zero, we can simply average over all values of θ mod π, noting that л has average zero since it is an odd function. $\square$

This function $\text{л}(\theta)$ is closely related to the *dilogarithm function*

$$L_2(z) = \sum z^n/n^2$$

which has been studied by Euler, Abel, and many others. The precise relationship is described by the identity

$$L_2(e^{2i\theta}) = \pi^2/6 - \theta(\pi - \theta) + 2i\text{л}(\theta)$$

for $0 \leqslant \theta \leqslant \pi$. (Compare Lewin [1958].) This identity can be verified by differentiating both sides for $0 < \theta < \pi$, using the limiting value as $\theta \to 0$ to compute the constant $\pi^2/6$.

Taking the imaginary part of both sides, we obtain the uniformly convergent Fourier series expansion

$$(2) \qquad \sum (\sin 2n\theta)/n^2 = 2\text{л}(\theta),$$

which is valid for all values of θ.

For actual computation, it is much better to work with the series

$$\text{л}(\theta) = \theta\left(1 - \log|2\theta| + \sum \frac{B_n(2\theta)^{2n}}{2n(2n + 1)!}\right),$$

which converges for $|\theta| \leqslant \pi$ and hence converges quite rapidly for small θ. Here

$$B_1 = \tfrac{1}{6}, \qquad B_2 = \tfrac{1}{30}, \ldots$$

are Bernoulli numbers. This equation can be proved by twice integrating the usual Laurent series expansion for $-\cot \theta$.

Here is an illustration of the utility of this function in computing hyperbolic volumes.

LEMMA 2. *Consider an ideal hyperbolic 3-simplex, that is a simplex Δ with all four vertices on the sphere of points at infinity. If α, β, γ are the dihedral angles along three edges meeting at a common vertex, then $\alpha + \beta + \gamma = \pi$, and*

$$\text{volume}(\Delta) = \text{л}(\alpha) + \text{л}(\beta) + \text{л}(\gamma).$$

It does not matter which particular vertex we choose, since it follows easily that *opposite dihedral angles are equal* so that we obtain the same three dihedral angles α, β, γ incident to any vertex.

This computation would follow easily from formulas of Lobachevsky [1836], although as far as I know he never worked out this particular example. (Compare Coxeter [1935].)

PROOF. We will use the Beltrami upper half-space model, with metric $ds^2 = (dx^2 + dy^2 + dz^2)/z^2$ and associated volume element $dx\,dy\,dz/z^3$, where $z > 0$. Let us position the ideal simplex Δ so that one of its faces lies in the hemisphere $z = (1 - x^2 - y^2)^{1/2}$, and its opposite vertex lies at the infinite point. Projecting Δ orthogonally to the unit disk in the x,y-plane, we obtain a triangle inscribed in this disk with angles α, β, γ. These are the angles of a euclidean triangle; so evidently $\alpha + \beta + \gamma = \pi$.

It follows easily from this picture that these angles determine Δ up to congruence (compare Andreev [1970]), and are subject to no other restrictions.

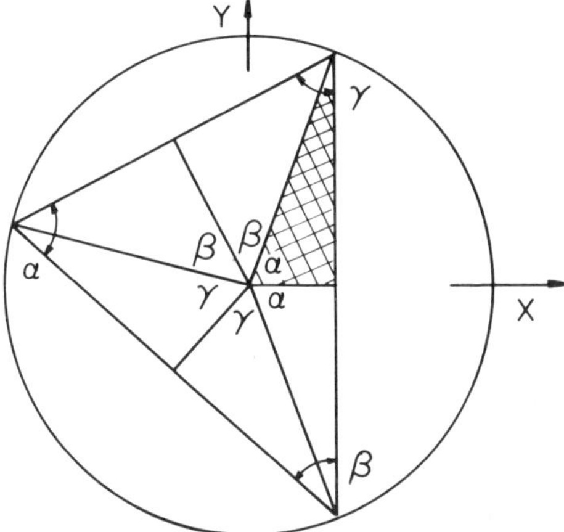

FIGURE 6. A triangle inscribed in the unit disk

If all three angles are acute, then barycentrically subdividing from the origin, we obtain six right triangles as illustrated in Figure 6. (The case of an obtuse angle can be handled by a similar argument, or by analytic continuation.) Take the region in the upper half space lying over just one of these triangles, namely the shaded one. We must integrate the Beltrami volume element $dx\,dy\,dz/z^3$ over the region defined by inequalities

$$z \geqslant \sqrt{1 - x^2 - y^2}, \qquad 0 \leqslant y \leqslant x \tan \alpha, \qquad 0 \leqslant x \leqslant \cos \alpha.$$

Integrating with respect to z, we obtain $\frac{1}{2}\,dx\,dy/(1 - x^2 - y^2)$. Integrating with respect to y, we obtain

$$\frac{dx}{4A}\left[\log(A + y) - \log(A - y)\right]_0^{x \tan \alpha} = \frac{dx}{4A}\log\frac{A \cos \alpha + x \sin \alpha}{A \cos \alpha - x \sin \alpha}$$

where $A = (1 - x^2)^{1/2}$. Substituting $A = \sin\theta$, $x = \cos\theta$, $dx = -A\,d\theta$, and integrating, we get

$$\text{volume} = \frac{1}{4}\int_\alpha^{\pi/2}\log\frac{2\sin(\theta + \alpha)}{2\sin(\theta - \alpha)}\,d\theta$$

$$= \left(-\text{Л}\left(\frac{\pi}{2} + \alpha\right) + \text{Л}(2\alpha) + \text{Л}\left(\frac{\pi}{2} - \alpha\right) - \text{Л}(0)\right)/4.$$

Now, if we substitute the identity

$$\text{Л}(2\alpha) = 2\text{Л}(\alpha) + 2\text{Л}(\alpha + \pi/2),$$

this reduces easily to   volume $= \text{Л}(\alpha)/2$. Adding the volumes of the other five regions, we obtain the required formula. $\square$

COROLLARY. *The maximum possible volume of a hyperbolic 3-simplex is* $3\text{Л}(\pi/3)$.

This works out as $1.0149416\ldots$.

PROOF. We must maximize the continuous function $\text{Л}(\alpha) + \text{Л}(\beta) + \text{Л}(\gamma)$ subject to the constraints $\alpha, \beta, \gamma \geqslant 0$ and $\alpha + \beta + \gamma = \pi$. Since the degenerate cases, where one dihedral angle is zero, all have zero volume, the maximum must occur at an interior point, where the derivative $\text{Л}'(\theta) = -\log(2\sin\theta)$ is defined and satisfies

$$\text{Л}'(\alpha) = \text{Л}'(\beta) = \text{Л}'(\gamma),$$

hence $\sin\alpha = \sin\beta = \sin\gamma$. Inspection shows that $\alpha = \beta = \gamma = \pi/3$ is the only interior solution. $\square$

REMARK. Haagerup and Monkholm [**1979**] have proved the analogous theorem in n-dimensional hyperbolic space. *The unique n-simplex of maximal volume is the regular ideal n-simplex.*

Now let us compare Lemma 2 with Humbert's formulas for the volumes of certain fundamental domains. The presentation will be based on Borel [**1981**], which contains much further information about arithmetic computations of volume.

Let $\mathbf{Z}[\alpha]$ be the ring of integers in an imaginary quadratic number field $F = \mathbf{Q}[\alpha]$, with discriminant $-D = (\bar\alpha - \alpha)^2 < 0$; and let

$$\Gamma = PSL_2\mathbf{Z}[\alpha] \subset PSL_2\mathbf{C}$$

be the corresponding Bianchi group of isometries of hyperbolic 3-space. Borel shows that a fundamental domain for Γ has

$$\text{volume} = D^{3/2}\zeta_F(2)/4\pi^2,$$

where ζ_F is the Dedekind zeta function. Using the standard identity

$$\zeta_F(s) = \zeta_\mathbf{Q}(s)\sum_{n=1}^\infty\left(\frac{-D}{n}\right)\frac{1}{n^s}$$

(see for example Hecke [**1923**, §49]), where $\zeta_\mathbf{Q}(2) = \pi^2/6$, we obtain the formula

$$(3)\qquad\qquad \text{volume} = \frac{D^{3/2}}{24}\sum\left(\frac{-D}{n}\right)\frac{1}{n^2}$$

of Humbert. Here $(-D/n)$ is a generalized quadratic residue symbol, with values ± 1 or 0. In order to express this infinite series in terms of the function $\text{л}(\theta)$, note first that the real-valued function $n \mapsto (-D/n)$ on $\mathbf{Z}/D$ is equal to its own Fourier transform, up to a factor of $\sqrt{-D} = i\sqrt{D}$. More precisely

$$\sum_{k \bmod D} \left(\frac{-D}{k}\right) e^{2\pi i k n/D} = \left(\frac{-D}{n}\right)\sqrt{-D}$$

for every positive integer n. (See Hecke §§52, 58.) Substituting this formula in (3), we obtain

$$\text{volume} = \frac{D}{24i} \sum_{k \bmod D} \left(\frac{-D}{k}\right) \sum_{n \geqslant 1} e^{2\pi i k n/D}/n^2.$$

Evidently the real parts of the exponentials must cancel out. Hence, using the Fourier series expansion (2) we obtain

$$\text{volume} = \frac{D}{12} \sum_{k \bmod D} \left(\frac{-D}{k}\right)\text{л}(k\pi/D).$$

Thus the volume of Humbert's fundamental domain is a finite linear combination, with rational coefficients, of values of the function $\text{л}(\theta)$ at rational multiples of π.

This suggests the conjecture that there may be some direct geometrical argument, describing such fundamental domains in terms of cutting and pasting of ideal simplices. (Compare Ash [**1977**].)

As an example, taking $D = 3$ we obtain

$$\text{volume} = \frac{1}{4}\left(\text{л}\left(\frac{\pi}{3}\right) - \text{л}\left(\frac{2\pi}{3}\right)\right) = \frac{1}{2}\text{л}\left(\frac{\pi}{3}\right).$$

According to Riley, the figure eight knot complement must have volume equal to twelve times this number, or $6\text{л}(\pi/3)$. This is precisely twice the volume $3\text{л}(\pi/3)$ of a maximal hyperbolic 3-simplex, thus checking Thurston's observation that the figure eight complement can be triangulated by two copies of a maximal 3-simplex.

Such results suggest the interest of determining number theoretical properties of the real numbers $\text{л}(\theta)$. Here is an explicit guess.

CONJECTURE. *If we consider only angles θ which are rational multiples of π, then every $\mathbf{Q}$-linear relation*

$$q_1\text{л}(\theta_1) + \cdots + q_n\text{л}(\theta_n) = 0$$

is a consequence of the relations

$$\text{л}(\pi + \theta) = \text{л}(\theta), \qquad \text{л}(-\theta) = -\text{л}(\theta),$$

$$\text{л}(n\theta) = n \sum_{k \bmod n} \text{л}(\theta + k\pi/n)$$

of Lemma 1.

38   JOHN MILNOR

For example л(π/3) and л(π/6) are subject to the **Q**-linear relation

$$2л(π/6) = 3л(π/3),$$

as one easily verifies. But the conjecture would imply that л(π/3) and л(π/4) are rationally independent. In other words, no finite covering manifold of the figure eight complement should have the same volume as some covering of the Whitehead link complement.

A completely equivalent conjecture would be the following. *For any fixed $n > 2$, the $\varphi(n)/2$ real numbers* л($k\pi/n$), *with* $0 < k < n/2$ *and k relatively prime to n, are linearly independent over* **Q**. Here φ is the Euler φ-function.

The author hopes to discuss this question further in a later paper.[2]

References

[1827] C. F. Gauss, *Disquisitiones generales circa superficies curvas*, Commentationes Gottingensis **6** (Werke **4**, 218–258; German summary, 341–347).

[1829–30] N. I. Lobachevsky, *On the foundations of geometry*, Kazan Messenger **25–28**; German transl., 'Zwei geometrische Abhandlungen', Leipzig, 1898.

[1832] T. Clausen, *Ueber die Function* sin φ + $(1/2^2)$ sin 2φ + $(1/3^2)$ sin 3φ + *etc.*, J. Reine Angew. Math. **8**, 298–300.

J. Bolyai, *The absolute science of space, independent of the truth or falsity of Euclid's Axiom 11 (which can never be proved a priori)*, Maros-Vásárhelyini.

[1836] N. I. Lobachevsky, *Imaginary geometry and its application to integration*, Kazan (German transl. Abh. Gesch. Math. **19**, 1904).

[1837] ______, *Géometrie imaginaire*, J. Reine Angew. Math. **17**, 295–320.

[1859] A. Cayley, *Sixth memoir upon quantics*, Phil. Trans. **149**, 61–91.

[1868] B. Riemann, *Ueber die Hypothesen welche der Geometrie zu Grunde liegen*, Abh. K. G. Wiss. Göttingen **13** (from his Inaugural Address of 1854).

E. Beltrami, *Saggio di interpetrazione della geometria non-euclidea*, Gior. Mat. **6**, 248–312 (Also Op. Mat. **1**, 374–405; Ann. École Norm. Sup. **6** (1869), 251–288).

______, *Teoria fondamentale degli spazii di curvatura costante*, Annali di mat. ser. II **2**, 232–255 (Op. Mat. **1**, 406–429; Ann. École Norm. Sup. **6** (1869), 345–375).

[1871] F. Klein, *Über die sogenannte Nicht-Euklidische Geometrie*, Math. Ann. **4**, 573–625 (cf. Ges. Math. Abh. **1**, 244–350).

E. Betti, *Sopra gli spazi di un numero qualunque di dimensioni*, Annali di mat. ser. II **4**, 140–158.

[1873] H. A. Schwarz, *Ueber diejenigen Fälle in welchen die Gaussische hypergeometrische Reihe eine algebraische Function ihres vierten Elementes darstellt*, J. Reine Angew. Math. **75**, 292–335 (Ges. Math. Abh. **2**, 211–259).

[1882] H. Poincaré, *Theorie des groupes fuchsiens*, Acta Math. **1**, 1–62; (Oeuv. **2**, 108–168).

[1883] ______, *Mémoire sur les groupes kleinéens*, Acta Math. **3**, 49–92; (Oeuv. **2**, 258–299).

[1884] E. Picard, *Sur un groupe de transformations des points de l'espace situés du même côté d'un plan*, Bull. Soc. Math. France **12**, 43–47.

[1890] W. Dyck, *Beiträge zur Analysis situs*. II, Math. Ann. **37** 273–316.

F. Klein, *Zur Nicht-Euklidischen Geometrie*, Math. Ann. **37**, 544–572 (Ges. Abh. **1**, 353–383).

[1891] L. Bianchi, *Geometrische Darstellung der Gruppen Linearer Substitutionen mit ganzen complexen Coefficienten nebst Anwendungen auf die Zahlentheorie*, Math. Ann. **38**, 313–333.

---

[2]"On polylogarithms, Hurwitz zeta functions and the Kubert identities" (in preparation).

W. Killing, *Ueber die Clifford-Klein'schen Raumformen*, Math. Ann. **39**, 258–278.

[1895] H. Poincaré, *Analysis situs*, J. École Polyt. **1**, 1–121 (Oeuv. **6**, 193–288).

[1898] J. Hadamard, *Les surfaces à courbures opposées et leur lignes géodésiques*, J. Math. Pures Appl. Ser. 5 **4**, 27–73 (Oeuv. **2**, 729–775).

[1900] C. F. Gauss, Werke **8**, 175–239.

[1908] H. Tietze, *Über die topologischen Invarianten mehrdimensionaler Mannigfaltigkeiten*, Monatshefte für Math. Phys. **19**, 1–118.

[1912] H. Gieseking, *Analytische Untersuchungen ueber topologische Gruppen*, Thesis, Muenster (Compare Magnus (1974), 153).

L. E. J. Brouwer, *Über Abbildung von Mannigfaltigkeiten*, Math. Ann. **71**, 97–115.

[1913] O. Veblen and J. W. Alexander, *Manifolds of n dimensions*, Ann. of Math. (2) **14**, 163–178.

H. Weyl, *Die Idee der riemannschen Fläche*, Teubner, Leipzig.

[1914] F. Hausdorff, *Grundzüge der Mengenlehre*, von Veit, Leipzig.

[1919] G. Humbert, *Sur la mesure des Classes d'Hermite de discriminant donné dans un corps quadratique imaginaire, et sur certains volumes non euclidiens*, Comptes Rendus (Paris) **169**, 448–454.

[1923] E. Hecke, *Vorlesungen über die Theorie der algebraischen Zahlen*, Akad. Verl., Leipzig.

[1925] H. Hopf, *Zum Clifford-Kleinschen Raumproblem*, Math. Ann. **95**, 313–339.

[1928] E. Cartan, *Leçons sur la géométrie des espaces de Riemann*, Gauthier-Villars, Paris.

[1931] H. Hopf and W. Rinow, *Über den Begriff der vollständigen differentialgeometrischen Fläche*, Comment. Math. Helv. **3**, 209–225.

[1935] H. S. M. Coxeter, *The functions of Schläfli and Lobatschefsky*, Quart. J. Math. **6**, 13–29 (pp. 3–20 of twelve geometric essays, S. Ill. Univ. Press, Carbondale, 1968).

[1937] J. H. C. Whitehead, *On doubled knots*, J. London Math. Soc. **12**, 63–71 (Works **2**, 59–67).

[1942] H. S. M. Coxeter, *Non-Euclidean geometry*, Univ. Toronto Press, Toronto.

[1958] L. Lewin, *Dilogarithms and associated functions*, MacDonald, London.

[1963] A. Borel, *Compact Clifford-Klein forms of symmetric spaces*, Topology **2**, 111–112.

[1970] G. A. Margulis, *Isometry of closed manifolds of constant negative curvature with the same fundamental group*, Soviet Math. Dokl. **11**, 722–723.

E. M. Andreev, *On convex polyhedra [of finite volume] in Lobačevskiĭ space*, Mat. Sbornik **81**, 445–478 and **83**, 256–260 (transl. **10**, 413–440 and **12**, 255–259).

[1971] G. D. Mostow, *The rigidity of locally symmetric spaces*, Proc. (1970) Internat. Congr. Math., vol. 2, pp. 187–197, Gauthier-Villars, Paris.

[1973] ______, *Strong rigidity of locally symmetric spaces*, Ann. Math. Studies, No. 78, Princeton Univ. Press, Princeton, N. J.

G. Prasad, *Strong rigidity of Q-rank 1 lattices*, Invent. Math. **21**, 255–286.

[1974] W. Magnus, *Noneuclidean tesselations and their groups*, Academic Press, New York.

[1975] R. Riley, *A quadratic parabolic group*, Math. Proc. Cambridge Philos. Soc. **77**, 281–288.

[1977] A. Ash, *Deformation retracts with lowest possible dimension of arithmetic quotients of self-adjoint homogeneous cones*, Math. Ann. **225**, 69–76.

[1978] W. Thurston, *Geometry and 3-manifolds*, Lecture Notes, Princeton University.

N. Wielenberg, *The structure of certain subgroups of the Picard group*, Math. Proc. Cambridge Philos. Soc. **84**, 427–436.

[1979] U. Haagerup and H. Munkholm, *Simplices of maximal volume in hyperbolic n-space*, preprint, Odense University.

[1980] H. J. Munkholm, *Simplices of maximal volume in hyperbolic space, Gromov's norm, and Gromov's proof of Mostow's rigidity theory* (following Thurston), pp. 109–124 of Topology Symposium—Siegen 1979, ed. Koschorke and Neumann, Springer Lecture Notes in Math., no. 788.

R. Rieley, *Applications of a computer implementation of Poincaré's theorem on fundamental polyhedra*, preprint, I.A.S.

______, *Seven excellent knots*, preprint, I.A.S.

M.-F. Vignéras, *Variétés riemanniennes isospectrales et non isométriques*, Ann. of Math. (2) **112**, 21–32.

[1981] A. Borel, *Commensurability classes and volumes of hyperbolic 3-manifolds*, Ann. Scuola Norm. Sup. Pisa Cl. Sci. (4) **8**, 1–33.

M. Gromov, *Hyperbolic manifolds according to Thurston and Jórgensen*, Sémin. Bourbaki **546**, Springer Lecture Notes in Math., no. 842, pp. 40–53.

D. Sullivan, *Travaux de Thurston sur les groupes quasi-fuchsiens et les variétés hyperboliques de dimension 3 fibrées sur S^1*, Sémin. Bourbaki **554**, Springer Lecture Notes in Math., no. 842, pp. 196–214.

W. Thurston, *Three dimensional manifolds, Kleinian groups and hyperbolic geometry* (to appear).

______, *Hyperbolic structures on 3-manifolds, I: Deformation of acylindrical manifolds* (to appear).

SCHOOL OF MATHEMATICS, INSTITUTE FOR ADVANCED STUDY, PRINCETON, NEW JERSEY 08540

Proceedings of Symposia in Pure Mathematics
Volume **39** (1983), Part 1

Completeness of the Kähler-Einstein Metric on Bounded Domains and the Characterization of Domains of Holomorphy by Curvature Conditions

NGAIMING MOK AND SHING-TUNG YAU

Abstract. In this article we show that a bounded domain is a domain of holomorphy if and only if it carries a complete Kähler-Einstein metric with negative Ricci curvature. The completeness theorem is generalized to a variety of subdomains on Kähler manifolds with curvature conditions. We also investigate the question of characterizing domains of holomorphy by holomorphic sectional curvature.

In [2] Cheng and Yau constructed complete Kähler-Einstein metrics with negative Ricci curvature on strictly pseudoconvex domains with C^k boundary, $k \geqslant 5$. For any bounded domain of holomorphy Ω, they show that the complete Kähler-Einstein metrics of scalar curvature $-C$ on any exhausting family Ω_i of strictly pseudoconvex domains (with C^k boundary, $k \geqslant 5$) converge uniformly on compact sets to a unique metric independent of the exhaustion Ω_i. They call such a metric almost complete. Moreover, under very mild boundary conditions on Ω, for example, assuming that Ω is the intersection of a (possibly infinite) family of pseudoconvex domains with C^2 boundary, the limit metric is actually complete. This naturally poses the problem whether in fact all bounded domains of holomorphy admit complete Kähler-Einstein metrics with negative Ricci curvature.

On the other hand, by the construction due to Grauert [6] of a complete Kähler metric on $\mathbf{C}^n - \{0\}$ and the embedding theorem of Bishop-Narasimhan, one can easily show that complete Kähler metrics exist on $X - V$, where X is a Stein manifold and V is a closed analytic subvariety of any dimension. However, a consequence of an extension theorem due independently to Griffiths [8] and Shiffman [16] shows that any Riemann domain admitting a complete Hermitian metric with nonpositive holomorphic sectional curvature is necessarily a domain of holomorphy. As a converse to the question in the preceding paragraph, one would ask if the existence of complete Kähler-Einstein metrics with negative Ricci curvature is sufficient to characterize bounded domains of holomorphy. In this article, we prove the following theorem.

Presented to the Symposium on the Mathematical Heritage of Henri Poincaré April 7–10, 1980; received by the editors January 9, 1982.

1980 *Mathematics Subject Classification*. Primary 53E, 32E.

MAIN THEOREM. *Any bounded domain of holomorphy Ω admits a complete Kähler-Einstein metric. Conversely, if a bounded domain Ω admits a complete Hermitian metric such that $-C \leqslant$ Ricci curvature $\leqslant 0$, then Ω is a domain of holomorphy.*

The proof of the Main Theorem will be given in the first two sections. In §3 we present a generalization to the first statement. We prove that a Stein open subset Ω of an ambient Kähler manifold X also admits a complete Kähler metric under some mild geometric assumptions on X and Ω. We remark that the Main Theorem is also true for bounded domains spread over Stein manifolds and certain unbounded domains in $\mathbf{C}^n$.

The second statement is sharp in the sense that in general one cannot weaken the hypothesis by assuming $-C \leqslant$ Ricci curvature $\leqslant C$. In fact, in §4 we construct a complete Kähler metric of bounded Ricci curvature on the punctured unit ball $B^n - \{0\}$. This metric is obtained by piecing together the Fubini-Study metric on $\mathbf{P}^{n-1}$ and the Poincaré metric on the punctured disc.

The proof of the first statement of the Main Theorem and the generalization in §3 depends essentially on the maximum principle for the Monge-Ampère equation, which was developed in Cheng and Yau [2] and used in the same paper to prove the completeness of the limit Kähler-Einstein metric in special cases, and the theorem of Oka that $-\log d$ (Euclidean distance to the boundary) is plurisubharmonic on any domain of holomorphy. The second statement depends strongly on one form of the Schwarz lemma due to Yau [22] and the Kontinuitätssatz of Oka for domains of holomorphy.

Since the maximum principle for the complex Monge-Ampère equation that we need is actually a version of the Schwarz lemmas which can be developed using techniques in Yau [22], we include a proof of it in this article. As another application of the Schwarz lemmas, we give an application to the problem of Behnke-Stein on increasing unions of Stein manifolds in the Appendix.

In §5, we study the question of characterizing Riemann domains of holomorphy by holomorphic sectional curvature. The natural question is whether one can allow a certain amount of positive holomorphic sectional curvature in the result of Griffiths [8] and Shiffman [16] quoted above. We show that if a Riemann domain Ω admits a complete Kähler metric such that an upper bound for the holomorphic sectional curvatures decreases asymptotically to zero, then Ω is a domain of holomorphy. This theorem is optimal because straightforward computations of the example in §4 show that the complete Kähler metric on $B^n - \{0\}$ has bounded holomorphic sectional curvature. The proof of the theorem depends on the Kontinuitätssatz of Oka, the area-minimizing property of complex submanifolds in a Kähler manifold and some elementary estimates of the inequality $\Delta(-\log g) \leqslant Cg$.

We remark that recently Ozawa [14] and later Diedrich and Pflug [3] have proved theorems on Steinness of domains carrying complete Kähler metrics by imposing boundary conditions in place of curvature conditions. Their results depend on L^2-estimates of $\bar{\partial}$ on complete Kähler manifolds due to Andreotti and Vesentini [1], Hörmander [9] and Skoda [17].

The Kähler-Einstein metric should be considered as a direct generalization of the Poincaré metric on Riemann surfaces. Thus it is not inappropriate to dedicate this paper to Poincaré. It should be mentioned that most of the material here was worked out half a year after the conference.

Table of Contents

1. A Schwarz lemma.

(1.1) We shall need a form of the Schwarz lemma for volume forms where the target manifold N carries only a volume form V_N. If

$$V_N = V_N^\alpha \left(\tfrac{i}{2}\right)^n dz_\alpha^1 \wedge d\bar{z}_\alpha^{-1} \wedge \cdots \wedge dz_\alpha^n \wedge d\bar{z}_\alpha^n$$

in local coordinates, then $\tfrac{i}{2}\partial\bar{\partial} \log V_N^\alpha$ is a globally defined $(1,1)$ form, called the Ricci form of V_N. We prove here a Schwarz lemma for volume forms, a weaker form of which was stated without proof in Yau [22].

SCHWARZ LEMMA FOR VOLUME FORMS. *Let M be a complete Hermitian manifold with scalar curvature bounded from below by $-K_1$ and let N be a complex manifold of the same dimension with a volume form (i.e., positive (n, n) form, $n = \dim N$) V_N such that the Ricci form is negative definite and $(\tfrac{i}{2}\partial\bar{\partial} \log V_n)^n \geqslant K_2 V_N (\tfrac{i}{2}\partial\bar{\partial} \log V_N$ is to be interpreted with local coordinates as above). Suppose $f\colon M \to N$ is a holomorphic map and the Jacobian is nonvanishing at one point. Then $K_1 > 0$ and*

$$\sup f^*V_N / V_M \leqslant K_1^n / n^n K^2$$

where V_M is the volume form associated to the given Kähler metric on M.

PROOF OF THE LEMMA. Let $u = f^*V_N / V_M$ be the volume ratio and let Δ denote the Laplacian on the manifold M. Let (w_j), (z_i) be local holomorphic coordinates on M and N respectively. In the coordinates (z_i), we shall write the form as $V_N = V_N'(\tfrac{i}{2})^n dz_1 \wedge d\bar{z}_1 \wedge \cdots \wedge dz_n \wedge d\bar{z}_n$. Then,

$$(1) \qquad \Delta \log u(x) = R(x) + \Delta \log \left| \det \frac{\partial f_i}{\partial w_j}(x) \right|^2 + \Delta \log V_N'(f(x))$$

where R stands for the scalar curvature of M. Hence,

$$(2) \qquad \Delta \log u(x) \geqslant R(x) + \Delta \log V_N'(f(x)).$$

Suppose f is nondegenerate at x. We choose local coordinates (w_j) at x such that $g_{ij}(x) = \delta_{ij}$, and such that the form $\frac{i}{2}\partial\bar\partial \log V_N'(f(x))$ is in diagonal form. Since f is nondegenerate at x, (w_j) also serve as local holomorphic coordinates for $f(x)$. In terms of w_j

$$\frac{i}{2}\partial\bar\partial \log V_N'(f(x)) = \frac{i}{2}\sum_j a_j\, dw_j \wedge d\bar w_j, \qquad a_j > 0.$$

By the arithmetic-geometric inequality

$$\Delta \log V_N'(f(x)) = \sum a_j = n\left(\frac{\sum a_j}{n}\right) \geq n(a_1 \cdots a_n)^{1/n}.$$

On the other hand, from $(\frac{i}{2}\partial\bar\partial \log V_N') \geq K_2 V_N$, it follows that

$$a_1 \cdots a_n \geq K_2 \frac{f^*(V_N)}{V_M} = K_2 u$$

and

(3) $$\Delta \log u(x) \geq R(x) + n(K_2 u)^{1/n}(x).$$

From $\Delta e^v = e^v(\Delta v + |\nabla v|^2)$ it follows that on all of M

(4) $$\Delta u \geq u\left(R + nK_2^{1/n}u^{1/n}\right).$$

Consider the function $1/(u + c)^{1/2n}$, $c > 0$, arbitrary

$$\Delta\left(\frac{1}{(u+c)^{1/2n}}\right) = \frac{-\Delta u}{2n(u+c)^{1/2n+1}} + \left(\frac{1}{2n}\right)\left(\frac{2n+1}{2n}\right)\frac{|\nabla u|^2}{(u+c)^{1/2n+2}}.$$

From $u \geq u(R + nK_2^{1/n}u^{1/n})$ (4), it follows that

(5) $$\Delta\left(\frac{1}{(u+c)^{1/2n}}\right) \leq \left(\frac{1}{2n}\right)\left(\frac{2n+1}{2n}\right)\frac{|\nabla u|^2}{(u+c)^{1/2n+2}}$$
$$- \frac{u}{2n(u+c)^{1/2n+1}}\left(-K_1 + nK_2^{1/n}u^{1/n}\right).$$

We quote the following maximum principle in Yau [21].

Maximum principle on complete Riemannian manifolds. Let M be a complete Riemannian manifold with Ricci curvature bounded from below. Let f be a C^2-function which is bounded from below on M. Then, for any $\varepsilon > 0$, there exists a point p in M such that

$$|\nabla f| < \varepsilon, \qquad \Delta f > -\varepsilon, \qquad f(p) < \inf f + \varepsilon.$$

Applying the maximum principle to our function $1/(u + c)^{1/2n}$, for any $\varepsilon > 0$, there exists p such that

$$\frac{1}{(u + c)^{1/n+2}} |\nabla u|^2 < \varepsilon, \qquad \Delta\left(\frac{1}{(u + c)^{1/2n}}\right) > -\varepsilon, \qquad u < \inf u + \varepsilon.$$

Multiplying (5) by $(u + c)^{-1/2n}$ and applying the maximum principle, at the point p we have

(6)

$$\frac{1}{(u + c)^{1/2n}}(-\varepsilon) \leqslant \left(\frac{1}{2n}\right)\left(\frac{2n + 1}{2n}\right)\varepsilon - \frac{u}{2n(u + c)^{1/n+1}}\left(-K_1 + nK_2^{1/n}u^{1/n}\right),$$

hence

$$\frac{K_1 u}{2n(u + c)^{1/n+1}} - \frac{nK_2^{1/n}u^{1+1/n}}{2n(u + c)^{1+1/n}} \geqslant \frac{-\varepsilon}{(u + c)^{1/2n}} - \frac{(2n + 1)\varepsilon}{(2n)^2}.$$

Fix c and let $\varepsilon \to 0$ so that $1/(u + c)^{1/2n} \to \inf 1/(u + c)^{1/2n}$ and $u + \sup u$, we see that $\sup u$ must, in fact, be bounded and

$$(\sup u)^{1/n} \leqslant \frac{K_1}{nK_2^{1/n}}, \qquad \sup u \leqslant \frac{K_1^n}{n^n K_2}$$

proving the lemma.

2. Proof of the Main Theorem.

(2.1) *Completeness of the Kähler-Einstein metric on bounded Stein domains.* In this section, we prove the first statement of the Main Theorem. Namely, we shall show that the almost complete Kähler-Einstein metric constructed by Cheng and Yau [2] is, in fact, complete on any bounded Riemann domain of holomorphy. Let $\pi\colon \Omega \to \mathbf{C}^n$ denote the spreading map with $\pi(\Omega)$ bounded and let $d(z) = d(z, \partial\Omega)$ denote the Euclidean distance to the boundary. Without loss of generality, we can assume that $\pi(\Omega)$ is contained in a Euclidean ball of radius $1/e$. Let $V_0'(\frac{i}{2})^n\, dz_1 \wedge d\bar{z}_1 \wedge \cdots \wedge dz_n \wedge d\bar{z}_n$ denote the volume form of the Kähler-Einstein metric of B with Ricci curvature $-(n + 1)$. Following Cheng and Yau [2], we consider a test function

$$u = \log \frac{1}{d^2(-\log d)^2} + \log V_0'.$$

Then,

$$\partial\bar{\partial}\left(c \log \frac{1}{d^2(-\log d)^2} + \log V_0'\right) \qquad (c > 0)$$

$$= 2c\partial\bar{\partial}(-\log d) - \frac{2c\partial\bar{\partial}(-\log d)}{-\log d} + \frac{2c\partial d \wedge \bar{\partial}d}{d^2(-\log d)^2} + \partial\bar{\partial}(\log V_0').$$

By the Theorem of Oka, $-\log d$ is plurisubharmonic on Ω. Since $-\log d < 1$, as a $(1,1)$ current

$$\partial\bar\partial u \geq \frac{2c\partial d \wedge \bar\partial d}{d^2(-\log d)^2} + \partial\bar\partial(\log V_0').$$

Clearly d is a Lipschitz function, and hence differentiable almost everywhere. Moreover, ∇d has unit length wherever differentiable. Since the Kähler-Einstein metric on B dominates the Euclidean metric up to a constant multiple by taking c sufficiently small

$$\det(u_{i\bar j}) \geq Ce^u$$

whenever $u_{i\bar j} = \partial^2 u/\partial z_i \partial\bar z_j$ exists.

To take care of the nondifferentiability of u and to be able to apply either the maximum principle for the complex Monge-Ampère equation (Cheng and Yau [2]) or the Schwarz lemma for volume forms $(1,1)$, we smooth v on compact subdomains. Let $\Omega = \bigcup \Omega_i$ be an exhaustion of Ω by relatively compact strictly pseudoconvex subdomains, and let V denote the Lipschitz volume form $e^u(\frac{i}{2})^n dz_1 \wedge d\bar z_1 \wedge \cdots \wedge dz_n \wedge d\bar z_n$. On Ω_i let $u_{i,\varepsilon}$ denote a smoothing of Ω_i on u by the standard symmetric smoothing kernel over balls of sufficiently small radius ε. Write $V_{i,\varepsilon} = e^{u_{i,\varepsilon}}(\frac{i}{2})^n dz_1 \wedge d\bar z_1 \wedge \cdots \wedge dz_n \wedge d\bar z_n$. It is clear that as (n,n) forms

$$\left(\tfrac{i}{2}\partial\bar\partial \log V_{i,\varepsilon}'\right)^n \geq (C+\varepsilon')V_{i,\varepsilon}$$

for some $\varepsilon' \geq 0$, where $V_{i,\varepsilon}' = e^{u_{i,\varepsilon}}$ is the coefficient of $V_{i,\varepsilon}$. Furthermore, ε' tends to zero as ε shrinks to zero. Since Ω_i is a bounded strictly pseudoconvex domain, it admits a complete Kähler-Einstein metric of Ricci curvature $-(n+1)$. We shall write $V_{E,i}$ for the Kähler-Einstein volume form. By the Schwarz lemma (1.1),

$$V_{E,i} \geq C'V_{i,\varepsilon}.$$

Since $V_{i,\varepsilon}$ converges pointwise to V on Ω_i, we have

$$V_{E,i} \geq C'V.$$

Since Ω is a bounded domain, it admits a Hermitian metric of Ricci curvature bounded from above by a negative constant. The complete Kähler-Einstein metric of Ricci curvature $-(n+1)$ converge, therefore, to a Kähler-Einstein metric $(g_{i\bar j})$ on Ω by the arguments of Cheng and Yau [2, Theorem 8.4]. We shall denote the volume form by V_E and its coefficient by V_E'. We have

$$V_E \geq C'V, \quad \text{so that } V_E' \geq \frac{C'}{d^2(-\log d)^2} \qquad (C' \geq 0).$$

Observe now on Ω with the Kähler-Einstein metric $(g_{i\bar j})$, $\Delta \log V_E' = n(n+1)$. Although $(\Omega, g_{i\bar j})$ is a priori not known to be complete, by the standard limit process we employed just now, one can still apply the gradient estimate on complete Riemannian manifolds of Yau [21] on manifolds with Ricci curvature bounded from below. In this case, we have

$$|\nabla \log V_E'| \leq C(\log V_E' - \inf \log V_E') \qquad (C > 0)$$

where $\inf \log V'_E > -\infty$ since V_E dominates a constant multiple of the Euclidean metric, by the Schwarz lemma. Write $c = \inf \log V'_E$

$$\frac{|\nabla(\log V'_E - c)|}{(\log V'_E - c)} \leq C, \qquad |\nabla \log(\log V'_E - c)| \leq C.$$

Fix a base point z_0 in Ω. Then,

$$\log(\log V'_E(z) - c) - \log(\log V'_E(z_0) - c)$$
$$= \int_{z_0}^z d\log(\log V'_E(z) - c) \leq C\delta(z; z_0)$$

where $\delta(z; z_0)$ denotes the distance function in the almost-complete Kähler-Einstein metric, so that

$$\delta(z; z_0) \geq \frac{\log(\log V'_E(z) - c) - \log(\log V'_E(z_0) - c)}{C}$$
$$\geq C_1 \log(-\log d(z, \partial\Omega)) + C_2.$$

In particular, the almost-complete Kähler-Einstein metric $(g_{i\bar{j}})$ is, in fact complete on Ω. Finally, the order of growth of both the volume form and the distance function are sharp as exemplified by the Poincaré-metric on products of punctured discs and polydiscs.

(2.2) *Steinness of bounded domains admitting Kähler metrics of nonpositive Ricci curvature.* In this subsection, we prove the second statement in the Main Theorem. Let $\omega = \frac{i}{2}\Sigma_{i,j}g_{i\bar{j}}dz^i \wedge d\bar{z}^j$ be the Kähler form of the given metric on Ω. By assumption

$$0 \leq \partial\bar{\partial}\log\det(g_{i\bar{j}}) \leq C\omega$$

as $(1, 1)$ forms. We are going to prove that the existence of such a metric implies that Ω satisfies the Kontinuitätssatz. To prove by absurdity, we shall need the fact that otherwise Ω would admit a one-parameter family of discs smooth up to the boundary which first touches $\partial\Omega$ with the interior of a disc, despite the fact that no boundary condition is put on Ω. This is an easy consequene of Oka's original proof [12, 13]. Nonetheless, we reproduce the proof here for the convenience of the reader. For notational convenience, we shall assume that Ω is a domain. Let v be a unit vector in $\mathbf{C}^n$. Denote by R_v the Hartogs radius on Ω in the direction of v. Namely, if $z \in \Omega$, $R_v(z) = d(z, \partial\Omega_v(z))$, where $\Omega_v(z)$ is the intersection of the complex plane passing through z in the direction v and the domain Ω. By the Theorem of Oka, Ω is a domain of holomorphy if and only if

$$-\log d = \sup_{\|v\|=1} -\log R_v$$

is plurisubharmonic. Recall now that Ω is equipped with a complete Kähler metric with $-C \leq$ Ricci curvature ≤ 0. Suppose Ω is not a domain of holomorphy; then for some unit vector v, $-\log R_v$ is not plurisubharmonic at some $x \in \Omega$. We write the complex coordinates on $\mathbf{C}^n$ in the form

$$(z_1,\ldots,z_n) = \left(x_1 + \sqrt{-1}\,y_1,\ldots,x_n + \sqrt{-1}\,y_n\right).$$

After a complex affine change of coordinates, we can assume that $v = \partial/\partial x_1$ and $-\log R_v$ is not subharmonic at $x = 0$ in the z_2 direction. (In the z_1 direction $-\log R_v$ is necessarily subharmonic.) The function $-\log R_v$ violates the mean-value inequality at 0 on the slice $\{z_1 = 0, z_3 = 0,\ldots,z_n = 0\} \cap \Omega = \Omega_v(0)$. Hence there exists γ_0 such that

$$-\log R_1(0) < \frac{1}{2\pi} \int_0^{2\pi} -\log R_1(\gamma_0 e^{i\theta})\, d\theta.$$

Let u be the harmonic function on $D(r_0) = \{z \in \mathbf{C}^n: z_1 = x_1^0, z_3 = x_3^0,\ldots,z_n = 0, |z_2| < \gamma_0\}$ defined by $u|_{\partial D(r_0)} = \log R_1$. Let v be a harmonic conjugate on $D(r_0)$. $u + iv = f$ is analytic. Consider the change of coordinates $(z_1, z_2,\ldots,z_n) \to (e^{f(z_2)}z_1, z_2,\ldots,z_n) = (w_1, w_2,\ldots,w_n)$. The set of complex coordinates $(w_1,\ldots,w_n)$ is defined only when $|z_2| < \gamma_0$. We write R_1' for the Hartogs radius in the w_1-direction. For any γ sufficiently close to γ_0, $\gamma < \gamma_0$,

$$-\log R_1'(0) = -\log R_1(0) + u(0) > \frac{1}{2\pi} \int_0^{2\pi} \left(-\log R_1(\gamma e^{i\theta}) + u(\gamma e^{i\theta})\right) d\theta$$

$$= \frac{1}{2\pi} \int_0^{2\pi} -\log R_1'(\gamma e^{i\theta}).$$

On the other hand, as $\gamma \to \gamma_0$, $-\log R_1'(\gamma e^{i\theta}) \to -\log R_1(r_0 e^{i\theta}) + u(\gamma_0 e^{i\theta}) = 0$. We can therefore assume $R_1'(\gamma e^{i\theta}) > 1 - \varepsilon > R_1'(0)$ for ε sufficiently small. Consider now the complex one-dimensional family of discs $\Phi(\xi_1; \xi_2)$ indexed by ξ_1 defined by $\Phi(\xi_1, \xi_2) = (\xi_1, \xi_2, 0,\ldots,0)$ in the w-coordinates for $|\xi_1| < 1 - \varepsilon$ and $|\xi_2| < \gamma$. Since $R_1'(\gamma e^{i\theta}) > 1 - \varepsilon > R_1'(0)$ there is some $\delta < 1 - \varepsilon$ such that

$$\Phi\left(\Delta(\delta) \times \overline{\Delta(r)}\right) \subset \Omega$$

and such that $\Phi(\overline{\Delta(r)} \times \overline{\Delta(r)}) \cap \partial\Omega \neq \varnothing$. We can assume, without loss of generality, that the point $Q = (\delta, p_2, 0,\ldots,0)$ lies on $\Phi(\overline{\Delta(\delta)} \times \overline{\Delta(r)}) \cap \partial\Omega$. The boundary of the disc $(\delta, \xi_2, 0,\ldots,0)$ parametrized by $|\xi_2| < \gamma$ stays in Ω but the interior hits the boundary at p.

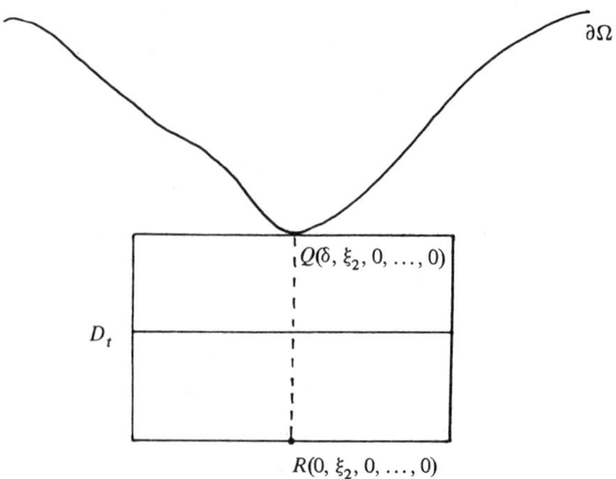

Section of some neighborhood of Q by the plane $\mathrm{Im}\, w_1 = \mathrm{Im}\, w_2 = w_3 = \cdots = w_n = 0$
after some local holomorphic change of coordinates.

The coefficient of the volume form of the metric ω, i.e., $\det(g_{i\bar{j}}) = e^{\log \det(g_{i\bar{j}})}$ is plurisubharmonic because

$$\partial\bar{\partial}\big(\log \det(g_{i\bar{j}})\big) = -(\text{Ricci curvature form}) \geq 0.$$

By the maximum principle on the disc D_t (see diagram)

$$\det\big(g_{i\bar{j}}(t, p_2)\big) \leq \sup_{\theta} \det\big(g_{i\bar{j}}(t, \gamma e^{i\theta})\big), \qquad 0 \leq t < \delta.$$

Since $\Phi(\Delta(\delta) \times \partial\Delta(r)) \subset \Omega$,

$$\det\big(g_{i\bar{j}}(t, p_2)\big) \leq C_1 \quad \text{for } 0 \leq t < \delta.$$

The Ricci curvature of the metric $(\Omega, g_{i\bar{j}})$ has a lower bound. Take a large Euclidean ball B containing Ω with the Poincaré metric $h_{i\bar{j}}$. Applying the Schwarz lemma of Yau [22] to the inclusion map $(\Omega, g_{i\bar{j}}) \to (B, h_{i\bar{j}})$ we obtain

$$(g_{i\bar{j}}) > c(\delta_{i\bar{j}}) \qquad (c > 0) \text{ as a matrix,}$$

so that

$$g_{1\bar{1}}(t, p_2) \leq C_2.$$

The length of the Euclidean line segment from $R(0, \xi_2, 0, \ldots, 0)$ to $Q(\delta, \xi_2, 0, \ldots, 0)$ is now given by

$$\int_0^{\delta} \sqrt{g_{1\bar{1}}(t, p_2)}\, dt \leq \sqrt{C_2}\,\delta,$$

contradicting the completeness of the metric ω. By contradiction, we have shown that Ω must, in fact, satisfy the Kontinuitätssatz and hence be a domain of holomorphy. The proof of the Main Theorem is complete.

3. Generalization to certain Stein manifolds.

(3.1) We have the following generalization of the first part of the Main Theorem.

THEOREM. *Suppose $\Omega \Subset X$ is a locally Stein manifold. Suppose X carries a (not necessarily complete) Kähler metric $(g_{i\bar{j}})$. Then, Ω admits a complete Kähler-Einstein metric with negative Ricci curvature under one of the following assumptions.*

(i) *The metric $(g_{i\bar{j}})$ has negative Ricci curvature and X carries a nonvanishing holomorphic n-form φ.*

(ii) *The metric $(g_{i\bar{j}})$ has positive Ricci curvature on X, and Ω carries a Hermitian metric $(h_{i\bar{j}})$ of negative holomorphic sectional curvature bounded from above by a negative constant such that $(h_{i\bar{j}}) > c(g_{i\bar{j}})$ for some $c > 0$.*

(iii) *The Kähler metric $(g_{i\bar{j}})$ on X is arbitrary. There exists a Hermitian metric $(h_{i\bar{j}})$ as in (ii) and Ω is furthermore assumed to be Stein.*

REMARK. The case of unbounded domains of holomorphy is covered by either (i), (ii), or (iii).

PROOF. We are going to construct a suitable volume form on Ω in order to apply the Schwarz lemma (1.1). We define a function ρ which is a ratio of two volume forms in the different cases:

(i)

$$\rho = \frac{\det(g_{i\bar{j}})(\frac{i}{2})^n \, dz_1 \wedge d\bar{z}_1 \wedge \cdots \wedge dz_n \wedge d\bar{z}_n}{\varphi \wedge \bar{\varphi}}.$$

(Recall that $(X, g_{i\bar{j}})$ has negative Ricci curvature and φ is a nonvanishing holomorphic n-form.)

(ii) and (iii)

$$\rho = \frac{\det(h_{i\bar{j}})}{\det(g_{i\bar{j}})}.$$

(Recall that $(\Omega, h_{i\bar{j}})$ has negative Ricci curvature and $(X, g_{i\bar{j}})$ has positive Ricci curvature.)

We define a Lipschitz volume form by

$$V = \frac{1}{d^a(-\log d)^b} \rho^k \det(g_{i\bar{j}})\left(\frac{i}{2}\right)^n dz_1 \wedge d\bar{z}_1 \wedge \cdots \wedge dz_n \wedge d\bar{z}_n$$

where $d(z) = d(z, \partial\Omega)$ is the distance function on Ω from the boundary in terms of the ambient metric on X. We scale the metric $(g_{i\bar{j}})$ on X so that $-\log d > 0$. Direct computation gives

$$\partial\bar{\partial} \log V' \qquad (V' = \text{local coefficient of the } (n, n) \text{ form } V)$$

$$= a\partial\bar{\partial}(-\log d) - \frac{b}{(-\log d)}\partial\bar{\partial}(-\log d) + \frac{b\partial d \wedge \bar{\partial}d}{d^2(-\log d)^2}$$

$$+ k\partial\bar{\partial} \log \rho + \partial\bar{\partial} \log \det(g_{i\bar{j}}).$$

Since Ω is locally Stein, by Ellencwajg [4] and Suzuki [19], $\partial\bar{\partial}(-\log d)$ has a lower bound on $(X, g_{i\bar{j}})$. We treat the three cases in two groups as below.

Case (i). We choose $a = 2$, $b = 2$. Scale the metric so that $-\log d < 1$. By choosing k large enough

$$k\partial\bar{\partial} \log \rho + \left(2 - \frac{2}{-\log d}\right)\partial\bar{\partial}(-\log d) + \partial\bar{\partial} \log \det(g_{i\bar{j}}) > (g_{i\bar{j}}).$$

Since $|\nabla d| = 1$ almost everywhere

$$\left(\frac{i}{2}\partial\bar{\partial} \log V'\right)^n \geq C\frac{1}{d^2(-\log d)^2} \det(g_{i\bar{j}})\left(\frac{i}{2}\right)^n dz_1 \wedge d\bar{z}_1 \wedge \cdots \wedge dz_n \wedge d\bar{z}_n \geq C'V$$

because ρ is bounded in case (i).

Cases (ii) *and* (iii). Since $(h_{i\bar{j}}) > c(g_{i\bar{j}})$ $(c > 0)$ and $(\Omega, h_{i\bar{j}})$ has negative Ricci curvature bounded from above by a negative constant,

$$\partial\bar{\partial} \log \det(h_{i\bar{j}}) > \varepsilon_0(g_{i\bar{j}}) \quad \text{for some } \varepsilon_0 > 0.$$

Since $\rho = \det(h_{i\bar{j}})/\det(g_{i\bar{j}})$,

$$\partial\bar{\partial}\log V' = k\partial\bar{\partial}\log\rho + \partial\bar{\partial}\log\det(g_{i\bar{j}})$$

$$+ \left(a - \frac{b}{-\log d}\right)\partial\bar{\partial}(-\log d) + \frac{b\partial d \wedge \bar{\partial}d}{d^2(-\log d)^2}$$

$$\geqslant \varepsilon_0(g_{i\bar{j}}) + (1-k)\partial\bar{\partial}\log\det(g_{i\bar{j}})$$

$$+ \left(a - \frac{b}{-\log d}\right)\partial\bar{\partial}(-\log d) + \frac{b\partial d \wedge \bar{\partial}d}{d^2(-\log d)^2}.$$

Fix $x \in \Omega$ and assume that the function $d(z) = d(z, \partial\Omega)$ is differentiable at x. Diagonalize the Hermitian matrices

$$(g_{i\bar{j}}) \quad \text{and} \quad \left(\frac{\partial^2}{\partial z_i \partial\bar{z}_j}\log\det(h_{i\bar{j}})\right)(x)$$

simultaneously so that $g_{i\bar{j}}(x) = \delta_{i\bar{j}}$ and

$$\left(\frac{\partial^2}{\partial z_i \partial\bar{z}_j}\log\det(h_{i\bar{j}})\right)(x) = \rho_i \delta_{i\bar{j}} \quad (\rho_1 \leqslant \rho_2 \leqslant \cdots \leqslant \rho_n).$$

Since $(\Omega, h_{i\bar{j}})$ has negative Ricci curvature bounded from above, $\rho_1,\ldots,\rho_n \geqslant C$ for some $C > 0$. Denote the matrix

$$\left(\frac{b}{d^2(-\log d)^2}\frac{\partial d}{\partial z_i}\frac{\partial d}{\partial\bar{z}_j}\right)(x)$$

by $(p_{i\bar{j}})$. Write

$$s = \frac{b}{d^2(-\log d)^2}.$$

Now choose k sufficiently close to 1, but less than it and set $a = b = 2 - 2k$, so that

$$\varepsilon_0(g_{i\bar{j}}) + (1-k)\partial\bar{\partial}\log\det(g_{i\bar{j}}) + (2-2k)\left(1 - \frac{1}{-\log d}\right)\partial\bar{\partial}(-\log d) \geqslant \varepsilon(g_{i\bar{j}})$$

for some $\varepsilon > 0$.

By diagonalization, it follows easily that

$$\det\left(\frac{\partial^2}{\partial z_i \partial\bar{z}_j}\log V'\right) \geqslant \det\begin{bmatrix} \varepsilon + \rho_1 & & 0 \\ & \ddots & \\ 0 & & \varepsilon + \rho_n \end{bmatrix} + p_{i\bar{j}}$$

$$\geqslant C\left[(\varepsilon + \rho_1)\cdots(\varepsilon + \rho_{n-1})(\varepsilon + \rho_n + s)\right]$$

$$\geqslant C\left[(\varepsilon + \rho_1)\cdots(\varepsilon + \rho_n)\right]^k s^{1-k} \quad \text{(arithmetic-geometric inequality)}$$

$$\geqslant C\rho^k \frac{b^{1-k}}{d^{2-2k}(-\log d)^{2-2k}} \geqslant C'V'.$$

(Recall that

$$V = \frac{1}{d^a(-\log d)^b} \rho^k \det(g_{i\bar{j}}) \left(\frac{i}{2}\right)^n dz_1 \wedge d\bar{z}_1 \wedge \cdots \wedge dz_n \wedge d\bar{z}_n.$$

Hence, we have shown that in all cases (i)–(iii),

$$\left((\tfrac{i}{2})\partial\bar{\partial} \log V'\right)^n \geqslant C'V \quad \text{for some } C' > 0.$$

Since by assumption Ω is locally Stein, in all cases $\partial\bar{\partial}(-\log d) > -c(g_{i\bar{j}})$ on Ω, by the result of Ellencwajg [4]. In cases (i) and (ii), $-\log d + C\log \rho$ is a smooth plurisubharmonic exhaustion function on Ω for C sufficiently large. It follows that Ω is a Stein manifold (Grauert [7]). In case (iii), Ω is Stein by assumption. One can, therefore, exhaust Ω by relatively compact Stein subdomains Ω_α with strictly pseudoconvex boundary. By Cheng and Yau [2, Theorem 8.4], each Ω_α carries a complete Kähler-Einstein metric with Ricci curvature $-(n + 1)$. On a coordinate patch $U \cap \Omega$, $\Delta \log V'_E = n(n + 1)$, where

$$V_E = V'_E \left(\tfrac{i}{2}\right)^n dz_1 \wedge d\bar{z}_1 \wedge \cdots \wedge dz_n \wedge d\bar{z}_n$$

denotes the volume form of the limit Kähler-Einstein metric $(p_{i\bar{j}})$ on Ω (Cheng and Yau [2, Theorem 8.4]) and Δ denotes the corresponding Laplacian. By standard smoothing arguments and the process of taking limits, we have $V_E \geqslant CV$. In all cases the Schwarz lemma of Yau [22, Theorem 1] and the modified version due to Royden [15] imply that the Kähler-Einstein metric on Ω_α dominates a constant multiple of the metric $(g_{i\bar{j}})$ on X, with a constant independent of α. In place of the global gradient estimate in the Main Theorem, we now use the gradient estimate on geodesic balls of Cheng and Yau [2]. Cover X by a finite number of open sets U_i such that there exists $U'_i \Subset U_i$ which already cover X. On $U'_i \cap \Omega$ apply the gradient estimate on geodesic balls to conclude that $|\nabla \log(\log V_i - a_i)| \leqslant C_i$ for the local coefficient V'_i of V_E on U_i and suitable a_i. On $U'_i \cap U'_j \log V'_j - \log V'_j$ are bounded. Let ρ_i be a partition of unity subordinate to (U''_i). The function $\Sigma \rho_i \log(\log V'_i - a_i)$ will then have bounded gradient. This gives by integration a lower bound for the distance function and proves the completeness of $(\Omega, p_{i\bar{j}})$.

(3.2) *Generalization to bounded domains on Stein manifolds.*

COROLLARY OF THE MAIN THEOREM. *Let $\pi: \Omega \to X$ be a domain in a Stein manifold X such that $\pi(\Omega)$ is bounded on X. Then the Main Theorem remains valid on Ω.*

PROOF. Suppose Ω is a bounded domain of holomorphy in X. To prove completeness of the limit Kähler-Einstein metric, it suffices to construct the volume form $V/d^2(-\log d)^2$ for comparison as in the proof of the Main Theorem and finally to integrate the metric using the gradient estimates of Cheng and Yau and the partition of unity argument at the end of (3.1). We can then take d as the distance to the boundary using any Kähler metric on X, take V to be $e^\varphi V_0$, where V_0 is the restriction of the Poincaré metric of some big Euclidean ball obtained by properly embedding X into some $\mathbf{C}^N$ and φ is an appropriate plurisubharmonic function to make sure that

$$\partial\bar\partial \log\left(\frac{e^{\varphi}V_0}{d^2(-\log d)^2} \right) > 0$$

in the interior. This is necessary because $-\log d$ is no longer plurisubharmonic in general, but $\partial\bar\partial(-\log d) \geq -c_i$ Kähler form on X. The restriction of the Poincaré metric to Ω carries negative bisectional curvature, so that $\partial\bar\partial \log V_0 \geq c_2$ is a Kähler form on X. The partition of unity argument for integrating the metric goes through because the limit Kähler-Einstein metric on Ω dominates a positive multiple of the Poincaré metric of the big Euclidean ball by the Schwarz lemma of Yau. The last statement also implies that the second statement of the Main Theorem is true for bounded domains in X.

Finally, we remark that the Main Theorem can be generalized to domains $\Omega \subset \mathbf{P}^n - V$ for divisors V such that $\mathbf{P}^n - V$ is uniformized by a bounded domain. One simply reduces the theorem to the covering domain by using the theorem of Stein [18] and the local nature of Steinness on domains in $\mathbf{P}^n$ (Kieselman [10], Takeuchi [20]). Examples of divisors V such that $\mathbf{P}^n - V$ is covered by a bounded domain can be obtained by the method of simultaneous uniformization due to Bers.

4. A complete Kähler metric on $B^n - \{0\}$ with bounded Ricci curvature.

(4.1) In this subsection, we shall construct an explicit complete Kähler metric with bounded Ricci curvature on $B^n - \{0\}$, which is non-Stein for $n > 1$. The particular role played by the Fubini-Study metric will be explained in the next section on characterizing domains of holomorphy by holomorphic sectional curvature. For simplicity of notation, we shall consider $n = 2$.

Construction of the metric. On $B^2 - \{0\}$ consider the Kähler metric whose Kähler form is given by $\omega = \partial\bar\partial f(r^2)$, $r^2 = |z_1|^2 + |z_2|^2$ (z_1, z_2 coordinates on $\mathbf{C}^2$), where

$$f(r^2) = \log \frac{r^2}{(-\log r^2)},$$

$$\omega = \partial\left(f'(r^2)\bar\partial r^2 \right) = f''(r^2)\partial r^2 \wedge \bar\partial r^2 + f'(r^2)\partial\bar\partial r^2.$$

At the point $x = (z_1, 0)$, $r = |z_1|$, the Kähler form has components

$$\begin{cases} g_{1\bar1}(x) = f'(r^2) + r^2 f''(r^2), \\ g_{2\bar2}(x) = f'(r^2), \\ g_{1\bar2}(x) = g_{2\bar1}(x) = 0. \end{cases}$$

Direct computation gives

$$g_{1\bar1}(x) = \frac{1}{r^2(\log r^2)^2}, \qquad g_{2\bar2}(x) = \frac{1}{r^2} + \frac{1}{r^2(-\log r^2)}.$$

The potential is asymptotically equivalent to $-\log(1 - r^2)$ (up to a function smooth up to ∂B^n) near the boundary of B^n. Direct computation shows that ω has asymptotically constant negative holomorphic bisectional curvature close to ∂B^n.

The volume form ω^2 is given by

$$\omega^2 = \left(\frac{1}{r^2} + \frac{1}{r^2(-\log r^2)} \right) \left(\frac{1}{r^2(\log r^2)} \right) \left(\frac{i}{2} \right)^2 dz_1 \wedge d\bar{z}_1 \wedge dz_2 \wedge d\bar{z}_2.$$

The Ricci form Ric ω is given by

$$\text{Ric } \omega(x) = \partial\bar{\partial} \log \frac{1}{(-\log r^2)} - 2\partial\bar{\partial}\log r^2 - \partial\bar{\partial}\log\left(1 - \frac{1}{\log r^2} \right)$$

$$= \frac{1}{r^2(\log r^2)^2} dz_1 \wedge d\bar{z}_1$$

$$+ \left(\frac{2}{r^2} + \frac{1}{r^2(-\log r^2)} \right) dz_2 \wedge d\bar{z}_2 - \partial\bar{\partial}\log\left(1 - \frac{1}{\log r^2} \right).$$

The first two terms when measured against the Kähler form ω will give negative eigenvalue -1 in the normal direction, and positive eigenvalue for x close to the origin, asymptotically equal to 2. Direct computation of $\partial\bar{\partial}\log(1 - 1/\log r^2)$ gives asymptotically

$$\frac{2}{r^2(-\log r^2)^3} dz_1 \wedge d\bar{z}_1 - \frac{1}{r^2(\log r^2)^2} dz_2 \wedge d\bar{z}_2$$

for x close to the origin, which is negligible when measured against the Kähler form ω. Hence the Kähler metric $(B^2 - \{0\}, \omega)$ has bounded Ricci curvature. Restricted to the punctured unit disc, the metric ω is nothing but the standard complete Poincaré metric for the punctured disc. Since ω is radially symmetric $(B^2 - \{0\}, \omega)$ is complete and gives the desired metric of bounded Ricci curvature.

5. Characterizing domains of holomorphy by holomorphic sectional curvature.

(5.1) As a consequence of an extension theorem due independently to Griffiths [8] and Shiffman [16], domains of holomorphy (not necessarily bounded) are characterized by the existence of a complete Hermitian metric with nonpositive holomorphic sectional curvature. On the other hand, it is easy to show that any domain admitting a complete Bergmann metric is a domain of holomorphy (see below). It is well known that the Bergmann metric has holomorphic bisectional curvature bounded from above by $+2$. Hence, it is natural to ask if positive holomorphic sectional curvature can be allowed in the result of Griffiths and Shiffman stated above. However, a lengthy but straightforward computation shows that $(B^2 - \{0\}, \omega)$ in (4.1), in fact, has bounded holomorphic sectional (even bisectional) curvature. On the positive side, we obtain

THEOREM. *Let Ω be a Riemann domain spread over $\mathbf{C}^n$ and $(g_{i\bar{j}})$ be a complete Kähler metric on Ω. For any holomorphic tangent vector ξ at $x \in \Omega$ we assume that the holomorphic sectional curvature $K(\xi, \bar{\xi})$ in the direction ξ satisfies*

$$K(\xi, \bar{\xi}) \leqslant f(d(x_0, x))$$

where $d(x_0, x)$ denotes the geodesic distance from some fixed point x_0 and $f(t)$ is a function decreasing to zero as $t \to \infty$.

REMARK. Contrary to the result of Griffiths [8] and Shiffman [16], the Kähler condition will be crucial in our argument because the area-minimizing property of complex submanifolds is only valid for the Kähler manifolds.

PROOF. As in the second statement of the Main Theorem (2.2), we shall show that Ω satisfies the Kontinuitätssatz. Otherwise one has the same picture as in §2 at some boundary point Q after some holomorphic change of coordinates (where the holomorphic coordinates w_i are replaced by z_i here).

In other words, we have a smooth one real parameter family of discs D_t such that the boundary of each disc lies in Ω and the interior D_t touches the boundary $\partial\Omega$ of Ω at Q, at $t = \delta$. We choose holomorphic local coordinates as in (2.2)

$$\text{Area of } D_t = \int_{D_t} \omega$$

where $\omega =$ Kähler form of the given metric.

We denote the cylindrical surface joining D_0 and D_t by S_t and the interior of the cylinder by C_t. Then, since ω is a closed form by the Stokes' Theorem

$$\int_{D_t} \omega + \int_{S_t} \omega - \int_{D_0} \omega = \int_{C_t} d\omega = 0$$

where C_t stands for the interior of the cylinder and D_t, S_t are given appropriate orientations. Hence,

$$\int_{D_t} \omega = \int_{D_0} \omega - \int_{S_t} \omega.$$

Since $\partial D_t \subset \Omega$ for all t,

$$\left| \int_{S_t} \omega \right| \leqslant C_1 \quad \text{for all } t, 0 \leqslant t < \delta,$$

so that Area of $D_t \leqslant C_2$ for all t, $0 \leqslant t < \delta$. Write $\omega = \frac{i}{2} \Sigma_{i,j} g_{i\bar{j}} dz_i \wedge d\bar{z}_j$. By restricting the Kähler metric to the disc D_t and using the fact that holomorphic sectional curvatures are nonincreasing after restricting to submanifolds we have on D_t

$$\frac{\partial^2}{\partial z_1 \partial \bar{z}_1}(-\log g_{1\bar{1}}(z)) \leqslant \left(\frac{R_{1\bar{1}1\bar{1}}(z)}{g_{1\bar{1}}^2(z)} \right) g_{1\bar{1}}(z) = K_{1\bar{1}}(z) g_{1\bar{1}}(z)$$

where $R_{i\bar{j}k\bar{l}}$ stands for the components of the sectional curvature tensor of the Kähler metric and the last equality gives the definition of $K_{1\bar{1}}$. Suppose V_t is a point on the horizontal disc $\overline{D}_t$, $0 \leqslant t < \delta$, where $g_{1\bar{1}}$ is maximum. Let $\{t_i\}$ be a sequence of real numbers such that $t_i \to \delta$. We observe first that the top disc D_δ may intersect $\partial\Omega$ at more than a single point. Write $S = D_\delta \cap \partial\Omega$, S is a closed subset of D_δ. We exhaust $\overline{D}_\delta - S$ by a sequence X_j such that

$$X_j = \left\{ p \in \overline{D}_\delta - S : \text{dist}(p, D_\delta \cap \partial\Omega) > \tfrac{1}{j} \right\}$$

where dist denotes the Euclidean distance. Suppose first that the points V_{t_i}, for the sequence $t_i \to \delta$ are such that the projection to the disc $\overline{D}_\delta$ all lie on some fixed X_j for i sufficiently large. Now regard X_j as an open subset of the z_1-plane. Then,

$$X_j \times [0, \delta] \Subset \Omega,$$

so that $M = \sup_{X_j \times [0,\delta]} g_{1\bar{1}} < \infty$.

By passing to the limiting disc D_δ, we see that

$$g_{1\bar{1}}(\delta, z_2, 0, \ldots, 0) \leqslant M \quad \text{for all } (\delta, z_2, 0, \ldots, 0) \in \Omega \cap D_\delta.$$

It follows that any Euclidean line segment joining a point on ∂D_δ to a point on $S = \partial\Omega \cap D_\delta$ would have bounded length, contradicting the completeness of the metric. By re-indexing $\{t_i\}$ if necessary, we can assume that $V_{t_i} \in X_i \times \{t_i\}$. By the same reasoning as above, we can furthermore assume that $g_{1\bar{1}}(V_{t_i})$ increases and tends to infinity as i tends to infinity. We assert the following lemma.

LEMMA. *The area of the disc D_{t_i} in the Kähler metric ω tends to infinity as $i \to \infty$.*

Given the lemma, the boundedness of the area of the horizontal discs D_t $(0 \leqslant t < \delta)$ will be contradicted, so that Ω must in fact satisfy the Kontinuitätssatz locally. Equivalently, this shows that Ω is a domain of holomorphy, proving the theorem.

To prove the lemma, consider the inequalities on D_{t_i}

$$\begin{cases} \Delta(-\log g_{1\bar{1}}) \leqslant K_{1\bar{1}} g_{1\bar{1}} \quad \text{and} \\ g_{1\bar{1}} \leqslant g_{1\bar{1}}(V_{t_i}) = k_i, \end{cases}$$

where Δ denotes $\partial^2/\partial z_1 \partial\bar{z}_1$.

Let $V_{t_i} = (x_i, t_i)$. Since holomorphic sectional curvature at a point x is bounded by $f(d(x_0, x))$, decreasing to zero at infinity, we can assume without loss of generality that the following statement holds: On the disc $\overline{D}_{t_i}$, $k_i = g_{1\bar{1}}(V_{t_i}) \geqslant i^3$, and the curvature $K_{1\bar{1}}$ (defined above) $\leqslant \tfrac{1}{i}$ on the Euclidean disc of radius $\tfrac{1}{i}$ centered at V_{t_i}, assumed to be in Ω. The second statement is possible after re-indexing the t_i's. (The points $\{V_{t_i}\}$ may converge to (different) points on $\partial\Omega \cap D_\delta = S$.)

Consider the function $(V_{t_i} = (x_i, t_i))$

$$h_i(z) = \log g_{1\bar{1}}(z + x_i, t_i) - \log g_{1\bar{1}}(x_i, t_i) + \frac{k_i}{i} r^2, \qquad r = |x|^2,$$

defined for z sufficiently small.

From the inequalities above,

$$
\begin{cases}
\Delta h_i \geqslant - (K_{1\bar{1}}g_{1\bar{1}})(z + x_i, t_i) + \dfrac{k_i}{i}, \\[2mm]
\quad\quad \geqslant 0 \quad \text{for } |z| < \tfrac{1}{i}, \text{ where } g_{1\bar{1}} \leqslant k_i,\ K_{1\bar{1}} \leqslant \tfrac{1}{i}, \\[2mm]
h_i(0) = 0.
\end{cases}
$$

Hence,

$$
g_{1\bar{1}}(z + x_i, t_i) = k_i e^{-k_i r^2/i} e^{h_i},
$$

$$
h_i \text{ subharmonic for } |z| < \tfrac{1}{i},\ h_i(0) = 0,
$$

$$
\int_{|z|<1/i} g_{1\bar{1}}(z + x_i, t_i)\frac{\sqrt{-1}}{2} dz_1 \wedge d\bar{z}_1 = \int_0^{1/i}\int_0^{2\pi} k_i r e^{-k_i r^2/i} e^{h_i}\, d\theta\, dr
$$

$$
\geqslant \int_0^{1/i} k_i r e^{-k_i r^2/i}\, dr = -\frac{i}{2}\int_0^{1/i} d\!\left(e^{-k_i r^2/i}\right)
$$

$$
= \frac{i}{2}\left(1 - e^{-k_i/i^3}\right)
$$

where the first inequality follows easily from the sub-mean value inequality for e^{h_i}. Since by assumption $k_i = g_{1\bar{1}}(V_{t_i}) \geqslant i^3$,

$$
\text{Area of } D_{t_i} \geqslant \int_{|z|<1/i} g_{1\bar{1}}(z + x_i, t_i)\frac{\sqrt{-1}}{2} dz_1 \wedge d\bar{z}_1
$$

$$
\geqslant \frac{i}{2}\left(1 - \frac{1}{e}\right) \to \infty \quad \text{as } i \to \infty.
$$

This finishes the proof of the lemma, hence proving the theorem by contradiction.

REMARKS. (1) Nontrivial examples of complete Kähler metrics on domains of holomorphy Ω satisfying the hypothesis in the theorem can be obtained by embedding Ω properly into some $\mathbf{C}^N$ and pulling back an appropriate complete Kähler metric with positive holomorphic sectional curvature decaying to zero at infinity. Examples of such metrics on $\mathbf{C}^N$ can be constructed by taking $\partial\bar{\partial}\varphi$ as the Kähler form, where $\varphi = \varphi(|z|)$ is an appropriate radial function on $\mathbf{C}^N$ (see, e.g., Klembeck [11]).

(2) Suppose ω is a complete Kähler metric on $B^2 - \{0\}$ with bounded holomorphic sectional curvature. Let $A_\delta(t) = $ area of $\{|z_1| < \delta,\ z_2 = t\}$, $\delta < 1$. By the area minimizing property of complex submanifolds in Kähler manifolds, $A_\delta(0) < \infty$ for all δ. For $\varepsilon > 0$, $A_\delta(0) < \varepsilon$ when δ is small enough. On the other hand, our arguments show that for any δ, there is a sequence $t_i \to 0$ such that $A_\delta(t_i) > c$, independent of δ, since the metric ω is complete. This means that for δ fixed, the function $A_\delta(t)$ is not continuous in t. This explains the role played by the Fubini-Study metric in our example in §4, which traps some mass at the origin (given by $\partial\bar{\partial}\log|z_1|^2$).

Appendix. An application of the Schwarz lemmas to the problem of Behnke-Stein.

THEOREM. *Let $(X, g_{i\bar{j}})$ be a Kähler manifold. Let $\Omega \Subset X$, $\Omega = \bigcup_\alpha \Omega_\alpha$, be a relatively compact subdomain which is the increasing union of a sequence of Stein subdomains (Ω_α). Suppose $(X, g_{i\bar{j}})$ has positive Ricci curvature in a neighborhood of $\partial\Omega$ and Ω_α admits a volume form V_α such that if $V_\alpha = V'_\alpha(\frac{i}{2})^h dz_1 \wedge d\bar{z}_1 \wedge \cdots \wedge dz_n \wedge d\bar{z}_n$ in local coordinates $(\frac{i}{2}\partial\bar\partial \log V'_\alpha)^n \geqslant C_1 V_\alpha$, $V'_\alpha \geqslant C_2 \det(g_{i\bar{j}})$ for positive constants C_1 and C_2 independent of α. Then Ω is a Stein manifold.*

PROOF. Fix any point $x \in \Omega$ and consider a coordinate ball B in a neighborhood of x. By the Schwarz lemma (1.1),

$$V_\alpha \leqslant C(\text{volume form of Poincaré metric on } B).$$

The functions $\varphi_\alpha = \log V'_\alpha - \log \det(g_{i\bar{j}})$ are defined on Ω_α. $\partial\bar\partial\varphi_\alpha \geqslant -\partial\bar\partial \log \det(g_{i\bar{j}}) > c(g_{i\bar{j}})$ $(c > 0)$ on a neighborhood of $\partial\Omega$ because $(X, g_{i\bar{j}})$ has positive Ricci curvature there. φ_α is lower bounded by assumption and upper bounded by the Schwarz lemma (1.1). The function $\varphi = (\limsup \varphi_\alpha)^*$, where the star sign denotes upper semi-continuous regularization is then defined in Ω and plurisubharmonic outside a compact set, with $\partial\bar\partial\varphi > c(g_{i\bar{j}})$ $(c > 0)$ in the generalized sense. The function $-\log d$, $d(z) = d(z, \partial\Omega)$ being the distance function in terms of the Kähler metric $(X, g_{i\bar{j}})$ satisfies $\partial\bar\partial(-\log d) > -c'(g_{i\bar{j}})$ since Ω, being locally a union of domains of holomorphy, is a locally Stein domain on X. The function $-\log d + C\varphi$ for C large enough then gives an upper semi-continuous function which is strictly plurisubharmonic outside a compact set. By the improved version of the Grauert solution of the Levi problem due to Fornaess and Narasimhan [5] (which allows discontinuity in the exhaustion function) Ω is therefore a strongly pseudoconvex manifold. The exceptional compact subvariety must, however, lie in some Ω_α for α large enough. Since Ω_α is Stein, it must be trivial, implying that Ω is, in fact, a Stein manifold.

BIBLIOGRAPHY

1. A. Andreotti and E. Vesentini, *Carleman estimates for the Laplace-Beltrami equation on complex manifolds*, Inst. Hautes Études Sci. Publ. Math. **25** (1965), 91–130.

2. S.-Y. Cheng and S.-T. Yau, *On the existence of a complete Kähler metric on non-compact complex manifolds and the regularity of Fefferman's equation*, Comm. Pure Appl. Math. **33** (1980), 507–544.

3. K. Diederich and P. Pflug, *Über Gebiete mit vollständiger Kählermetrik*, Math. Ann. **257** (1981), 191–198.

4. G. Ellencwajg, *Pseudoconvexité locale dans les variétés kählériennes*, Ann. Inst. Fourier (Grenoble) **25** (1975), 295–314.

5. J. E. Fornaess and R. Narasimhan, *The Levi problem on complex spaces with singularities*, Math. Ann. (1980), 47–72.

6. H. Grauert, *Characterisierung der Holomorphiegebiete durch die vollständige Kählersche Metrik*, Math. Ann. **131** (1956), 38–75.

7. ______, *On Levi's problem and the imbedding of real-analytic manifolds*, Ann. of Math. (2) **68** (1958), 460–472.

8. P. A. Griffiths, *Two theorems on extension of holomorphic mappings*, Invent. Math. **14** (1971), 27–62.

9. L. Hörmander, L^2-*estimates and existence theorems of the $\bar\partial$-operator*, Acta Math. **113** (1965), 89–152.

10. C. O. Kieselman, *On entire functions of exponential type and indicators of analytic functionals*, Acta Math. **117** (1967), 1–35.

11. P. Klembeck, *Complete Kähler metrics with positive holomorphic bisectional curvature* (preprint).

12. K. Oka, *Domaines pseudoconvexes*, Tôhoku Math. J. **49** (1942), 15–52.

13. ______, *Domaines finis sans point critique intérieur*, Japan. J. Math. **23** (1953), 97–115.

14. T. Ozawa, *On complete Kähler domains with C^1-boundary*, Publ. Res. Inst. Math. Sci. Kyoto Univ. **16** (1980), 929–940.

15. H. Royden, *The Ahlfors-Schwarz lemma in several complex variables* (preprint).

16. B. Shiffman, *Extension of holomorphic maps into hermitian manifolds*, Math. Ann. **194** (1971), 249–258.

17. H. Skoda, *Application des techniques L^2 à la théorie des ideaux d'une algèbre de fonctions holomorphes avec poids*, Ann. Sci. École Norm. Sup. **5** (1972), 545–580.

18. K. Stein, *Überlagerungen holomorph-vollständiger komplexer Räume*, Arch. Math. **7** (1956), 354–361.

19. Suzuki, *Pseodoconvex domains on a Kähler manifold with positive holomorphic bisectional curvature*, Publ. Res. Inst. Math. Sci. **12** (1976/77), 191–214; ibid. **12** (1976/77), 439–445.

20. A. Takeuchi, *Domaines pseudoconvexes infinis et la métrique riemannienne dans un espace projectif*, J. Math. Soc. Japan **16** (1964), 159–181.

21. S.-T. Yau, *Harmonic functions on complete Riemannian manifolds*, Comm. Pure Appl. Math. **28** (1975), 201–228.

22. ______, *A general Schwarz lemma for Kähler manifolds*, Amer. J. Math. **100** (1978), 197–203.

DEPARTMENT OF MATHEMATICS, STANFORD UNIVERSITY, STANFORD, CALIFORNIA 94305

Current address (S.-T. Yau): Department of Mathematics, Institute for Advanced Study, Princeton, New Jersey 08540

Current address (Ngaiming Mok): Department of Mathematics, Princeton University, Princeton, New Jersey 08544

Proceedings of Symposia in Pure Mathematics
Volume 39 (1983), Part 1

Symplectic Geometry

ALAN WEINSTEIN

0. Introduction. Classical mechanics in the time of Huygens (1629–1695) and Newton (1642–1727) was very geometrical. Although Newton invented the calculus in order to formulate and solve physical problems, many of his arguments made heavy use of euclidean geometry. After Newton, there came a period of "mécanique analytique," during which Lagrange (1736–1813) could boast that his treatise on mechanics contained no pictures.[1] Following the path of Euler (1707–1783) and Lagrange, Jacobi (1804–1851) and Hamilton (1805–1865) continued the development of analytic techniques for the explicit solution of the differential equations describing mechanical systems. Finally, geometry has taken a new role in mechanics through the contributions of Poincaré (1854–1912) and Birkhoff (1884–1944). Now, though, the geometry is the more flexible geometry of canonical (in particular, area preserving) transformations instead of the rigid geometry of Euclid; accordingly, the conclusions of the geometrical arguments are often qualitative rather than quantitative.

In this lecture (and paper), I would like to explain what symplectic geometry is and to describe its role in contemporary mathematics. I think it is not unfair to say that symplectic geometry is of interest today, not so much as a theory in itself, but rather because of a series of remarkable "transforms" which connect it with various areas of analysis.[2]

The *lagrangian submanifolds* play an especially important part in symplectic geometry and its applications. In §3 of this lecture, I will outline an approach to symplectic geometry in terms of a "category" in which the morphisms are exactly the lagrangian submanifolds; this approach suggests some interesting

---

Reprinted from Bulletin Amer. Math. Soc. (N.S.) 5 (1981), 1–13.

1980 *Mathematics Subject Classification.* Primary 58F05.

[1] Poinsot (1777–1859) reacted strongly against this analytical tradition. Referring to "l'illustre Lagrange" in his famous study of rigid body rotation, Poinsot wrote that in Lagrange's treatment of the subject, "on ne voit guère que des calculs, sans aucune image nette de la rotation du corps." (I would like to thank J. J. Duistermaat for calling my attention to Poinsot's vivid critique of analytical mechanics, as well as for his comments on this manuscript.)

[2] G. D. Birkhoff's "disturbing secret fear that geometry may ultimately turn out to be no more than the glittering intuitional trappings of analysis" [BI] may be especially appropriate when applied to symplectic geometry. I learned of Birkhoff's statement from Chern [C], who tends to dismiss the fear on the grounds that Birkhoff was an analyst. The recent success of symplectic geometric methods in linear partial differential equations (see §5 for an example) suggests that one might need the glitter to find the gold. This opinion is also expressed by Poinsot (see previous footnote) who suggests that calculations are merely a tool in the service of geometrical and mechanical reasoning.

new problems as well as unifying previous results. Some of the ideas in this section were developed independently by Guillemin and Sternberg, implicitly in [**G-S1**] and explicitly in §§9 and 10 of [**G-S2**].

1. Symplectic spaces and manifolds. The model space for symplectic geometry is the $2n$-dimensional cartesian space $\mathbf{R}^{2n}$ with *symplectic structure* provided by the differential form $\Omega_n = \Sigma_{j=1}^{n} dq_i \wedge dp_i$ in coordinates $(q_1, \ldots, q_n, p_1, \ldots, p_n)$. Eventually, we shall consider differentiable transformations preserving this structure, but for a moment let us look at the linear ones, from which the name "symplectic" comes.

The planes (through the origin) in $\mathbf{R}^{2n}$ on which Ω_n vanishes (as a bilinear form) form a hypersurface $\mathcal{C}$ in the Grassmann manifold $G_{2,2n}(\mathbf{R})$. The Grassmann manifold may be identified with the space of lines in $\mathbf{R}P^{2n-1}$, and a hypersurface in this space of lines is called a *line complex* in projective geometry. These complexes were studied extensively by Plücker (1801–1868) and others in the 19th century. The complex $\mathcal{C}$ is called a linear complex, since it is defined by linear equations on the exterior product $\mathbf{R}^{2n} \wedge \mathbf{R}^{2n}$, so the group of projective transformations leaving $\mathcal{C}$ invariant (equal to the projectivization of the linear group leaving Ω_n invariant) was called the *linear complex group*. In 1946, Hermann Weyl [**WL**] decided that the terminology was too confusing to perpetuate, so he took the Latin roots in com-plex (meaning "plaited together") and replaced them by the Greek roots symplectic.[3] The name *symplectic group* is now universally used for the group of linear transformations preserving Ω_n.

The mathematical point to be retained from this linguistic digression is that the symplectic structure is essentially determined by the subspaces on which it vanishes. In addition to the planes in $\mathbf{R}^{2n}$ on which Ω_n vanishes, we may consider larger subspaces with this property, called *isotropic*. The maximal isotropic subspaces have dimension n and are called *lagrangian* (following Maslov [**MAS**]).

A submanifold L of $\mathbf{R}^{2n}$ is called lagrangian if its tangent space at each point is lagrangian, i.e. if L is n-dimensional, and the pullback of Ω_n to L vanishes. If L projects diffeomorphically onto the subspace $\{(q_1, \ldots, q_n, 0, \ldots, 0)\}$, so that L is defined by equations of the form $p_i = \theta_i(q_i, \ldots, q_n)$, then it is easy to verify that L is lagrangian if and only if $\partial\theta_i/\partial q_j = \partial\theta_j/\partial q_i$ on $\mathbf{R}^n$, i.e. if and only if there is a function $S(q_1, \ldots, q_n)$ such that $\theta_i = \partial S/\partial q_i$. Such a function S is called a *generating function* for the lagrangian submanifold L. One of the most important recent developments in the applications of symplectic geometry is the discovery that, if we identify this L with the function e^{iS}, then we can extend this identification so that more general lagrangian submanifolds (not necessarily projecting diffeomorphically onto $\mathbf{R}^{2n}$) correspond to generalized functions on $\mathbf{R}^n$. (See [**G-S1**] for a detailed account of this correspondence.) In a sense, we may think of the

---

[3] According to the Oxford English Dictionary, the only use of "symplectic" in English is with reference to a certain bone in the head of fish. The dictionary gives a citation from Todd's Cyclopedia of Anatomy (1839–47). A picture of the symplectic bone may be found on p. 67 of [**BO**]; it appears that the bone is quite small, but that it may serve to hold things together.

lagrangian submanifolds themselves as being a generalization of functions.

The symplectic group is too rigid for many purposes. The class of lagrangian submanifolds, for instance, is invariant under the group of all diffeomorphisms $f: \mathbf{R}^{2n} \to \mathbf{R}^{2n}$ for which $f^*\Omega_n = \Omega_n$. More generally, we may consider diffeomorphisms between open subsets of $\mathbf{R}^{2n}$ which preserve Ω_n; these diffeomorphisms are called canonical transformations, symplectic diffeomorphisms, or (following Souriau [**SO**]), *symplectomorphisms*. We may define a symplectic manifold to be a differentiable manifold with an atlas for which the coordinate changes are symplectomorphisms. Alternately, we may define a symplectic structure on a manifold P as a closed 2-form Ω which is nondegenerate in the sense that the bundle map $\tilde{\Omega}: TP \to T^*P$ defined by $[\tilde{\Omega}(v)](w) = \Omega(v, w)$ is an isomorphism. It is a theorem of Darboux that, near any point in a manifold with a symplectic structure, there are local coordinates $(q_1, \ldots, q_n, p_1, \ldots, p_n)$ for which $\Omega = \Sigma_i \, dq_i \wedge dp_i$, so the two definitions are equivalent.

An important class of symplectic manifolds consists of the cotangent bundles T^*X. Every local coordinate system $(q_1, \ldots, q_n)$ on X gives rise to local coordinates $(q_1, \ldots, q_n, p_1, \ldots, p_n)$ on T^*X, and the form $\Sigma \, dq_1 \wedge dp_i$ is independent of the choice of such coordinates; we denote the symplectic structure on T^*X by Ω_X. The image of a section $\sigma: X \to T^*X$ is a lagrangian submanifold of (T^*X, Ω_X) if and only if $d\sigma = 0$; if $\sigma = dS$, S is called a generating function for $\sigma(X)$.

2. Symplectic geometry as Lagrange did it. The first symplectic manifold was introduced by Lagrange [**LA1**] in 1808.[4] In studying the motion of the planets under the influence of their mutual gravitational interaction, he took as a starting point the elliptical motion of a single planet around the sun. (Actually, the focus of the ellipse is the center of the mass of the sun-planet system, but we will ignore this point here.) This ellipse was then considered to "drift" under the disturbing influence of the other planets. The drift was described by a system of differential equations, and Lagrange sought to put these equations in the simplest possible form.

Let $\mathscr{E}$ be the set of possible elliptical motions of a planet. A member of $\mathscr{E}$ may be described by six real numbers, called *elements* of the orbital motion. For instance, one may take the major axis and eccentricity of the ellipse, two variables describing the orientation of the plane containing the ellipse, one variable giving the orientation of the ellipse within this plane, and a final variable specifying the time at which the planet passes a designated point on the ellipse. In modern terms, we may say that $\mathscr{E}$ is a six-dimensional manifold, with the elements forming a local coordinate system. (Chapter 9 of [**A-M**] contains a nice discussion of various systems of elements.)

Let $a_1, \ldots, a_6$ be a system of elements. The problem faced by Lagrange and his contemporaries, and solved by them in many different ways, was to write down a system of first-order differential equations for the a_j's. Finally, in 1808, Lagrange came up with a formulation of these equations which was

---

[4] In this section, we take the symposium title "Mathematical Heritage of Poincaré" to refer to what Poincaré inherited, rather than what he left for us.

much simpler than any which had been found before. He first constructed, for any system of elements, and each pair of indices with $1 \leqslant i, j \leqslant 6$, a "bracket" $[a_i, a_j]$ which, like each of the a_i's, is a function on $\mathcal{E}$. The bracket expressions (now called *Lagrange brackets*) satisfy the *antisymmetry* condition

$$[a_i, a_j] = -[a_j, a_i]. \tag{1}$$

Lagrange also constructed a real valued function Φ on $\mathcal{E}$, called the *disturbing function*, which depends on the perturbing forces but is independent of the choice of elements, such that the drift caused by the perturbation satisfies the equations:

$$\frac{\partial \Phi}{\partial a_i} = \sum_{j=1}^{6} [a_i, a_j] \frac{da_j}{dt}. \tag{2}$$

The matrix of brackets satisfies the *nondegeneracy* condition

$$\det([a_i, a_j]) \neq 0 \tag{3}$$

so there is an inverse matrix (b_{ij}), and the equations (2) become

$$\frac{da_i}{dt} = \sum_{j=1}^{6} b_{ij} \frac{\partial \Phi}{\partial a_j} \tag{4}$$

in which the derivatives of the elements are expressed in terms of one function instead of six. At the time of Lagrange, this advance was of advantage largely because it shortened by a factor of six some rather complicated computations. The theoretical properties of equations of the form (4) were yet to be explored, beyond the following further results due to Lagrange.

The antisymmetry of the matrix (b_{ij}) implies that the total derivative of Φ with respect of time along a solution of (4) is zero; i.e. the disturbing function Φ is a conserved quantity for the drift motion. This conservation law could be used to check computations, as well as in discussions of stability.

Lagrange also found another invariant for the solutions of (4), which we may describe as follows. Let $f_t \colon \mathcal{E} \to \mathcal{E}$ be the mapping which associates to each elliptical orbit e the elliptical orbit $f_t(e)$ to which the planet will have drifted from e after the perturbing forces have acted for t units of time. For any value of t, the six functions $a_i \circ f_t$ are a new set of elements, and the brackets $[a_i \circ f_t, a_j \circ f_t]$ may be formed. Lagrange's last invariance principle states that the functional form of the expression of $[a_i \circ f_t, a_j \circ f_t]$ in terms of the $a_i \circ f_t$'s is independent of time, i.e.

$$[a_i \circ f_t, a_j \circ f_t] = [a_i \circ a_j] \circ f_t. \tag{5}$$

Finally, in [**LA2**], Lagrange found special elements $(q_1, q_2, q_3, p_1, p_2, p_3)$ for which $[q_i, q_i] = [p_i, p_j] = 0$ and $[p_i, q_j] = \delta_{ij}$. In terms of these elements, the equations (4) become

$$dq_i/dt = \partial \Phi/\partial p_i, \qquad dp_i/dt = -\partial \Phi/\partial q_i \tag{6}$$

which are known today as *Hamilton's equations*.

In modern differential geometric terms, we may identify the Lagrange brackets $[a_i, a_j]$ as the coefficients $\Omega(\partial/\partial a_j, \partial/\partial a_i)$ of a differential 2-form Ω.

Condition (1) means that Ω is a skew-symmetric form, while condition (3) means that Ω is nondegenerate in the sense defined in §1. Equation (2) is now expressed in the form $d\Phi = \tilde{\Omega} \circ \xi$, where ξ is the vector field describing the drift motion, and equation (4) becomes $\xi = \tilde{\Omega}^{-1} \circ d\Phi$. With f_t defined as the map at time t of the flow generated by ξ, equation (5) states that $f_t^*\Omega = \Omega$, i.e. Ω is invariant under the flow. It turns out that this invariance condition (for all functions Φ) is equivalent to the form Ω being *closed*, i.e. $d\Omega = 0$. Note that if Σ is any 2-cycle in $\mathfrak{G}$, then $\int_\Sigma \Omega = \int_{f_t(\Sigma)} \Omega$. The invariance principle in this form was discovered by Poincaré, who called Ω an *integral invariant*. Finally, in terms of the special elements $(q_1, q_2, q_3, p_1, p_2, p_3)$, Ω is equal to $\Sigma_i\, dq_i \wedge dp_i$, as in the conclusion of Darboux's theorem.

Since the time of Lagrange, the general form (4) of the equations of drift (or motion) seems largely to have been abandoned in favor of the "canonical" form (6). Recently, inspired by the geometric point of view, R. Littlejohn [**LI**] has found that the use of (6) is well suited to some computations concerning the stability of motion of a charged particle in a magnetic field, a problem of interest in the physics of accelerators, plasmas, and the earth's atmosphere. This application of geometrically based ideas to a concrete physical problem is very much in the spirit of Poincaré's work in classical mechanics.

3. The symplectic "category". The constructions described in the last section can be carried out on any symplectic manifold (P, Ω). If ξ is a vector field on P, then the flow of ξ leaves Ω invariant if and only if $\tilde{\Omega} \circ \xi$ is *closed* 1-form. As we saw in §1, $(\tilde{\Omega} \circ \xi)(P)$ is a lagrangian submanifold of T^*P, so we have a correspondence between infinitesimal symplectomorphisms of P and certain lagrangian submanifolds of T^*P. The correspondence is an illustration of what I might call the "symplectic creed":

EVERYTHING IS A LAGRANGIAN SUBMANIFOLD

In practice, the symplectic creed means that one should try to express objects and constructions in symplectic geometry in terms of lagrangian submanifolds. This approach, which lends a certain unity to symplectic geometry, is particularly fruitful in view of the transforms of Hörmander [**HR**], Maslov [**MAS**], and Sato [**S-K-K**] which assign classes of generalized functions to lagrangian submanifolds of cotangent bundles.

After the images of closed 1-forms, the next most important lagrangian submanifolds are the graphs of symplectomorphisms. If (P_1, Ω_1) and (P_2, Ω_2) are symplectic manifolds, the product $P_1 \times P_2$ may be given the symplectic structure $\pi_1^*\Omega_1 - \pi_2^*\Omega_2$, where $\pi_j\colon P_1 \times P_2 \to P_j$ are the cartesian projections. Now a diffeomorphism $f\colon P_2 \to P_1$ is a symplectomorphism if and only if its graph is lagrangian in $(P_1 \times P_2, \pi_1^*\Omega_1 - \pi_2^*\Omega_2)$. Since a symplectomorphism is also called a canonical transformation, and an arbitrary subset of $P_1 \times P_2$ is a relation, the lagrangian submanifolds of $(P_1 \times P_2, \pi_1^*\Omega_1 - \pi_2^*\Omega_2)$ have been named by Hörmander [**HR**] *canonical relations* from P_2 to P_1. (They were also studied by Sniatycki and Tulczejew [**SN-T**], who called them symplectic relations.) It was proven by Hörmander that the set-theoretic composition of canonical relations is again a canonical relation, provided that a transversality condition is satisfied. (A weaker form of the transversality condition was found to be sufficient by Guillemin [G] and the author [WE1].)

To systematize the application of the symplectic creed, we may define the symplectic "category" $\mathcal{S}$, in which the objects are symplectic manifolds, the morphisms are canonical relations, and composition of morphisms is set-theoretic composition of relations. $\mathcal{S}$ is not a true category, since not all compositions are morphisms in $\mathcal{S}$, but we can still use the language of category theory to study canonical relations from this point of view.[5]

A *point* in a category is an object which has only one endomorphism; in $\mathcal{S}$, any point is isomorphic to the zero-dimensional manifold $\mathbf{R}^0$; the only 2-form on $\mathbf{R}^0$ is 0, but this is a symplectic structure. Now we may define the *elements* of an object in a category to be the morphisms from a point to the object. In $\mathcal{S}$, the elements of (P, Ω) are the lagrangian submanifolds of $(P \times \mathbf{R}^0, \pi_1^*\Omega - \pi_2^*0)$, which is naturally isomorphic to (P, Ω); i.e. the elements of (P, Ω) are its lagrangian submanifolds.

The symplectic manifold $(P_1 \times P_2, \pi_1^*\Omega_1 + \Omega_2^*\pi_2)$ plays the role in $\mathcal{S}$ of the product $(P_1, \Omega_1) \times (P_2, \Omega_2)$; a pair of morphisms from (P_j', Ω_j') to (P_j, Ω_j) $(j = 1, 2)$ gives rise to a morphism from $(P_1', \Omega_1') \times (P_2', \Omega_2') \to (P_1, \Omega_1) \times (P_2, \Omega_2)$ (note that the projections are not in $\mathcal{S}$). The disjoint union of (P_1, Ω_1) and (P_2, Ω_2) acts as a sum.

A reflexive duality in $\mathcal{S}$ is given by the contravariant functor which assigns to each (P, Ω) the manifold $(P, \Omega)^* = (P, -\Omega)$ and to each canonical relation its inverse. Notice that the morphisms from (P_2, Ω_2) to (P_1, Ω_1) are precisely the elements of $(P_1, \Omega_1) \times (P_2, \Omega_2)^*$; it is reasonable, therefore, to denote the latter product by $\mathrm{Hom}((P_2, \Omega_2), (P_1, \Omega_1))$. (Compare this with the isomorphism $\mathrm{Hom}(V_2, V_1) \simeq V_1 \otimes V_2^*$ in the category of vector spaces.)

For any object (P, Ω) in $\mathcal{S}$, the object $\mathrm{Hom}((P, \Omega), (P, \Omega))$, which we denote by $\mathrm{End}(P, \Omega)$, has the structure of a "*-algebra" in the category $\mathcal{S}$. Namely, we have:

(i) A multiplication morphism from $\mathrm{End}(P, \Omega) \times \mathrm{End}(P, \Omega)$ to $\mathrm{End}(P, \Omega)$, i.e. from $(P, \Omega) \times (P, \Omega)^* \times (P, \Omega) \times (P, \Omega)^*$ to $(P, \Omega) \times (P, \Omega)^*$, given by

$$\{(p_1, p_2, p_3, p_4, p_5, p_6) \in P^6 | p_1 = p_3, p_2 = p_6, p_4 = p_5\};$$

(ii) An identity element of $\mathrm{End}(P, \Omega)$ given by the diagonal in $P \times P$;

(iii) An involution from $\mathrm{End}(P, \Omega)$ to its dual given by

$$\{(p_1, p_2, p_3, p_4) | p_1 = p_4, p_2 = p_3\}.$$

Another important example of a *-algebra in $\mathcal{S}$ is given by (T^*G, Ω_G), where G is a Lie group; the operations are:

(i) the multiplication morphism

$$M \subseteq (T^*G, \Omega_G) \times (T^*G, -\Omega_G) \times (T^*G, -\Omega_G)$$

given by $\{((g_1, \gamma_1), (g_2, \gamma_2), (g_3, \gamma_3)) | \gamma_j \in T_{g_j}^*G, g_1 = g_2 g_3, \gamma_2 = r_{g_3}^*\gamma_1, \gamma_3 = l_{g_2}^*\gamma_1\}$, where l_g^* and r_g^* are the pullbacks by left and right translation by g in

---

G. M may also be described as the normal bundle in $T^*(G \times G \times G)$ of the graph $\{(g_1, g_2, g_3) | g_1 = g_2 g_3\} \subseteq G \times G \times G$, with the sign of the last two covectors reversed; analytically, M is the wavefront set (see **[HR]**) of the convolution operator on the group algebra of G.

(ii) The identity element given by the cotangent space at the identity $T_e^* G \subseteq T^* G$.

(iii) The involution from $(T^* G, \Omega_G)$ to its dual given by $\{((g_1, \gamma_1), (g_2, \gamma_2)) | g_1 = g_2^{-1}, \gamma_1 = -\text{inv}^* \gamma_2\}$, where inv: $G \to G$ is the inversion mapping.

For every manifold X, the symplectic manifold $(T^* X, \Omega_X)$ is a commutative *-algebra in S, the operations being:

(i) The multiplication morphism $\{(x_1, \xi_1), (x_2, \xi_2), (x_3, \xi_3) | x_1 = x_2 = x_3, \xi_1 = \xi_2 + \xi_3\}$, which is the wavefront set of the ordinary multiplication operator for functions on X.

(ii) The identity element given by the zero section in $T^* X$.

(iii) The involution from $(T^* X, \Omega_X)$ to its dual given by $\{((x_1, \xi_1), (x_2, \xi_2)) | x_1 = x_2, \xi_1 = -\xi_2\}$.

Now we may consider homomorphisms between *-algebras in S, i.e. morphisms which are compatible with the three parts of the *-algebra structure. A simple example is given by the "Fourier transform", as follows. Let $G = X = \mathbf{R}^n$, considered first as a Lie group and then as a manifold, so that $T^* \mathbf{R}^n$ is a *-algebra in two different ways. A homomorphism from the "convolution" to the "multiplication" structure is given by

$$\mathcal{B} = \{((x, \xi), (g, \gamma)) | x \in \mathbf{R}^n, g \in \mathbf{R}^n, \xi \in (\mathbf{R}^n)^*, \gamma \in (\mathbf{R}^n)^*,$$

$$\gamma = Bx, \xi = -Bg\},$$

where $B: \mathbf{R}^n \to \mathbf{R}^{n*}$ is any symmetric linear transformation. (B need not be invertible for $\mathcal{B}$ to be a canonical relation, but only for B invertible is $\mathcal{B}$ a homomorphism from the multiplication to the convolution structure.)

Another important class of homomorphisms in S arises from actions of Lie groups on symplectic manifolds. Whenever a Lie group G acts differentiably by symplectomorphisms on a symplectic manifold (P, Ω), there is an induced Lie algebra homomorphism α from the Lie algebra $\mathfrak{g}$ of G to the Lie algebra $\Xi(P, \Omega)$ of infinitesimal symplectomorphisms of (P, Ω). It is known that, for purposes of quantization, some extra structure should be brought in; namely, the homomorphism α should be lifted to a homomorphism β: $\mathfrak{g} \to C^\infty(P, \Omega)$; here, $C^\infty(P, \Omega)$ has the Lie algebra structure given by the *Poisson bracket* operation $\{\Phi_1, \Phi_2\} = \Omega(\xi_{\Phi_1}, \xi_{\Phi_2})$, where $\xi_\Phi = \tilde{\Omega}^{-1} \circ d\Phi$. (The Jacobi identity for this operation is equivalent to the equation $d\Omega = 0$.) Another way to look at the homomorphism β is as a map $\mu: P \to \mathfrak{g}^*$ defined by $[\mu(p)](v) = [\beta(v)](p)$, for $p \in P$ and $v \in \mathfrak{g}$. If G is connected, the map μ, called a *momentum map* for the action, is equivariant with respect to the coadjoint representation of G on $\mathfrak{g}^*$; for general G, this equivariance must be taken as an additional condition.

All this extra structure associated with a group action becomes quite natural from the point of view of the symplectic category. Given an action of

G on P and a mapping $\mu: P \to \mathfrak{g}^*$, we may define a mapping $A: P \times G \to P \times P \times T^*G$ by the formula

$$A(p, g) = \left(gp, p, l_{g^{-1}}\mu(p) \right).$$

It turns out (see p. 21 of [WE3], or [WE4]) that $L = A(P \times G)$ is a homomorphism from (T^*G, Ω_G) to $\mathrm{End}(P, \Omega)$ if and only if μ is an equivariant momentum mapping coming from a Lie algebra homomorphism $\mathfrak{g} \to C^\infty(P, \Omega)$; the action of G on P, as well as μ, can then be recovered from L. In terms of $\mathcal{S}$, therefore, L *is the action*; it is an action of the symplectic $*$-algebra T^*G on P and contains all the extra structure which is usually obtained from quantization theory.

4. Quantization. The mathematical quantization problem arises from the physical problem of assigning quantum-mechanical observables (self-adjoint operators on a Hilbert space) to corresponding classical-mechanical ones (real valued functions on $\mathbf{R}^{2n}$). If one requires that Poisson brackets go over to (a multiple of) commutator brackets, then the quantization problem has no solution. (Van Hove's theorem; see [A-M].) One way to make progress on the quantization problem is to make it simultaneously harder and easier. We enlarge the class of classical objects which are to be quantized, but we relax the requirements by which the quantum and classical objects are to correspond.

The first step in enlarging the problem is to replace the observables by the groups of which they are the infinitesimal generators. Our basic classical objects are then symplectomorphisms, which should be represented by unitary operators on quantum Hilbert spaces. This idea was introduced already by Weyl and von Neumann as a way of avoiding some of the analytical difficulties connected with unbounded operators.

The category $\mathcal{S}$ suggests an approach to the quantization problem by a still further extension. First of all, to every symplectic manifold (P, Ω) we try to assign a Hilbert space $H(P, \Omega)$. This should be done in such a way that $H(P, -\Omega)$ is the dual of $H(P, \Omega)$, and $H[(P_1, \Omega_1) \times (P_2, \Omega_2)]$ is a tensor product of $H(P_1, \Omega_1)$ and $H(P_2, \Omega_2)$. Now, for every lagrangian submanifold $L \subseteq (P, \Omega)$, we should try to assign an element of $H(P, \Omega)$. In particular, for every morphism in $\mathcal{S}$ from (P_2, Ω_2) to (P_1, Ω_1), we will get an operator from $H(P_2, \Omega_2)$ to $H(P_1, \Omega_1)$. Simple examples show that it may take more data than just L to specify an element of H; furthermore, the analytic object constructed from L may be in some extension of H (e.g. a distribution rather than an L^2 function). Nevertheless, the general idea outlined here has proven extremely fruitful when complemented by hard analysis in specific situations.

An excellent example of the quantization procedure just outlined is the theory of Fourier integral operators. (See the lecture of Leray in this symposium for another example.) Here the symplectic manifolds considered are all obtained from cotangent bundles with their zero sections deleted by taking duals and products; the Hilbert space assigned to $(\dot{T}X, \Omega_X)$ is the space $|X|^{1/2}$ of L^2 half-densities on X; the lagrangian submanifolds quantized are those which are invariant under multiplication by real positive scalars. If Γ_f is the graph of a homogeneous symplectomorphism $f\colon \dot{T}^*X \to \dot{T}^*Y$, then Γ_f is quantized by a whole class C_f of operators from $|X|^{1/2}$ to $|Y|^{1/2}$. Barring a possible obstruction of Fredholm index type (see [WE2]) one can choose a unitary operator U_f in C_f. Ideally, one would like to choose the U_f's such that U_{fg} is always equal to $U_f U_g$, but such a choice is probably impossible. (A rigorous proof of this impossibility may not have been published anywhere, but the ideas behind van Hove's theorem should apply.) Instead, one finds that $U_{fg} - U_f U_g$ is generally a nonzero pseudodifferential operator of order -1. It is unknown to what extent (for specific sets or groups of f's) the error terms $U_{fg} - U_f U_g$ can be made to have lower order, or be smoothing operators, or be zero. One result in the positive direction is that any compact group of homogeneous symplectomorphisms can be represented homomorphically by unitary Fourier integral operators. This was proven by the author (unpublished) by interpreting the operator-theoretic constructions of [H-K] in terms of $\mathbb{S}$, and by Boutet de Monvel with the aid of his abstract Toeplitz structures [BU].

The theory of Fourier integral operators illustrates an important feature of quantization, which is also stated quite explicitly by Maslov [MAS]. The Hilbert spaces associated to symplectic manifolds usually carry some sort of filtration, and the quantization is only "correct" to within a certain degree of accuracy as measured by the filtrations. It is only in special situations that approximate or asymptotic correspondences can be made into exact ones, and in those cases there are usually some ideas from outside of symplectic geometry which are brought into play (e.g. representation theory in [AU-K], and the theory of hyperbolic equations in [D-H]).

To close this section, we shall briefly describe the "geometric quantization" scheme of Kostant [KO1] and Souriau [SO], motivated by earlier work of Segal [SE] and Kirillov [KI1]. To quantize a symplectic manifold (P, Ω) by this scheme, one begins by finding a principal circle bundle $Q \to P$ whose curvature form is Ω. One then defines $H(P, \Omega)$ to be, roughly, a space of sections of the associated complex line bundle which are covariant constant along the leaves of a foliation by lagrangian manifolds; such a foliation is called a *polarization*. We highly recommend the recent monograph by Sniatycki [SN] for a thorough discussion of this quantization theory.

The category $\mathbb{S}$ might prove useful in geometric quantization; here are a few suggestions. First of all, the circle bundle and polarization constructions can be interpreted in $\mathbb{S}$. Instead of looking at the space Q, which with the connection form ϕ is a contact manifold, we may consider its "symplectification" $\hat{Q} = \{r\phi(q) \in T^*Q | q \in 0, r > 0\}$ with the symplectic structure induced by the embedding in (T^*Q, Ω_Q). There is a natural action of the symplectic algebra T^*S^1 on Q, coming from a symplectic action of S^1 on Q, and one of the corresponding reduced manifolds (see [AR], [MA-W], or [MEY] for the theory of *reduction*) is (P, Ω). The other reduced manifolds are the manifolds $(P, r\Omega)$ for $r > 0$; since one can now let $r \to \infty$, the circle bundle construction introduces asymptotics into the situation. (Some of the ideas expressed here are borrowed from Boutet de Monvel and Guillemin [BU-G], who develop a quantization procedure for contact manifolds which leads to a remarkable unification of spectral theory with the theory of the Hilbert polynomial in algebraic geometry.)

Polarizations can be interpreted in $\mathcal{S}$, too. If $\mathcal{L}$ is a foliation of (P, Ω) by lagrangian submanifolds, we consider the submanifold $C_{\mathcal{L}} \subseteq P \times P$ defined as $\{(p_1, p_2) | p_1 \text{ and } p_2 \text{ are in the same leaf of } \mathcal{L}\}$. This is an immersed *coisotropic* submanifold of $\mathrm{End}(P, \Omega)$. Now every coisotropic submanifold C of a symplectic manifold (P_1, Ω_1) carries the *characteristic foliation* by lagrangian submanifolds, and we may define an immersed lagrangian submanifold $L(C) \subseteq \mathrm{End}(P_1, \Omega_1)$ by

$$L(C) = \{(c_1, c_2) | c_1 \in C, c_2 \in C,$$

$$c_1 \text{ and } c_2 \text{ are in the same leaf of the characteristic foliation}\}.$$

$L(C)$, considered as a morphism from $\mathrm{End}(P_1, \Omega_1)$ to itself, is idempotent, so it is like a projection operator in $\mathcal{S}$. (This fact may be "quantized" in Fourier integral operator theory; see [G-S2]. $L(C)$ has also been used in [MO].) Putting together the two observations above, we find that a polarization of (P, Ω) leads to a "projection operator" on $\mathrm{End}(P, \Omega)$ and thus corresponds to a subspace (subalgebra?) of "operators" on (P, Ω).

Finally, we propose the following synthesis of geometric quantization and Fourier integral operator theory. Given a symplectic manifold (P, Ω) together with circle bundle and polarization, and a lagrangian submanifold $L \subseteq (P, \Omega)$, construct an element or class of elements of $H(P, \Omega)$. Use of the fact, clear from Fourier integral operator theory, that some extra data besides L may be necessary might help to resolve some of the nonunitarity problems described in [G-S1] and [SN].

5. Some applications of symplectic geometry. In this section, I shall describe very briefly a (by no means complete) sample of the recent applications of symplectic geometry. In each case, the application begins with a "transform" which brings symplectic geometry into the area of interest.

A. *Conservative mechanics.* The law of motion of a system of bodies is commonly given as a system of first order differential equations in the positions and velocities of the bodies. The *Legendre transform* replaces the velocities by momenta and puts the equations in hamiltonian form. Methods of symplectic geometry such as the Birkhoff normal form, the Kolmogorov-Arnold-Moser theorem on the persistence of invariant tori, the intersection theory of lagrangian submanifolds, and the classification of group actions on symplectic manifolds may now be applied to study questions of stability, periodicity, and symmetry in conservative mechanics. (A good general reference in this area is [AR]. Also see [A-M], [G-S1], [SO], and [ST].)

B. *Completely integrable systems.* The nice structure of the solutions of the Korteweg-de Vries partial differential equation $u_t = -u_{xxx} + 6uu_x$ originally discovered by Gardner, Green, Kruskal and Miura is now explained neatly in terms of the *Faddeev-Zaharov symplectic structure* [F-Z] on the space of functions on $\mathbf{R}$, which makes the *KdV* equation into an infinite-dimensional hamiltonian system. A closely related finite dimensional system is the Toda lattice; this example has been studied and generalized by, among many others, Kostant [**KO2**], who uses as a basic tool the technique of reduction of symplectic manifolds with symmetry (see [**K-K-S**], [**M**], [**MA-W**], [**MEY**]). Another application of reduction to completely integrable systems is given by Mishchenko and Fomenko in [**MI-F**].

C. *Representation theory*. The geometric quantization scheme of Segal-Kirillov-Kostant-Souriau (see §4) leads to a correspondence between irreducible unitary representations of a Lie group G and transitive symplectic actions of these groups. The correspondence is essentially bijective for nilpotent, compact, and certain solvable groups. Although there are examples where the correspondence fails (see [**R-W**]), the symplectic viewpoint, often called the *orbit method*, since the homogeneous symplectic manifolds are coadjoint orbits, is at least a useful heuristic tool for investigating the representations of more general groups. The most extensive discussion available of the orbit method can be found in [**KI2**]. The nilpotent case is also treated in detail in [**WA**]. The coadjoint orbits have also appeared recently in the work of Simon [**SI**] on the statistical mechanics of representations of compact groups.

D. *Linear partial differential equations*. The relevant "transforms" here are the *Fourier integral operator calculus* of Hörmander [**HR**], the *canonical operator* of Maslov [**MAS**], and the *quantized contact transformation theory* of Sato [**S-K-K**]. In each case, when a differential operator on a manifold X is represented by a function on T^*X (its *principal symbol*), a symplectomorphism which simplifies the function may be realized as a conjugation which simplifies the operator. Problems for general operators may thereby be reduced to certain model cases. See [**D**] for a survey through 1974. A more recent example of this is contained in the work of R. Melrose. In [**MEL1**], he finds a normal form under symplectomorphisms for pairs of hypersurfaces which are tangent to one another at a point in a certain "nondegenerate" way. This theorem in symplectic geometry can then be applied to the problem of diffraction by a convex obstacle, in which the obstacle is represented by the first hypersurface and a family of light rays by the second. Among the results of this work is a theorem to the effect that, when a ray of light meets an obstacle in a "glancing" way (i.e. tangent to the obstacle at a convex point), the ray continues as if the obstacle were not there–here, a ray represents a path along with the solution is not C^∞. (See also [**T**].) This is in contrast with the situation for the propagation of analyticity where a ray, after grazing an obstacle, produces supplementary rays which "creep" around the boundary of the obstacle and then fly off into space once again at any moment they like. Melrose's approach also leads to very refined results in scattering theory (see [**MEL2**].)

References

[**A-M**] R. Abraham and J. Marsden, *Foundations of mechanics*, 2nd ed., Benjamin/Cummings, Reading, Mass., 1978.

[**AR**] V. I. Arnold, *Mathematical methods of classical mechanics*, Graduate Texts in Math., vol. 60, Springer-Verlag, Berlin and New York, 1978.

[**AU-K**] L. Auslander and B. Kostant, *Polarization and unitary representations of solvable Lie groups*, Invent. Math. **14** (1971), 255–354.

[BI] G. D. Birkhoff, *Fifty years of American mathematics*, Semicentennial Addresses of Amer. Math. Soc.,1938, p. 307.

[BO] C. E. Bond, *Biology of fishes*, Sanders, Philadelphia, Pa., 1979.

[BU] L. Boutet de Monvel, private communication.

[BU-G] L. Boutet de Monvel and V. Guillemin, *The spectral theory of Toeplitz operators*, preprint.

[C] S. S. Chern, *From triangles to manifolds*, Amer. Math. Monthly **86** (1979), 339–349.

[D] J. J. Duistermaat, *Applications of Fourier integral operators*, Proc. Internat. Congr. Math. (Vancouver, 1974), Canadian Mathematical Congress, 1975, pp. 263–268.

[D-H] J. J. Duistermaat and L. Hörmander, *Fourier integral operators*. II, Acta Math. **128** (1972), 183–269.

[F-Z] L. Faddeev and V. Zaharov, *Korteweg-de Vries equation: a completely integrable Hamiltonian system*, Funkcional. Anal. i Priložen **5** (1971), 18–27.

[G] V. W. Guillemin, *Clean intersection theory and Fourier integral operators*, Lecture Notes in Math., vol. 459, Springer-Verlag, Berlin and New York, 1975, pp. 23–35.

[G-S1] V. W. Guillemin and S. Sternberg, *Geometric asymptotics*, Amer. Math. Soc., Providence, R. I., 1976.

[G-S2] ______, *Some problems in integral geometry and some related problems in micro-local analysis*, Amer. J. Math. **101** (1979), 915–955.

[H-K] P. de la Harpe and M. Karoubi, *Perturbations compactes des representations d'un groupe dans un espace de Hilbert*, Bull. Soc. Math. France Mém. **46** (1976), 41–65.

[HR] L. Hörmander, *Fourier integral operators*. I, Acta Math. **127** (1971), 79–183.

[K-K-S] D. Kazhdan, B. Kostant, and S. Sternberg, *Hamiltonian group actions and dynamical systems of Calogero Type*, Comm. Pure Appl. Math. **31** (1978), 481–508.

[KI1] A. A. Kirillov, *Unitary representations of nilpotent Lie groups*, Uspehi Mat. Nauk. **17** (1962), 57–110. (English translation in Russian Math. Surveys, 53, 104).

[KI2] ______, *Elements of the theory of representations*, Springer-Verlag, Berlin and New York, 1976.

[KO1] B. Kostant, *Quantization and unitary representations*: Part I, *Prequantization*, Lecture Notes in Math., vol. 170, Springer-Verlag, Berlin and New York, 1970, pp. 87–208.

[KO2] ______, *The solution to a generalized Toda lattice and representation theory*, Advances in Math. **34** (1979), 195–338.

[LA1] J. L. Lagrange, *Mémoire sur la théorie des variations des éléments des planètes*, Mémoires de la classe des sciences mathématiques et physiques de l'institut de France, 1808, pp. 1–72.

[LA2] ______, *Second mémoire sur la théorie de la variation des constantes arbitraires dans les problèmes de mécanique*, Mémoires de la classe des sciences mathématiques et physiques de l'institut de France 1809, pp. 343–352.

[LI] R. G. Littlejohn, *A guiding center Hamiltonian: A new approach*, J. Math. Phys. **20** (1979), 2445–2458.

[M] G.-M. Marle, *Symplectic manifolds, dynamical groups, and Hamiltonian mechanics*, in Differential Geometry and Relativity (Cahen and Flato (eds.)), D. Reidel, Dordrecht, 1976, pp. 249–269.

[MA-W] J. Marsden and A. Weinstein, *Reduction of symplectic manifolds with symmetry*, Reports on Math. Phys. **5** (1974), 121–130.

[MAS] V. P. Maslov, *Théorie des perturbations et méthodes asymptotiques*, Dunod, Gauthier-Villars, Paris, 1972. (Translation of 1965 Russian edition.)

[MEL1] R. B. Melrose, *Equivalence of glancing hypersurfaces*, Invent. Math. **37** (1976), 165–191.

[MEL2] ______, *Forward scattering by a convex obstacle*, Comm. Pure Appl. Math. **33** (1980), 461–499.

[MEY] K. Meyer, *Symmetries and integrals in mechanics*, in Dynamical Systems (M. Peixoto (ed)), Academic Press, New York, 1973, pp. 259–273.

[MI-F] A. S. Miščenko and A. T. Fomenko, *Euler equations on finite-dimensional Lie groups*, Math. USSR-Izv. **12** (1978), 371–389.

[MO] J. Moser, *A fixed point theorem in symplectic geometry*, Acta Math. **141** (1978), 17–34.

[R-W] L. Rothschild and J. Wolf, *Representations of semi-simple groups associated to nilpotent orbits*, Ann. Sci. École Norm. Sup. (4) **7** (1974), 155–174.

[S-K-K] M. Sato, T. Kawai and M. Kashiwara, *Microfunctions and partial differential equations*, Lecture Notes in Math., vol. 287, Springer-Verlag, Berlin and New York, 1973, pp. 265–529.

[SE] I. E. Segal, *Quantization of nonlinear systems*, J. Math. Phys. **1** (1960), 468–488.

[SI] B. Simon, *The classical limit of quantum partition functions*, Comm. Math. Phys. **71** (1980), 247–276.

[SN] J. Sniatycki, *Geometric quantization and quantum mechanics*, Springer-Verlag, New York, 1980.

[SN-T] J. Sniatycki and W. M. Tulczyjew, *Generating forms of lagrangian submanifolds*, Indiana Univ. Math. J. **22** (1972), 267–275.

[SO] J.-M. Souriau, *Structure des systemes dynamiques*, Dunod, Paris, 1970.

[ST] S. Sternberg, *Celestial mechanics*. II, W. A. Benjamin, New York, 1969.

[T] M. E. Taylor, *Grazing rays and reflection of singularities of solutions to wave equations*, Comm. Pure Appl. Math. **24** (1976), 1–38.

[WA] N. R. Wallach, *Symplectic geometry and Fourier analysis*, Math. Sci. Press, Brookline, Mass., 1977.

[WE1] A. Weinstein, *On Maslov's quantization condition*, Lecture Notes in Math., vol. 459, Springer-Verlag, Berlin and New York, 1975, pp. 341–372.

[WE2] ______, *Fourier integral operators, quantization, and the spectra of riemannian manifolds*, Colloques Internationaux du CNRS **237** (1976), 289–298.

[WE3] ______, *Lectures on symplectic manifolds*, CBMS Regional Conf. Series, no. 29, Amer. Math. Soc., Providence, R. I., 1977.

[WE4] ______, *The symplectic "category,"* Proc. Conf. Differential Geometric Methods in Mathematical Physics (Clausthal-Zellerfeld, 1980) (in preparation).

[WL] H. Weyl, *The classical groups*, Princeton Univ. Press, Princeton, N. J., 1946.

DEPARTMENT OF MATHEMATICS, CALIFORNIA INSTITUTE OF TECHNOLOGY, PASADENA, CALIFORNIA 91106

DEPARTMENT OF MATHEMATICS, UNIVERSITY OF CALIFORNIA, BERKELEY, CALIFORNIA 94720

Section 2
TOPOLOGY

Proceedings of Symposia in Pure Mathematics
Volume **39** (1983), Part 1

Graeme Segal's Burnside Ring Conjecture

J. FRANK ADAMS

1. Introduction. My theme will be that algebraic topology still offers problems which reveal the present state of our art as inadequate. Such problems make us feel that there could be something good going on; but if there is, we have yet to understand it.

I shall devote §2 to explaining Segal's original conjecture. In §3–§6 I shall review what is proved about it so far. Broadly, for those few groups we have been able to handle the conjecture is found to be true, and there is no group for which the conjecture is known to be false. In §7–§9 I shall explain further conjectures related to the original one; the fact that these also are not yet disproved contributes to the impression that there could be something good going on. In §10 I shall comment briefly on our chances of going further.

2. Statement of the conjecture. In this section I shall explain Segal's original conjecture. I must begin by explaining cohomotopy.

Let X, Y be a good pair of spaces, for example, a finite-dimensional CW-pair. Then we have

$$\pi^n(X, Y) = \operatorname*{Lim}_{m \to \infty} [S^m X/Y, S^{m+n}].$$

Here maps and homotopies preserve the base-point; and if $Y = \phi$, I use Atiyah's convention that X/ϕ means X with a disjoint base-point. We can rewrite the right-hand side as

$$\operatorname*{Lim}_{m \to \infty} [X/Y, \Omega^m S^{m+n}]$$

or as

$$\left[X/Y, \operatorname*{Lim}_{m \to \infty} \Omega^m S^{m+n}\right].$$

The last expression gives us a definition of $\pi^n(X, Y)$ valid whether the pair X, Y is finite dimensional or not.

Cohomotopy is a generalised cohomology theory, namely the one corresponding to the sphere spectrum; and its coefficient groups are the stable homotopy groups of spheres. So it is like stable homotopy; if you could compute it, it would give you a lot of information, but unfortunately it is hard to compute.

However, there is one case in which we have a conjecture, and it is due to Graeme Segal. He asks for the analogue of a well-known theorem of Atiyah [3].

Reprinted from Bulletin Amer. Math. Soc. (N.S.) 6 (1982), 207–270.

1980 *Mathematics Subject Classification.* Primary 55Q55, 55R35.

Atiyah wished to compute the K-theory of BG, the classifying space of G, where G was originally a finite group. So Atiyah took the representation ring $R(G)$ and constructed a map

$$\alpha\colon R(G) \to K(BG).$$

He then completed $R(G)$ for the topology defined by the powers of the augmentation ideal. He extended α by continuity to get a map

$$\hat{\alpha}\colon R(G)\hat{\ } \to K(BG).$$

And finally he proved that this map $\hat{\alpha}$ is an isomorphism.

Segal proposed that we should replace K-theory by cohomotopy; so on the right we put $\pi^0(BG)$. On the left he proposed to replace the representation ring $R(G)$ by the Burnside ring $A(G)$ of G. To define this, instead of taking actions of G on vector spaces, we take actions of G on finite sets. (That is, assuming G is finite, of course.) We divide these actions into isomorphism classes, and make these classes into a semigroup under disjoint union. We take the corresponding Grothendieck group, and that is $A(G)$. We make it a ring so that the product of two finite G-sets is their Cartesian product. I emphasise that $A(G)$ should be regarded as something computable and known.

We have again a map

$$\alpha\colon A(G) \to \pi^0(BG).$$

I will not stop to define it here; the reader may consult §9 for a more general construction. We complete $A(G)$ for the topology defined by the powers of the augmentation ideal; and again α extends by continuity to give a map

$$\hat{\alpha}\colon A(G)\hat{\ } \to \pi^0(BG).$$

Segal's conjecture is that this map $\hat{\alpha}$ is an isomorphism.

For the trivial group $G = 1$ the conjecture is trivially true; both sides are Z and the map is an isomorphism. This seems to be the only group for which matters are trivial.

3. Results of Lin; the case $G = Z_2$. The group $G = Z_2$ already presents a problem. On the left, $A(Z_2)\hat{\ }$ comes to $Z \oplus \hat{Z}_2$, the direct sum of the integers and the 2-adic integers. On the right we can take real projective space RP^∞ as our model for BZ_2, and so the problem is to compute $\pi^0(RP^\infty)$; but for ten years nobody could get hold of it. This case was finally solved by W. H. Lin [11, 12], using a method which I once heard Graeme Segal call "that damn spectral sequence of yours". Lin found that the conjecture is true for $G = Z_2$.

Lin also found that $\pi^n(RP^\infty) = 0$ for $n > 0$. It seems fair to add to our list of conjectures the conjecture

$$\pi^n(BG) = 0 \quad \text{for } G \text{ finite}, n > 0.$$

Let me emphasise that this is not a statement which is true for trivial reasons, in the way that the homotopy groups of a space or spectrum vanish below the Hurewicz dimension. If you try to get hold of $\pi^n(BG)$ by obstruction-theory, or by an Atiyah-Hirzebruch spectral sequence, you find an infinity of nonzero groups

$$H^i\big(BG; \pi^j(pt)\big) \qquad (i+j=n).$$

What happens is that these groups all conspire to cancel each other out for $n > 0$; for $G = Z_2$ this phenomenon was noticed long ago by Mahowald, and later by me [1], although neither of us had a proof.

Various sorts of further argument seem likely to need information about the groups $\pi^n(RP^\infty)$ also in the range $n < 0$, where they are nontrivial. Lin obtained such information; but to describe the result will require some preliminary explanation.

4. Functional duals. In this section I will explain about functional duals, and so reach the results on $\pi^n(RP^\infty)$ for $n < 0$.

Let us work in a suitable category of spectra where we can do stable homotopy theory. Consider $[W \wedge X, S^0]$, where X is fixed and W varies. This is a contravariant functor of W, and satisfies the hypotheses of Brown's Representability Theorem, so we get a natural (1-1) correspondence

$$[W \wedge X, S^0] \leftrightarrow [W, DX],$$

where the representing object DX is by definition the "functional dual" of X. In particular,

$$\pi^{-n}(X) = [S^n \wedge X, S^0] \leftrightarrow [S^n, DX] = \pi_n(DX).$$

So if we know DX, we get information about the cohomotopy of X.

Let's use the same symbol for a space with base-point and for its suspension spectrum. Then we want to know $D(RP^\infty/\phi)$, and we have

$$RP^\infty/\phi \simeq S^0 \vee (RP^\infty/pt),$$

so

$$D(RP^\infty/\phi) \simeq (DS^0) \vee D(RP^\infty/pt) \simeq S^0 \vee D(RP^\infty/pt).$$

Thus it is sufficient to know $D(RP^\infty/pt)$.

Let $M\hat{Z}_2$ be a Moore spectrum for the 2-adic integers $\hat{Z}_2$ (in dimension 0). Let

$$M\hat{Z}_2 \wedge (RP^\infty/pt) \to S^0$$

be any (stable) map for which the functional Sq^2 is nonzero; since the left-hand side is equivalent to RP^∞/pt, such maps exist. Let

$$(RP^\infty/pt) \wedge (RP^\infty/pt) \to S^0$$

be any map for which the functional Sq^4 is nonzero; again, such maps exist. Then by adjointness we get a map

$$(M\hat{Z}_2) \vee (RP^\infty/pt) \to D(RP^\infty/pt).$$

PROPOSITION. *Any map*

$$\left(M\hat{Z_2} \right) \vee \left(RP^\infty / pt \right) \to D\left(RP^\infty / pt \right)$$

constructed in this way is an equivalence.

This is a result of Lin and myself which is not yet written up for publication; I sketch the proof. We have an induced map of homotopy

$$\pi_n\left(\left(M\hat{Z_2} \right) \vee \left(RP^\infty / pt \right)\right) \to \pi_n\left(D\left(RP^\infty / pt \right)\right) \cong \pi^{-n}\left(RP^\infty / pt \right).$$

In fact we can obtain an induced map between the Adams spectral sequences which should converge to these groups; we have one spectral sequence converging to

$$\left[S^0, \left(M\hat{Z_2} \right) \vee \left(RP^\infty / pt \right)\right]_*$$

and one converging to

$$\left[\left(RP^\infty / pt \right), S^0 \right]_*.$$

It is necessary to insist, however, that this map of spectral sequences carries $E_r^{s,t}$ to $E_r^{s+1,t+1}$; that is, it raises the filtration degree by 1. By quoting the algebraic result of [**12**], it is shown that this comparison map of spectral sequences is an isomorphism.

The reader will notice that the result about the functional dual implicitly contains the result that $\pi^n(RP^\infty) = 0$ for $n > 0$ and the result that $\pi^0(RP^\infty)$ is $Z \oplus \hat{Z_2}$, as well as information about the groups $\pi^n(RP^\infty)$ for $n < 0$. Of course, it is still necessary to take the isomorphism $\pi^0(RP^\infty) \cong Z \oplus \hat{Z_2}$ given by the proposition and reconcile it with the map $\hat{\alpha}$. Similar remarks will apply whenever we describe a functional dual $D(BG)$.

This completes my account of the results for $G = Z_2$.

5. Results of Gunawardena and Ravenel; the case $G = Z_n$. The next case should be the case $G = Z_p$, where p is an odd prime. Here Gunawardena has shown that with suitable care Lin's results go over [**9**].

The next case should be the case $G = Z_{p^e}$, p prime, $e \geqslant 1$. In this case D. C. Ravenel finds

$$D\left(BZ_{p^e} / pt \right) \simeq \left(\overset{e}{\underset{1}{\vee}} M\hat{Z_p} \right) \vee \left(\overset{e}{\underset{i=1}{\vee}} BZ_{p^i} / pt \right).$$

This is in agreement with the conjecture. An announcement of Ravenel's work is in course of publication [**14**]. The proof sketched lacks only a treatment of the convergence of a spectral sequence constructed for the purposes of the proof; it is plausible that this should not present an essential obstacle. The proof is by induction over e, and it relies on the previous work of Lin and Gunawardena to deal with the case $e = 1$ and start the induction.

The case of a general cyclic group Z_n follows immediately. After all, for a cyclic group Z_n we have a stable equivalence

$$BZ_n / pt \simeq \underset{p}{\vee} BS_p / pt,$$

where S_p is the Sylow p-subgroup of Z_n. This gives

$$D(BZ_n/pt) \simeq \bigvee_p D(B(S_p/pt)),$$

where the right-hand side is known by the work of Ravenel.

6. Other results. One should expect that suitable statements for a finite group G will follow from the corresponding statements about the Sylow subgroups S_p of G. For the conjecture "$\hat{\alpha}$ mono" this was proved by Laitinen [**10**]. For the conjectures "$\hat{\alpha}$ iso" and "$\pi^n(BG) = 0$ for $n > 0$" it has been proved by May and McClure; but I do not have details about publication. In particular, the conjectures should be true for any finite group G all of whose Sylow subgroups are cyclic.

Results which show that $\hat{\alpha}$ is mono are perhaps less impressive than ones which show that $\hat{\alpha}$ is iso. However, Laitinen [**10**] has proved that $\hat{\alpha}$ is mono when G is an elementary abelian p-group. This result was later extended to all finite abelian groups by Segal and Stretch. They use BP-theory; again I do not know details about publication.

At this point I more or less run out of positive results to report. Noncyclic groups present a substantial problem. To do calculations (corresponding to those which Lin did for $G = Z_2$) for so small a group as $G = Z_2 \times Z_2$ seems to be something contemplated only by a select few.

7. Could the functional duals be doing something good? At this point it is natural to step back and see if we can guess some larger pattern into which the pieces might fit. In this section I will do so for the functional duals.

Let G be a finite group, and let H be a subgroup of G. Then the coset space G/H is a finite G-set. We can also consider the set of G-maps $G/H \to G/H$, and this turns out to be $W_H = N_G(H)/H$, where $N_G(H)$ is the normaliser of H in G. So we have $W_H \times G$ acting on the finite set G/H. Thus we get a finite covering associated to the universal bundle over $B(W_H \times G) \simeq BW_H \times BG$, say

$$G/H \to E \xrightarrow{p} BW_H \times BG.$$

Take the unit element $1 \in \pi^0(E)$ and apply the transfer [**2, 5, 7, 8**]; we get an element

$$p_!1 \in \pi^0(BW_H \times BG) = \left[(BW_H \times BG)/\phi, S^0\right]$$
$$= \left[(BW_H/\phi) \wedge (BG/\phi), S^0\right].$$

This corresponds to a map

$$BW_H/\phi \to D(BG/\phi).$$

Conjugate subgroups H lead to the same outcome. So I get a map

$$\bigvee_{\{H\}} (BW_H)/\phi \to D(BG/\phi)$$

where in the wedge-sum I take one subgroup H from each conjugacy class. The left-hand side contains a summand S^0 which arises for $H = G$ ($W_G = N_G(G)/G = 1$). This maps to the obvious summand S^0 of $D(BG/\phi)$ (the constant map $G \to 1$ gives $BG/\phi \to B1/\phi$, which dualises to $D(B1/\phi) \to D(BG/\phi)$). As for the complementary summands, one may conjecture that this map gives an equivalence provided that G is a p-group and on the left you complete at the prime p. This appears consistent with what is known for cyclic p-groups and can be guessed for other groups. If G were not a p-group life would become more complicated.

8. Could equivariant cohomotopy be doing something good? In this section I will discuss some generalisations of the original conjecture which involve equivariant cohomotopy.

In a seminar in Oxford in February 1980 I proposed the conjecture in §7; and they were very polite and told me that this was indeed the obvious conjecture, and that it could be incorporated into further conjectures which were better and brighter and even more conjectural. These conjectures are again due to Segal, and again he follows an analogy.

Atiyah and Segal long ago generalised Atiyah's theorem

$$\hat{\alpha}: R(G)\hat{\ } \overset{\cong}{\to} K(BG)$$

in the following direction [4]. Let X be a suitable G-space, for example, a finite G-CW-complex in the sense of Matumoto [13]. Then there are two ways to define the equivariant K-theory of X. First, we can form $E_G \times_G X$ and take ordinary K of it, so as to get $K(EG \times_G X)$. This way works with K replaced by any generalised cohomology theory k; you can form $k(EG \times_G X)$. The second way only works if you have a good geometrical construction for k; then you go through the construction again with G acting on everything. In particular, Atiyah and Segal define $K_G(X)$ in terms of G-vector-bundles over the G-space X. Then they get

$$K_G(X)\hat{\ } \overset{\cong}{\to} K(EG \times_G X).$$

Taking X to be a point, they recover as a special case the theorem

$$R(G)\hat{\ } \overset{\cong}{\to} K(BG).$$

When one comes to replace K-theory by cohomotopy, one replaces $K(EG \times_G X)$ on the right by $\pi^n(EG \times_G X)$. On the left we should replace $K_G(X)$ by the equivariant cohomotopy of X according to some direct definition. For simplicity I will begin with the case $G = Z_2$. In this case we have two ways to suspend a given G-space: one, say S, in which Z_2 preserves the suspension coordinate, and another, say T, in which Z_2 reverses the suspension coordinate. So let X, Y be a finite-dimensional G-CW-pair, where $G = Z_2$. We can form

$$\pi_G^{i,j}(X, Y) = \varinjlim_{m,n \to \infty} \left[S^m T^n X/Y, S^{m+i} T^{n+j} \right]_G.$$

Here maps and homotopies are G-maps and G-homotopies preserving the base-point. Now just as I did in §2, I can rewrite this in the following form.

$$\operatorname*{Lim}_{m,n\to\infty}\left[X/Y,\Omega_S^m\Omega_T^n S^{m+i}T^{n+j}\right]_G.$$

Here the function-space on the right has the obvious action of G; an element $g \in G$ acts on a function f to give $g \circ f \circ g^{-1}$. I can rewrite things again in the following form.

$$\left[X/Y,\ \operatorname*{Lim}_{m,n\to\infty}\Omega_S^m\Omega_T^n S^{m+i}T^{n+j}\right]_G.$$

In this form the definition is valid whether the pair X, Y is finite-dimensional or not.

For a general finite group G of course I have to allow for suspension using any finite-dimensional real representation of G, and I get groups $\pi_G^\rho(X, Y)$ indexed by elements ρ in the real representation ring $RO(G)$. It is clear enough how the generalisation goes.

On the left of our conjecture we can now put the groups $\pi_G^\rho(X, Y)$. But on the right we still have ordinary cohomotopy groups

$$\pi^n(EG \times_G X, EG \times_G Y)$$

which are indexed over the integers. What is to be done?

It seems the best move is the following. EG is usually considered as a space with G acting on its right, but we can make G act on its left by using $e.g.^{-1}$. Then we can form

$$\pi_G^\rho(EG \times X, EG \times Y).$$

The constant map $c: EG \to pt$ of course induces

$$(c \times 1)^*: \pi_G^\rho(X, Y) \to \pi_G^\rho(EG \times X, EG \times Y).$$

If ρ happens to be an integer n, we can prove

$$\pi_G^n(EG \times X, EG \times Y) \cong \pi^n(EG \times_G X, EG \times_G Y).$$

The proof is not quite the same as it was for K-theory, but still it can be done. Whether ρ is an integer or not, we can conjecture that $(c \times 1)^*$ becomes an isomorphism if we complete the left-hand side. (Notice that $c: EG \to pt$ may be a homotopy equivalence, but it is not a G-homotopy-equivalence.)

On the face of it, it is not clear how the left-hand side $\pi_G^\rho(X, Y)$ just considered is related to the sort of left-hand side, involving groups W_H, which we saw in the last section. Experts in this area see a relation, but I am not ready to report on it with confidence.

I believe that when G is cyclic the conjecture in this section can probably be deduced from the results I have described above.

9. Could something make sense for compact Lie groups in general? Up to this point G has been a finite group; I now seek to generalise so that G can be any compact Lie group. Let $(BG)^m$ be the m-skeleton of BG; then we have the following exact sequence due to Milnor.

$$0 \to \operatorname*{Lim_m^1}\pi^*\!\left((BG)^m\right) \to \pi^*(BG) \to \operatorname*{Lim_m^0}\pi^*\!\left((BG)^m\right) \to 0.$$

When G is finite the Lim^1 term is zero, but this need no longer happen when G has positive dimension. It seems advisable to replace $\pi^n(BG)$ in our conjecture by $\varprojlim_m{}^0 \pi^n((BG)^m)$. Once this is done, hope does not seem to be ruled out.

At this point I should give briefly some perspective on the Burnside ring in this context. Take some good class of compact G-spaces X, for example, finite G-CW-complexes. By a G-Euler characteristic, I shall mean a pair (A, χ) of the following sort. A is a given abelian group; χ is a function which assigns to each finite G-CW-complex X an element $\chi(X)$ in A; and χ satisfies the following axioms.

(0) (Zero) $\chi(\phi) = 0$.

(i) (Invariance). If X and Y are G-homotopy-equivalent, then $\chi(X) = \chi(Y)$.

(iii) (Mayer-Vietoris) Under the obvious assumptions,

$$\chi(X_1 \cup X_2) + \chi(X_1 \cap X_2) = \chi(X_1) + \chi(X_2).$$

(Axiom (0) seems to be needed; one can replace $\chi(X)$ by $\chi(X) + c$, where c is a constant, without affecting axioms (i) and (ii).)

Among such G-Euler-characteristics, there is one which is universal. In this, the group A is a free abelian group with one generator $a_{\{H\}}$ for each conjugacy class $\{H\}$ of closed subgroups $H \subset G$. The function χ is defined as follows. Given X, let $r(\{H\}, n)$ be the number of G-cells in X of type $(G/K) \times E^n$ with $K \in \{H\}$; set

$$\chi(X) = \sum_{\{H\},n} (-1)^n r(\{H\}, n) a_{\{H\}}.$$

This is the most naive generalisation of the Euler characteristic; we count the G-cells with a sign $(-1)^n$ depending on their dimension, but we keep accounts separately for G-cells of the different symmetry types. For all this, see [6].

We now proceed as follows. For any such G-space X we can form the bundle with fibres X associated to the universal G-bundle, that is

$$EG \times_G X \xrightarrow{p} BG.$$

We can take the unit element

$$1 \in \pi^0(EG \times_G X)$$

and apply transfer [5, 7, 8] getting an "index"

$$p_! 1 \in \pi^0(BG).$$

We check that this satisfies the axioms for a G-Euler-characteristic of X; by the universal property, we get a homomorphism

$$\alpha: A \to \pi^0(BG).$$

(If we use a version of the transfer which is only defined when the base is finite dimensional, or even when the base is a finite complex, we just replace $\pi^0(BG)$ by a suitable inverse limit.)

At this point the general case has one new twist which doesn't show up when G is finite. Let H be a closed subgroup of G; then by definition, $\alpha a_{\{H\}}$ is $p_!(1)$ in the following fibering.

$$G/H \to EG \times_H pt \to BG.$$

Let N be the normaliser of H, and assume that N/H is of positive dimension. Then we can find a 1-parameter subgroup θ_t in N which leads from the identity element to a point not in H. Acting on $EG \times_H pt$ by right translation with θ_t, we can deform the identity map

$$1 \colon EG \times_H pt \to EG \times_H pt$$

to a map without fixed-points. So the fixed-point index $p_!1$ is zero in this case. Thus we can pass to a quotient of our previous universal A in which we map to zero each generator $a_{\{H\}}$ such that N/H is of positive dimension; we leave only the generators $a_{\{H\}}$ such that N/H is finite. We thus reach tom Dieck's version of the Burnside ring $A(G)$ [6], and we still have a homomorphism

$$\alpha \colon A(G) \to \pi^0(BG).$$

The one case which seems accessible to checking at the present time is the case $G = S^1$, for one can approximate to S^1 by its cyclic subgroups Z_n. As our model for BS^1 we can take the complex projective space CP^∞. When we audit the conjectures about

$$\varprojlim_m \pi^n(CP^m)$$

for $n \geqslant 0$, they seem to stand up.

10. What are our chances? Finally, I owe you some comments about our chances of going further. To begin with, it will be prudent to allow for a chance of say 5% or 10% that the conjectures fail when you go beyond cyclic groups. If so, then "calculation is the way to the truth"; one can always hope that a sufficiently heroic effort will settle one more group. On the other hand, if the conjectures are true, then we must ask what sort of a proof to seek. From the attractive nature of some of the formulations, you might be encouraged to hope for a fairly conceptual proof. I don't think it's all that likely, myself. After all, the original formulation was copied from K-theory, and in that case the best proof we have is the proof by Atiyah and Segal [4], which works by slowly building up the results for a succession of well-chosen particular groups. It would seem more reasonable to look for a proof like that, and the question is, how to build up.

The first obvious suggestion is to follow the approach used by Atiyah in [3]; try to prove something for p-groups G by induction, applying the inductive hypothesis either to a suitable normal subgroup $H \subset G$, or to the corresponding quotient G/H, or to both. This is hard and nobody can do it yet; the reader will find that the difficulty is already present in the case $G = Z_2 \times Z_2$.

The second obvious suggestion is to follow the approach used by Atiyah and Segal in [4]; go via compact Lie groups, even if you only want a conclusion about finite groups. This approach also does not seem realistic yet.

My conclusion is that this area deserves further study. But in order to work in it, those who love conceptual theories had better not scorn calculation, while those who love horrendous calculations should not neglect such help and guidance as may be had from conceptual theory.

REFERENCES

1. J. F. Adams, *Operations of the nth kind in K-theory, and what we don't know about RP^{∞}*, London Math. Soc. Lecture Notes No. 11, Cambridge Univ. Press, 1974, pp. 1–9.

2. ______, *Infinite loop spaces*, Princeton Univ. Press, Princeton, N.J., 1978, especially Chapter 4.

3. M. F. Atiyah, *Characters and cohomology of finite groups*, Inst. Hautes Études Sci. Publ. Math., No. 9 (1961), 23–64.

4. M. F. Atiyah and G. B. Segal, *Equivariant K-theory and completion*, J. Differential Geometry **3** (1969), 1–18.

5. J. C. Becker and D. Gottlieb, *Transfer maps for fibrations and duality*, Compositio Math. **33** (1976), 107–133.

6. T. tom Dieck, *Transformation groups and representation theory*, Lecture Notes in Math., vol. 766, Springer-Verlag, Berlin and New York, 1979.

7. A. Dold, *The fixed-point index of fibre-preserving maps*, Invent. Math. **25** (1974), 281–297.

8. ______, *The fixed point transfer of fibre-preserving maps*, Math. Z. **148** (1976), 215–244.

9. J. H. C. Gunawardena, *Segal's conjecture for cyclic groups of (odd) prime order*, J. T. Knight Prize Essay, Cambridge, 1980.

10. E. Laitinen, *On the Burnside ring and stable cohomotopy of a finite group*, Math. Scand. **44** (1979), 37–72.

11. W. H. Lin, *On conjectures of Mahowald, Segal and Sullivan*, Math. Proc. Cambridge Philos. Soc. **87** (1980), 449–458.

12. W. H. Lin, D. M. Davis, M. E. Mahowald and J. F. Adams, *Calculation of Lin's Ext groups*, Math. Proc. Cambridge Philos. Soc. **87** (1980), 459–469.

13. T. Matumoto, *On G-CW-complexes and a theorem of J. H. C. Whitehead*, J. Fac. Sci. Univ. Tokyo **18** (1971), 363–374.

14. D. C. Ravenel, *The Segal conjecture for cyclic groups*, Bull. London Math. Soc. **13** (1981), 42–44.

PURE MATHEMATICS DEPARTMENT, UNIVERSITY OF CAMBRIDGE, CAMBRIDGE, ENGLAND

Proceedings of Symposia in Pure Mathematics
Volume 39 (1983), Part 1

Three Dimensional Manifolds, Kleinian Groups and Hyperbolic Geometry

WILLIAM P. THURSTON

1. A conjectural picture of 3-manifolds. A major thrust of mathematics in the late 19th century, in which Poincaré had a large role, was the uniformization theory for Riemann surfaces: that every conformal structure on a closed oriented surface is represented by a Riemannian metric of constant curvature. For the typical case of negative Euler characteristic (genus greater than 1) such a metric gives a hyperbolic structure: any small neighborhood in the surface is isometric to a neighborhood in the hyperbolic plane, and the surface itself is the quotient of the hyperbolic plane by a discrete group of motions. The exceptional cases, the sphere and the torus, have spherical and Euclidean structures.

Three-manifolds are greatly more complicated than surfaces, and I think it is fair to say that until recently there was little reason to expect any analogous theory for manifolds of dimension 3 (or more)—except perhaps for the fact that so many 3-manifolds are beautiful. The situation has changed, so that I feel fairly confident in proposing the

1.1. CONJECTURE. *The interior of every compact 3-manifold has a canonical decomposition into pieces which have geometric structures.*

In §2, I will describe some theorems which support the conjecture, but first some explanation of its meaning is in order.

For the purpose of conservation of words, we will henceforth discuss only oriented 3-manifolds. The general case is quite similar to the orientable case.

1. The decomposition referred to really has two stages. The first stage is the prime decomposition, obtained by repeatedly cutting a 3-manifold M^3 along 2-spheres embedded in M^3 so that they separate the manifold into two parts neither of which is a 3-ball, and then gluing 3-balls to the resulting boundary components, thus obtaining closed 3-manifolds which are "simpler". Kneser [**Kn**] proved that this process terminates after a finite number of steps. The resulting pieces, called the prime summands of M^3, are uniquely determined by M^3 up to homeomorphism; cf. Milnor, [**Mil 1**].

The second stage of the decomposition involves cutting along tori. This was discovered much more recently, by Johannson [**Joh**] and Jaco and Shalen [**Ja, Sh**], even though the elementary theory of the torus decomposition does not

---

Reprinted from Bulletin Amer. Math. Soc. (N.S.) 6 (1982), 357–381.

1980 *Mathematics Subject Classification.* Primary 57M99, 30F40, 57S30; Secondary 57M25, 20H15.

require any deep techniques. In the torus decomposition, one cuts along certain tori embedded nontrivially in M^3, thus obtaining a 3-manifold whose boundary consists of tori. There is no canonical procedure to close off the boundary components, so they are left alone. The interested reader can learn the details elsewhere.

2. One way to think of a geometric structure on a manifold M is that it is given by a complete, locally homogeneous Riemannian metric. It is better, however, to define a geometric structure to be a space modelled on a homogeneous space (X, G), where X is a manifold and G is a group of diffeomorphisms of X such that the stabilizer of any point $x \in X$ is a compact subgroup of G. For example, X might be Euclidean space and G the group of Euclidean isometries. M is equipped with a family of "coordinate maps" into X which differ only up to elements of G. We make the assumption that M is complete. If X is simply-connected, this condition says that M must be of the form X/Γ, where Γ is a discrete subgroup of G without fixed points.

There are precisely eight homogeneous spaces (X, G) which are needed for geometric structures on 3-manifolds. These eight homogeneous spaces are determined by the following conditions:

(a) The space X is simply-connected. A manifold modelled on a non-simply-connected homogeneous space is also modelled on its universal cover.

(b) The group G is unimodular, that is, there is a measure on G invariant by multiplication on the right or the left. Otherwise, X would possess a vector field invariant by G which expands volume, so there could be no (X, G)-manifolds which are compact or which even have finite volume.

(c) G is a maximal group of homeomorphisms of X with compact stabilizers. If G were contained in a larger group G', then any (X, G)-manifold would be an (X, G')-manifold, so (X, G) would be redundant.

We will describe these eight geometries in §4. For the moment, it will suffice to say that of these eight, hyperbolic geometry is by far the most interesting, the most complex, and the most useful. The other seven come into play only in exceptional cases.

Conjecture 1.1 subsumes the Poincaré conjecture, which was posed by Poincaré not as a conjecture but as a question: Is every 3-manifold with trivial fundamental group homeomorphic to the 3-sphere? Poincaré raised the question in [**Poin**], but he did not pursue it, for as he said, "cette question nous entraînerait trop loin". Conjecture 1.1, just as the Poincaré conjecture, is likely not to be resolved quickly, but I hope it will be a more productive guide to research on 3-manifolds than Poincaré's question has proven to be. My hope is based on the fact that it applies to all 3-manifolds, so there are many examples which arise. To find a geometric structure for a particular manifold is a great help in understanding that manifold.

2. Supporting evidence. As I hinted earlier, Conjecture 1.1 can be proven in many cases. For instance, I will state a simple necessary and sufficient topological condition for the interior of a 3-manifold with nonempty boundary to have a hyperbolic structure. This theorem will imply Conjecture 1.1 for prime manifolds with nonempty boundary. There is a similar theorem for closed manifolds which satisfy a technical hypothesis (which is not necessary but frequently true).

A surface N^2 embedded in a 3-manifold M^3 is *two-sided* if N^2 cuts a regular neighborhood of N^2 into two pieces, i.e., the normal bundle to N^2 is oriented. Since we are assuming that M^3 is oriented, this is equivalent to the condition that N^2 is oriented. A two-sided surface is *incompressible* if every simple curve on N^2 which bounds a disk in M^3 with interior disjoint from N^2 also bounds a disk on N^2.

A 3-manifold M^3 is *geometrically atoroidal* if every incompressible torus embedded in M^3 is isotopic to a boundary component.

A 3-manifold M^3 is *homotopically atoroidal* if every map of T^2 to M^3 which acts injectively on the fundamental group is homotopic to ∂M. Homotopically atoroidal implies geometrically atoroidal; the converse is not quite true.

The condition that M is homotopically atoroidal can be rephrased in terms of the fundamental group. Assuming that ∂M is incompressible, the condition that M is homotopically atoroidal is equivalent to the condition that, first, $\pi_1(M)$ is not expressible as a free product, (this is guaranteed if M is prime), and second, that any subgroup of $\pi_1(M)$ isomorphic to $\mathbf{Z}^2$ is conjugate to a subgroup of the fundamental group of some torus boundary component.

Here are two simple examples of homotopically atoroidal manifolds. They are nonrepresentative because their interiors have Euclidean structures.

2.1. EXAMPLE. $M^3 = T^2 \times I$. The interior of M^3 is homeomorphic to $\mathbf{E}^3/\Gamma$, where $\mathbf{E}^3$ is Euclidean 3-space and Γ is generated by

$$(x, y, z) \to (x + 1, y, z) \quad \text{and} \quad (x, y, z) \to (x, y + 1, z).$$

2.2. EXAMPLE. $M^3 = T^2 \times I/(\mathbf{Z}/2)$, where the $\mathbf{Z}/2$ action flips I and acts as a covering transformation of T^2 over the Klein bottle. Its interior is the quotient of $\mathbf{E}^3$ by the group Γ generated by

$$(x, y, z) \to (x + 1, y, z) \quad \text{and} \quad (x, y, z) \to (-x, y + 1, -z).$$

2.3. THEOREM [**Th 2**]. *The interior of a compact 3-manifold M^3 with nonempty boundary has a hyperbolic structure iff M^3 is prime, homotopically atoroidal and not homeomorphic to Example 2.2.*

A hyperbolic structure on the interior of a compact manifold M^3 has finite volume iff ∂M^3 consists of tori, with the single exception of Example 2.1, which has no hyperbolic structure of finite volume.

2.4. COROLLARY. *Conjecture 1.1 is true for all compact, prime 3-manifolds with nonempty boundary.*

A good class of examples arises from knots in S^3. (A knot is an embedding of S^1 in S^3.) The complement of a knot is homeomorphic to the interior of a manifold whose boundary is a torus.

A *torus knot* is a knot which can be placed on an ordinary torus in S^3.

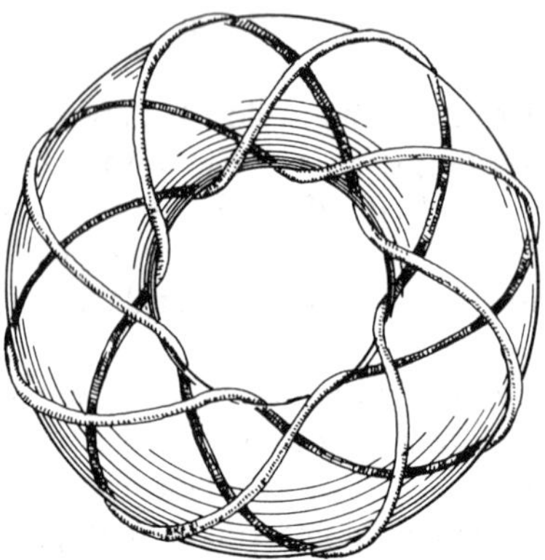

FIGURE 1. A torus knot of type $(3, 8)$. It can be placed on a torus so that it winds 3 times around the short way while going 8 times around the long way.

With any nontrivial knot K there is associated a whole collection of other knots, known as *satellites of* K; these are knots which are obtained by a nontrivial embedding of a circle in a small solid torus neighborhood of K. Here, "nontrivial" means that the embedding is not isotopic to K itself and is not contained within a ball inside the solid torus. A knot is a *satellite knot* if it is a satellite of a nontrivial knot.

FIGURE 2. A knot and a satellite of it.

2.5. COROLLARY. *If $K \subset S^3$ is a knot, $S^3 - K$ has a geometric structure iff K is not a satellite knot. It has a hyperbolic structure iff, in addition, K is not a torus knot.*

Indeed, the complement of a knot is always prime, and the torus decomposition is nontrivial exactly when K is a satellite.

Corollary 2.5 was first conjectured by R. Riley [**Ri 1**] based on his construction of a number of beautiful examples, with the aid of the computer. His work gave a big impetus to me to prove Theorem 2.3.

In order to give the statement for closed manifolds, we need some more terminology.

A 3-manifold M^3 is called a *Haken* manifold if it is prime and it contains a 2-sided incompressible surface (whose boundary, if any, is on ∂M) which is not a 2-sphere. A prime 3-manifold whose boundary is not empty is always Haken (with the trivial exception of the 3-ball, which is often considered to be Haken anyway). Any prime 3-manifold whose first homology has positive rank is Haken.

2.5. THEOREM [**Th 2**]. *Conjecture 1.1 is true for Haken manifolds. A closed Haken manifold has a hyperbolic structure iff it is homotopically atoroidal.*

It is hard to say how general the class of Haken manifolds is. There are many closed manifolds which are Haken and many which are not. Haken manifolds can be analyzed by inductive processes, because as Haken proved [**Hak**], a Haken manifold can be cut successively along incompressible surfaces until one is left with a collection of 3-balls. The condition that a 3-manifold has an incompressible surface is useful in proving that it has a hyperbolic structure (when it does), but intuitively it really seems to have little to do with the question of existence of a hyperbolic structure.

A *link L* in a 3-manifold M^3 is a 1-dimensional compact submanifold. A 3-manifold N^3 is said to be obtained from M^3 by Dehn surgery along L if N^3 is obtained by removing a regular neighborhood of L, and gluing it back in by some new identification. The new identification is determined by choosing a diffeomorphism of the torus for each component of L. Two choices of diffeomorphisms ϕ and ψ give rise to diffeomorphic manifolds if $\psi^{-1}\phi$ extends to a diffeomorphism of a solid torus (the regular neighborhood of the component of L).

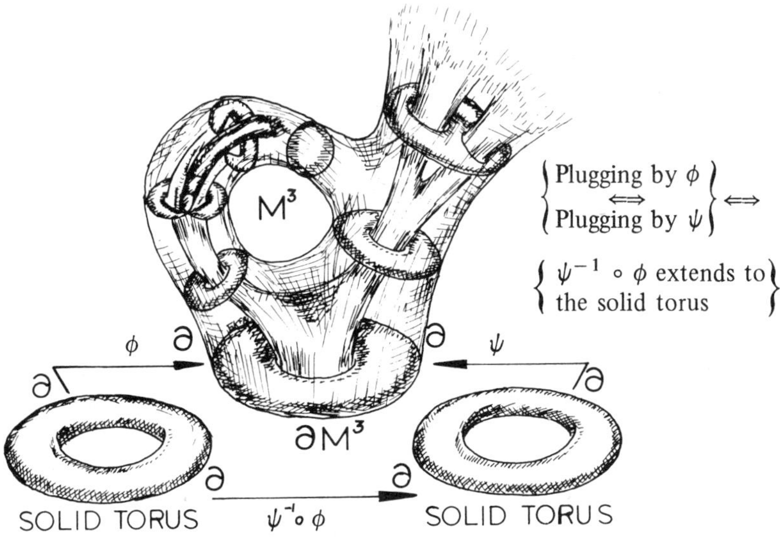

FIGURE 3. Modifying a 3-manifold by Dehn surgery. Plugging in a solid torus by φ gives a result diffeomorphic to plugging by ψ iff the diffeomorphism of the torus $\psi^{-1} \circ \varphi$ extends to a diffeomorphism of the solid torus.

There remains, for each component of L, a countably infinite set of essentially distinct choices.

2.6. THEOREM [**Th 1**]. *Suppose $L \subset M^3$ is a link such that $M - L$ has a hyperbolic structure. Then most manifolds obtained from M by Dehn surgery along L have hyperbolic structures. In fact, if we exclude, for each component of L, a finite set of choices of identification maps (up to the appropriate equivalence relation as mentioned above), all the remaining Dehn surgeries yield hyperbolic manifolds.*

It has been proven that for many knots or links, most Dehn surgeries give rise to non-Haken manifolds (see Thurston [**Th 1**], Hatcher and Thurston [**H, T**], Culler, Jaco and Rubinstein [**C, J, R**], Floyd and Hatcher [**F, H**], Hatcher [**Hat**]), so Theorem 2.6 yields many hyperbolic manifolds not covered by Theorem 2.5.

Every closed 3-manifold is obtained from the three-sphere S^3 by Dehn surgery along some link whose complement is hyperbolic, so in some sense Theorem 2.6 says that most 3-manifolds are hyperbolic. The most promising approach to Conjecture 1.1 seems to be to try to eliminate the vagueness from Theorem 2.6, and analyze exactly what happens under all Dehn surgeries. This can be studied by analytic continuation through families of geometric structures on $M - L$ which become singular at L. For certain examples (see [**Th 1**]) this has been successfully accomplished, so that here we have geometric structures for all manifolds obtained by Dehn surgeries along L.

It is quite feasible to use computers to study Conjecture 1.1 for classes of manifolds obtained by Dehn surgery. In fact, I have undertaken such a project in order to gain more feeling for the evolution of geometric structures on 3-manifolds. The examples I studied were arbitrary torus bundles over the circle, with Dehn surgery performed along a section of this bundle. Troels Jørgensen was the first to prove that the complement of such a section has a hyperbolic structure, in most cases. I wrote a computer program which found hyperbolic structures for particular Dehn surgeries on particular torus bundles over the circle. With the aid of the program, I found the pattern of which Dehn surgeries give rise to hyperbolic manifolds, and I could then show by hand that those examples for which the program was not producing hyperbolic structures in fact had other geometric structures. For sufficiently complicated torus bundles, the empirical observation is that all but the trivial Dehn surgery gives a hyperbolic manifold.

The geometric structures turn out to be very beautiful when you learn to see them. Often, the information which determines a geometric structure can be expressed in terms of some construction in plane Euclidean geometry. For instance, the output from my computer program which performs Dehn surgery on torus bundles over the circle is a tesselation of the plane minus the origin by triangles. The combinatorial pattern is predetermined, together with rules that certain triangles are similar. If such a pattern exists, then the manifold in question has a hyperbolic structure.[1]

---

[1]*Added in proof.* I can now prove that 1.1 is true for all prime 3-manifolds with a symmetry having fixed point set of dimension ≥ 1. This includes the examples above.

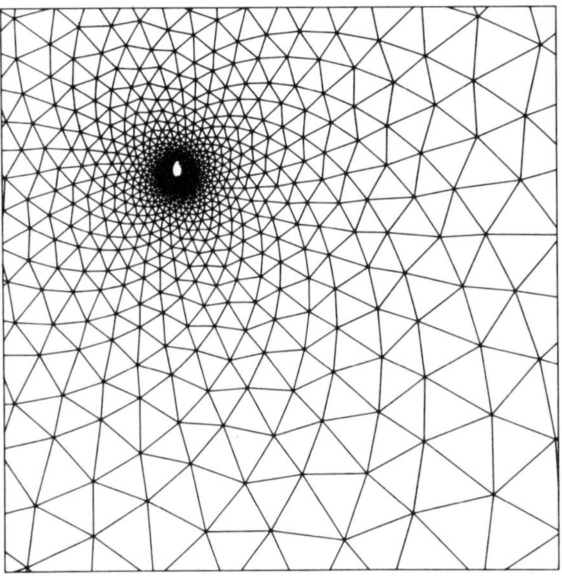

FIGURE 4. Three o'clock sky.

3. Applications. Theorems 2.3 and 2.5, or better yet, the conviction that Conjecture 1.1 is true, have many applications to the understanding of 3-manifolds.

First, we should point out the well-known Mostow rigidity theorem, which asserts that hyperbolic structure on closed manifolds, or more generally, any hyperbolic structure of finite volume, is canonical, provided the dimension is at least 3. (The original theorem, which does not include the case of noncompact manifolds, was proven by Mostow [**Mos**]. The proof for the noncompact case was completed by Prasad [**Pra**].)

3.1. MOSTOW RIGIDITY THEOREM. *Suppose M_1 and M_2 are hyperbolic manifolds of finite volume for which there is an isomorphism*

$$\phi: \pi_1(M_1) \to \pi_1(M_2).$$

Then there is an isometry $F: M_1 \to M_2$ which induces the isomorphism ϕ (up to conjugacy) between the fundamental groups.

The homotopy type of a hyperbolic manifold is determined by its fundamental group. Theorem 3.1 says that any invariant of the geometry of a hyperbolic manifold is actually an invariant of its homotopy type. This gives a powerful tool for distinguishing 3-manifolds.

3.2. COROLLARY. *Let M be a homotopically atoroidal Haken manifold. Then there are only a finite number of isotopy classes of homeomorphisms $\phi: M \to M$. The group of isotopy classes of self-homeomorphisms of M lifts to a group of actual homeomorphisms.*

OUTLINE OF PROOF. The group of isometries of any Riemannian manifold of finite volume is compact. An isometry of a hyperbolic manifold of finite volume which is near the identity actually equals the identity, so the group of isometries is finite. An isometry homotopic to the identity is equal to the identity, so the group of isometries is isomorphic to the group of automorphisms of the fundamental group. Waldhausen [**Wa 1**] showed that a homeomorphism inducing the identity on π_1 (up to conjugacy) is isotopic to the identity, so the group of isometries is also isomorphic to the group of isotopy classes of homeomorphisms.

REMARK. When M is not geometrically atoroidal, there are self-homeomorphisms of M supported in a neighborhood of a torus which have infinite order up to isotopy.

A group G is called *residually finite* if for each $g \in G$ there is some finite quotient map $f\colon G \to F$ such that $f(g) \neq 1$. If $G = \pi_1(M)$, an equivalent condition is that every compact subset of the universal covering space of M maps homeomorphically to some finite sheeted covering space.

3.3. THEOREM. *The fundamental group of a Haken manifold, or of any manifold for which Conjecture* 1.1 *holds, is residually finite.*

It is a standard fact that a finitely generated subgroup of $GL_n(\mathbf{C})$ is residually finite. Using this, one easily sees that the fundamental group of any geometric 3-manifold is residually finite. After a certain amount of fussing, one can assemble finite quotients of the fundamental groups of pieces of a geometric decomposition of a 3-manifold to obtain finite quotients of the fundamental group of the entire manifold.

One difficulty in the study of 3-manifolds has been the lack of any good invariants, such as the Euler characteristic for 2-manifolds. (The Euler characteristic of a closed 3-manifold is always 0.) Just as the area of a complete hyperbolic 2-manifold is a topological invariant (proportional to the Euler characteristic), so the volume of a hyperbolic 3-manifold is a topological invariant, which in some sense is a single measure of complexity of the manifold. One indication of this is the following theorem, whose basic core is due to Gromov, and which was sharpened in Thurston [**Th 1**] to give parts (b) and (c).

3.4. THEOREM. (a) *Suppose* $f\colon M \to N$ *is a map of nonzero degree between closed hyperbolic 3-manifolds. Then* $\mathrm{volume}(M) \geqslant |\,\mathrm{degree}(f)\,|\,\mathrm{volume}(N)$.

(b) *If equality holds in* (a), *then f is homotopic to a covering map which is a local isometry.*

(c) *Suppose M is a hyperbolic manifold of finite volume, and N is a hyperbolic manifold containing a* (*nonempty*) *link L such that $M \approx N - L$. Then* $\mathrm{volume}(N) < \mathrm{volume}(M)$.

The situation in part (c) is not unusual.

3.5. THEOREM (JØRGENSEN). *For any constant C, let $\mathcal{K}_C$ denote the set of hyperbolic 3-manifolds of volume* $\leqslant C$. *Then there is a finite subset* $\mathfrak{M} \subset \mathcal{K}_C$ *such that any element $N \in \mathcal{K}_C$ contains a link $L \subset N$ whose components consist of short geodesics such that $N - L$ is homeomorphic to some element $M \in \mathfrak{M}$.*

Intuitively, there is a finite set of great-grandmothers whose offspring are all the hyperbolic manifolds of volume $\leq C$.

If the link L has a short total length, one can show that the volumes of M and N are close. With this information, Theorems 3.4 and 3.5 imply

3.6. THEOREM. *The set of volumes of hyperbolic manifolds is a well-ordered subset of* **R**. *The set of manifolds with any given volume is finite.*

Using the Existence Theorem, 2.6, for hyperbolic Dehn surgeries, one deduces

3.7. THEOREM (SEE [**Th 1**]). *The order type of the set of all volumes of hyperbolic manifolds is* ω^ω.

Note. The notation ω^ω is that of ordinals, not cardinals: ω^ω is the countable ordinal which describes the order type of polynomials (of finite degree) in the symbol ω with natural number coefficients, ordered by the limiting order of the values when higher and higher integers are substituted for ω.

In other words, there is some lowest volume v_1; some second lowest volume v_2; a third lowest volume v_3, and so on until the first accumulation point of volumes v_ω, which is the smallest volume of a hyperbolic manifold with one cusp. There is a next highest volume $v_{\omega+1}$, and so on until the second accumulation point $v_{2\omega}$. The first accumulation point of accumulation points is called v_{ω^2}, and it is the smallest volume of a hyperbolic manifold with two cusps. And so on.

The *structure* of the set of volumes is significant, but it seems impractical to actually compute the real numbers v_α, for α a typical ordinal, and probably silly to try for very many ordinals. It would be interesting to know a few simple cases, like v_1 and v_ω, however. Based on a not at all exhaustive examination I made of a few examples and some computer computations of likely classes of examples by Bob Meyerhoff, it seems plausible that v_ω could be the volume of the complement of the figure eight knot, 2.0...,

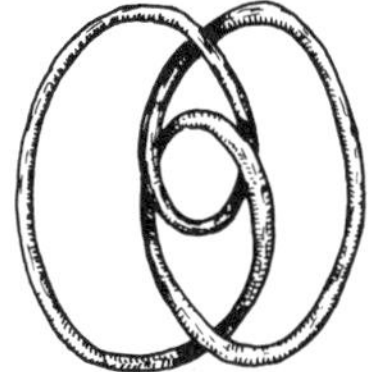

Is $v_\omega = 2.029883 \ldots$
the volume
of $S^3 - \S$

FIGURE 5.

The best candidate for v_1 so far is the volume of a manifold obtained by Dehn surgery along the figure eight knot; its volume is about .98.

There is another promising invariant of oriented hyperbolic 3-manifolds, about which practically nothing has been known until recently. This is the eta invariant, or the Chern-Simons invariant which is obtained by reducing the eta invariant mod a constant. The eta invariant should probably be thought of as the imaginary part of a complex constant, of which the volume is the real part. The eta invariant changes sign under reversal of orientation, while the volume is fixed.

Unfortunately, the eta invariant, unlike the volume, has been very difficult to compute, basically because in its definition some very arbitrary choices are involved, and these choices are difficult to make in any reasonable way. Recently, however, Bob Meyerhoff [**Mey**] has developed some good formulas for the Chern-Simons invariant. Among other things, he has shown that the set of values of the Chern-Simons invariant is dense in the circle—in fact, the manifolds obtained by Dehn surgery on any link with hyperbolic complement have Chern-Simons invariants whose values are dense.

The solution of the Smith conjecture is a development in which several mathematicians were fairly directly involved and in which Theorem 2.3 played an essential part, along with the equivariant loop theorem of Meeks and Yau [**M, Y**].

3.4. THEOREM. *Suppose* $\phi: S^3 \to S^3$ *is an orientation-preserving diffeomorphism such that* $\phi^n = 1$ *and for some* x, $\phi x = x$. *Then the fixed point set of* ϕ *is an unknotted circle, and* ϕ *is conjugate to an isometry.*[2]

For an explanation of the various parts of the proof and the somewhat complex way they fit together, see the proceedings of the Smith conjecture conference [**Smi**].

Conjecture 1.1 leads to a very concrete method of analyzing 3-manifolds. Eventually, it should be practical to begin with any of the usual descriptions of a 3-manifold M (e.g., a description of M using Dehn surgery along a link) and by a routine procedure on the computer, calculate its geometric decomposition (assuming that it has one). The first case to implement should be the case that M is hyperbolic. Robert Riley has, in fact, found hyperbolic structures for a variety of knot complements by computer, but his calculations are not routine except in special cases. Riley's work makes it clear that there is a rigorous, but not generally practical, algorithm for computing hyperbolic structures. The first step in this algorithm is to calculate representations of $\pi_1(M)$ in $PSL_2(\mathbf{C})$, which is the group of isometries of $\mathbf{H}^3$. One wants a discrete, faithful representation; such a representation is unique up to conjugacy if it exists. Conjugacy classes of representations correspond to solutions of a system of polynomials in a number of variables; one need check only isolated solutions, of which there are only a finite number. An algorithm certainly exists for finding them, but unfortunately, it is not practical for a complicated 3-manifold since the number of variables becomes large and the degree of the polynomials grows exponentially large with the complexity of the 3-manifold. For many less complicated 3-manifolds, however, the set of all representations can be computed practically.

The second step is the part which at first sounds dubious. However, it works theoretically and Riley has written a computer program which handles it quite nicely and practically. This step is to check whether a given representation is discrete and faithful. Riley's method (which originates from Poincaré) is to

---

[2]*Added in proof.* The result alluded to in the previous footnote generalizes this and gives a new, purely geometric proof. It implies for instance, that a diffeomorphism of finite order of a hyperbolic manifold has a fixed point set isotopic to a finite union of geodesics.

find a fundamental domain for the group action, or else find enough group elements close to the identity to imply the group is not discrete. If a fundamental domain is obtained, that implies that the image of the representation is discrete; if all the relations which come from the combinatorics of the fundamental domain are implied by the relations in the original group, the representation is faithful.

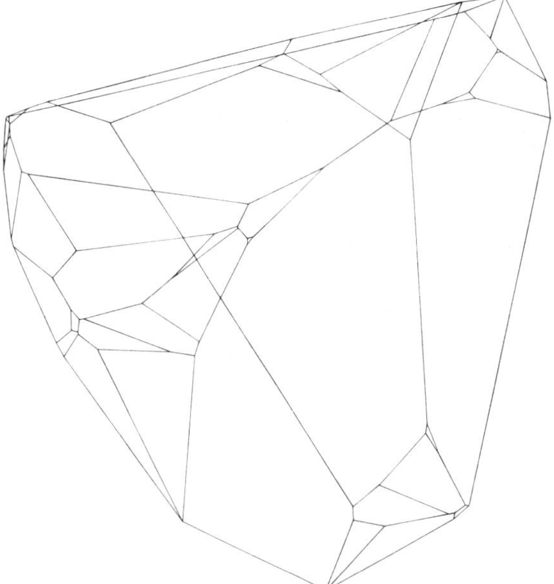

FIGURE 6. A fundamental domain for a discrete group of hyperbolic motions with compact quotient, generated by R. Riley's computer program. This is a projectively correct picture, using the Klein model for hyperbolic space.

To make the calculation of hyperbolic structures routine, what remains to be done is to find a practical general method of computing a likely candidate for the right representation of $\pi_1(M)$ in $PSL_2\mathbf{C}$. An approach using analytic continuation, depending on the theory of hyperbolic Dehn surgery, seems promising, but we will know whether it is successful only after it has been implemented on the computer.

There is an alternate approach to the second step, provided for the first step one calculates *all* representations of $\pi_1(M)$ in $PSL_2\mathbf{C}$ and provided also that one knows ahead of time that M has a hyperbolic structure. There is a characteristic number associated to a representation, fairly readily computable, which equals the volume of the quotient space when the representation is discrete and faithful. This characteristic number takes its unique maximal value for a discrete, faithful representation.

Successful completion of the two steps outlined gives a hyperbolic manifold whose fundamental group is $\pi_1(M)$. Conjecture 1.1, or simply the validity of Conjecture 1.1 for M, would imply that the hyperbolic manifold obtained is homeomorphic to M. If one doubts Conjecture 1.1, then, in general, a third step remains: to check whether M is homeomorphic to the hyperbolic manifold. The method of hyperbolic Dehn surgery often guarantees this. There are also very tedious algorithms which will eventually construct a homeomorphism provided it exists, but unfortunately there is no known algorithm which will show that the two manifolds are not homeomorphic.

There are indications that computations of other kinds of geometric structures may come as a byproduct of a method of computing hyperbolic structures: often there is a way to interpret these structures as degenerate hyperbolic structures.

4. Eight 3-dimensional geometries. As we promised in §1, we will now describe briefly the eight 3-dimensional homogeneous spaces, or geometries, which are needed to give geometric structures for 3-manifolds. We will in each case give a description of a simply-connected 3-manifold X, and a group G of orientation-preserving diffeomorphisms of X with compact stabilizers G_x. We will group them by the identity component of G_x: it is isomorphic either to $SO(3)$, to $SO(2)$, or to the trivial group $SO(1)$. For the first three geometries, G_x is $SO(3)$.

1. SPHERICAL GEOMETRY. X is the 3-sphere S^3 and G is $SO(4)$.

All 3-dimensional spherical manifolds have been classified. Cf. Seifert [**Sei**], for a complete enumeration.

The three-sphere S^3 has the structure of a group: the group of unit quaternions. Therefore $S^3 \times S^3$ acts on S^3 by the formula $x \overset{(g,h)}{\to} gxh^{-1}$. The kernel of this action is $\mathbf{Z}/2$, generated by $(-1, 1)$ in quaternionic notation, and this produces an incredible isomorphism

$$S^3 \times S^3/\mathbf{Z}/2 = SO(4).$$

It is with the aid of this isomorphism that the structure of 3-dimensional spherical manifolds can be analyzed. The most beautiful is the Poincaré dodecahedral space, obtained from a dodecahedron by identifying opposite faces with a $1/10$ right-handed rotation. This identification pattern forces the edges to be identified in triples. A regular dodecahedron in Euclidean space has dihedral angles somewhat less than $120°$. Regular convex polyhedra in S^3 have angles anywhere between the Euclidean value and $180°$. To put a geometric structure on the Poincaré dodecahedral space, find the regular spherical dodecahedron with dihedral angles of $120°$, and identify them in the given pattern. Note that the universal cover of the Poincaré dodecahedral space is S^3, with a tiling by 120 dodecahedra. (Poincaré first discussed this example in [**Poin**], in a very nongeometric form.)

It is a strange fact that all spherical manifolds also have a stronger structure, based on $X = S^3$ and $G = S^3 \times S^1/\mathbf{Z}/2$.

2. EUCLIDEAN GEOMETRY. $X = E^3$, $G = R^3 \times SO(3)$, the group of Euclidean isometries. There are only 10 nonhomeomorphic 3-dimensional Euclidean manifolds. Each one is finitely covered by the 3-torus.

3. HYPERBOLIC GEOMETRY (cf. the article by Milnor in this volume). One good picture of hyperbolic space is the Poincaré upper half space, $X = \{x, y, z \mid z > 0\}$. The group G is the group $PGL_2(\mathbf{C})$ of nonsingular 2×2 complex matrices of determinant 1, up to multiplication by $\pm I$ scalars. The action of G on X may be defined as fractional linear transformation using quaternionic

notation, $q = x + yi + zj$. The formula is

$$q \xrightarrow{\left[\begin{smallmatrix} a & b \\ c & d \end{smallmatrix}\right]} (aq + b)(cq + d)^{-1}.$$

One easily-described example of a hyperbolic manifold is the Seifert-Weber dodecahedral space, obtained by identifying opposite faces of a dodecahedron by $3/10$ right-handed rotations. One checks that edges of the dodecahedron are identified in quintuples. To form a geometric model, use the regular hyperbolic dodecahedron whose dihedral angles are $72°$.

For the next four cases, the identity component of G_x is $SO(2)$.

4. $X = S^2 \times \mathbf{E}^1 = G$ consists of (spherical isometries) $\times$ (isometries of the Euclidean line). Note that an isometry which reverses the orientation of both factor preserves the total orientation. There are only two nonhomeomorphic examples of compact manifolds with this geometry; can the reader find them?

5. $X = \mathbf{H}^2 \times \mathbf{E}^1$, G consists of (isometries of $\mathbf{H}^2$) $\times$ (isometries of $\mathbf{E}^1$). Every manifold modelled on this geometry is finitely covered by the product of a surface and a circle. Can the reader find an example which is not itself a product?

6. The space X is the universal covering space set of unit length vectors in the hyperbolic plane, $T_1(\mathbf{H}^2)$, and the group G is $\mathbf{R} \times$ (the universal covering of the group of isometries of $\mathbf{H}^2$). The isometries of $\mathbf{H}^2$ act by their derivatives, and $\mathbf{R}$ acts as simultaneous rotations of all vectors, keeping their based points fixed. Note that even orientation-reversing isometries of $\mathbf{H}^2$ preserve the orientation of $T_1(\mathbf{H}^2)$. The space of unit tangent vectors to any hyperbolic surface is an example of a manifold with this geometry. Another example is the quotient of $T_1(\mathbf{H}^2)$ by, say, the group of orientation-preserving automorphisms of the regular $(2, 3, 7)$ tiling.

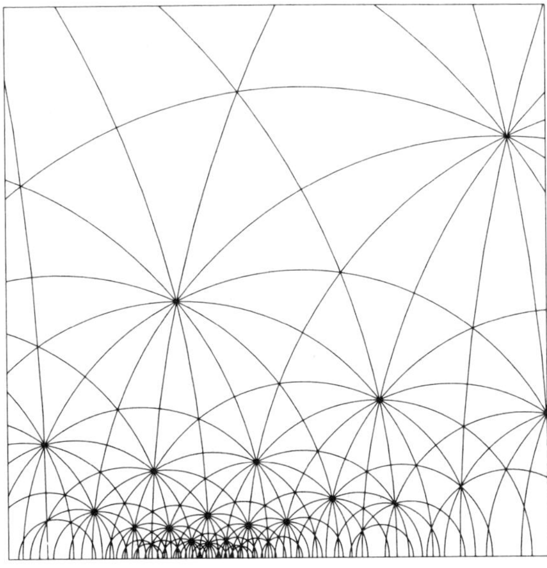

FIGURE 7. A tiling of the hyperbolic plane by congruent triangles with angles $\pi/2$, $\pi/3$, $\pi/7$. Upper half-space projection.

7. X is a twisted product of $\mathbf{E}^1$ with $\mathbf{E}^2$, and G is an extension of the group of isometries $\mathbf{E}^2$ by $\mathbf{R}$,

$$\mathbf{R} \to G \to \text{isometries of } \mathbf{E}^2.$$

To see a picture of X, consider $\mathbf{R}^3$ with a certain 2-plane field τ. Along the z-axis, τ is horizontal (orthogonal to the axis). Along any ray emanating from the z-axis and orthogonal to it, τ always contains the tangent to the ray, and its slope in the direction orthogonal to the ray increases linearly, so this slope always equals one half the distance from the z-axis. In coordinates, if e_1, e_2 and e_3 are the three coordinate vector fields in $\mathbf{R}^3$, τ is spanned by

$$e_1 - (1/2)y\, e_3, \qquad e_2 + (1/2)x\, e_3.$$

Any isometry of the x-y plane lifts to an affine map of $\mathbf{R}^3$ which preserves τ and also preserves distances along lines parallel to the z-axis. In fact, there is a whole 1-parameter family of such lifts. (These affine maps may be thought of as obtained by an isometry of $\mathbf{E}^3$, followed by some shearing map to adjust τ.) These affine transformations of $\mathbf{R}^3$ constitute G. Note that an orientation-reversing isometry of $\mathbf{E}^2$ lifts only to maps which reverse the z-direction.

Any oriented circle bundle over a 2-torus which is not the 3-torus has this kind of geometric structure. The simplest circle bundle over the 2-torus is the one with Euler class 1, and it is obtained as the quotient of X by the discrete, nilpotent subgroup Γ of G generated by

$$(x, y, z) \to (x + 1, y, z + y/2) \quad \text{and} \quad (x, y, z) \to (x, y + 1, z - x/2).$$

Notice that the commutator of these two generators is the vertical translation

$$(x, y, z) \to (x, y, z + 1).$$

Every other example of a manifold with this geometry is finitely covered by a manifold homeomorphic to X/Γ.

An alternate description of this geometry is to define X as the Heisenberg group,

$$H = \left\{ \begin{bmatrix} 1 & x & z \\ & 1 & y \\ & & 1 \end{bmatrix} \right\}$$

and G as the semidirect product of H by S^1 acting as a group of automorphisms of H which rotate the x-y plane. In this form, the group Γ can be taken as the subgroup of H where x, y and z are integers.

8. Finally, there is one example for which the identity component of G_x is trivial. In this case, X is very naturally a Lie group, the solvable Lie group

$$\mathbf{R}^2 \to X \to \mathbf{R}$$

where $\mathbf{R}$ acts on $\mathbf{R}^2$ (by conjugation) with the formula

$$(x, y) \to (e^t x, e^{-t} y).$$

The group G is an extension of X by $(\mathbf{Z}_2)^2$ acting as a group of automorphisms, whose 3 nontrivial elements are the 180° rotations

$$(x, y, t) \to (-x, -y, t), \quad (x, y, t) \to (y, x, -t) \quad \text{and} \quad (x, y, t) \to (-y, -x, -t).$$

Any torus bundle over S^1 whose monodromy is a linear map with distinct real eigenvalues has a geometric structure of this form.

5. The varying geometry of Kleinian groups. The Mostow rigidity theorem applies only to hyperbolic 3-manifolds with finite volume, which are interiors of compact 3-manifolds whose boundary consists of tori. Hyperbolic structures for the interior of a manifold with more complicated boundary are not rigid, and there is a wonderful deformation theory for them. The proofs by induction of Theorems 2.3 and 2.5 involve a study of the varying geometry of such manifolds. The theory of this varying geometry is tricky and rich, and a number of intriguing and significant questions remain. Here we will outline a few of the considerations.

In §4 we mentioned the Poincaré upper half-space picture for $\mathbf{H}^3$. If one adjoins a single point at ∞ in $\mathbf{R}^3$, then the action of the group $PGL_2(\mathbf{C})$ extends to the closure of upper half-space, which is a ball. The boundary of the ball is the Riemann sphere

$$\mathbf{C} = \mathbf{C} \cup \{\infty\} = \mathbf{CP}^1,$$

and the action of $PGL_2(\mathbf{C})$ is the usual action as Moebius transformations or complex projective transformations

$$z \xrightarrow{\left[\begin{smallmatrix} a & b \\ c & d \end{smallmatrix}\right]} \frac{az + b}{cz + d}.$$

A *Kleinian group* Γ is a discrete subgroup of $PGL_2(\mathbf{C})$, that is, a subgroup for which each element has a neighborhood in the group containing only that point. If Γ has no elements of finite order, then it acts freely on $\mathbf{H}^3$, and $\mathbf{H}^3/\Gamma$ is a complete hyperbolic manifold. Henceforth we will assume this to be the case.

Each orbit of Γ acting on $\mathbf{C}$ has accumulation points (provided Γ is not finite). The set of all accumulation points of orbits is called the limit set L_Γ of Γ, while its complement is called the domain of discontinuity, D_Γ. The Kleinian group Γ acts nicely on D_Γ (properly discontinuously) so that its quotient D_Γ/Γ is a surface, which inherits a conformal structure (= complex structure) from D_Γ. In fact, D_Γ/Γ is the boundary of the three-manifold $(H^3 \cup D_\Gamma)/\Gamma$. Sometimes this "Kleinian" 3-manifold is compact, and sometimes it isn't.

The basic deformation theorem, developed by Ahlfors, Bers, Mostow, Sullivan and others, says that a certain class of *relatively* mild deformations of Γ, the quasi-conformal deformations of Γ, are controlled precisely by conformal structures on D_Γ/Γ. Except in degenerate cases (when Γ is abelian) the conformal structure on D_Γ/Γ is represented by a unique hyperbolic metric, by the classical uniformization theorem.

5.1. AHLFORS FINITE AREA THEOREM. *The total hyperbolic area of D_Γ/Γ is finite.*

Thus one has a correspondence between the space of quasi-conformal deformations of Γ and the space of finite-area hyperbolic metrics on D_Γ/Γ, which is called the Teichmüller space of D_Γ/Γ.

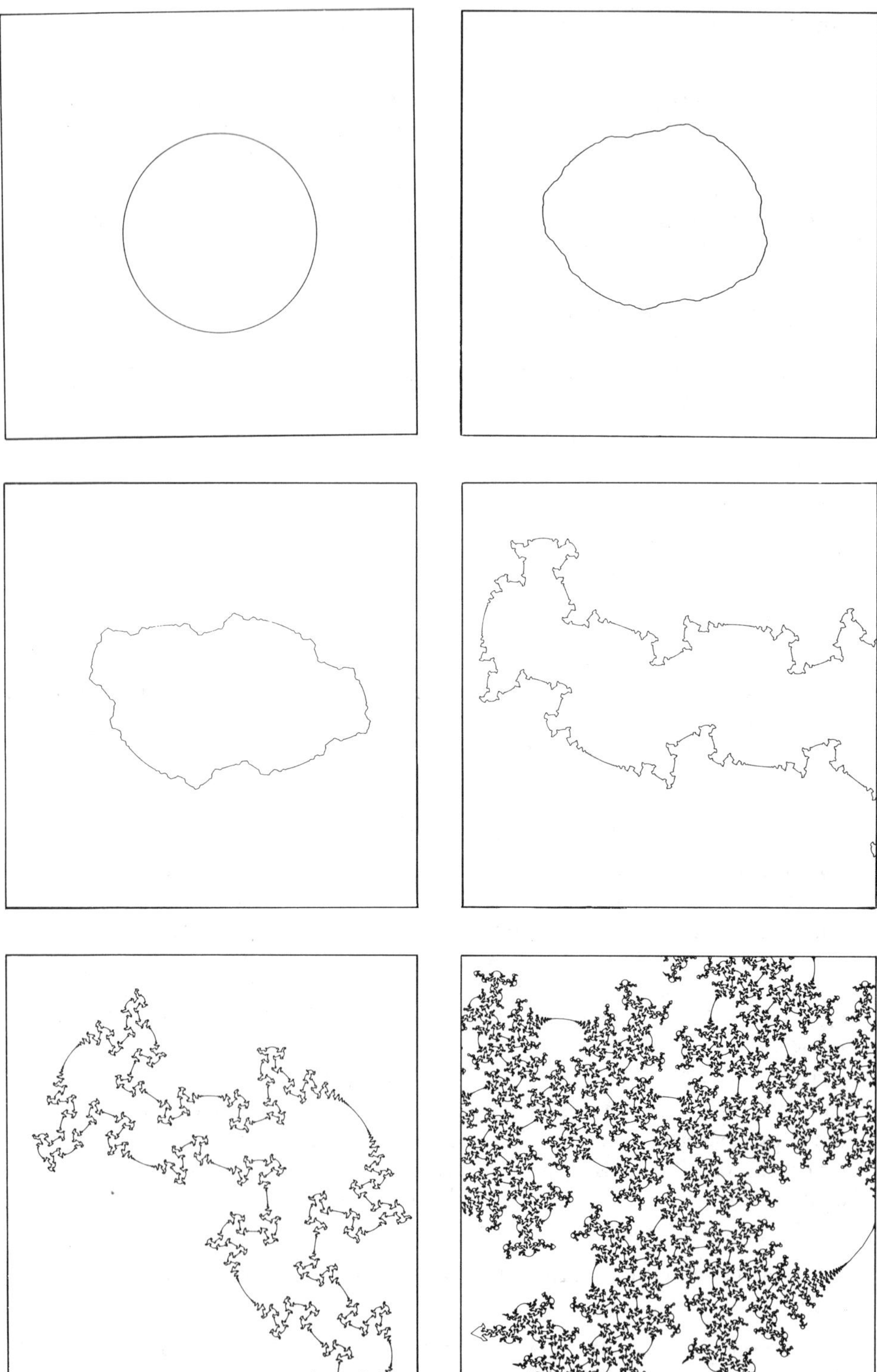

FIGURE 8. Six limit curves. The last curve is omnipresent.

Here are some computer sketches of examples in which the limit set is a curve. Poincaré was taken by the nondifferentiable nature of the curves which even have Hausdorff dimension > 1, as described in Sullivan's article [**Sul 1**].

The first limit set is a circle. This means the group leaves invariant a hyperbolic plane inside $\mathbf{H}^3$; in Poincaré upper half-space, hyperbolic planes appear as Euclidean hemispheres resting on $\mathbf{C}$. The other groups are obtained by bending the original group. All the curves except the last are Jordan curves; the last curve actually fills $\mathbf{C}$, but to make the computer actually fill $\mathbf{C}$ would require an amount of computer money (and foolishness) approximating the "defense" budget.

A group whose limit set is a circle is called the Fuchsian group (because of Poincaré's modesty), and groups whose limit sets are Jordan curves are called quasi-Fuchsian groups; they are obtained by quasi-conformal deformations of Fuchsian groups. The final group in our sequence of pictures cannot be obtained by a quasi-conformal deformation of a Fuchsian group, since the topology of the limit set has changed. How can we explain limiting phenomena such as this?

First we need to know something about the limiting geometry of hyperbolic surfaces, as the hyperbolic structure goes to infinity in Teichmüller space. One way that hyperbolic structures can go to infinity is that a curve, or system of curves, can be pinched:

FIGURE 9. A surface with a pinched waist.

The hyperbolic metric develops a long, skinny waist to accomplish this.

To describe more generally how surfaces can go to infinity in Teichmuller space one needs a generalization of the notion of a simple closed curve.

A *geodesic lamination* λ of a hyperbolic surface S is a closed subset of S which is a disjoint union of complete geodesics on S (called the leaves of λ). A simple closed geodesic is one example of a geodesic lamination. To get other examples, consider a sequence of longer and longer simple closed geodesics. There is always a subsequence so that the pictures "converge", usually to an uncountable set of geodesics. A typical local cross-section of a geodesic lamination is a Cantor set, but other behavior can also occur. The 2-dimensional Lebesgue measure of a geodesic lamination is always 0.

A *transverse invariant measure*, μ, for a geodesic lamination can be thought of as a rule which assigns to each transverse arc α to λ a measure that is supported on $\lambda \cap \alpha$ and invariant under maps from one arc α to another arc β which take each point of intersection of α with a leaf of λ to a point of intersection of β with the same leaf. One can think of μ as a weighted counting of the leaves of λ, or as a measure of the amount of exertion required to cross a certain set of leaves. A lamination equipped with a transverse invariant measure of full support is a *measured lamination*. We also include here the trivial lamination consisting of no leaves and having 0 measure.

There is a good topology for the set of measured laminations, making it a Hausdorff space. In this topology, a bunch of leaves can continuously get smaller and smaller measure until they vanish. Let $\mathfrak{ML}(S)$ denote the space of measured laminations on S, and $\mathfrak{ML}_0(S)$ denote the space of compactly supported measured laminations (in the case S is noncompact). Despite the apparently erratic behavior of laminations, we have

5.2. THEOREM [**Th 1**]. *The space $\mathfrak{ML}_0(S)$ is homeomorphic to Euclidean space of the same dimension as the Teichmüller space for S.*

There are also projectivized versions of the lamination spaces. That is, any transverse invariant measure may be multiplied by a positive constant to give a new transverse measure. We define the projective lamination spaces

$$P\mathfrak{L}(S) = (\mathfrak{ML}(S) - 0)/\text{multiplication by scalars}$$

and

$$P\mathfrak{L}_0(S) = (\mathfrak{ML}_0(S) - 0)/\text{multiplication by scalars}.$$

5.3. THEOREM. *There is a natural topology on $\mathfrak{T}(S) \cup P\mathfrak{L}_0(S)$ which makes it a ball, where $\mathfrak{T}(S)$ is the Teichmuller space for S.*

Intuitively, the interpretation is that a sequence of hyperbolic structures on S can go to infinity by "pinching" a certain geodesic lamination λ; then it converges to λ. As a lamination is pinched toward 0, lengths of paths crossing it are forced toward infinity. The ratios of these lengths determine the transverse invariant measure. A good exposition of this theory may be found in the book by Fathi, Laudenbach, Poénaru et al [**F, L, P**], although this work deals with the closely related theory of measured foliations.

Geodesic laminations enter into the theory of hyperbolic 3-manifold in a number of guises, but we will describe only one result. Two compactly supported laminations λ_1 and λ_2 *fill up* S if $S - \lambda_1 \cup \lambda_2$ consists of chunks which, when lifted to S, are bounded if S is compact, or possibly contained in a horoball neighborhood of a cusp.

5.4. DOUBLE LIMIT THEOREM [**Th 3**]. *Let λ_1 and λ_2 be a pair of projective laminations which fill up S. Then if $\{g_1^i\} \to \lambda_1$ and $\{g_2^i\}$ are sequences in $\mathfrak{T}(S)$ tending toward λ_1 and λ_2, the sequence of quasi-Fuchsian groups whose two components of D_Γ/Γ have conformal structures given by g_1^i and g_2^i has a subsequence which converges algebraically to some Kleinian group Γ isomorphic to $\pi_1(S)$.*

This was proven first in the case S is a punctured torus by Jørgensen, [**Jør**].

This theorem implies the existence of many geometrical complicated Kleinian groups isomorphic to $\pi_1(S)$, since from the theory developed in [**Th 1**] the laminations λ_1 and λ_2 can be reconstructed from Γ although the information of the projective class of measures is lost.

A geodesic lamination is *arational* if it intersects every simple closed geodesic. When λ_1 and λ_2 are arational (which is the likely case), then the limit set of Γ is the entire 2-sphere.

There is one construction for 3-manifolds which relates very directly to the theory of surfaces, and which gave me considerable difficulty for a time in the proof of Theorem 2.5. In fact, when I first heard the suggestion that Theorem 2.5 was true (by one mathematician giving a distorted quote of another), I immediately came up with this class of 3-manifolds as "obvious" counterexamples.

This construction is the mapping torus construction, which produces a 3-manifold M_ϕ depending on a diffeomorphism $\phi \colon S \to S$ of a surface S. One simply forms the product $S \times I$ and identifies each point $(x, 1)$ to $(\phi(x), 0)$. It is easy to construct quite complicated diffeomorphisms ϕ by taking compositions of simple diffeomorphisms, called Dehn twists, which are the identity outside the neighborhood of a simple closed curve on S.

The diffeomorphism ϕ of S gives rise to a natural transformation of $\mathfrak{I}(S)$, which extends continuously to the ball $\overline{\mathfrak{I}}(S) = \mathfrak{I}(S) \cup \mathscr{P}\mathscr{L}_0(S)$. By the Brouwer fixed point theorem, ϕ has at least one fixed point in $\overline{\mathfrak{I}}(S)$.

5.5 THEOREM. *One of the following 3 alternatives holds:*

(a) *There is a fixed point in $\mathfrak{I}(S)$, and ϕ is isotopic to a diffeomorphism of finite order.*

(b) *There is a finite system of disjoint simple curves invariant (up to isotopy) by ϕ.*

(c) *There are precisely two points in $P\mathscr{L}_0(S)$ fixed by ϕ. These are arational laminations which together fill up S.*

The deduction in part (a) that if ϕ fixes a point in Teichmüller space, ϕ is isotopic to a diffeomorphism of finite order is the same as the deduction of Corollary 3.2 from the Mostow rigidity Theorem 3.1.

Theorem 5.5 is part of the classification of conjugacy classes of diffeomorphisms of surfaces up to isotopy [**Th 4**] or [**F, L, P**], analogous to Jordan form. In case (b), one can cut the surface along the system of curves and analyze the diffeomorphism into diffeomorphisms of simpler surfaces. In case (c), there is a very nice, canonical representative of the isotopy class of ϕ (up to conjugacy), called a pseudo-Anosov homeomorphism.

5.6. THEOREM [**Th 3**]. *The mapping torus M_ϕ has a hyperbolic structure if and only if ϕ satisfies condition* (c) *(ϕ is isotopic to a pseudo-Anosov homeomorphism).*

This was proven first for the case S is a punctured torus by Jørgensen, [**Jør**]. This is a special case of Theorem 2.5 or 2.3. An exposition of this can also be found in [**Su 2**]. The proof in [**Th 3**] is by applying the double limit Theorem 5.4 to the two laminations in $P\mathscr{L}_0(S)$ fixed by ϕ. One obtains an action of $\pi_1(S)$ on $\mathbf{H}^3$, which by an extension of the Mostow rigidity theorem from [**Th 1**] or [**Su 3**] can be shown to be conjugate to the action obtained by composing with the automorphism ϕ. The conjugating isometry, when adjoined to the isometries coming from $\pi_1(S)$, gives a discrete, faithful action of $\pi_1(M_\phi)$. The quotient manifold is homeomorphic to M_ϕ by 3-manifold theory. This was first proven by Stallings [**Sta**].

Consider now a copy of the surface S inside the 3-manifold M_ϕ. The universal covering space S sits inside the universal covering space of M_ϕ, which is $\mathbf{H}^3$. The closure of S must contain all of $\mathbf{C}$, because as we have indicated the limit set of $\pi_1(S)$ is all of $\mathbf{C}$. (This is also easy to deduce directly from the existence of a hyperbolic structure on M_ϕ.) If we consider any hyperbolic metric on S, this gives a diffeomorphism of S to $\mathbf{H}^2$. Does the map of $\mathbf{H}^2$ to $\mathbf{H}^3$ so defined extend continuously to a map of the 2-disk to the 3-ball? This question is more subtle than it first appears, but it was answered affirmatively by J. Cannon and me:

5.7. THEOREM [**Can-Th**]. *The circle at infinity in the universal cover of a fiber of a mapping torus maps continuously to the sphere at infinity in $\mathbf{H}^3$, to give a sphere-filling curve.*

The topology of the map of S^1 to S^2 can be described exactly, as follows. Let λ_1 and λ_2 be the two geodesic laminations invariant by ϕ. Put two copies of S on a 2-sphere, one filling the northern hemisphere and one filling the southern hemisphere, with corresponding points on the circles at infinity glued together. Put a copy of λ_1 lifted to S on the northern hemisphere, and λ_2 lifted to S on the southern hemisphere.

FIGURE 10. A pattern of identifications of a circle, here represented as the equator, whose quotient space is topologically a sphere. This defines, topologically, a sphere-filling curve.

Now, form the identification space of S^2 obtained by identifying the closure of each leaf of a lamination to a point and the closure of each component of the complement of the laminations to a point. It can be deduced readily from a theorem of R. L. Moore that the identification space is homeomorphic to the 2-sphere.

Since each equivalence class meets the equator, the image of the equator is the entire 2-sphere. This is the topological model we promised.

There are similar models for the possible behaviour of the boundary of S in *all* the cases covered by the double limit theorem. It is not hard to show that these are the correct topological models provided the maps of S^1 are continuous, but continuity is not known in the general case.

We have already given one picture of a space-filling curve, which, indeed, came from this construction. The surface in question was a punctured torus in the figure eight knot complement.

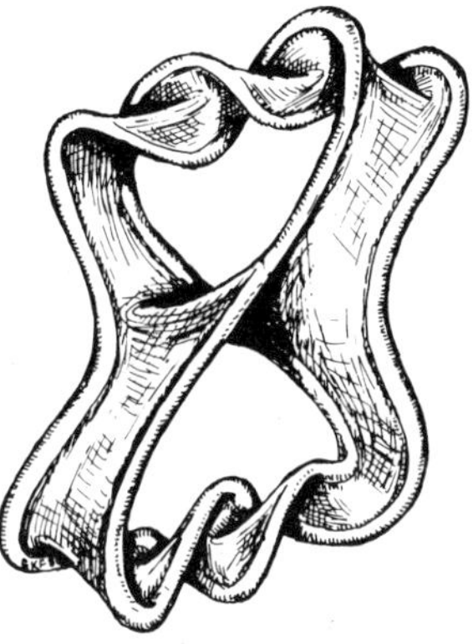

FIGURE 11. The figure eight knot is spanned by a surface (topologically, a punctured torus) which can be swept around through all of S^3-(the figure eight knot) and brought back to its starting position!

This knot complement is homeomorphic to the mapping torus of the diffeomorphism of the punctured torus determined by the linear map $\left[\begin{smallmatrix}2&1\\1&1\end{smallmatrix}\right]$ of the torus. Here is another example. Note the definite sense of the spirals. This reflects the fact that, unlike in the case of the figure eight knot complement, M_ϕ is not diffeomorphic to its mirror image. In this case, ϕ is the diffeomorphism of the punctured torus coming from the linear map $\left[\begin{smallmatrix}5&4\\1&1\end{smallmatrix}\right]$ of the torus.

FIGURE 12. The sphere-filling curve determined by the punctured torus bundle over the circle with monodromy $\left[\begin{smallmatrix}5&4\\1&1\end{smallmatrix}\right]$.

The double limit theorem is a powerful result for analyzing surface groups, which are among the most flexible of Kleinian groups. There are several other significant theorems which back up the double limit theorem by deriving information about the geometry of the quotient manifolds in the limits which the double limit theorem produces (see [**Th 1**] and [**Th 2**]). There are still many basic conjectures which have not been proven in general, however. Information about the infinitely complicated manifolds obtained by the double limit theorem has relevance because it relates to the geometry of closed manifolds, especially to the mapping tori M_ϕ, but also to much more general manifolds.

We will now give a sample of a theorem which complements the double limit theorem by analyzing a certain class of Kleinian groups which are more complicated and more rigid than surface groups.

5.8. THEOREM [**Th 2**]. *Let Γ be a Kleinian group such that* (a) *the Kleinian manifold $M = (\mathbf{H}^3 \cup D_\Gamma)/\Gamma$ is compact,*

(b) *the components of D_Γ are simply connected and*

(c) *the closures of any two components of D_Γ are disjoint.*

Then the space of discrete, faithful representations of Γ in $PSL_2\mathbf{C}$, up to conjugacy, is compact.

Condition (b) is equivalent to the topological condition that ∂M is incompressible, and condition (c) is equivalent to the topological condition that M is acylindrical: there are no essential cylinders in M.

The space of conjugacy classes of discrete faithful representations of Γ is called the algebraic deformation space. For a precise definition of this and other topologies or sets of Kleinian groups, see [**Th 2**].

Since the quasi-conformal deformation space is noncompact, Theorem 5.8 produces many limits of sequences of quasi-conformal deformations of Γ, which are not themselves quasi-conformal deformations. Just as for the double limits of surface groups produced by Theorem 5.4, the limit set of most of the limiting groups is the entire sphere $\mathbf{C}$, and there is an "ending lamination" associated to each component of ∂M such that the associated components of D_Γ collapse in the limit.

The limiting groups produced by Theorem 5.8, unlike the ones produced by 5.4, do not occur as subgroups of groups with compact quotients. Nonetheless, 5.8 is very important in the proof of Theorem 2.5 because it implies a good deal about what can happen, before passing to the limit. In the inductive proof of Theorem 2.5, this enables one to show that manifolds such as M can be deformed until they fit together to give closed hyperbolic manifolds.

Kleinian groups are very beautiful and rich, with an amazing variety of productive ways to think about them. (Compare the articles of Bers [**Bers**] and Sullivan [**Sul 1**] for two different points of view. These samples are not exhaustive.) We have seen a great deal of progress in understanding them, but we are still in the midst of a number of significant and intriguing problems. Perhaps by the year 2000 our understanding of 3-manifolds and Kleinian groups will be solid, and the phenomena we now expect will be proven.

6. Some open questions. Here are a few questions and projects concerning 3-manifolds and Kleinian groups which I find fascinating.

1. Do all 3-manifolds have decompositions into geometric pieces?

2. Is every finite group action on a 3-manifold equivalent to an action respecting the geometry? In particular, let $\phi\colon M \to M$ be a diffeomorphism of finite order of a hyperbolic manifold. Is there an equivariant isotopy of ϕ to an isometry? In particular, is the fixed point set of ϕ isotopic to a union of geodesics?[3]

3. Does every 3-dimensional orbifold which contains no bad 2-dimensional suborbifolds admit a geometric decomposition? (For terminology, see [**Th 1**]. This question contains 2.)[3]

4. Develop a global theory of hyperbolic Dehn surgery. Give specific general upper bounds for nonhyperbolic Dehn surgeries. Describe the limiting geometry which occurs when hyperbolic Dehn surgery breaks down. (See [**Th 1**]. From a number of examples, we know that the behavior in the limit can be very beautiful, but there is no general theory. This is a possible approach to solving 1 and 3.)

5. Are all Kleinian groups geometrically tame? (See [**Th 1**] for a definition, which should be extended appropriately to the general case.)

6. Is every Kleinian group a limit of geometrically finite groups? (In many cases, this implies 5.)

7. Develop a theory of Schottky groups and their limits analogous to the theory of quasi-fuchsian groups and their limits developed in [**Th 1**].

8. Analyze limits of quasi-fuchsian groups with accidental parabolics. (6, 7 and 8 would combine to prove 5 in a very satisfying way.)

9. Is $\mathbf{H}^3/\Gamma$, where Γ is finitely generated, always homeomorphic to the interior of a compact manifold? (This was proven in many cases for geometrically tame groups in [**Th 1**].)

10. (AHLFORS MEASURE 0 PROBLEM). Does the limit set of a finitely-generated Kleinian group always have full measure or 0 measure, and in the former case does Γ act ergodically? (This was proven for many cases of geometrically tame groups in [**Th 1**].)

11. Classify geometrically tame representations of a given group. Are they parametrized by their ending laminations and their parabolics, together with the conformal structure on the domain of discontinuity?

12. Describe the quasi-isometry type of an arbitrary Kleinian group. In other words, give a formula for a Riemannian manifold that has a diffeomorphism to $\mathbf{H}^3/\Gamma$ such that the metrics have a bounded ratio. (A good deal of information about the quasi-isometry types is already known, but it is not yet complete.)

13. If the limit set of a finitely-generated Kleinian group has Hausdorff dimension less than 2, is it geometrically finite? (This would probably be a consequence of 5.)

14. Suppose Γ has the property that $(\mathbf{H}^3 \cup D_\Gamma)/\Gamma$ is compact. Then is it true that the limit set of any other Kleinian group Γ' isomorphic to Γ is the

---

[3]*Added in proof.* This is now proven, provided, for (3), the complement of the singular locus is irreducible.

homeomorphic image of the limit set of Γ, by a homeomorphism taking the fixed point of an element γ to the fixed points of the corresponding element γ'? (Theorems 5.7 is a special case of this.) There are examples to show that there is no continuous map

$$L_\Gamma \times \{\text{algebraic deformation space of } \Gamma\} \to S^2$$

which parametrizes the limit sets. Perhaps, though, there is a parametrization which is continuous separately in the two factors.

15. Can finitely-generated subgroups of a finitely-generated Kleinian group be residually separated from the group? In other words, given a subgroup $H \subset \Gamma$ and $\gamma \in \Gamma - H$, is there a finite quotient of Γ in which the image of γ is not in the image of H? Peter Scott proved this property for surface groups. It is useful for a number of topological arguments, even for special subgroups H.

16. Does every aspherical 3-manifold, or every hyperbolic 3-manifold, have a finite-sheeted cover which is Haken? This is related to (15). By applying Mostow's theorem, and (2.5), it is easy to see that a homotopically atoroidal manifold with a finite-sheeted cover which is Haken is homotopy equivalent to a hyperbolic manifold. Unfortunately, there seems to be little prospect of finding such finite-sheeted coverings without first knowing the manifold is hyperbolic.

17. Does every aspherical 3-manifold have a finite-sheeted cover with positive first Betti number? This is stronger than 16.

18. Does every hyperbolic 3-manifold have a finite-sheeted cover which fibers over the circle? This dubious-sounding question seems to have a definite chance for a positive answer.

19. Find topological and geometric properties of quotient spaces of arithmetic subgroups of $PSL_2 \mathbf{C}$. These manifolds often seem to have special beauty.

20. Develop a computer program to calculate the canonical form for a general diffeomorphism of a surface, and to calculate the action of the group of diffeomorphisms on $P\mathcal{L}_0(S)$. Use this to implement an algorithm for the word problem and the conjugacy problem in the group of isotopy classes of diffeomorphisms of a surface.

21. Develop a computer program to calculate hyperbolic structures on 3-manifolds.

22. Tabulate the volumes and the Chern-Simons invariants and other simple information for a bunch of 3-manifolds: for instance, the knots in the knot tables. Try to develop a practical feel for the well ordering.

23. Show that volumes of hyperbolic 3-manifolds are not all rationally related. Cf. [**Mil 2**].

24. Show that most 3-manifolds with Heegard diagrams of a given genus have hyperbolic structures—in analogy to the hyperbolic Dehn surgery theorem. This would be the next step after 7.

ACKNOWLEDGMENT.   I would like to thank George Francis for the illustrations.

Bibliography

[**Bers**] L. Bers, *Finite dimensional Teichmüller spaces and generalizations*, Bull. Amer. Math. Soc. (N.S.) **5** (1981), 131–172.

[**Can, Th**] J. Cannon and W. Thurston, *Sphere-filling curves and limit sets of Kleinian groups* (to appear).

[**C, J, S**] M. Culler, W. Jaco and H. Rubinstein, *Incompressible surfaces in once-punctured torus bundles* (to appear).

[**F, H**] W. Floyd and A. Hatcher, *Incompressible surfaces in 2-bridge link complements* (to appear).

[**F, L, P**] A. Fathi, F. Laudenbach, V. Poénaru et al., *Travaus de Thurston sur les surfaces*, Astérisque **66-67**, Société Mathématique de France, 1979.

[**Hak**] W. Haken, *Theorie der Normal Flachen*, Acta Math. **105** (1961), 5.

[**Hat**] A. Hatcher, *Incompressible surfaces in once-punctured torus bundles* (to appear).

[**H, T**] A. Hatcher and W. Thurston, *Incompressible surfaces in 2-bridge knot complements*, Invent. Math. (to appear).

[**Ja, Sh**] W. H. Jaco and P. B. Shalen, *Seifert fibered spaces in 3-manifolds*, Mem. Amer. Math. Soc. No. 2 (1979).

[**Joh**] K. Johannson, *Homotopy equivalence of 3-manifolds with boundaries*, Lecture Notes in Math., no. 7761, Springer-Verlag, Berlin, Heidelberg, New York, 1979.

[**M, Y**] W. Meeks and S-T Yau, *Topology of three dimensional manifolds and the embedding problems in minimal surface theory*, Ann. of Math. **112** (1980), 441–484.

[**Mey**] R. Meyerhoff, *The Chern-Simons invariant for hyperbolic 3-manifolds*, thesis, Princeton University, 1981.

[**Mil 1**] J. Milnor, *A unique factorization theorem for 3-manifolds*, Amer. J. Math. **84** (1962), 1–7.

[**Mil 2**] ______, *Hyperbolic geometry: the first 150 years*, proceedings.

[**Mos 1**] G. D. Mostow, *Strong rigidity of locally symmetric spaces*, Ann. Math. Studies, No. 78, Princeton Univ. Press, Princeton, N.J., 1973.

[**Mos 2**] G. D. Mostow, Inst. Hautes Études Sci. Publ. Math.

[**Poin**] H. Poincaré, *Cinquième complèment a l'analysis situs*, Rend. Circ. Mat. Palermo **18** (1904), 45–110, or *Oeuvres*, t. VI, pp. 435–498.

[**Pras**] G. Prasad, *Strong rigidity of Q-rank 1 lattices*, Invent. Math., 5–6.

[**Ril**] R. Riley, *Discrete parabolic representations of link groups*, Mathematika **22** (1975), 141–150.

[**Seif**] H. Seifert, *Topologie drei dimensionales gefasertei Raume*, Acta Math. **60** (1933), 147–288.

[**Smi**] Proceedings of the Smith Conjecture Conference at Columbia University, (to appear).

[**Sta**] J. Stallings, *On fibering certain 3-manifolds*, Prentice-Hall, Engelwood Cliffs, N.J., 1962, pp. 95–100.

[**Sul 1**] D. Sullivan, *Discrete conformal groups and measurable dynamics*, these proceedings.

[**Sul 2**] ______,

[**Th 1**] W. Thurston, *The geometry and topology of 3-manifolds*, preprint, Princeton Univ. Press (to appear).

[**Th 2**] ______, *Hyperbolic structures on 3-manifolds*, I: *deformations of acylindrical manifolds*, preprint.

[**Th 3**] ______, *Hyperbolic structures on 3-manifolds*, II: *surface groups and 3-manifolds which fiber over the circle*.

[**Th 4**] ______, *On the geometry and dynamics of diffeomorphisms of surfaces*, I, preprint.

[**Wal**] F. Waldhausen, *On irreducible 3-manifolds which are sufficiently large*, Ann. of Math. (2) **87** (1968), 56–88.

DEPARTMENT OF MATHEMATICS, UNIVERSITY OF COLORADO, BOULDER, COLORADO 80309

Section 3

RIEMANN SURFACES, DISCONTINUOUS GROUPS AND LIE GROUPS

Proceedings of Symposia in Pure Mathematics
Volume **39** (1983), Part 1

Finite Dimensional Teichmüller Spaces and Generalizations

LIPMAN BERS[1]

CONTENTS

0. Background
1. Introduction
2. Teichmüller spaces and modular groups
3. Teichmüller's theorem
4. Real boundaries
5. Classification of modular transformations
6. Embedding into complex number space
7. Metrics in Teichmüller space
8. Quasi-Fuchsian groups
9. Fiber spaces over Teichmüller spaces
10. Complex boundaries
11. b-groups
12. Ahlfors' problem, Sullivan's theorem and the complex boundary
13. The Maskit embedding
14. Riemann surfaces with nodes
15. Strong deformation spaces
16. Complex structure of deformation spaces
17. Moduli spaces

APPENDIX

18. Some auxiliary results.
19. Infinite dimensional Teichmüller spaces
References

0. Background.

(A) This paper is an expanded version of one read at the Poincaré Symposium of the American Mathematical Society at Bloomington, Indiana, in April 1980. The subject belongs to the "higher" theory of Riemann surfaces, and some readers may not object to being reminded of the main facts in the "standard" theory.

A *Riemann surface S* is a connected surface on which one can do complex function theory which is, locally, not distinguishable from ordinary complex function theory in a domain of the complex number plane **C**. More precisely, it is required that S be a connected Hausdorff space, that certain continuous complex valued functions on subdomains of S be designated as holomorphic, and that the following propositions be valid. (i) For every point P of S there

Reprinted from Bulletin Amer. Math. Soc. (N.S.) 5 (1981), 131–172.

[1] This material is based upon work supported by the National Science Foundation under Grant No. MCS 78-27119.

1980 *Mathematics Subject Classification.* Primary 14H15, 30C60, 30F40, 32G15.

115

is an open neighborhood N and a holomorphic function $\zeta\colon N \to \mathbf{C}$ which is a homeomorphism of N onto a domain in $\mathbf{C}$; such a ζ is called a *local parameter* defined in N. (ii) A function $f\colon D \to \mathbf{C}$ defined in a domain $D \subset S$ is holomorphic if and only if for every local parameter defined in $N \subset D$, the function $f \circ \zeta^{-1}$ is holomorphic (in the ordinary sense) in $\zeta(N) \subset \mathbf{C}$.

It follows that S is a surface (locally homeomorphic to $\mathbf{R}^2$), orientable and, which is not quite obvious, *triangulable*.

(B) The simplest examples of Riemann surfaces are the complex z-plane $\mathbf{C}$ (z is a local parameter defined everywhere), the *Riemann sphere* $\hat{\mathbf{C}} = \mathbf{C} \cup \{\infty\}$ (z is a local parameter in $\mathbf{C}$, $\zeta = 1/z$, with $\zeta = 0$ at ∞, is a local parameter in $\hat{\mathbf{C}} - \{0\}$) and every domain $D \subset \mathbf{C}$, in particular, the *upper half-plane* $U = \{z | \operatorname{Im} z > 0\}$. More generally, every subdomain of a Riemann surface is a Riemann surface.

A bijective homeomorphism h between two Riemann surfaces S_1 and S_2 is called *conformal* if f and f^{-1} take (germs of) holomorphic functions into (germs of) holomorphic functions. Conformal Riemann surfaces are identical *qua* Riemann surfaces. Note that on a Riemann surface one can measure angles (use any local parameter). Conformal maps are angle and orientation preserving homeomorphisms.

Nonconstant globally defined holomorphic functions exist on any noncompact Riemann surface. On every Riemann surface, compact or not, there exist globally defined nonconstant *meromorphic* functions, i.e. functions which are holomorphic except for isolated poles. (The proofs always involve some variant of the Dirichlet principle.)

The ring of holomorphic functions on a noncompact Riemann surface S, and the field of meromorphic functions on any Riemann surface S, each determine S uniquely, up to a conformal mapping and a reflection. (A reflection takes a Riemann surface S into its mirror image $\bar{S}$; S and $\bar{S}$ are identical surfaces but the local parameters on $\bar{S}$ are the conjugates of the local parameters on S.)

(C) In a certain sense the theory of Riemann surfaces goes back to Gauss who proved in 1822 that a sufficiently small piece on a sufficiently smooth oriented surface Σ in Euclidean space can be mapped conformally, i.e. preserving angles and orientation, onto a plane domain. Therefore any such Σ can be made into a Riemann surface by declaring a nonconstant continuous complex-valued function $g\colon \Delta \to \mathbf{C}$, defined on a domain $\Delta \subset \Sigma$, to be holomorphic if it is locally, except at isolated points, a conformal mapping.

Indeed, the embedding $\Sigma \hookrightarrow \mathbf{R}^N$ is irrelevant; all that matters is the existence of a positive definite Riemannian metric

$$ds^2 = E(x, y)\, dx^2 + 2F(x, y)dx\, dy + G(x, y)\, dy^2$$

(which, in case $\Sigma \hookrightarrow \mathbf{R}^N$ can be taken as induced by the metric in $\mathbf{R}^N$). A local parameter on a part of Σ is a complex-valued function

$$\zeta = u(x, y) + iv(x, y)$$

such that

$$\frac{\partial(u, v)}{\partial(x, y)} > 0, \qquad ds^2 = \rho(x, y)(dx^2 + dy^2).$$

The real-valued functions u, v are called *isothermal parameters*.

Gauss proved the existence of isothermal parameters u, v under the hypothesis that E, F, G are real analytic functions of the local coordinates (x, y). Later this condition has been considerably weakened; finally Morrey (1938) showed that it is enough to assume that E, F, G are measurable and the ratio

$$\frac{E + G}{\sqrt{EG - F^2}}$$

is bounded.

It is not difficult to verify that every Riemann surface can be defined by putting a properly chosen Riemannian metric onto an orientable sufficiently smooth surface. It is surprising that one may require that this metric be complete (every geodesic can be continued arbitrarily far) and have constant Gaussian curvature $+1$, 0, or -1; then the metric is determined uniquely (except for an arbitrary constant factor in the case of zero curvature). Except in a few cases the canonical curvature is -1. Let us recall how one finds this canonical metric.

(D) A deep theorem (the *uniformization theorem* of Klein-Poincaré-Koebe, 1882–1907) asserts that every simply connected Riemann surface is conformal to one of three:

$$(0.1) \qquad\qquad \hat{\mathbf{C}}, \mathbf{C} \text{ and } U.$$

If S is any Riemann surface, then the general theory of covering spaces (created just for this purpose) shows that S may be identified with $\tilde{S}/G$ where $\tilde{S}$ is the universal covering space of S, and G the covering group. Now $\tilde{S}$ is simply connected and can be made into a Riemann surface by lifting local parameters from S to $\tilde{S}$. Then $\tilde{S}$ becomes one of the three surfaces (0.1) and G becomes a discrete and fixed point free group of conformal self-mappings of $\tilde{S}$, hence a discrete group of *Möbius transformations*

$$(0.2) \qquad\qquad z \mapsto \frac{az + b}{cz + d}, \qquad ad - bc = 1$$

of $\tilde{S}$ onto itself.

One sees easily that if $\tilde{S} = \hat{\mathbf{C}}$, G can contain only the identity so that $S = \hat{\mathbf{C}}$. If $\tilde{S} = \mathbf{C}$, then G can contain only Euclidean translations. One verifies that in that case either $G = 1$ (the trivial group), or G is generated by one translation (say by $z \mapsto z + 1$) or by two (say by $z \mapsto z + 1$ and $z \mapsto z + \tau$, $\operatorname{Im} \tau > 0$). In the first case S is conformal to $\mathbf{C}$, in the second to $\mathbf{C} - \{0\}$, and in the third S is homeomorphic to a torus.

In all other cases $\tilde{S} = U$ and G is a discrete group of Möbius transformations (0.2) subject to the condition that a, b, c, d are real. Such groups have been named *Fuchsian* by Poincaré.

Now $\hat{\mathbf{C}}$ carries a complete metric of Gaussian curvature 1 obtained by mapping $\hat{\mathbf{C}}$ stereographically onto the sphere $x_1^2 + x_2^2 + x_3^2 = 1$, and $\mathbf{C}$ has the canonical complete metric $|dz| = \sqrt{dx^2 + dy^2}$ of curvature 0. This metric is inherited by the infinite cylinder ($\mathbf{C}$ identified under the group $z \mapsto z + n$, $n \in \mathbf{Z}$) and the torus ($\mathbf{C}$ identified under $z \mapsto n + m\tau$; $n, m \in \mathbf{Z}$, Im $\tau > 0$). Finally, U has a complete metric of curvature (-1), the *Poincaré metric* $|dz| y^{-1}$ which makes it into a model of the non-Euclidean plane, and every Riemann surface distinct from a sphere, a sphere with one or two punctures and a torus is of the form $S = U/G$, G a fixed point free Fuchsian group, and carries a Poincaré metric inherited from U.

(E) If G is any Fuchsian group, which need not be torsion free, i.e. may have fixed points in U, then $S = U/G$ is again, in a natural way, a Riemann surface. The (globally defined) holomorphic and meromorphic functions on S can be identified with holomorphic and meromorphic functions on U which are invariant (automorphic) with respect to G. Similarly, for every integer q the holomorphic and meromorphic *q-differentials* on S can be identified with holomorphic and meromorphic functions φ on U with $\varphi(z)\, dz^q$ G-invariant, i.e. satisfying

$$\varphi(g(z))g'(z)^q = \varphi(z), \qquad g \in G$$

(automorphic forms). (A special convention is needed at elliptic and parabolic fixed points.)

(F) Fuchsian groups are special cases of *Kleinian* groups (this name is also due to Poincaré). A Kleinian group G is a discrete group of not necessarily real Möbius transformations (0.2) with the following property: the *limit set* Λ of G, defined as the set of accumulation points of trajectories $\{ g(z), g \in G \}$ is not all of $\hat{\mathbf{C}}$. If so, the open dense set $\Omega = \hat{\mathbf{C}} - \Lambda$ is called the *region of discontinuity* of G and every component of Ω is called a *component* of G. The group acts properly discontinuously on Ω and the quotient Ω/G is, in a natural way, a disjoint union of Riemann surfaces.

While the theory of Kleinian groups goes back to Klein, Poincaré and Koebe (and even earlier, to Schottky) a flowering of this theory occurred during the past fifteen years. A landmark in this development was Ahlfors' *finiteness theorem*: if a Kleinian group G is finitely generated, the quotient Ω/G has finitely many components, each component is a compact Riemann surface or obtained from one by removing finitely many points, and the natural projection $\Omega \to \Omega/G$ is ramified over at most finitely many points.

(G) A compact Riemann surface S is, topologically, a sphere with $p \geqslant 0$ handles, the number p is called the *genus* of S. The holomorphic function on S are, of course, constants (by the maximum principle). The meromorphic functions form a field of algebraic functions of one variable (over $\mathbf{C}$). This means that any two meromorphic functions, z and w, on S are connected by an irreducible polynomial equation with complex coefficients

$$(0.3) \qquad P(z, w) = \sum_{\nu=1}^{n} \sum_{\mu=1}^{m} a_{\nu\mu} z^\nu w^\mu = 0,$$

and that, if these two functions are chosen suitably, any third is a rational function of the two.

In the latter case one says that S is the Riemann surface of the *plane algebraic curve* (0.3). This does *not* mean, of course, that S is isomorphic to the set of points (z, w) in $\mathbf{C}^2$ satisfying (0.2) or even to the set of points (t_1, t_2, t_3) in the complex projective plane P_2 satisfying the corresponding homogeneous equation

$$(0.3')\qquad \sum_{\nu=1}^{n} \sum_{\mu=1}^{m} a_{\nu\mu} t_1^{\nu} t_2^{\mu} t_3^{n+m-\nu-\mu} = 0.$$

Indeed, the curve (0.3) will in general have singularities. However, every compact Riemann surface is isomorphic to a *nonsingular algebraic curve* in complex projective 3-space.

(H) The theory of compact Riemann surfaces and the (more general) theory of algebraic curves over $\mathbf{C}$ are very well developed and exceedingly rich. Closely connected with compact Riemann surfaces are those with *finitely generated fundamental groups*. Such a surface S is obtained from a compact surface $\hat{S}$, of some genus p by removing $r \geqslant 0$ disjoint continua. If n of those are points and $m = r - n$ are nondegenerate continua, we say that S has *type* (p, n, m), type (p, n) if $m = 0$.

A Riemann surface S of type (p, n, m), with $m > 0$, is said to have m *ideal boundary curves*. Such an S can be always doubled, i.e., represented as a subdomain of a surface S^d, of type $(2p + m - 1, 2n, 0)$, which admits an anticonformal (angle preserving and orientation reversing) involution j which leaves precisely m disjoint Jordan curves $C_1, \ldots, C_m$ fixed and such that $S^d - \{C_1 \cup \cdots \cup C_m\}$ as two components, one of which is S. The curves $C_1, \ldots, C_m$ are the ideal boundary curves of S and S^d is its *Schottky double*. (Example: the double of U is $\hat{\mathbf{C}}$.)

(I) The uniformization theorem implies that every compact Riemann surface of genus 0 is conformal to a sphere (i.e. to $\hat{\mathbf{C}}$). The conformal type of a compact surface of genus 1 depends on one complex parameter $\tau \in U$. Already Riemann computed that the conformal type of a compact Riemann surface of genus $p > 1$ depends on $3p - 3$ complex parameters (*moduli*) and this implies that the number of complex parameters for the conformal type of a surface of type (p, n) is $3p - 3 + n$, provided that number is not negative. Similarly, the conformal type of a surface of type (p, n, m) with $m > 0$ depends on $6p - 6 + 2n + 3m$ real parameters, provided this number is not negative.

The main aim of Teichmüller theory is to make the dependence of a Riemann surface (of finite type) on the complex or real moduli as explicit and as transparent as possible. (The theory for type (p, n, m) with $m > 0$ can be reduced, at least in principle, to that of type (p, n) by using doubling.)

(J) One of Teichmüller's great contributions was to recognize that the problem becomes more accessible if one considers not only conformal mappings between Riemann surfaces, but also quasiconformal mappings.

Let S and $\hat{S}$ be two homeomorphic Riemann surfaces and $f\colon S \to \hat{S}$ an orientation preserving diffeomorphism. If P is a point on S, $\hat{P} = f(P) \in \hat{S}$, $z = x + iy$ a local parameter on S defined near P and $\hat{z} = \hat{x} + i\hat{y}$ one on $\hat{S}$ defined near $\hat{P}$, then the map $\hat{z} = h(z)$ defined by $h = \hat{z} \circ f \circ z^{-1}|N$ (N a sufficiently small neighborhood of P) is called a *local representer* of f at P.

The map f is called *K-quasiconformal* if there is a constant $K \geqslant 1$ such that, for every representer, at every point of S,

$$(0.4) \qquad \left| \frac{\partial \hat{z}}{\partial \bar{z}} \right| \leqslant \frac{K-1}{K+1} \left| \frac{\partial \hat{z}}{\partial z} \right|$$

or, in real terms,

$$(0.4') \qquad \frac{\left(\dfrac{\partial \hat{x}}{\partial x}\right)^2 + \left(\dfrac{\partial \hat{x}}{\partial y}\right)^2 + \left(\dfrac{\partial \hat{y}}{\partial x}\right)^2 + \left(\dfrac{\partial \hat{y}}{\partial y}\right)^2}{\dfrac{\partial \hat{x}}{\partial x}\dfrac{\partial \hat{y}}{\partial y} - \dfrac{\partial \hat{x}}{\partial y}\dfrac{\partial \hat{y}}{\partial x}} \leqslant \frac{1}{2}\left(K + \frac{1}{K}\right).$$

Geometrically this means that f takes every infinitesimal circle into an infinitesimal ellipse whose major axis is at most K times the minor axis.

The smallest K for which f is K-quasiconformal is called the *dilatation* of f and is denoted by $K(f)$. If $K(f) = 1$, f is conformal.

The dilatation, or better, its logarithm, measures the deviation of f from conformality and thus gives an upper bound for the difference between the conformal structures of S and $\hat{S}$.

For technical reasons it is convenient not to insist that the orientation preserving homeomorphism $f\colon S \to \hat{S}$ be a diffeomorphism but only that the partial derivatives $\partial \hat{x}/\partial x$, $\partial \hat{y}/\partial x$, $\partial \hat{x}/\partial y$, $\partial \hat{y}/\partial y$ taken in the sense of distribution theory be locally square integrable measurable functions satisfying $(0.4')$ almost everywhere.

1. Introduction.

(A) Volume 1 of *Acta Mathematica* appeared in 1882; it begins with Poincaré's paper on Fuchsian groups, the first of a series of memoirs which contain his work on discontinuous groups, automorphic functions and uniformization. Already on p. 6 Poincaré defines the *Poincaré line element*

$$(1.1) \qquad ds = \frac{|dz|}{y}$$

in the upper half-plane $U = \{z = x + iy \,|\, y > 0\}$ and explains, somewhat reluctantly, its geometric meaning: "Je ne puis passer sous silence le lien qui rattache les notions précédentes a la géométrie noneuclidienne de Lobatschewsky . . . Cette terminologie m'a rendie de grands services de mes recherches, mais je ne l'employerai pas ici pour éviter toute confusion."

The memoir in question is concerned with constructing *Fuchsian groups*, i.e., discrete groups of plane non-Euclidean motions or, in the Poincaré half-plane model, discrete group of real Möbius transformations, i.e. discrete subgroups of $PSL(2, \mathbf{R})$. One such group, the elliptic modular group $PSL(2, \mathbf{Z})$ was already well known at that time. For reasons which will become apparent later we shall denote this group by $\mathrm{Mod}_{1,1}$.

(B) Now every point $\tau \in U$ defines a lattice $\mathcal{L} = \{\omega = n + m\tau \mid n, m \in \mathbf{Z}\}$ and a punctured torus $(\mathbf{C} - \mathcal{L})/\mathcal{L}$ where we regard $\mathcal{L}$ as a subset and as an additive subgroup of $\mathbf{C}$. This torus is *marked*, i.e. it comes with a distinguished sequence of generators of the fundamental group, defined by the segments going from $z = 0$ to $z = 1$ and to $z = \tau$, respectively. Thus U may be thought of as the space $T_{1,1}$ of isomorphism classes of marked Riemann surfaces of genus 1 with 1 puncture. Two distinct points τ_1 and τ_2 of U represent conformally equivalent but differently marked Riemann surfaces if and only if $\tau_2 = g(\tau_1)$ for some $g \in PSL(2, \mathbf{Z})$.

Therefore $X_{1,1} = T_{1,1}/\mathrm{Mod}_{1,1}$ is the space of conformal equivalence classes (moduli) of once punctured Riemann surfaces of genus 1. Of course, $X_{1,1}$ is $\mathbf{C}$, the canonical mapping $T_{1,1} \to X_{1,1}$ being ramified over the points corresponding to $\tau = i$ and $\tau = (1 + \sqrt{-3})/2$. The moduli space $T_{1,1}$ can be compactified by adding to it the point $\tau = \infty$ representing a thrice punctured sphere (obtained by 'degenerating' a punctured torus).

The part of Teichmüller space theory and an addition to this theory, on which we report here, deal with extending the statements made above to Riemann surfaces of genus p with n punctures, cf. [5, 14, 26, 30, 35, 66].

(C) We remark that the Teichmüller space $T_{p,n}$ is one of three distinct but related generalizations of the Poincaré half-plane connected with the theory of Riemann surfaces. The other two are the Poincaré model of the non-Euclidean 3-space:

$$\mathbf{H}_3 = \{(x, y, t) \in \mathbf{R}^3 \mid t > 0\} = \{(z, t) \mid t \in \mathbf{C}, t \in \mathbf{R}, t > 0\}$$

with the Poincaré metric

$$(1.2) \qquad ds^2 = \frac{dx^2 + dy^2 + dt^2}{t^2} = \frac{|dz|^2 + dt^2}{t^2},$$

and the generalized Siegel upper half-plane H_p consisting of complex symmetric $p \times p$ matrices $Z = X + iY$ with positive definite imaginary parts, with the Siegel metric

$$ds^2 = \mathrm{trace}(Y^{-1}(dX + idY)Y^{-1}(dX - idY)).$$

(D) We assume acquaintance with the standard theory of Riemann surfaces, including uniformization theory and with the definition and basic properties of quasiconformal mappings in two dimensions, cf. [13, 15, 66, 81]. We also assume the elementary properties of Kleinian groups [55].

In particular, we shall use without comment the theorem that a Riemann surface S, distinct from a sphere with 0, 1 or 2 punctures and from a torus, can be represented as U/G where G is a torsion-free Fuchsian group determined by S up to a conjugacy in $PSL(2, \mathbf{R})$, and of the equivalent theorem that S carries a uniquely determined *Poincaré metric*, i.e. a complete Riemannian metric of constant Gaussian curvature (-1), which respects the conformal structure of S. The latter condition means that near every point of S the metric can be written as $ds = \lambda(z)|dz|$, z being a local parameter.

The two equivalent statements just made constitute the main case of the so-called *limit circle theorem* the truth of which was perceived by Klein and by Poincaré in 1882, though a rigorous proof was given, by Poincaré and by Koebe, only in 1907.

(E) No claim of completeness is made for this report or for the bibliography. Analytic methods are emphasized throughout. The powerful geometric methods invented by Thurston [**54, 117, 118**] could not be described here and only a few of his important results are mentioned. Important topics not covered also include the Grothendieck [**62, 44, 53**] and the Earle and Eells [**48, 49**] appoaches to Teichmüller theory, the Jenkins and Strebel quadratic differentials [**69, 74, 107, 109**], connections with the theory of Jacobians [**47**], and others [**57, 74**]. Teichmüller spaces of Fuchsian groups and infinite dimensional Teichmüller spaces are described briefly in the Appendix.

2. Teichmüller spaces and modular groups.

(A) Let p and n be nonnegative integers with $2p - 2 + n > 0$. In order to define $T_{p,n}$ precisely, we choose a topological oriented surface $\Sigma = \Sigma_{p,n}$ obtained by removing n distinct points from a compact surface $\hat{\Sigma}$ of genus p (with $\Sigma = \hat{\Sigma}$ if $n = 0$) and call two orientation-preserving homeomorphisms of Riemann surfaces onto Σ

$$f: S \to \Sigma \quad \text{and} \quad f': S' \to \Sigma$$

equivalent if there is a commutative diagram

$$
\begin{array}{ccc}
S & \xrightarrow{f} & \Sigma \\
\varphi\downarrow & & \downarrow\psi \\
S' & \xrightarrow[f']{} & \Sigma
\end{array}
$$

where φ is an isomorphism (conformal mapping onto) and ψ a homeomorphism homotopic to the identity. The equivalence classes

$$[f] = [f: S \to \Sigma]$$

of homeomorphisms of compact Riemann surfaces of genus p with n punctures onto Σ are called *marked Riemann surfaces* of type (p, n) or points of the Teichmüller space $T_{p,n}$.

(B) Every orientation-preserving homeomorphism ω of Σ onto itself induces a mapping ω_* of $T_{p,n}$ onto itself, by the rule

$$\omega_*([f]) = [\omega \circ f].$$

Clearly ω_* depends only on the homotopy type of ω, $\mathrm{id}_* = \mathrm{id}$ and $(\omega_1 \circ \omega_2)_* = \omega_{1*} \circ \omega_{2*}$. The transformations ω_* form the (Teichmüller) *modular group* $\mathrm{Mod}_{p,n}$.

Two points $[f_1]$ and $[f_2]$ of $T_{p,n}$ are equivalent under $\mathrm{Mod}_{p,n}$ if and only if the Riemann surfaces $f_1^{-1}(\Sigma)$ and $f_2^{-1}(\Sigma)$ are conformally equivalent. Therefore $X_{p,n} = T_{p,n}/\mathrm{Mod}_{p,n}$ is the space of *moduli* (conformal equivalence classes) of Riemann surfaces of type (p, n).

(C) There are several ways of topologizing $T_{p,n}$. They are all equivalent and lead to the following theorem.

The Teichmüller space $T_{p,n}$ is homeomorphic to $\mathbf{R}^{6p-6+2n}$ (*and the modular group* $\mathrm{Mod}_{p,n}$ *is a properly discontinuous group of homeomorphisms*).

The first proof of this result is, in effect, due to Fricke [**56**], cf. [**70**].

(D) The topology in $T_{p,n}$ can be derived from the so-called *Teichmüller metric*. The Teichmüller distance between two points $[f_1]$ and $[f_2]$ of $T_{p,n}$ is defined as

$$\langle [f_1], [f_2] \rangle = \tfrac{1}{2} \log \inf K(f)$$

where f ranges over all quasiconformal mappings in the homotopy class of $f_1^{-1} \circ f_2$, and $K(f)$ is the (global) *dilatation* of f.

If f is a diffeomorphism, $K(f)$ is the supremum of the local dilatation $K(P; f)$, $P \in f_2^{-1}(\Sigma)$, where $K(P, f)$ is the ratio of the major axis to the minor axis of the infinitesimal ellipse into which f maps an infinitesimal circle centered at P. For a general quasiconformal map the definition is the same, except that $K(P; f)$ is defined only almost everywhere.

The Teichmüller space $T_{p,n}$ is a complete metric space, and the modular group $\mathrm{Mod}_{p,n}$ *is a group of isometries* (with respect to the Teichmüller metric).

(E) The space $T_{1,1}$ with the Teichmüller metric is isometric to U with the Poincaré metric; under this isometry $\mathrm{Mod}_{1,1}$ is the elliptic modular group.

The space $T_{0,3}$ is, of course, a point.

(F) Now let $\Sigma_{p,n,m}$ be a topological oriented surface obtained from a closed surface $\hat{\Sigma}$ of genus p by removing $n + m$ distinct points which we divide into two sets consisting of n and $m > 0$ points, respectively. We assume that $6p - 6 + 2n + 3m > 0$.

The *reduced Teichmüller space* $T_{p,n,m}^{\#}$ consists of equivalence classes $[f\colon S \to \Sigma]$ of orientation-preserving homeomorphisms of Riemann surface of type (p, n, m), i.e., of genus p, with n punctures and m ideal boundary curves, onto Σ. It is required that f take the punctures of S into the first set of punctures of Σ.

The space $T_{p,n,m}^{\#}$ is again a *complete metric* space (under the Teichmüller metric) and is *homeomorpic* to $\mathbf{R}^{6p-6+2n+3m}$.

The definition of the (reduced) modular group $\mathrm{Mod}_{p,n,m}^{\#}$ is obvious. It is again a group of isometries and it acts properly discontinuously.

3. Teichmüller's theorem.

(A) Given two marked Riemann surfaces of type (p, n),

$$[f_2\colon S_1 \to \Sigma] \quad \text{and} \quad [f_2\colon S_2 \to \Sigma],$$

we want to find an *extremal map* in the homotopy class of $f_2^{-1} \circ f_1$, that is a map $f\colon S_1 \to S_2$ homotopic to $f_2^{-1} \circ f_1$ and with the smallest possible dilatation, namely

$$K(f) = e^{2\langle [f_1], [f_2] \rangle}.$$

The existence of an extremal map follows easily from general compactness properties of quasiconformal maps. Teichmüller discovered that the extremal map is unique and of a very special form: if it is not conformal, which can happen only if $[f_1] = [f_2]$, the extremal f is, in the neighborhoods of all but finitely many points, a conformal map followed by an affine map followed by another conformal map.

(B) More precisely, a homeomorphism $f: S_1 \to S_2$ is called a *Teichmüller map* if there are holomorphic quadratic differentials, q_1 on S_1 and q_2 on S_2, with

$$(3.1) \qquad 0 < \int\int_{S_1} |q_1| = \int\int_{S_2} |q_2| < \infty$$

such that for every point $P \in S_1$, with $q_1 \neq 0$ at P, we have that $q_2 \neq 0$ at $f(P)$ and near P the map may be written as

$$(3.2) \qquad \int q_2^{1/2} = K^{1/2} \operatorname{Re} \int q_1^{1/2} + \sqrt{-1}\ K^{-1/2} \operatorname{Im} \int q_1^{1/2}$$

(for a proper choice of the signs of the square roots) where $K > 1$ is constant, and is the dilatation of f. We call q_1 and q_2 the *initial* and *terminal differentials* of f.

(We recall that a holomorphic quadratic differential q on a Riemann surface S is a holomorphic section of twice the canonical bundle; locally, in terms of a local parameter ζ we have $q = \varphi(\zeta)d\zeta^2$ where $\varphi(\zeta)$ is a holomorphic function. Note that $|\varphi(\zeta)|d\xi\,d\eta$ is a density, so that condition (3.1) is meaningful. It implies that the singularities of q_j at the punctures of S_j are at worst simple poles.)

Relation (3.2) may be written as

$$(3.3) \qquad \xi_2 = K^{1/2}\xi_1, \qquad \eta_2 = K^{-1/2}\eta_1$$

where $\zeta_1 = \xi_1 + \sqrt{-1}\ \eta_1$ and $\zeta_2 = \xi_2 + \sqrt{-1}\ \eta_2$ are local parameters on S_1 and S_2, respectively, such that $\zeta_1 = 0$ at P, $\zeta_2 = 0$ at $f(P)$ and, near these points, $q_1 = d\zeta_1^2$, $q_2 = d\zeta_2^2$.

(C) The inverse f^{-1} of a Teichmüller map is again a Teichmüller map, with the same dilatation K, with initial differential $(-q_2)$ and with terminal differential $(-q_1)$.

(D) Teichmüller's theorem asserts that (α) *every Teichmüller map is the unique extremal in its homotopy class and* (β) *that every extremal map is either conformal or a Teichmüller map.*

Teichmüller first proved, in his famous 1939 paper [**114**], statement (α); in 1943 he derived (β) from (α) by a continuity argument [**115**]. But his eccentric style (and unsavory political reputation) delayed the acceptance of his bold ideas until Ahlfors' paper [**6**] which contained a new proof of (β). Later a modernized version of Teichmüller's own proof was given by the author [**17**], and a variational proof of (β) was achieved by Schiffer [**106**] (in a special case), Krushkal [**79, 80**] and Hamilton [**65**].

(E) Kravetz [**78**] showed that through any two distinct points of $T_{p,n}$ there is a unique *Teichmüller line*, i.e. an isometric image of **R**, and that three distinct points in $T_{p,n}$ are collinear if and only if they can be denoted by τ_1, τ_2, τ_3 in such a way that $\langle \tau_1, \tau_2 \rangle + \langle \tau_2, \tau_3 \rangle = \langle \tau_1, \tau_3 \rangle$.

At one time it was believed that the Teichmüller metric (which is not a Riemannian metric but can be derived from a Finsler metric (O'Byrne [97])) has negative curvature in the following sense: if τ_0, τ_1, τ_1', τ_2 and τ_2' are points in Teichmüller space such that

$$\langle \tau_0, \tau_1 \rangle = \langle \tau_1, \tau_2 \rangle = \tfrac{1}{2}\langle \tau_0, \tau_2 \rangle, \qquad \langle \tau_0, \tau_1' \rangle = \langle \tau_1', \tau_2' \rangle = \tfrac{1}{2}\langle \tau_0, \tau_2' \rangle$$

then

$$\langle \tau_2, \tau_2' \rangle > 2\langle \tau_1, \tau_1' \rangle.$$

However, Masur [89] showed that this is not so if $\dim T_{p,n} \geq 2$.

Nevertheless, there are certain similarities between $T_{p,n}$ and spaces of negative curvature.

(F) The extremal problem considered in (A) can be generalized as follows. Let $f_0\colon S_1 \to S_2$ be a given quasiconformal map between two Riemann surfaces, and if S_1 and S_2 have ideal boundary curves, let σ be a closed set on the union of the ideal boundary curves of S_1. Consider all quasiconformal maps in the homotopy class of f_0 modulo σ, and find in it an extremal, i.e. one with the smallest possible dilatation.

(We recall that the quasiconformal map f_0 is defined, by continuity, also on the ideal boundary curves of S_1. Another quasiconformal map f is homotopic to f_0 modulo σ if $f_0^{-1} \circ f|\sigma = \mathrm{id}$ and $f_0^{-1} \circ f$ can be continuously deformed into the identity keeping every point of σ fixed.)

We assume, in order to exclude trivial cases, that the only conformal map $h\colon S_1 \to S_1$ with $h|\sigma = \mathrm{id}$ is the identity.

The existence of an extremal map follows again from compactness properties of quasiconformal maps.

A *Teichmüller map* $f\colon S_1 \to S_2$ is defined as in (B), except that we now require that q_1 (and q_2) be *continuous and real* at all point of the ideal boundary curves of S_1 (of S_2) except at the points belonging to σ (to $f_0(\sigma)$).

If the fundamental group of S_1 is *finitely generated* and σ is *finite*, *Teichmüller's theorem* (statements (α) and (β) in (D)) *is valid*.

This can be proved, for instance, by doubling.

(G) Without any finiteness assumptions, *statement (α) is always valid* (Reich and Strebel [102, 103, 109, 112]). Also Hamilton [65] gave a *necessary condition* for an extremal map which, according to Reich and Strebel, is also *sufficient*.

On the other hand, *statement (β) is in general false*. There are extremal maps which are not unique (Strebel [108]), and there are unique extremal maps which are not Teichmüller maps (in the sense defined above).

4. Real boundaries.

(A) Teichmüller's theorem gives a new proof of Fricke's result, see §2, (C). Indeed it yields a representation, or rather infinitely many representations, of $T_{p,n}$ as a ball in $\mathbf{R}^{6p-6+2n}$.

We choose a point $[f_0: S_0 \to \Sigma]$ as an origin of $T_{p,n}$ and we choose on S_0 a real basis of integrable quadratic (holomorphic) differentials $q_1, \ldots, q_{6p-6+2n}$ (the number $6p - 6 + 2n$ comes from the Riemann-Roch theorem). The points $\tau = [f: S \to \Sigma]$ of $T_{p,n}$ distinct from the origin are in a one-to-one correspondence with Teichmüller maps $f_\tau: S_0 \to S$, f_τ being the extremal map in the homotopy class of $f \circ f_0^{-1}$. The initial quadratic differential of f_τ can be written as

$$\Lambda(\alpha_1 q_1 + \cdots + \alpha_{6p-6+2n} q_{6p-6+2n}),$$

with

$$\alpha_1^2 + \cdots + \alpha_{6p-6+n}^2 = 1, \qquad \Lambda > 0,$$

and the dilatation K of f_τ can be rewritten as

$$K = (1 + k)/(1 - k), \qquad 0 < k < 1.$$

We now assign to the origin in $T_{p,n}$ the origin in $\mathbf{R}^{6p-6+n}$ and to the point τ the point in $\mathbf{R}^{6p-6+2n}$ with the coordinates $k\alpha_1, \ldots, k\alpha_{6p-6+n}$. The resulting bijection of $T_{p,n}$ onto the open unit ball is a homeomorphism.

(B) This homeomorphism yields a *compactification* of $T_{p,n}$; the resulting boundary, called the *Teichmüller boundary*, is homeomorphic to the $(6p - 7 + 2n)$-dimensional sphere and can be canonically identified with the set of rays in the space $Q(S_0)$ of integrable quadratic differentials on S_0.

It is natural to ask whether this compactification is essentially independent of the choice of the origin $[S_0 \to \Sigma]$, and whether the action of $\mathrm{Mod}_{p,n}$ on $T_{p,n}$ extends continuously to the Teichmüller boundary. According to Kerckhoff [71] the answer to both questions is *no*.

(C) A compactification of $T_{p,n}$, which does *not* depend on the choice of an origin is due to Thurston ([117], cf. the presentation in [54]).

Let $\tau = [f: S \to \Sigma]$ be a point in $T_{p,n}$, C a Jordan curve on Σ homotopic neither to a point nor a puncture of S and $l_\tau(C)$ the length of the unique Poincaré geodesic on S freely homotopic to $f_\tau^{-1}(C)$. It is known that the knowledge of all numbers $l_\tau(C)$ determines τ; as a matter of fact, it suffices to know these numbers for a properly chosen finite sequence $C_1, \ldots, C_N$. Also, since there are nonhomogeneous relations between the various numbers $l_\tau(C)$, it is enough to know these numbers up to a common positive factor.

Thus there is a bijection between $T_{p,n}$ and a subset M of an infinitely dimensional (or, if one prefers, a finite dimensional) real projective space. This bijection is a homeomorphism and Thurston proves, in a highly imaginative way, that the closure $\overline{M}$ of M is homeomorphic to a ball, while the *Thurston boundary* $\overline{M} - M$ is homeomorphic to a sphere, both of the "right" dimension.

For reasons which will become apparent later the Teichmüller and Thurston boundaries are called *real boundaries*.

The action of $\mathrm{Mod}_{p,n}$ *extends continuously to the Thurston boundary* (because it amounts to a permutation of the Jordan curves C on Σ).

(D) In Thurston's proof $\overline{M} - M$ appears as the set of *measured foliations* on Σ, determined up to equivalence and up to a common positive factor. It is strange that just as the various Teichmüller boundaries, the Thurston boundary can be identified with the space of rays of $Q(S)$, for every $\tau = [S \to \Sigma]$ in $T_{p,n}$.

This follows from the fact, proved by Hubbard and Masur [68], and later by Kerckhoff [71], that every measured foliation on a Riemann surface is equivalent to one defined by a quadratic differential q, the leaves being the horizontal trajectories $q > 0$, and the transverse measure being derived from the metric $ds = |q|^{1/2}$.

(E) The Teichmüller theorem stated in §2, (F) leads to a representation of $T_{p,n,m}^{\#}$ as a ball in $\mathbf{R}^{6p-6+2n+3m}$. The argument is the same as in (A).

5. Classification of modular transformations.

(A) Thurston [117], cf. [54], used his compactification of Teichmüller space to classify orientation preserving homeomorphisms ω of a surface Σ onto itself up to homotopy (or, which is the same, up to isotopy) and discovered that in the general case ω is isotopic to a mapping of a very special form. He himself indicated that these results could be obtained from those obtained by Nielsen many years ago; this was made explicit by J. Gilman [60] and by R. Miller [93]. In [31] the author showed how the classification can be obtained by a generalization of Teichmüller's extremal problem. We outline this approach below. (Further developments and applications may be found in Kra [75].)

(B) The elements γ of $\mathrm{Mod}_{p,n}$ can be classified as follows: we call $\gamma \neq \mathrm{id}$ *elliptic* if it has a fixed point, *parabolic* if it has no fixed point but

$$(5.1) \qquad \inf\langle \tau, \gamma(\tau)\rangle = 0 \qquad (\tau \in T_{p,n}).$$

If γ is neither elliptic nor parabolic, then

$$(5.2) \qquad a = \inf\langle \tau, \gamma(\tau)\rangle > 0 \qquad (\tau \in T_{p,n})$$

and γ is called *hyperbolic* if there is a point τ_0 with $\langle \tau_0, \gamma(\tau_0)\rangle = a$, *pseudohyperbolic* if not.

For $p = n = 1$ this is the usual classification of real Möbius transformations (and the pseudohyperbolic case does not occur).

In all cases γ is elliptic if and only if it is of finite order, and γ is hyperbolic if and only if it leaves a Teichmüller line fixed.

(C) Every $\gamma \in \mathrm{Mod}_{p,n}$ is of the form $\gamma = \omega_*$ where ω is a topological orientation preserving self-mapping of Σ (which may be chosen as a diffeomorphism if Σ is given a differentiable structure) and the metric properties of ω_* reflect the topological properties of ω. Thus ω_* is elliptic if and only if ω is homotopic to a periodic mapping.

A self-mapping $\omega\colon \Sigma \to \Sigma$ is called *reduced* by a finite nonempty set $C_1, \ldots, C_k$ of Jordan curves on Σ such that the C_j are disjoint, no C_j is freely homotopic to a curve C_i, $i \neq j$, and no C_j can be continuously deformed into a point or puncture of Σ, if

$$\omega(C_1 \cup \cdots \cup C_k) = C_1 \cup \cdots \cup C_k.$$

It turns out that $\gamma = \omega_*$ *is hyperbolic if and only if ω is not homotopic to a periodic map and irreducible* (not homotopic to a reduced map).

(D) A reducible map ω is always homotopic to a ω_0 which is reduced by curves $C_1, \ldots, C_k$ and has the following property. For every component Π of $\Sigma - (C_1 \cup \cdots \cup C_k)$ set $\omega_\Pi = \omega_0 | \Pi$ and let r_Π denote the smallest positive integer such that $\omega_\Pi^{r_\Pi}(\Pi) = \Pi$. Then $g_\Pi = \omega_\Pi^{r_\Pi}$ is irreducible. It turns out that if ω is reducible and not homotopic to a periodic map then $\gamma = \omega_*$ is parabolic or pseudohyperbolic according to whether all or not all maps g_Π are homotopic to periodic maps or the identity.

(E) If $\gamma = \omega_*$ is hyperbolic then, by definition, there are points $\tau = [f: S \to \Sigma]$ of $T_{p,n}$ with the following property: the self-map $f^{-1} \circ \omega \circ f$ of S is homotopic to a *Teichmüller map g which is absolutely extremal*, i.e. its dilatation cannot be decreased by changing g within its homotopy class *and* by changing the conformal structure of S.

It can be shown that *a Teichmüller map $g: S \to S$ is absolutely extremal if and only if its initial differential q is also its terminal differential*.

If so, g is what Thurston calls a *pseudo-Anosov diffeomorphism*: there are two transversal measured foliations, the horizontal trajectories of q and $(-q)$, respectively, such that g maps them onto themselves while multiplying the distance between the leaves by e^{2a} and e^{-2a}, respectively; here a is the number defined by (5.2).

Thus a self-mapping $\omega: \Sigma \to \Sigma$ which is not homotopic to a periodic map is either homotopic to a pseudo-Anosov map or reducible, but not both, as has been first proved by Thurston.

(F) The elements of $\mathrm{Mod}^{\#}_{p,n,m}$ can be classified similarly. However, it turns out that if $m > 0$, *there are no hyperbolic elements*.

6. Embedding into complex number space.

(A) Already Teichmüller guessed [**116**], and claimed to have proved, that the Teichmüller space $T_p = T_{p,0}$ is, *in a natural way, a complex manifold*, but the first complete proof was given by Ahlfors in [**7**].

Note that there are certain functions on T_p which must be holomorphic under a "natural" definition of the complex structure in T_p.

Choose a "canonical" integral homology basis $\alpha_1, \ldots, \alpha_p, \beta_1, \ldots, \beta_p$ on Σ, so that $\alpha_i \times \beta_i = -(\beta_i \times \alpha_i) = 1$, $i = 1, \ldots, p$, and "all" other intersection numbers vanish. For every $\tau = [f: S \to \Sigma] \in T_{p,0}$ there are p uniquely determined holomorphic Abelian differentials $a_1, \ldots, a_p$ on S such that

$$\int_{f^{-1}(\alpha_j)} a_i = \begin{cases} 1 & \text{if } i = j, \\ 0 & \text{if } i \neq j. \end{cases}$$

The $p \times p$ matrix $\Pi(\tau)$, with entries

$$\Pi_{ij}(\tau) = \int_{f^{-1}(\beta_j)} a_i$$

depends on τ only and is an element of the generalized Siegel upper half-plane H_p (see §1, (C)). It is natural to require that the map $T_{p,0} \to H_p$ just defined be holomorphic.

This requirement leads to a complex manifold structure on T_2 and on the part T_p' of T_p, $p > 2$, corresponding to nonhyperelliptic surfaces; one can use as complex coordinates, near every point of T_2 or T_p', an appropriately chosen set of $3p - 3$ entries of the period matrix Π, as was observed by Rauch [100, 101]. The most delicate part of Ahlfors' argument consists in showing that the complex structure does not become singular on the set $T_p - T_p'$.

(B) Actually there are in $T_{p,n}$ natural global complex coordinates: $T_{p,n}$, *with its complex structure, can be realized as a bounded domain in complex number space* [19].

Indeed, this realization is, to a certain extent, canonical; if we choose a point $\tau_0 = [f_0\colon S_0 \to \Sigma]$ as the "origin" in $T_{p,n}$, then $T_{p,n}$ can be represented canonically as a bounded domain in the $(3p - 3 + n)$-dimensional complex vector space $Q(\bar{S}_0)$ of integrable quadratic differentials on the mirror image $\bar{S}_0$ of S_0. This is accomplished as follows.

(C) We represent S_0 as U/G_0 where G_0 is a torsion-free Fuchsian group (under our assumptions on S_0, G_0 is finitely generated and every real point is an accumulation point of an orbit). Note that $\bar{S}_0 = L/G_0$ where L is the lower half-plane.

The space $Q(\bar{S}_0)$ can be identified with the space $B_2(L, G_0)$ of holomorphic functions $\varphi(z)$, $z \in L$, satisfying the functional equation of automorphic forms of weight (-4)

$$(6.1) \qquad \varphi(g(z))g'(z)^2 = \varphi(z) \quad \text{for } g \in G_0$$

and the two equivalent growth conditions:

$$(6.2) \qquad \iint_{L/G_0} |\varphi(z)|\, dx\, dy < \infty,$$

$$(6.3) \qquad \sup|y^2\varphi(z)| < +\infty,$$

which characterize the so-called *cusp forms*.

Now let $\tau = [f\colon S \to \Sigma]$ be any point of $T_{p,n}$. Without loss of generality we assume that f_0 and f are quasiconformal, so is then the mapping $f^{-1} \circ f_0$. The Beltrami differential m of $f^{-1} \circ f_0$ on S_0 can be lifted to a bounded (measurable) function $\mu(z)$ satisfying the inequality

$$(6.4) \qquad \|\mu\|_\infty = \text{ess sup}|\mu(z)| < 1$$

and the functional equation

$$(6.5) \qquad \mu(g(z))\frac{\overline{g'(z)}}{g'(z)} = \mu(z), \qquad g \in G_0.$$

Let w_μ be the unique μ-conformal mapping of U onto itself which preserves the points 0, 1, ∞. (A univalent map $z \mapsto w(z)$ is called μ-conformal if it satisfies the Beltrami equation

$$(6.6) \qquad \frac{\partial w}{\partial \bar{z}} = \mu \frac{\partial w}{\partial z}.$$

A μ-conformal map of U onto itself can be continued to a topological self-map of $U \cup \mathbf{R} \cup \{\infty\}$.)

One computes that for every $g \in G_0$ the map $w_\mu \circ g$ is again μ-conformal, hence of the form $g_\mu \circ w_\mu$ with a holomorphic g_μ, and since g_μ maps U onto itself it is a real Möbius transformation. Thus $G_\mu = w_\mu G_0 w_\mu^{-1}$ is again a torsion-free Fuchsian group, and one can show that there exists a commutative diagram

$$(6.7) \qquad \begin{array}{ccc} U & \overset{w_\mu}{\to} & U \\ \downarrow & & \downarrow \\ S_0 = U/G_0 & \underset{f^{-1} \circ f_0}{\to} & U/G_\mu = S \end{array}$$

where the vertical arrows represent canonical projection.

(D) One can prove that $w_\mu|\mathbf{R}$ as well as the isomorphism $g \mapsto w_\mu \circ g \circ w_\mu^{-1}$ of G_0 onto G_μ depend not on the choice of f_0 and f but *only* on τ_0 and τ. On the other hand, if τ_0 is given, any μ satisfying (6.4) and (6.5) defines a τ in $T_{p,n}$.

Now extend the definition of $\mu(z)$ to all of $\mathbf{C}$ by setting $\mu|L = 0$. Let $w^\mu(z)$ denote the unique μ-conformal map of $\mathbf{C}$ onto itself which leaves $0, 1, \infty$ fixed. Then

$$(6.8) \qquad w_\mu = h_\mu \circ w^\mu|U$$

where $h_\mu \colon w^\mu(U) \to U$ is conformal, and $w^\mu|L$ is also conformal. Furthermore, one can show that $w^\mu|L$ depends only on τ and τ_0 and not on the particular choice of μ. So does therefore the *Schwarzian derivative*

$$(6.9) \qquad \varphi^\mu(z) = \frac{du}{dz} - \frac{1}{2} u^2 \quad \text{where } u = \frac{d}{dz} \log \frac{d(w^\mu|L)}{dz}.$$

One computes that $\varphi^\mu(z)dz^2$ is G-invariant, i.e., φ^μ satisfies (5.1). Also, by the Kraus-Nehari inequality for the Schwarzian derivatives of univalent functions [77, 96],

$$(6.10) \qquad \sup|y^2 \varphi^\mu(z)| \leqslant \tfrac{3}{2}.$$

Thus $\varphi^\mu \in B_2(L, G)$.

(E) The mapping $\tau \mapsto \varphi^\mu$ of $T_{p,n}$ into $B_2(L, G_0)$ is *continuous* (this follows from the fact that $w^\mu(z)$ depends holomorphically on $\mu \in L_\infty(\mathbf{C})$). It is also *injective*. Indeed knowing φ^μ one can find $w^\mu|L$ as a quotient of two linearly independent solutions of the ordinary differential equation $2\eta'' = -\varphi^\mu \eta$. Knowing $w^\mu|L$ one can find h_μ as the conformal map of the Jordan domain complementary to $w^\mu(L)$ onto U which keeps $0, 1, \infty$ fixed, and knowing h_μ one obtains from (6.3) the function $w_\mu|\mathbf{R}$ which determines the isomorphism $G_0 \to G_\mu$ and τ.

(F) We conclude that the mapping $\tau \mapsto \varphi^\mu$ is a *homeomorphism of* $T_{p,n}$ *onto a domain in* $B_2(L, G_0) \approx \mathbf{C}^{3p-3+n}$, a *bounded* domain, by (6.10). We shall often *identify* $T_{p,n}$ with its image.

One can either use this identification to define the complex structure in $T_{p,n}$ (then one must check that this structure does not depend on the choice of τ_0) or, if this structure is already defined, one can verify that the map $\tau \mapsto \varphi^\mu$ is holomorphic.

(G) REMARK. It is easy to see that $T_{0,4}$ is isomorphic to $T_{1,1}$, $T_{0,5}$ to $T_{1,2}$ and $T_{0,6}$ to $T_{2,0}$. D. Patterson [98] showed that there are no other isomorphisms.

(H) The modular group $\mathrm{Mod}_{p,n}$ acts on $T_{p,n}$ as a group of holomorphic self-mapping. Since it also acts properly discontinuously, the *moduli space*

$$(6.11) \qquad\qquad X_{p,n} = T_{p,n}/\mathrm{Mod}_{p,n}$$

is a normal complex space. It has singularities which come from the fixed points of $\mathrm{Mod}_{p,n}$ corresponding to Riemann surfaces admitting conformal self-mappings.

However, Grothendieck and Serre [62] observed that a subgroup of $\mathrm{Mod}_{p,n}$ of finite index operates freely, for instance the subgroup induced by self-mapping of Σ which induce the identity on $H_1(\Sigma, \mathbf{Z}_m)$ where $m \geqslant 3$ is a fixed integer. Thus $X_{p,n}$ is a complex manifold divided by a finite group.

By methods of algebraic geometry Baily, Mumford and others [16, 40, 95] showed that $X_{p,n}$ is a quasi-projective algebraic variety.

(I) Since the (complex) dimension of $T_{1,1}$ is 1, this Teichmüller space is conformally equivalent to a disc. Numerical calculations show, however, that the image of $T_{1,1}$ in $B_2(L, G) \approx \mathbf{C}$ is *not* a disc (Porter [99]).

(J) In all cases $T_{p,n}$ *is a domain of holomorphy*. There are several proofs of this fact, the first (by Bers and Ehrenpreis [33]) is based on Ahlfors' characterization of quasi-circles, see §18, (C).

(K) On the other hand, if $\dim T_{p,n} > 1$, $T_{p,n}$ is not a symmetric domain and even *not homogeneous*. This was proved by Royden [104, 105] who showed that, under the condition stated, *the full group of holomorphic self-mappings* of $T_{p,n}$ *coincides with* $\mathrm{Mod}_{p,n}$ or (in a few cases) contains it as a subgroup of finite index. Hence it is discrete.

(L) The space $T_{p,n,m}^{\#}$ has a natural real-analytic structure but, if $m > 0$, *no natural complex structure*.

7. Metrics in Teichmüller space.

(A) Royden's result, stated in §6, (K), is closely connected with another theorem also due to him.

The Teichmüller distance between two points P and Q in $T_{p,n}$ can be defined, without any reference to quasiconformal mappings, by the formula

$$(7.1) \qquad\qquad \langle P, Q \rangle = \inf \langle F^{-1}(P), F^{-1}(Q) \rangle$$

where $\langle\ ,\ \rangle$ on the right denotes the Poincaré distance in U (= the Teichmüller distance in $T_{1,1}$) and F runs over all holomorphic maps $U \to T_{p,n}$. Royden's result can also be stated as follows: *the Teichmüller metric coincides with the Kobayashi metric.*

(The Kobayashi pseudo-metric [72] can be defined in every complex manifold; the general definition is more complicated than (7.1).)

(B) We mention at this occasion that given two distinct points P and Q in $T_{p,n}$ there is a holomorphic isometry $F: U \to T_{p,n}$ (an isometry with respect to the Poincaré metric in U and the Teichmüller metric is $T_{p,n}$) such that $F(U)$ contains P and Q and hence also the Teichmüller line through these points. This was known already to Teichmüller.

Also, for every fixed $P \in T_{p,n}$, the distance $\langle P, Q \rangle$ is a *differentiable* function of $Q \neq P$ (Earle [46]).

(C) In every bounded domain D in a complex number space one can define two metrics which are determined by the complex structure of D and hence invariant under any holomorphic self-mapping of the domain. The *Bergman metric* is a Kähler metric defined by the line element

$$ds^2 = \Sigma \frac{\partial^2 \log K(z; z)}{\partial z_i \partial \bar{z}_j}\, dz_i \wedge d\bar{z}_j$$

where $K(z; \zeta) = K(z_1, \ldots, z_\delta; \zeta_1, \ldots, \zeta_\delta)$ is the Bergman kernel of $D \subset \mathbf{C}^\delta$. The *Carathéodory* distance between two points P and Q of D is defined as

$$d_{\mathrm{Car}}(P, Q) = \sup\langle f(P), f(Q)\rangle,$$

where $\langle\ ,\ \rangle$ is again the Poincaré distance in U and f runs over all holomorphic maps $D \to U$. It is not known whether the Carathéodory metric in $T_{p,n}$ coincides with the Teichmüller metric, but Earle [45] proved that it is *complete*; so is the Bergman metric since it always dominates the Carathéodory metric (Hahn [63]). The completeness of the Carathéodory metric implies again that $T_{p,n}$ is a domain of holomorphy. (This also follows from Royden's theorem.)

(D) Another Hermitian metric in $T_{p,n}$ invariant under $\mathrm{Mod}_{p,n}$, has been defined by Weil [120]. The tangent space to $T_{p,n}$ at a point $\tau = [f: S \to \Sigma]$ can be identified with $Q(\bar{S}) \approx B_2(L, G_0)$, as is seen from §6, and in $B_2(L, G_0)$ there is a canonical scalar product defined many years ago by Petersson:

$$(\varphi, \psi) = \int\int_{L/G} y^2 \varphi(z)\, \overline{\psi(z)}\ dx\, dy.$$

The Hermitian *Weil-Petersson metric* derived from this scalar product clearly has the required invariance. It has been shown to be Kählerian (Weil, Ahlfors [8, 9]) and to have *negative Ricci and holomorphic sectional curvatures* (Ahlfors). But it is *incomplete* (Chen [38], Wolpert [121]) and therefore distinct from the Bergman metric. (Cf. also §16, (E).)

8. Quasi-Fuchsian groups.

(A) We return to the mapping w^μ and form the group

$$(8.1) \qquad\qquad G^\tau = w^\mu G_0 (w^\mu)^{-1}.$$

It is (α) a group of not necessarily real Möbius transformations (this is established as the similar statement about G_μ in §6, (C)), (β) it preserves a directed Jordan curve C which is also its limit set, and set of accumulation points of orbits (namely the curve $C = w^\mu(\mathbf{R} \cup \{\infty\})$) and ($\gamma$) it acts properly discontinuously on the two domains complementary to C (namely $w^\mu(U)$ and $w^\mu(L)$). Actually (γ) follows from (α) and (β).

A group with the properties (α), (β), (γ) is called a *quasi-Fuchsian group of the first kind* (it is Fuchsian if C is a circle or a line).

(B) Such a group always represents *two* Riemann surfaces, the quotients of the two Jordan domains by the group, and if the group is finitely generated, the two are homeomorphic. If the group is torsion free, it can be identified with the fundamental groups of both surfaces and hence (by Nielsen's theorem) the group determines a homotopy class of *orientation-reversing* homeomorphisms between the two surfaces.

In the case of the group G^τ defined by (8.1) the two surfaces are the mirror image $\overline{S}_0$ of $S_0 = f_0^{-1}(\Sigma)$ and $S = f^{-1}(\Sigma)$, and the homotopy class is that of the mapping

$$f^{-1} \circ f_0 \circ j^{-1}$$

where j is the canonical mapping of S_0 onto $\overline{S}_0$ (and $\tau_0 = [f_0 \colon S_0 \to \Sigma]$, $\tau = [f \colon S \to \Sigma]$). The verification is left to the reader.

(C) The *simultaneous uniformization theorem* [18], a generalization of the limit circle theorem by Klein-Poincaré-Koebe, asserts that *given two Riemann surfaces of the same type (p, n), and a homotopy class of orientation-reversing homeomorphisms between them, there exists a quasi-Fuchsian group representing these surfaces and this class. The group is determined uniquely*, up to a conjugation in $PSL(2, \mathbf{C})$.

The proof of the existence part of the theorem is essentially the construction of the group G^τ above. Briefly speaking, the desired group is obtained as a quasiconformal deformation of a Fuchsian group G_0, i.e., by conjugating G_0 by a μ-conformal automorphism of $\mathbf{C}$, μ being chosen appropriately but so as to satisfy (6.5). The uniqueness proof is rather straightforward but relies on delicate properties of quasiconformal mappings. (For the proof, and a more general version of the theorem see [20].)

(D) Quasi-Fuchsian groups had been known to Klein and Poincaré. The classical construction starts with a sequence of circles $C_1, \ldots, C_N, C_{N+1} = C_1$, such that C_j touches or intersects C_{j-1} and C_{j+1} and no other circle, $j = 1, \ldots, N$. The desired group is obtained as the subgroup of orientation-preserving transformations in the group generated by N reflections about the circles $C_1, \ldots, C_N$.

Also, Poincaré stated without proof that if one changes a little the generators of a Fuchsian group, preserving their types and all relations between them, but permitting the generators to become complex, one obtains a quasi-Fuchsian group.

The simultaneous uniformization theorem, however, was found only when the theory of quasiconformal mappings was sufficiently developed. As a matter of fact, no proof of the theorem not relying on quasiconformal mappings is known.

(E) The fixed curve of a quasi-Fuchsian group of the type considered here has many interesting properties. It admits a parametric representation by Hölder continuous functions and has measure zero. But it is not rectifiable, Bowen [37]. Indeed, Bowen proved that its *Hausdorff dimension is strictly between 1 and 2.*

(F) An important theorem by Maskit [37] states that *any finitely generated Kleinian group whose region of discontinuity has precisely two components is either quasi-Fuchsian and a quasiconformal deformation of a Fuchsian group, or a $\mathbf{Z}_2$ extension of a quasi-Fuchsian group.*

For a simplified proof, see Kra and Maskit [76].

(G) The set of conjugacy classes of quasi-Fuchsian groups representing two Riemann surfaces of type (p, n) is easily seen to be a $(6p - 6 + 2n)$-dimensional complex manifold. By the results of §6 it is isomorphic to $T_{p,n} \times T_{p,n}$.

9. Fiber spaces over Teichmüller spaces.

(A) We return to the embedding $T_{p,n} \subset B_2(L, G_0)$ of §6 and note that to every $\tau \in T_{p,n}$ there belongs a Jordan domain

$$D_\tau = w^\mu(U)$$

bounded by the Jordan curve consisting of $z = \infty$ and the arc

$$\zeta = w^\mu(x), \qquad -\infty < x < +\infty.$$

For every fixed $x \in \mathbf{R}$, ζ depends holomorphically on $\mu \in L_\infty(\mathbf{C})$; it is not changed if μ is replaced by, say, μ' as long as μ' defines the same τ. We express this by saying that *the Jordan domain D_τ depends holomorphically on τ.*

(B) The *fiber space* $F_{p,n}$ *over* $T_{p,n}$ is the set of pairs

$$(\tau, z) \quad \text{with } \tau \in T_{p,n}, z \in D_\tau.$$

It is a $(3p - 2 + n)$-dimensional domain of holomorphy and is *canonically isomorphic to* $T_{p,n+1}$, the projection $F_{p,n} \to T_{p,n}$ corresponding to "forgetting one puncture" [27].

(C) Set

$$\delta(1) = p \quad \text{if } n = 0, \qquad \delta(1) = p + n - 1 \quad \text{if } n > 0,$$
$$\delta(r) = (2r - 1)(p - 1) + n(r - 1) \quad \text{if } r > 1.$$

For every r there are $\delta(r)$ holomorphic functions $\Psi_1^r(\tau, z)$, $\Psi_2^r(\tau, z), \ldots$ in $F_{p,n}$ such that for every fixed $\tau \in T_{p,n}$ the function $\Psi_1^r, \ldots, \Psi_{\delta(r)}^r$ form a basis of the space of cusp forms of weight $(-2r)$ in D_τ for the group G^τ. (A cusp form of weight $(-2r)$ is a holomorphic function $\psi(z)$, $z \in D_\tau$ such that $\psi(z)dz^r$ is G^τ-invariant, i.e.

$$\psi(g(z))g'(z)^r = \psi(z), \qquad g \in G^\tau,$$

and

$$\int\int_{D_\tau/G^\tau} \lambda_\tau(z)^{2-2q}|\psi(z)|\, dx\, dy < \infty,$$

$$\sup|\lambda_\tau(z)^{-q}\psi(z)| < \infty,$$

where $\lambda_\tau(z)|dz|$ is the Poincaré line element in D_τ.)

(D) The existence of the functions Ψ_j^r can be obtained from Grauert's theorem on holomorphic triviality of topologically trivial holomorphic vector bundles over Stein manifolds. For $r > 1$ there is also an *explicit integral formula* which we will write down without discussing its derivation (which can be found in [21, 22, 24]).

If $\psi_1, \psi_2, \ldots, \psi_{\delta(r)}$ is a basis of cusp forms of weight $(-2r)$ in U for the group G_0, then (setting $\zeta = \xi + \sqrt{-1}\ \eta$)

$$\Psi_j^r(\tau, z) = \int\int_L \frac{\eta^{2r-2}\psi_j(\bar\zeta)[\partial w^\mu(\zeta)/\partial\zeta]^r}{[w^\mu(\zeta) - z]^{2r}}\, d\xi\, d\eta.$$

Here w^μ is the function defined in §6, (D); recall that $w^\mu|L$ is a holomorphic function depending only on τ and not on the particular choice of μ.

(E) For $n = 0$ and $r = 3$ the functions $\Psi_1^3(\tau, z), \ldots, \Psi_{5p-5}^3$ are holomorphic functions on $F_{p,0}$ with the property: *every algebraic curve of genus p admits a parametric representation by rational functions of these functions, for some fixed τ.*

(F) The group G_0 operates on $F_{p,n}$ by the rule $g_0((\tau, z)) = (\tau, g^\mu(z))$ where $g^\mu = w^\mu \circ g \circ (w^\mu)^{-1}$. The quotient $F_{p,n}/G_0$ is a complex manifold and a fiber space over $T_{p,n}$, the fiber over a point $\tau = [f: S \to \Sigma]$ being isomorphic to the Riemann surface S. The manifold $V_{p,n} = F_{p,n}/G_0$ has been called the *universal Teichmüller curve.*

Except for some special and rather obvious cases the universal Teichmüller curve admits *no holomorphic sections.* This has been first shown by Hubbard [67], for $n = 0$, and then by Earle and Kra [50, 51] for the general case. The proofs use essentially Royden's theorem.

(G) One can define [27] an *extended modular group* $\mathrm{mod}_{p,n}$ of holomorphic self-mappings of $F_{p,n}$ which acts properly discontinuously and contains G_0 as a normal subgroup with $\mathrm{Mod}_{p,n} = \mathrm{mod}_{p,n}/G_0$. The quotient

$$Y_{p,n} = F_{p,n}/\mathrm{mod}_{p,n}$$

is a normal complex space fibered over the moduli space $X_{p,n}$, the fiber over a point $x \in X_{p,n}$ being $S/\mathrm{Aut}(S)$ where S is a Riemann surface (unmarked) of type (p,n) represented by x and $\mathrm{Aut}(S)$ the finite group of conformal self-mappings of S. (This statement must be slightly modified if $(p, n) = (1, 1)$ $(1, 2)$ or $(2, 0)$.)

The isomorphism theorem in (B) implies that $X_{p, n+1}$ is a ramified $(n + 1)$-sheeted covering space of $Y_{p, n}$.

(H) Earle [4] constructed a fiber space over $T_{p, 0}$ whose fibers are Jacobian varieties of genus p. We refer to his paper for details.

10. Complex boundaries.

(A) We proceed to discuss the *complex boundary* $\partial T_{p, n}$ defined by the embedding $T_{p, n} \subset B_2(L, G_0)$, see §6. This boundary depends, of course, on the choice of the origin τ_0, cf. Bers [25], Maskit [84], Abikoff [2].

It is now convenient to associate with every cusp form $\varphi \in B_2(L, G)$ the function

$$(10.1) \qquad W_\varphi(z) = \eta_1(z)/\eta_2(z), \qquad z \in L,$$

where η_1 and η_2 satisfy

$$(10.2) \qquad 2\eta_j''(z) + \varphi(z)\eta_j(z) = 0, \qquad j = 1, 2, \quad z \in L,$$

and

$$(10.3) \qquad \eta_1(-i) = \eta_2'(-i) = 1, \qquad \eta_1'(-i) = \eta_2(-i) = 0.$$

This function, just as the function $w^\mu|L$ defined in §6, (C) has the Schwarzian derivative $\varphi = \varphi^\mu$. But while $w^\mu|L$ is defined only for $\varphi \in T_{p, n} \subset B_2(L, G_0)$ and is normalized by the "boundary conditions" $w^\mu(0) = 0$, $w^\mu(1) = 1$, $w^\mu(\infty) = \infty$, the function W_φ is defined for all $\varphi \in B_2(L, G)$ and is normalized by the initial condition $W_\varphi(t - i) = t^{-1} + 0(t)$, $t \to 0$. Of course, for $\varphi \in T_{p, n}$, the function W_φ is of the form $\alpha \circ w^\mu|L$ where $\alpha \in PSL(2, \mathbf{C})$, but α depends on φ and need not have a limit as φ converges to $\partial T_{p, n}$.

(B) For *every* $\varphi \in B_2(L, G_0)$ there is a homomorphism

$$G_0 \ni g \mapsto \chi_\varphi(g) \in PSL(2, \mathbf{C})$$

such that $W_\varphi \circ g = \chi_\varphi(g) \circ W_\varphi$. The group $\chi_\varphi(G_0) = G^\varphi$ is the so called *monodromy group* of the differential equation (10.2).

For $\varphi \in T_{p, n}$ the function W_φ is univalent meromorphic; its image contains the point at infinity. The same is true for $\varphi \in \partial T_{p, n}$, but $W_\varphi(L)$ is a Jordan domain *only* if $\varphi \in T_{p, n}$.

For $\varphi \in T_{p, n} \cup \partial T_{p, n}$ the homomorphism χ_φ is an *isomorphism* and

$$(10.4) \qquad G^\varphi = W_\varphi G_0 W_\varphi^{-1}.$$

For $\varphi \in T_{p, n}$ the group G^φ is quasi-Fuchsian, with precisely one unbounded component (a component of a Kleinian group is a component of its domain of discontinuity). The quotient of the unbounded component by G^φ is always $\bar{S}_0$, that of the bounded component the "variable" Riemann surface $S = f_\tau^{-1}(\Sigma)$, τ being the point in Teichmüller space represented by φ. (For the sake of brevity we often say "is" instead of "is conformally equivalent to".)

(C) It turns out that for $\varphi \in \partial T_{p, n}$, the group G^φ is always *Kleinian* and has a *single invariant component* $\Delta_0 = W_\varphi(L)$; this component contains the point at infinity and Δ_0 / G^φ is $\bar{S}_0$.

Any other component Δ of G^φ is simply connected, not invariant (i.e., there is a $g \in G^\varphi$ with $g(\Delta) \neq \Delta$), and the stabilizer G_Δ^φ of Δ in G^φ always contains an *accidental parabolic element*, i.e., a parabolic element $\chi_\varphi(g)$ with $g \in G_0$ hyperbolic. Also, $\Delta / G_\Delta^\varphi$ is a Riemann surface of some type $(\hat{p}, \hat{n})$.

(D) It is not difficult to verify that points $\varphi \in B_2(L, G_0)$, with $\chi_\varphi(g)$ parabolic for some hyperbolic $g \in G_0$, lie on countably many holomorphic hypersurfaces in $B_2(L, G_0)$. The union of these hypersurfaces has real codimension 2, while $\partial T_{p,n}$ has real codimension 1. Hence $W^\varphi(L)$ is the *only* component of G^φ for all $\varphi \in \partial T_{p,n}$, except for a subset of positive real codimension.

(E) Now let $\varphi \in \partial T_{p,n}$ belong to this small set. Then there is a finite nonempty set of noninvariant components $\Delta_1, \ldots, \Delta_N$ of G^φ such that $g(\Delta_j) \neq \Delta_k$ for $j \neq k$ and $g \in G^\varphi$, and every component of G^φ, distinct from the invariant component Δ_0, is of the form $g(\Delta_j)$ for some $g \in G^\varphi$ and some j, $1 \leq j \leq N$. If Ω denotes the region of discontinuity of G^φ, then

$$\Omega / G^\varphi = \Delta_0 / G^\varphi + \Delta_1 / G_{\Delta_1}^\varphi + \cdots + \Delta_N / G_{\Delta_N}^\varphi = \bar{S}_0 + S_1 + \cdots + S_N.$$

It can be shown that there are on S_0 finitely many Jordan curves, which can be chosen as Poincaré geodesics $C_1, \ldots, C_r$, $0 < r \leq 3p - 3 + n$, with the following property. The complement of $(C_1 \cup \cdots \cup C_n)$ in S_0 has $M \geq N$ components $D_1, \ldots, D_M$ and these can be ordered so that D_j is homeomorphic to S_j, $j = 1, \ldots, N$.

(F) A boundary point $\varphi \in \partial T_{p,n}$ with $M = N > 0$ is called *regular* and so is the group G^φ.

Such a group may be said to represent the mirror image $\bar{S}_0$ of the reference surface S_0 and one (if $N = 1$) or several (if $N > 1$) surfaces obtained by drawing $r > 0$ disjoint geodesic Jordan arc on S_0 and then "squeezing" each of those into a point (which may be thought of as two punctures joined into a node).

All combinatorially possible regular G^φ actually occur (Maskit, Abikoff). Furthermore, if φ is a regular point on $\partial T_{p,n}$ and the surfaces $S_1, \ldots, S_N$ have types $(p_1, n_1), \ldots, (p_N, n_N)$, then there is a complex manifold $A \subset \partial T_{p,n}$, containing φ, of dimension

$$3(p_1 + \cdots + p_N) - 3N + n_1 + \cdots + n_N,$$

which is isomorphic to the product $T_{p_1 n_1} \times T_{p_2 n_2} \times \cdots \times T_{p_N n_N}$.

(G) A boundary point $\varphi \in \partial T_{p,n}$ is called *partially degenerate* if $M > N > 0$, and so in the group G^φ. The group represents, besides $\bar{S}_0$, some but not all surfaces obtained by squeezing $r > 0$ disjoint geodesic Jordan curves on S_0 into punctures. It is worth noting that no thrice punctured sphere may be "thrown away".

Abikoff showed that all combinatorially possible types of partially degenerate groups G^{φ} occur as boundary points of the *boundary Teichmüller spaces A* described above.

(H) A point $\varphi \in \partial T_{p,n}$ and the group G^{φ} are called *totally degenerate* if G^{φ} has only the component $W_{\varphi}(L)$ and thus represent only $\bar{S}_0$. We already noted that a boundary group G^{φ} without accidental parabolic elements must be totally degenerate. However, there are also totally degenerate groups with accidental parabolic elements.

11. *b*-groups.

(A) A finitely generated Kleinian group G is called a *b-group* if it has precisely one invariant component Δ_0, and Δ_0 is simply connected. Groups G^{φ}, $\varphi \in T_{p,n}$, are *b*-groups and the name "*b*-group" was chosen in the hope that each such group appears on the boundary of some Teichmüller space. If G contains elliptic elements, the Teichmüller space in question must be a Teichmüller space of a Fuchsian group with torsion (see §19). Torsion requires no essentially new ideas and we will consider here only torsion-free *b*-groups.

(B) By Ahlfors' *finiteness theorem* [**11**, **73**] a *b*-group represents finitely many Riemann surfaces, each with a finite Poincaré area. By the "*second area inequality*", see [**73**], the total Poincaré area of Ω/G, where Ω is the region of discontinuity of G, is at most twice the Poincaré area of Δ_0/G. (In §10, (E) we encountered this inequality in the guise $M \geqslant N$.) If equality holds, G is called *regular*, and Abikoff proved that *a regular b-group is always a boundary group*.

(C) The existence of finitely generated Kleinian groups with a connected and simply connected region of discontinuity (*totally degenerate groups*) seems not to have been known to the founding fathers of the theory, Klein and Poincaré. The existence proof for such groups sketched in §10, (D) actually yields the following results: *given a Riemann surface S of type* (p, n), *with* $3p - 3 + n > 0$, *there exist continuously many nonconjugate torsion-free totally degenerate Kleinian groups G, with region of discontinuity* Ω *and* Ω/G *conformally equivalent to S*.

(D) A remarkable property of degenerate groups has been discovered by L. Greenberg. It refers to the interpretation of Kleinian groups as discrete groups of non-Euclidean motions in space. This interpretation, due to Poincaré plays an increasingly important part in recent work on Kleinian groups. In the hands of Thurston it became an amazingly powerful tool for investigating discrete groups and hyperbolic structures on 3-manifolds.

Call a quaternion $x + yi + tj + sk$ special if $t > 0$, $s = 0$, and write a special quaternion as $Z = z + tj$, $z = x + iy \in \mathbf{C}$. The Poincaré half-space $\mathbf{H}_3$ is the set of special quaternions (see §1, (C)); with the metric (1.2) it becomes a model of non-Euclidean space. Every Möbius transformation

$$z \mapsto g(z) = (az + b)/(cz + d),$$

where a, b, c, $d \in \mathbf{C}$ and $ad - bc = 1$, gives rise to an isometry of $\mathbf{H}_3$ which we denote by the same letter

$$Z \mapsto g(Z) = (aZ + b)(cZ + d)^{-1},$$

and every orientation-preserving isometry of $\mathbf{H}_3$ can be so written. (Similarly, every conformal self-mapping of $\mathbf{H}_n$ can be interpreted as an orientation-preserving isometry of $\mathbf{H}_{n+1}$, but the elegant formula involving quaternions is restricted to $n = 2$.)

(E) It is classical that every discrete finitely generated group of non-Euclidean motions in the plane, i.e., every finitely generated Fuchsian group, has a fundamental domain which is a non-Euclidean polygon with finitely many sides. Greenberg [**61**] showed that this is not so in space. *A degenerate Kleinian group, viewed as a group of motions in $\mathbf{H}_3$, does not have a finitely-many-sided fundamental polyhedron.* It is as one says, *geometrically infinite.* So also is every partially degenerate boundary group.

(F) The difference between geometrically finite and geometrically infinite Kleinian groups dominates the whole theory. In general, geometrically finite groups are easier to investigate. Maskit [**87, 88**] gave a complete classification of all such groups with an invariant component.

All finitely generated quasi-Fuchsian groups and regular boundary groups are geometrically finite.

12. Ahlfors' problem, Sullivan's theorem and the complex boundary.

(A) In a seminal paper [**11**] on Kleinian groups Ahlfors asked whether the limit set $\Lambda(G)$ of a finitely generated Kleinian group has area (2-dimensional Lebesgue measure) zero. Later [**12**] he proved that this is so if the group is geometrically finite. (The condition of being finitely generated is essential; Abikoff [**1**] constructed an infinitely generated quasi-Fuchsian group whose fixed curve has positive area.) Thurston showed that limit sets of so-called strong limits of (finitely generated) quasi-Fuchsian groups also have vanishing area (cf. [**118**] §8.12.4 and §9.2). This includes all groups on $\partial T_{p,n}$ without accidental parabolic elements.

(B) The importance of Ahlfors' problem for function theory derives from the following question. Suppose G is a finitely generated Kleinian group, w a quasiconformal self-mapping of $\mathbf{C} \cup \{\infty\}$ which is compatible with G, that is such that wGw^{-1} is again a Kleinian group, and $\mu = (\partial w/\partial \bar{z})/(\partial w/\partial z)$ the Beltrami coefficient of w. Is w determined, except for a conjugation, by knowing $\mu|\Omega$, that is by knowing how w changes the conformal structure of the Riemann surfaces represented by G?

The answer is yes if mes $\Lambda(G) = 0$. In an important recent paper [**113**] Sullivan showed that the answer is *always yes.*

(C) We ask now (as we asked for the real boundaries in §4) whether the modular group $\mathrm{Mod}_{p,n}$ acts (by continuity) on $\partial T_{p,n} \subset B_2(L, G_0)$. A partial answer [**32**] can be given using Sullivan's theorem.

A boundary group G^φ, $\varphi \in \partial T_{p,n}$, represents the fixed Riemann surface $\bar{S}_0$ and $N \geq 0$ other Riemann surfaces of types $(p_1, n_1), \ldots, (p_N, n_N)$. We say that φ *has no moduli* if either $N = 0$ or $(p_j, n_j) = (0, 3)$ for $j = 1, \ldots, N$.

The modular group acts (by continuity) at all boundary points with no moduli.

This includes all degenerate, some regular and some partially degenerate boundary points. (It includes the whole boundary if $p = n = 1$.)

(D) If $\varphi \in \partial T_{p,n}$ has moduli, $\{\varphi_\nu\} \subset T_{p,n}$ and $\varphi_\nu \to \varphi$, and if $\gamma \in \mathrm{Mod}_{p,n}$ and a subsequence $\{\varphi_{\nu_i}\}$ are such that $\gamma(\varphi_{\nu_i}) \to \varphi_1 \in \partial T_{p,n}$, then G^{φ_1} represents as many surfaces as G^φ does, and of the same types.

(E) If $\gamma \in \mathrm{Mod}_{p,n}$ leaves the origin $\tau_0 = [f_0: S_0 \to \Sigma]$ fixed, that is, if $\gamma = \omega_*$ where $\omega: \Sigma \to \Sigma$ can be written as $f_0 \circ h \circ f_0^{-1}$ with $h: S_0 \to S_0$ conformal, then γ *acts continuously on all of* $\partial T_{p,n}$. Indeed, in this case γ is the restriction to $T_{p,n}$ of a linear transformation on $B_2(L, G_0)$.

(F) According to a forthcoming paper by Kerckhoff and Thurston *some* $\gamma \in \mathrm{Mod}_{p,n}$ *must fail to act on some regular points on* $\partial T_{p,n}$ (provided that $\dim T_{p,n} > 1$).

13. The Maskit embedding.

(A) Besides the injection $T_{p,n} \hookrightarrow \mathbf{C}^{3p-3+n}$ described in §6 there is another injection due to Maskit [**86**] which depends not on the choice of an origin in $T_{p,n}$, which involves continuously varying parameters, but on finitely many choices. We describe it for the case $T_p = T_{p,0}$, the case considered by Maskit. The extension to $n \neq 0$ involves neither new ideas nor technical difficulties.

In order to define the Maskit embedding one chooses $3p - 3$ disjoint and homotopically independent Jordan curves $C_1, \ldots, C_{3p-3}$ on Σ (which is now a compact surface of genus p); this can be done in only finitely many topologically distinct ways.

The idea of the Maskit embedding is to represent a point $\tau = [f: S \to \Sigma]$ of T_p by a regular b-group G with invariant component Δ_0, which represents $S = \Delta_0/G$, and $2p - 2$ thrice punctured spheres. This group is to have the property that the lift of each curve $f^{-1}(C_j)$, $j = 1, \ldots, 3p - 3$, to Δ_0 is invariant under an accidental parabolic element of G.

(B) The Maskit coordinates are $(3p - 3)$-tuples of numbers $Z_j \in U$, belonging to a domain $\mathfrak{D} \subset U^{3p-3}$. The numbers $Z_1, \ldots, Z_{3p-3}$ determine a sequence of generators for the group G. The technical details are somewhat complicated and will be explained later (cf. §16, especially (F)), in a more general context.

(C) The boundary of T_p in the Maskit embedding, and the action of $\mathrm{Mod}_p = \mathrm{Mod}_{p,0}$ on this boundary, have not yet been investigated sufficiently. We mention, however, that if ω is a Dehn twist about C_j, ω_* leaves all Z_k, $k \neq j$, alone and replaces Z_j by $Z_j + 1$.

14. Riemann surfaces with nodes.

(A) We shall now describe an extension of Teichmüller space theory in which compact Riemann surfaces are replaced by Riemann surfaces with nodes. The theory had been announced almost ten years ago [28, 29] but the promised detailed presentation has . ot yet appeared. The author uses this occasion to present his apologies and to express the hope that the long delayed obligation will be fulfilled in the not too distant future.

(B) A compact surface with nodes Σ is a compact Hausdorff space such that every point $P \in \Sigma$ has a fundamental system of neighborhoods homeomorphic either to the disc $|z| < 1$ in $\mathbf{C}$ or to the set $|z_1| < 1, |z_2| < 1, z_1 z_2 = 0$ in $\mathbf{C}^2$. In the second case P is called a *node*.

Each component of the complement of the set of nodes is called a *part* of Σ, and Σ is called *oriented* if every part is, *stable* if no part has an Abelian fundamental group, *nonsingular* if there are no nodes.

We shall consider only stable oriented Σ.

The *genus* p of Σ is the genus of the nonsingular surface obtained by thickening each node; the stability condition implies that $p > 1$.

(C) A (compact stable) *Riemann surface with nodes S* is a (compact stable) surface with nodes such that every part S_j of S is a Riemann surface of some type (p_j, n_j). The stability condition implies that $2p_j - 2 + n_j > 0$ so that each part of S has a Poincaré metric. The total Poincaré area of S is $2\pi(2p - 2)$ as for a nonsingular surface.

(D) A *regular r-differential* on S is a rule which assigns a holomorphic r-differential (holomorphic section of r times the canonical line bundle) to every part S_j of S, with the provision that at the punctures the r-differential have poles at most of order r, and at any two punctures joined in a node the residues are equal, if r is even, or opposite, if r is odd.

(DEFINITION OF RESIDUE. Let q be an r-differential which has at P a pole of order at most r. Either the order is less than r and the residue at P is 0, or there is a local parameter ζ with $\zeta = 0$ at P and $q = a(d\zeta/\zeta)^r$ near P and the residue at P is a.)

The dimension $\delta(r)$ of the space of regular r-differentials on S is

$$\delta(r) = \begin{cases} p & \text{if } r = 1, \\ (2r - 1)(p - 1) & \text{if } r > 0, \end{cases}$$

as for nonsingular surfaces. For $r > 2$ one can use a basis of regular r-differentials to embed S into $\mathbf{P}_{\delta(r)-1}$ and thus realize it as an algebraic curve of arithmetic genus p, with no singularities except for simple nodes.

(E) A *strong deformation* of a surface with nodes Σ_1 onto a surface with nodes Σ_2, both of the same genus p, is a continuous surjection $\Sigma_1 \to \Sigma_2$ such that the image of a node of Σ_1 is a node Σ_2, the inverse image of a node of Σ_2 is a node of Σ_1 or a Jordan curve on a part of Σ_1, and the mapping restricted to the complement of the inverse images of all nodes is an orientation-preserving homeomorphism.

(F) We say that Σ_1 is *terminal* if it has the largest possible number of nodes, namely $3p - 3$. There are only finitely many topologically distinct terminal surfaces of a given genus p, and every strong deformation of a terminal surface is a homeomorphism.

(G) From now on the letter Σ (or S), with or without subscripts and superscripts will denote a compact surface (Riemann surface) with nodes, of genus p.

15. Strong deformation spaces.

(A) Two strong deformations $f\colon S \to \Sigma$ and $f'\colon S' \to \Sigma$ will be called *equivalent* if there is a commutative diagram

$$
(15.1) \qquad
\begin{array}{ccc}
S & \xrightarrow{\;f\;} & \Sigma \\
\varphi\downarrow & & \downarrow\psi \\
S' & \xrightarrow[f']{} & \Sigma
\end{array}
$$

where φ and ψ are bijective homeomorphisms homotopic to an isomorphism and to the identity, respectively (cf. the definition in §2, (A)).

If $f\colon S \to \Sigma$ and $g\colon \Sigma \to \Sigma_0$ are strong deformations, so is $g \circ f$ and the equivalence class of $g \circ f$ depends only on the equivalence class of f and the homotopy class of g.

(B) The *strong deformation* space $\mathfrak{D}(\Sigma)$ of Σ is the set of all equivalence classes $[f\colon S \to \Sigma]$ of strong deformations $f\colon S \to \Sigma$.

Every strong deformation $g\colon \Sigma \to \Sigma_0$ induces a mapping $g_*\colon \mathfrak{D}(\Sigma) \to \mathfrak{D}(\Sigma_0)$ which takes $[f\colon S \to \Sigma]$ into $[g \circ f\colon S \to \Sigma_0]$. The mappings ω_* induced by homeomorphisms $h\colon \Sigma \to \Sigma$ form the modular group $\mathrm{Mod}(\Sigma)$.

From now on we shall omit the adjective "strong" when there is no danger of misunderstanding.

(C) In order to topologize $\mathfrak{D}(\Sigma)$, let us set, for every closed curve C on a part of S,

$$
l_S(C) = \inf(\text{Poincaré length of } C')
$$

where C' runs over the free homotopy class of C. If $l_S(C) > 0$, then $l_S(C)$ equals the length of the unique Poincaré geodesic freely homotopic to C. We also set $l_S(P) = 0$ if P is a point on S.

A set $A \subset \mathfrak{D}(\Sigma)$ is called *open* if for every point $[f\colon S \to \Sigma]$ in A there are finitely many closed curves $C_1, \ldots, C_r$ on parts of S, and an $\varepsilon > 0$, such that whenever the deformation $h\colon S_1 \to S$ satisfies the conditions

$$
|l_{S_1}\!\left(h^{-1}(C_j)\right) - l_S(C_j)| < \varepsilon \quad \text{for } j = 1, \ldots, r,
$$

$$
|l_{S_1}(h^{-1}(Q))| < \varepsilon \quad \text{for all nodes } Q \text{ of } S,
$$

the point $[f \circ h\colon S_1 \to \Sigma]$ of $\mathfrak{D}(\Sigma)$ belongs to A.

This definition implies that induced maps are continuous and open.

It also implies that if Σ *is nonsingular*, $\mathfrak{D}(\Sigma)$ *may be identified with* T_p.

(D) Let Σ_0 be terminal, let $P_1, \ldots, P_{3p-3}$ be the nodes of Σ_0 and $\sigma_1, \ldots, \sigma_{2p-2}$ the parts of Σ_0, each of which is homeomorphic to a thrice punctured sphere.

Each node P_j has two banks (we forego a formal definition of this term) and each part σ_i has three boundary banks. For each i we order these three banks cyclically.

Let $[f\colon S \to \Sigma_0]$ be a point of $\mathfrak{D}(\Sigma_0)$. Without loss of generality we assume that $C_j = f^{-1}(P_j)$ is either a node of S or a Poincaré geodesic and we set

$$2\pi r_j = \text{Poincaré length of } C_j.$$

If $r_j > 0$ there is on each bank of C_j a *distinguished point* determined as follows. Assume the bank in question is $C_j^+ = f^{-1}(b')$ where b' is a bank of a node of Σ_0, and that C_j^+ and two inverse images $f^{-1}(b'')$ and $f^{-1}(b''')$ of banks of nodes "bound" the inverse image $f^{-1}(\sigma_i)$ of a part of Σ_0. Assume also that b''' precedes b' in the cyclic ordering chosen above. There is a unique Poincaré geodesic K in $f^{-1}(\sigma_i)$ which joins C_j^+ to $f^{-1}(b'')$, is orthogonal to C_j, and is orthogonal to C_k if $f^{-1}(b''')$ is a bank of C_k (rather than a node of S). The endpoint of K on C_j is the distinguished point on C_j^+.

If $r_j > 0$, let θ_j be the smallest nonnegative number with the property: if we approach the distinguished point on one bank of C_j along a geodesic arc in this bank turn left and proceed along C_j, we reach the second distinguished point after having covered the distance θr_j. Clearly $0 \leqslant \theta < 2\pi$.

Now set $z_j = 0$ if $r_j = 0$ and

$$z_j = r_j e^{i\theta} \quad \text{if } r_j > 0.$$

The numbers $z_1, \ldots, z_{3p-3}$ depend only on the equivalence class $[f\colon S \to \Sigma_0]$; we call them the *Fenchel-Nielsen coordintes* of this point in $\mathfrak{D}(\Sigma_0)$. It can be shown that the map

$$[f\colon S \to \Sigma_0] \mapsto \{z_1, \ldots, z_{3p-3}\}$$

is a homeomorphism of $\mathfrak{D}(\Sigma_0)$ onto $\mathbf{C}^{3p-3}$.

Note that the inverse image of a coordinate plane $z_k = 0$ consists of all $[f\colon S \to \Sigma_0]$ such that $f^{-1}(P_j)$ is a node.

(E) It can be shown that for every $g\colon \Sigma \to \Sigma_0$, Σ_0 as before, the induced map $g_*\colon \mathfrak{D}(\Sigma) \to \mathfrak{D}(\Sigma_0)$ is a universal covering of its image, the image being the set of those $[f\colon S \to \Sigma_0]$ for which $f^{-1}(P_j)$ is not a node whenever $g^{-1}(P_j)$ is not a node.

Hence $\mathfrak{D}(\Sigma)$ *is homeomorphic to* $\mathbf{C}^{3p-3}$ and one can define Fenchel-Nielsen coordinates $w_1, \ldots, w_{3p-3}$ in $\mathfrak{D}(\Sigma)$, defined by the deformation g, as complex numbers $w_1, \ldots, w_{3p-3}$ such that

$$w_j = z_j \quad \text{if } g^{-1}(P_j) \text{ is a node of } \Sigma,$$

$$e^{w_j} = z_j \quad \text{if } g^{-1}(P_j) \text{ is not a node of } \Sigma.$$

If Σ is nonsingular, one obtains the coordinates in T_p used by Fenchel and Nielsen in the famous manuscript.

(F) One now verifies easily that *every* induced map of one deformation space into another is a universal covering of the image.

(G) It is important to note that the Fenchel-Nielsen coordinates are written as complex numbers merely for convenience. They are *real coordinates* and do not reflect the natural complex structure of the deformation spaces, which we proceed to define.

16. Complex structure of deformation spaces.

(A) Let Σ_1 be a nonsingular surface of genus p and Σ any other surface. Since $\mathfrak{D}(\Sigma_1)$ is identical with T_p, see §15, (C), it is a complex manifold. Since there always is an induced map $g_*: \mathfrak{D}(\Sigma_1) \to \mathfrak{D}(\Sigma)$ which is a universal covering of the image $g_*(\mathfrak{D}(\Sigma_1))$, see §15, (F), the set $g_*(\mathfrak{D}(\Sigma_1)) \subset \mathfrak{D}(\Sigma)$ has a natural structure of a complex manifold, and since $\mathfrak{D}(\Sigma) - g_*(\mathfrak{D}(\Sigma_1))$ is nowhere dense (cf. §15, (D)), $\mathfrak{D}(\Sigma)$ has a *ringed structure*: a continuous function on an open set $A \subset \mathfrak{D}(\Sigma)$ is called holomorphic if its restriction to $A \cap g_*(\mathfrak{D}(\Sigma_1))$ is.

Note that induced maps are holomorphic with respect to this ringed structure.

Actually $\mathfrak{D}(\Sigma)$ *is a complex manifold which can be realized as a bounded domain in* $\mathbf{C}^{3p-3}$.

We sketch the (lengthy) proof of this theorem below.

(B) First we consider a terminal surface Σ_0 with nodes $P_1, \ldots, P_{3p-3}$ and parts $\sigma_1, \ldots, \sigma_{2p-2}$.

We draw $2p - 2$ circular discs $\Delta_1, \ldots, \Delta_{2p-2}$ in $\mathbf{C}$ and choose on each boundary $\partial\Delta_j$ points $Q_{j^1}, Q_{j^2}, Q_{j^3}$ which follow each other in this order, in the counterclockwise direction. We denote by $\gamma_{j^\nu}, j = 1, \ldots, 2p - 2, \nu = 1, 2, 3$, parabolic Möbius transformations with fixed points at Q_{j^ν} such that

$$\gamma_{j^3} \circ \gamma_{j^2} \circ \gamma_{j^1} = \text{id.}$$

Then $\gamma_{j^1}, \gamma_{j^2}, \gamma_{j^3}$ generate a torsion-free Fuchsian group Γ_j with limit set $\partial\Delta_j$. Both Δ_j/Γ_j and $\bar\Delta_j/\Gamma_j$, where $\bar\Delta_j$ is the domain exterior to $\partial\Delta_j$, are triply punctured spheres.

We assume that the Δ_j are so far apart that there are fundamental domains F_j for Γ_j in $\bar\Delta_j$ such that the closure of the complement of F_j lies in the intersection of all $F_k, j \neq k$. Then, according to Klein's combination theorem, the group Γ generated by $\Gamma_1, \ldots, \Gamma_{2p-2}$ is their free product and Kleinian. It has precisely one invariant component A_0, not simply connected, while every noninvariant component is of the form $\gamma(\Delta_j)$ for some $j = 1, \ldots, 2p - 2$ and $\gamma \in \Gamma$, and has $\gamma^{-1}\Gamma_j\gamma$ as stabilizer. Also, A_0/Γ is a sphere with $6p - 6$ punctures.

For later use we denote by α_{j^ν} the Möbius transformation which maps Δ_j onto U taking Q_{j^ν} into ∞ and the other two points Q on $\partial\Delta_j$ into 0 and 1. Then $\alpha_{j^\nu}\Gamma_j\alpha_{j^\nu}^{-1}$ is the principal congruence subgroup modulo 2 of $PSL(2, \mathbf{Z})$, and

$$\alpha_{j^\nu} \circ \gamma_{j^\nu} \circ \alpha_{j^\nu}^{-1}(z) = z - 2.$$

Next we divide the $6p - 6$ points Q_{j^ν} into $3p - 3$ disjoint ordered pairs, each pair being assigned to a node P_i of Σ_0, in such a way that to a P_i which lies in the closures of σ_j and $\sigma_{j'}, j < j'$, belongs a pair $(Q_{j^\nu}, Q_{j'^\nu})$, and to a P_i which lies in the closure of one part σ_j only belongs a pair $(Q_{j^\nu}, Q_{j^{\nu'}}), \nu < \nu'$.

Note that if, for every ordered pair $(Q_{j'}, Q_{j''})$, we join the punctures of the Riemann surfaces $\Delta_{j'}/\Gamma_j$ and $\Delta_{j''}/\Gamma_j$ corresponding to $Q_{j'}$ and $Q_{j''}$, respectively, into a node, we obtain a Riemann surface with nodes S_0 homeomorphic to Σ_0.

We shall now thicken some of the $3p - 3$ nodes of S_0, in a way which depends on $3p - 3$ complex parameters.

(C) Let $\zeta_1, \ldots, \zeta_{3p-3}$ be complex numbers with $|\zeta_i| < 1$, and set

$$\delta_i(z) = z \quad \text{if } \zeta_i = 0,$$

$$\delta_i = \alpha_{j'}^{-1} \circ \beta_i \circ \alpha_{j''} \quad \text{if } \zeta_i \neq 0$$

where $(Q_{j'}, Q_{j''})$ belongs to P_i and

$$(16.1) \qquad \beta_i(z) = -z - \sqrt{-1} \log \zeta_i$$

where we use *some* determination of the logarithm. Note that for $\zeta_i \neq 0$ the Möbius transformation δ_i conjugates the cyclic group generated by $\gamma_{j''}$ into that generated by $\gamma_{j'}$.

We say that $(\zeta_1, \ldots, \zeta_{3p-3})$ belongs to $\mathfrak{D}'(\Sigma_0)$ if for every P_i with $\zeta_i \neq 0$ to which there is associated the pair $(Q_{j'}, Q_{j''})$ there are Jordan curves $K_{j'}, K_{j''}$ with the following properties: (a) $K_{j'}$ contains the point $Q_{j'}$ (and $K_{j''}$ contains the point $Q_{j''}$), lies otherwise in Δ_j (in $\Delta_{j'}$) and is invariant under $\gamma_{j'}$ (under $\gamma_{j''}$); (b) the transformation δ_i maps the domain exterior to $K_{j''}$ onto the domain interior to $K_{j'}$, and (c) the curves K lying in a fixed Δ_j are taken, by the canonical projection $\Delta_j \to \Delta_j/\Gamma_j$, into disjoint Jordan curves.

One can verify that the condition

$$(16.2) \qquad (\zeta_1, \ldots, \zeta_{3p-3}) \in \mathfrak{D}'(\Sigma_0)$$

does *not* depend on the choice of the discs Δ_j or the points $Q_{j'}$. Nor does it depend on the determination of the logarithms in (16.1). Finally, (16.2) always holds if all $|\zeta_i|$ are sufficiently small for then the curves K can be chosen as horocycles.

The group $\mathfrak{G} = \mathfrak{G}(\zeta_1, \ldots, \zeta_{3p-3})$ generated by γ and $\delta_1, \ldots, \delta_{3p-3}$ also does not depend on the determination of the logarithms.

(D) Under condition (16.2) the group $\mathfrak{G}$ is Kleinian, in view of Maskit's second combination theorem, cf. [**83, 85**].

Assume that this is so and let N denote the number of components of the space obtained from Σ_0 by removing all nodes P_i with $\zeta_i = 0$.

One can show that $\mathfrak{G}$ has precisely $N + 1$ nonconjugate components A_0, $A_1, \ldots, A_n$, none of which is invariant; the component A_0 is the domain bounded by the $2p - 2$ circles $\partial \Delta_j$. Let $\mathfrak{G}_t$ denote the stabilizer of A_t in $\mathfrak{G}$. Then $\mathfrak{G}_0 = \Gamma$ and for $t > 0$ each $\mathfrak{G}_t$ is a regular b-group and if (p_t, n_t) is the type of $A_t/\mathfrak{G}_t$, then $n_1 + \cdots + n_N = 2k$, where k is the number of zeros among $\zeta_1, \ldots, \zeta_{3p-3}$.

If we join any two punctures on the surfaces $A_1/\mathfrak{G}_1, \ldots, A_N/\mathfrak{G}_N$, which correspond to points $Q_{j'}$ and $Q_{j''}$ belonging to a P_i with $\zeta_i = 0$, into a node, we obtain a compact Riemann surface S of genus p with k nodes.

The curves K project into $3p - 3 - k$ disjoint Jordan curves on S. Together with the nodes they divide S into $2p - 2$ triply connected domains $s_1, \ldots, s_{2p-2}$ which are in a natural bijection with the parts $\sigma_1, \ldots, \sigma_{2p-2}$ of Σ_0. There is a deformation $f: S \to \Sigma_0$ such that $f(s_i) = \sigma_i$ for all i, and the point $[f: S \to \Sigma_0]$ in $\mathfrak{D}(\Sigma)$ depends only on the numbers $\zeta_1, \ldots, \zeta_{3p-3}$ (and not on the choice of the curves K).

We have thus defined a mapping $\mathfrak{D}'(\Sigma_0) \to \mathfrak{D}(\Sigma_0)$. It turns out that this is a *bijective holomorphic* (with respect to the ringed structure of $\mathfrak{D}(\Sigma_0)$) *homeomorphism*.

(E) We know now that $\mathfrak{D}(\Sigma_0)$, for a terminal Σ_0, is a bounded domain in $\mathbf{C}^{3p-3}$. For every node P_i of Σ_0 the set R_i of points $[f: S \to \Sigma_0]$ in $\mathfrak{D}(\Sigma_0)$ such that $f^{-1}(P_i)$ is a node is the intersection of $\mathfrak{D}(\Sigma_0)$ with the hyperplane $\zeta_j = 0$.

For every Σ there is a terminal Σ_0 and a subset I of $\{1, \ldots, 3p - 3\}$ such that there is a deformation $g: \Sigma \to \Sigma_0$ which induces a universal covering

$$g_*: \mathfrak{D}(\Sigma) \to \mathfrak{D}(\Sigma_0) - \bigcup_{i \in I} R_i.$$

We conclude that $\mathfrak{D}(\Sigma)$ *is a complex manifold* and that there are complex coordinates $Z_1, \ldots, Z_{3p-3}$ in $\mathfrak{D}(\Sigma)$ such that

$$Z_i = \zeta_i \quad \text{if } i \in I,$$
$$e^{2\pi\sqrt{-1}Z_i} = \zeta_i \quad \text{if } i \notin I.$$

Thus $\mathfrak{D}(\Sigma)$ is a domain in the product of $|I|$ unit discs and $3p - 3 - |I|$ half-planes. ($|I|$ is the cardinality of I.)

In particular, if Σ is nonsingular, $\mathfrak{D}(\Sigma) = T_p$, $|I| = 0$ and $Z_1, \ldots, Z_{3p-3}$ are the *Maskit coordinates* (cf. §13, (B)).

(F) It can be shown that for every Σ, $\mathfrak{D}(\Sigma)$ *is a domain of holomorphy*. Using this one can prove that for every $r = 1, 2, \ldots$, there is a trivial complex vector bundle over $\mathfrak{D}(\Sigma)$ such that the fiber over a point $[f: S \to \Sigma]$ is the space of regular r-differentials on S.

17. Moduli spaces.

(A) Let $\hat{X}_p$ denote the *space of moduli* (isomorphism classes) of compact Riemann surfaces with nodes, of genus p. This space can be topologized the same way as $\mathfrak{D}(\Sigma)$ was topologized in §15, (C). It follows easily that $X_p = X_{p,0}$ the space of moduli of nonsingular Riemann surfaces of genus p, is an open dense subset of $\hat{X}_p$. Since X_p is a normal complex space (cf. §6, (H)), $\hat{X}_p$ has a natural ringed structure.

Studying the natural maps of all deformation spaces of genus p into $\hat{X}_p$ one can show that $\hat{X}_p$ *is a normal complex space* (and a V-manifold).

(B) We now state a useful *à priori inequality*.

Let S be a compact Riemann surface of genus p with $k \geqslant 0$ nodes. *One can draw $3p - 3 - k$ disjoint* (Poincaré) *geodesic Jordan curves on the parts of S, none of which has length exceeding a constant α_p depending only on p.*

This was first conjectured by Mumford. The author's proof uses the

Keen-Halpern [**64, 92**] collar lemma. A different proof, and a numerical estimate for α_p, was recently given by Buser [**38**].

(C) Now let $\Sigma^{(1)}$, $\Sigma^{(2)}, \ldots, \Sigma^{(r)}$ be all topologically distinct terminal surfaces of genus p. The inequality in (B) implies that there are compact sets $F^{(j)} \subset \mathfrak{D}(\Sigma^{(j)})$ such that the images of $F^{(1)}, \ldots, F^{(r)}$ under the natural maps $\mathfrak{D}(\Sigma^{(j)}) \to \hat{X}_p$ cover $\hat{X}_p$. We conclude that $\hat{X}_p$ *is compact.*

That X_p can be compactified by adding to it moduli of algebraic curves of genus p with nodes, and that the resulting space $\hat{X}_p$ is a normal complex space has been established by methods of algebraic geometry. It is also known that $\hat{X}_p$ *is projective algebraic* (Mumford and others [**40, 94, 95**]); this result has not yet been obtained by methods described here.

(D) A different construction of $\hat{X}_p$ has been indicated by Earle and Marden [**52**] and by Abikoff [**3**]. These authors work with the standard embedding $T_p \subset B_2(L, G_0)$, cf. §6, and introduce a new topology in the *augmented Teichmüller space* $\hat{T}_p$ consisting of T_p and all its regular boundary points, cf. §10, (F). The modular group Mod_p acts continuously on all of $\hat{T}_p$ and $\hat{X}_p$ is obtained as $\hat{T}_p/\mathrm{Mod}_p$. ($\hat{T}_p$ is not a manifold.)

(E) On the other hand, Masur [**90**] proved that $\hat{X}_p$ can be obtained by completing X_p with respect to the metric induced in it by the Weil-Petersson metric in T_p, cf. §7.7.

APPENDIX

18. Some auxiliary results.

(A) Teichmüller spaces of Riemann surfaces which cannot be compactified (as Riemann surfaces) by adding finitely many points are infinite dimensional. We shall summarize the theory of such spaces in §19. Here we state some results from the theory of quasiconformal mappings and from functional analysis which will be needed later, and which are of independent interest.

(B) The following theorem is due to Ahlfors and Beurling [**36**].

An increasing continuous function $x \mapsto f(x)$ with $f(-\infty) = -\infty$ and $f(+\infty) = +\infty$ is the restriction to $\mathbf{R}$ of a quasiconformal self-mapping F of $\mathbf{C}$ if and only if it is *quasisymmetric*, i.e., if and only if there is a constant $M > 0$ such that

$$\frac{1}{M} \leqslant \frac{f(x + y) - f(x)}{f(x) - f(x - y)} \leqslant M.$$

Futhermore, $K(F)$ can be estimated in terms of M and vice versa.

One may always assume that $F(\bar{z}) = \overline{F(z)}$. Quasisymmetric functions are the restrictions to the real axis of continuous extensions of quasiconformal self-mappings of U which fix ∞.

(C) A *quasicircle* C is a Jordan curve in $\mathbf{C} \cup \{\infty\}$ which is the image of a circle under a quasiconformal self-mapping F of $\mathbf{C} \cup \{\infty\}$. The dilatation $K(C)$ is the infinum of $K(F)$ where F runs over all such maps.

Ahlfors [10] showed that a Jordan curve C containing the point ∞ is a quasicircle if and only if there is a constant M such that *for any 3 finite points a, b, c on C with b between a and c,*

$$\left| \frac{b - a}{c - a} \right| \leqslant M$$

and if and only if there is *Lipschitz continuous quasireflection about C,* i.e., a topological orientation reversing self-map Φ of $\mathbf{C}$, such that $\Phi \circ \Phi = \mathrm{id}$, $\Phi | C = \mathrm{id}$ and there is a constant M' such that

$$|\Phi(z_1) - \Phi(z_2)| \leqslant M'|z_1 - z_2|.$$

Furthermore $K(C)$ can be estimated in terms of M (or of M') and vice versa.

(D) Let D be a domain in $\mathbf{C} \cup \{\infty\}$ with at least two boundary points, $\lambda(z)|dz|$ the Poincaré metric in D, Γ a discrete group of conformal self-mappings of D, which may be the trivial group $1 = \{\mathrm{id}\}$, and $q > 1$ an integer. An automorphic form of weight $(-2q)$ for Γ in D is a holomorphic function $\varphi(z)$, $z \in D$, such that $\varphi(z) = O(|z|^{-2q})$, $z \to \infty$, if $\infty \in D$ and

$$\varphi(\gamma(z))\gamma'(z)^q = \varphi(z) \quad \text{for } \gamma \in \Gamma.$$

The automorphic forms φ with

$$\|\varphi\|_A = \int\!\!\int_{D/\Gamma} \lambda(z)^{2-q}|\varphi(z)|\, dx\, dy < \infty$$

form the Banach space $A_q(D, \Gamma)$ of *integrable forms*, those with

$$\|\varphi\|_B = \sup|\lambda(z)^{-q}\varphi(z)| < \infty$$

form the Banach space $B_q(D, \Gamma)$ of *bounded forms*. The main properties of these spaces are as follows [**23, 73**].

There is a bounded linear map Θ of $A_q(D, 1)$ onto $A_q(D, \Gamma)$ obtained by forming the *Poincaré theta series*

$$(\Theta\Phi)(z) = \sum_{\gamma \in \Gamma} \Phi(\gamma(z))\gamma'(z)^q,$$

and there is a pairing (*Petersson's scalar product*) between $A_q(D, \Gamma)$ and $B_q(D, \Gamma)$ given by

$$(\varphi, \psi)_{D,\Gamma} = \int\!\!\int_{D/\Gamma} \lambda(z)^{2-2q}\varphi(z)\,\overline{\psi(z)}\; dx\, dy$$

(for $\varphi \in A$, $\psi \in B$) which establishes a *topological bijection between $B_q(D, \Gamma)$ and $A_q(D, \Gamma)^*$.*

The spaces $A_q(D, \Gamma)$ and $B_q(D, \Gamma)$ are finite dimensional if and only if the group Γ is finitely generated and the Fuchsian group $h\Gamma h^{-1}$ (where h is a conformal map of the universal covering surface of D onto U) has every point of $\mathbf{R}$ as a limit point.

If so, $A_q(D, \Gamma) = B_q(D, \Gamma)$.

(E) While the above statments have nothing to do with quasiconformal mappings the proof of the following theorem [**24**] depends on Ahlfors' characterization of quasicircles.

Let C be an oriented Jordan curve in $\hat{\mathbf{C}} \cup \{\infty\}$, D_1 and D_2 the domain interior and exterior to C, respectively, $\lambda_j(z)|dz|$ the Poincaré metric in D_j, $j = 1, 2$, and $q > 1$ an integer. Let Γ be a discrete group of Möbius transformations with map D_1 (and D_2) onto itself. Set

$$(\mathcal{L}_C\varphi)(z) = \int\int_{D_1} \frac{\lambda_1(\zeta)^{2-2q}\overline{\varphi(\zeta)}\,d\xi\,d\eta}{(\zeta - z)^{2q}}, \qquad z \in D,$$

and let $\mathcal{L}_{-C}$ be defined similarly, with D_1 and D_2 interchanged. *If C is a quasicircle, $\mathcal{L}_C$ is a topological bijection of $A_q(D_1, \Gamma)$ onto $A_q(D_2, \Gamma)$ and of $B_q(D_1, \Gamma)$ onto $B_q(D_2, \Gamma)$.* Furthermore

$$(\mathcal{L}_C\varphi, \psi)_{D_2, \Gamma} = (\varphi, \mathcal{L}_{-C}\psi)_{D_1, \Gamma}$$

for $\varphi \in A_q(D_1, \Gamma)$ and $\psi \in B_q(D_2, \Gamma)$.

19. Infinite dimensional Teichmüller spaces.

(A) In the theory of finite dimensional Teichmüller spaces, summarized in §§2–13, quasiconformal mappings are a tool for proving theorems. In the theory of infinite dimensional Teichmüller spaces [21, 22, 27] quasiconformal mappings necessarily appear in the basic definitions.

Let S be an arbitrary Riemann surface (without nodes) which is not a sphere with 0, 1 or 2 punctures and not a torus and let $f\colon S \to S_0$ be a quasiconformal bijection onto another such surface. Then f extends to a topological bijection of $S \cup b(S)$, where $b(S)$ is the union of all ideal boundary curves of S, onto $S_0 \cup b(S_0)$. We shall denote this extension by the same letter f.

Two quasiconformal maps, $f_1\colon S_1 \to S$ and $f_2\colon S_2 \to S$ and called *weakly equivalent* if there is a commutative diagram

$$\begin{array}{ccc} S_1 & \xrightarrow{f_1} & S \\ \varphi\downarrow & & \downarrow\psi \\ S_2 & \xrightarrow[f_2]{} & S \end{array}$$

where φ is conformal and ψ homotopic to the identity. The maps f_1 and f_2 are called *equivalent* if ψ, or rather its continuous extension, can be continuously deformed into the identity leaving the image of every point in $b(S)$ fixed.

The equivalence classes

$$[f] = [f\colon S' \to S]$$

of quasiconformal maps are the points of the *Teichmüller space* $T(S)$, the weak equivalence classes

$$(f) = (f\colon S' \to S)$$

are the points of the *reduced Teichmüller space* $T^{\#}(S)$. There is an obvious canonical surjection

$$(19.1) \qquad\qquad T(S) \to T^{\#}(S)$$

which is the identity if $b(S) = \varnothing$; cf. §2, (A) and (F).

The *Teichmüller distance* between two points $[f_1]$ and $[f_2]$ in $T(S)$ is defined as

$$\tfrac{1}{2}\log \inf K(f), \qquad f \in \left[f_1^{-1} \circ f_2 \right],$$

where K denotes the dilatation. The Teichmüller distance between two points in $T^{\#}(S)$ can be defined similarly, or by using the map (19.1). Both spaces $T(S)$ and $T^{\#}(S)$ are *complete metric spaces*.

Every quasiconformal bijection $\omega\colon S_0 \to S_1$ induces *allowable mappings*

$$\omega_*\colon T(S_0) \to T(S_1), \qquad T^{\#}(S_0) \to T^{\#}(S_1)$$

which take $[f]$ into $[\omega \circ f]$ and (f) into $(\omega \circ f)$. These mappings are *isometries*.

An allowable self-mapping of $T(S)$ or of $T^{\#}(S)$ is called a *modular transformation*.

(B) It is clear that if S is compact, except for n punctures, and of genus p, then $T(S)$ can be identified with $T_{p,n}$, and if S is compact except for n punctures and m "holes", and of genus p, then $T^{\#}(S)$ can be identified with $T^{\#}_{p,n,m}$. It turns out that in all other cases $T(S)$ and $T^{\#}(S)$ are *infinite dimensional*.

The theory of $T(S)$ is richer than that of $T^{\#}(S)$. In particular, $T(S)$ has a natural complex structure, whereas $T^{\#}(S)$ has only a real analytic structure; $T^{\#}(U)$ is a point, whereas $T(U)$ is a domain in an infinite dimensional complex Banach space (and is, in some sense, the "universal" Teichmüller space, see below). Also, the theory of $T^{\#}(S)$ can be reduced to that of $T(\hat{S})$ where $\hat{S}$ is obtained from S by doubling, cf. §2, (F).

From now on we shall be concerned with $T(S)$. It is convenient to extend the definitions further.

(C) Let Q denote the group of all conformal self-mappings of the upper half-plane U. Every $\omega \in Q$ extends to a homeomorphism of $U \cup \mathbf{R} \cup \{\infty\}$ which we denote by the same letter. We denote by Q_n the subgroup of those $\omega \in Q$ which fix 0, 1 and ∞, and by Q_0 the normal subgroup of those which leave every $x \in \mathbf{R}$ fixed. We may also consider $PSL(2, \mathbf{R})$ as a subgroup of Q. If $G \subset PSL(2, \mathbf{R})$ is discrete, we denote by $Q(G)$ the set of those $\omega \in Q$ for which

$$wGw^{-1} \subset PSL(2, \mathbf{R})$$

and we set $Q_n(G) = Q(G) \cap Q_n$.

The *Teichmüller space* $T(G)$ is defined as the image of $Q_n(G)$ under the canonical map

$$Q_n \to Q_n / Q_0.$$

If $\omega \in Q(G)$, ω induces an *allowable mapping* $T(G) \to T(\omega^{-1}G\omega)$ which takes the point $[w] \in T(G)$ defined by $w \in Q_n(G)$ into $[\alpha \circ w \circ \omega]$ where α is an element of $PSL(2, \mathbf{R})$ chosen so that $\alpha \circ w \circ \omega \in Q_n$. Allowable self-mappings of $T(G)$ are called *modular transformations*, they are induced by elements of the normalizer of G in Q.

(D) The group Q_n is a complete metric space (but *not* a topological group) under the Teichmüller metric

$$\langle w, \hat{w} \rangle = \tfrac{1}{2} \log K(\hat{w} \circ w^{-1})$$

(where K is the dilatation). This metric defines a Teichmüller metric δ_G in $T(G)$ via the canonical mapping $Q_n(G) \to T(G)$.

The Teichmüller spaces $T(G)$ are *complete* and allowable mappings are *isometries*.

(E) For $G = 1$ the space $T(G) = T(1)$ is the group Q_n/Q_0, that is, the group of orientation-preserving quasisymmetric (cf. §18, (B)) self-maps of **R** which fix 0 and 1.

By our definitions

$$T(G) \subset T(G_1) \quad \text{if } G_1 \subset G.$$

In particular, since $\{\mathrm{id}\} = 1 \subset G$, we have that $T(G) \subset T(1)$ for all Fuchsian groups G. For this reason $T(1)$ is called the *universal Teichmüller space*.

On every $T(G)$ the Teichmüller metric δ_1 and δ_G are topologically equivalent and $\delta_1 \leqslant \delta_G$. Recently Strebel [111] showed that, in general, $\delta_1 < \delta_G$.

(F) The connection between the Teichmüller spaces defined in §2 and those defined above is given by an isomorphism theorem below.

For every Fuchsian group $G \subset PSL(2, \mathbf{R})$ let U_G denote the complement in U of the fixed points of elliptic elements of G, so that U_G/G is a Riemann surface (and $U_G = U$ if G is torsion free).

Every conformal bijection $h\colon U_G/G \to S$ induces a homeomorphism of $T(G)$ onto $T(S)$. These homeomorphisms respect Teichmüller distances and allowable mappings.

In particular, $T(U)$ may be identified with $T(1)$.

If G has no torsion, the theorem follows directly from the definitions. Given G, $S = h(U/G)$ and $w \in Q_n(G)$, set $S' = U/(wGw^{-1})$ and let F be the (quasiconformal) map which makes the diagram

$$
\begin{array}{ccc}
U & \overset{w}{\to} & U \\
\downarrow & & \downarrow \\
\end{array}
$$
$$
S \overset{h}{\leftarrow} U/G \overset{F}{\to} U/(wGw^{-1}) = S'
$$

commutative (where vertical arrows denote canonical projections). Then the desired homeomorphism takes $[w] \in T(G)$ into $[f] \in T(S)$ where $f = h \circ F^{-1}$. The original proofs of the isomorphism theorem for the general case (by Marden [82] and by Bers-Greenberg [34]) were rather complicated, but Kra [74] observed that a short proof can be based on Teichmüller's theorem.

REMARK. One can also define reduced Teichmüller spaces $T^{\#}(G)$ and prove for them a corresponding isomorphism theorem.

(G) We can now repeat the construction in §6 and embed $T(G)$ into a Banach space, namely into the space $B_2(L, G)$ where L is the lower half-plane.

If $\tau = [w] \in T(G)$, there is a quasiconformal homeomorphism $\hat{w}$ of $C \cup \{\infty\}$ onto itself such that $\hat{w} \circ w^{-1}|U$ and $\hat{w}|L$ are conformal, and $\hat{w}$ keeps 0, 1, ∞ fixed. The Schwarzian derivative φ^τ of $\hat{w}|L$ depends only on $\tau = [w]$ and not on the particular choice of w, and φ^τ lies in a ball of radius $\frac{3}{2}$ in $B_2(L, G)$. The mapping

$$(19.1) \qquad T(G) \ni \tau \mapsto \varphi^\tau \in B_2(L, G)$$

is a homeomorphism onto a domain.

The proof of the latter statement is somewhat involved. It was first given by Ahlfors [10] for the case $G = 1$. The author's proof [21] used Ahlfors' theorems stated in §18, (C) to derive the theorem in §18, (E) and from it the desired result, for all G. Later Earle [41] found a simple way to go from $G = 1$ to an arbitrary G.

The embedding theorem gives a complex structure to $T(G)$ and, in view of the isomorphism theorem of (F), also to $T(S)$. It turns out that all allowable mappings are holomorphic.

(H) We may identify $T(G)$ with its image in $B_2(L, G)$ under (19.1). It has been shown that $T(G)$ is holomorphically convex for every G.

Also

$$T(G) = T(1) \cap B_2(L, G).$$

This is almost evident if $\dim T(G) < \infty$, due to Kra if G is finitely generated, and has been recently proved by Tukia [119] in the general case.

(I) Some results in §§9 and 10 can be extended to the general case. We omit the details and refer the reader to the literature.

(J) We conclude by some remarks about the connection between the universal Teichmüller space $T(1)$ and *univalent functions*.

Let S denote the set of Schwarzian derivatives of univalent functions $f(z)$, $z \in L$. Then S is a bounded closed set in $B = B_2(L, 1)$ and $T(1)$ may be described as the set of those $\varphi \in S$ for which $f(L)$ is a quasidisc (i.e., a domain bounded by a quasicircle, cf. §18, (C)).

The boundary $\partial T(1)$ of $T(1)$ in B lies in S. Gehring [58, 59] proved that $T(1)$ *is the interior of* S, but that *the complement of* $T(1) \cup \partial T(1)$ *in* S *is not empty*.

It seems interesting to determine which simply connected domains D have the property that the Schwarzian derivatives of the conformal maps $L \to D$ belong to $\partial T(1)$. It is also interesting to describe precisely the set $\partial S - \partial T(1)$. At present we do not know whether $S \subset B$ is connected.

REFERENCES

1. W. Abikoff, *Some remarks on Kleinian groups*, Advances in the Theory of Riemann Surfaces, Ann. of Math. Studies, No. 66, Princeton Univ. Press, Princeton, N. J., 1971, pp. 1–5.

2. ______, *On boundaries of Teichmüller spaces and on Kleinian groups*. III, Acta Math. **134** (1975), 212–237.

3. ______, *Augmented Teichmüller spaces*, Bull. Amer. Math. Soc. **82** (1976), 333–334.

4. ______, *Degenerating families of Riemann surfaces*, Ann. of Math. (2) **105** (1977), 29–44.

5. ______, *Topics in the real analytic theory of Teichmüller space*, Lecture Notes in Math., vol. 820, Springer-Verlag, Berlin and New York, 1980.

6. L. V. Ahlfors, *On quasiconformal mappings*, J. d'Analyse Math. **3** (1953/54), 1–58.

7. ______, *The complex analytic structure of the space of closed Riemann surfaces*, R. Nevanlinna et al., Analytic Functions, Princeton Univ. Press, Princeton, N. J., 1960, pp. 45–66.

8. ______, *Curvature properties of Teichmüller's space*, J. d'Analyse Math. **9** (1961), 161–176.

9. ______, *Some remarks on Teichmüller's space of Riemann surfaces*, Ann. of Math. (2) **74** (1961), 171–191.

10. ______, *Quasiconformal reflections*, Acta Math. **109** (1963), 291–301.

11. ______, *Finitely generated Kleinian groups*, Amer. J. Math. **86** (1964), 413–423; **87** (1965), 759.

12. ______, *Fundamental polyhedrons and limit sets of Kleinian groups*, Proc. Nat. Acad. Sci. U.S.A. **55** (1966), 251–254.

13. ______, *Lectures on quasiconformal mappings*, Van Nostrand, New York, 1966.

14. ______, *Quasiconformal mappings, Teichmüller spaces, and Kleinian groups*, Proc. Internat. Congr. Math. (Helsinki 1978), pp. 71–84.

15. L. V. Ahlfors and L. Bers, *Riemann's mapping theorem for variable metrics*, Ann. of Math. (2) **72** (1960), 385–404.

16. W. L. Baily, Jr., *On the theory of θ-functions, the moduli of abelian varieties, and the moduli of curves*, Ann. of Math. (2) **75** (1962), 342–381.

17. L. Bers, *Quasiconformal mappings and Teichmüller's theorem*, R. Nevanlinna et. al., Analytic Functions, Princeton Univ. Press, Princeton, N. J., 1960, pp. 89–119.

18. ______, *Simultaneous uniformization*, Bull. Amer. Math. Soc. **66** (1960), 94–97.

19. ______, *Correction to Spaces of Riemann surfaces as bounded domains*, Bull. Amer. Math. Soc. **67** (1961), 465–466.

20. ______, *Uniformization by Beltrami equations*, Comm. Pure. Appl. Math. **14** (1961), 215–228.

21. ______, *On moduli of Riemann surfaces*, Lecture notes, E.T.H., Zurich, 1964.

22. ______, *Automorphic forms and general Teichmüller spaces*, Proc. Conf. Complex Analysis (Minneapolis, 1964) pp. 109–113, Springer-Verlag, Berlin and New York, 1965.

23. ______, *Automorphic forms and Poincaré series for infinitely generated Fuchsian groups*, Amer. J. Math. **87** (1965), 196–214.

24. ______, *A non-standard integral equation with applications to quasiconformal mappings*, Acta Math. **116** (1966), 113–134.

25. ______, *On boundaries of Teichmüller spaces and on Kleinian groups*. I, Ann. of Math. (2) **91** (1970), 570–600.

26. ______, *Uniformization, moduli and Kleinian groups*, Bull. London Math. Soc. **4** (1972), 257–300.

27. ______, *Fiber spaces over Teichmüller spaces*, Acta Math. **130** (1973), 89–126.

28. ______, *On spaces of Riemann surfaces with nodes*, Bull. Amer. Math. Soc. **80** (1974), 1219–1222.

29. ______, *Deformations and moduli of Riemann surfaces with nodes and signatures*, Math. Scand. **36** (1975), 12–16.

30. ______, *Quasiconformal mappings, with applications to differential equations, function theory and topology*, Bull. Amer. Math. Soc. **83** (1977), 1083–1100.

31. ______, *An extremal problem for quasiconformal mappings and a theorem by Thurston*, Acta Math. **141** (1978), 73–98.

32. ______, *The action of the modular group on the complex boundary*, Riemann Surfaces and Related Topics: Proceedings of the 1978 Stony Brook Conference, Ann. of Math. Studies, No. 97 1981, pp. 33–52.

33. L. Bers and L. Ehrenpreis, *Holomorphic convexity of Teichmüller spaces*, Bull. Amer. Math. Soc. **70** (1964), 761–764.

34. L. Bers and L. Greenberg, *Isomorphisms between Teichmüller spaces*, Advances in the theory of Riemann surfaces, Ann. of Math. Studies, No. 66, 1971, pp. 53–79.

35. L. Bers and I. Kra, (eds.), *A crash course on Kleinian groups*, Lecture Notes in Math., vol. 400, Springer-Verlag, Berlin and New York, 1974.

36. A. Beurling and L. V. Ahlfors, *The boundary correspondence under quasiconformal mappings*, Acta Math. **96** (1956), 125–142.

37. R. Bowen, *Hausdorff dimension of quasicircles*, Inst. Hautes Études Sci. Publ. Math. **50** (1979), 153–170.

38. P. Buser, *Riemannsche Flächen und Längenspektrum vom trigonometrischen Standpunkt aus Habilitationsschrift*, Universität Bonn, 1980.

39. T. Chu, *The Weil-Petersson metric in the moduli space*, Chinese J. Math. **4** (1976), 29–51.

40. P. Deligne and D. Mumford, *The irreducibility of the space of curves of a given genus*, Inst. Hautes Études Sci. Publ. Math. **36** (1969), 75–109.

41. C. J. Earle, *The Teichmüller space of an arbitrary Fuchsian group*, Bull. Amer. Math. Soc. **70** (1964), 699–701.

42. ______, *Teichmüller spaces of groups of the second kind*, Acta Math. **112** (1964), 91–97.

43. ______, *Reduced Teichmüller spaces*, Trans. Amer. Math. Soc. **126** (1967), 54–63.

44. ______, *On holomorphic families of pointed Riemann surfaces*, Bull. Amer. Math. Soc. **79** (1973), 163–166.

45. ______, *On the Carathéodory metric in Teichmüller spaces*, Discontinuous Groups and Riemann Surfaces, Ann. of Math. Studies, No. 79, Princeton Univ. Press, Princeton, N. J., 1974, pp. 99–103.

46. ______, *The Teichmüller distance is differentiable*, Duke Math. J. **44** (1977), 389–397.

47. ______, *Families of Riemann surfaces and their Jacobi varieties*, Ann. of Math. **107** (1978), 255–286.

48. C. J. Earle and J. Eells, *On the differential geometry of Teichmüller spaces*, J. Analyse Math. **19** (1967), 35–52.

49. ______, *A fibre bundle description of Teichmüller theory*, J. Differential Geom. **3** (1969), 19–43.

50. C. J. Earle and I. Kra, *On holomorphic mappings between Teichmüller spaces*, Contributions to Analysis, Academic Press, New York, 1974, pp. 107–124.

51. ______, *On sections of some holomorphic families of closed Riemann surfaces*, Acta Math. **137** (1976), 49–79.

52. C. J. Earle and A. Marden (to appear).

53. M. Engber, *Teichmüller spaces and representability of functors*, Trans. Amer. Math. Soc. **201** (1975), 213–226.

54. A. Fathi, F. Laudenbach and V. Poénaru, *Travaux de Thurston sur les surfaces*, Astérisque **66–67** (1979), 1–284.

55. L. Ford, *Automorphic functions* (second ed.), Chelsea, New York, 1951.

56. R. Fricke and F. Klein, *Vorlesungen über die Theorie der automorphen Funktionen* (two volumes), B. G. Teubner, 1889 and 1926.

57. F. P. Gardiner, *On relative Teichmüller spaces and a globalization principle in Riemann surface theory*, J. d'Analyse Math. **35** (1979), 1–12.

58. F. W. Gehring, *Univalent functions and the Schwarzian derivative*, Comment. Math. Helv. **52** (1977), 561–572.

59. ______, *Spirals and the universal Teichmüller space*, Acta Math. **141** (1978), 99–113.

60. J. Gilman, *On the Nielsen type and the classification for the mapping-class group*, Advances in Math. **40** (1981), 68–96.

61. L. Greenberg, *Fundamental polyhedra for Kleinian groups*, Ann. of Math. **84** (1966), 433–441.

62. A. Grothendieck, *Techniques de construction en géometrie analytique*, Sém. Cartan (1961/62) exp. 17, (with an appendix by J. P. Serre).

63. K. T. Hahn, *On the completeness of the Bergman metric and its subordinate metrics*, Proc. Nat. Acad. Sci. U.S.A. **73** (1976), 42–44.

64. N. Halpern, *Some contributions to the theory of Riemann surfaces*, Thesis, Columbia University, 1978.

65. R. S. Hamilton, *Extremal quasiconformal mappings with prescribed boundary values*, Trans. Amer. Math. Soc. **138** (1969), 399–406.

66. W. J. Harvey, (ed.), *Discrete groups and automorphic functions*, Academic Press, New York, 1977.

67. J. H. Hubbard, *Sur les sections analytique de la courbe universelle de Teichmüller*, Mem. Amer. Math. Soc. No. 166, 1976, pp. 1–137.

68. J. Hubbard and H. Masur, *Quadratic differentials and foliations*, Acta Math. **142** (1979), 221–274.

69. J. A. Jenkins, *On the existence of certain general extremal metrics*, Ann. of Math. **66** (1957), 440–453.

70. L. Keen, *On Fricke moduli*, Advances in the Theory of Riemann Surfaces, Ann. of Math. Studies, No. 66, 1971, pp. 205–224.

71. S. P. Kerckhoff, *The asymptotic geometry of Teichmüller space*, Topology **19** (1980), 23–41.

72. S. Kobayashi, *Hyperbolic manifolds and holomorphic mappings*, M. Dekker, New York, 1970.

73. I. Kra, *Automorphic forms and Kleinian groups*, Benjamin, New York, 1972.

74. ______, *On new kinds of Teichmüller spaces*, Israel J. Math. **16** (1973), 237–257.

75. ______, *On the Nielsen-Thurston-Bers type of some self-maps of Riemann surfaces*, Acta Math. 146 (1981), 231–270.

76. I. Kra and B. Maskit, *Involutions on Kleinian groups*, Bull. Amer. Math. Soc. **78** (1972), 801–805.

77. W. Kraus, *Über den Zusammenhang einiger Charakteristiken eines eintach zusammehangenden Bereiches mit der Kreisabbildung*, Mitt. Math. Sem. Giessen 21 (1932), 1–28.

78. S. Kravetz, *On the geometry of Teichmüller spaces and the structure of their modular groups*, Ann. Acad. Sci. Fenn. **278** (1959), 1–35.

79. S. L. Krushkal, *On Teichmüller's theorem on extremal quasiconformal mappings*, Mat. Sb. **8** (1967), 313–332.(in Russian)

80. ______, *Quasiconformal mappings and Riemann surfaces*, V. H. Winston and Sons, 1979.

81. O. Lehto and K. I. Virtanan, *Quasiconformal mappings in the plane*, Springer-Verlag, Berlin, 1973.

82. A. Marden, *On homotopic mappings of Riemann surfaces*, Ann. of Math. **90** (1969), 1–8.

83. B. Maskit, *On Klein's combination theorem. II*, Trans. Amer. Math. Soc. **131** (1968), 32–39.

84. ______, *On boundaries of Teichmüller spaces and on Kleinian groups*. II, Ann. of Math. **91** (1970), 607–639.

85. ______, *On Klein's combination theorem*. III, Advances in the Theory of Riemann Surfaces, Ann. of Math. Studies, No. 66, 1971, pp. 297–316.

86. ______, *Moduli of marked Riemann surfaces*, Bull. Amer. Math. Soc. **80** (1974), 773–777.

87. ______, *On the classification of Kleinian groups*. I. *Koebe groups*, Acta. Math. **135** (1975), 249–270.

88. ______, *On the classification of Kleinian groups*. II. *Signatures*, Acta. Math. **138** (1977), 17–42.

89. H. Masur, *On a class of geodesics in Teichmüller space*, Ann. of Math. **102** (1975), 205–221.

90. ______, *The extension of the Weil-Petersson metric to the boundary of Teichmüller spaces*, Duke Math. J. **43** (1976), 623–635.

91. ______, *The Jenkins-Strebel differentials with one cylinder are dense*, Comment. Math. Helv. **54** (1979), 179–184.

92. J. P. Matelski, *A compactness theorem for Fuchsian groups of the second kind*, Duke Math. J. **43** (1976), 829–840.

93. R. T. Miller, *Nielsen's view point on geodesic laminations*, Advances in Math. (to appear).

94. D. Mumford, *The structure of the moduli spaces of curves and abelian varieties*, Proc. Intern. Congr. Math. (Nice 1970), vol. 1, pp. 457–465.

95. ______, *Curves and their Jacobians*, Univ. of Michigan Press, 1975.

96. Z. Nehari, *Schwarzian derivatives and schlicht functions*, Bull. Amer. Math. Soc. **55** (1949), 545–551.

97. B. O'Byrne, *On Finsler geometry and applications to Teichmüller spaces*, Advances in the Theory of Riemann surfaces, Ann. of Math. Studies, No. 66, 1971, pp. 317–328.

98. D. B. Patterson, *The Teichmüller spaces are distinct*, Proc. Amer. Math. Soc. **35** (1972), 179–182; **38** (1973), 668.

99. R. M. Porter, *Computation of a boundary point of Teichmüller space*, Bol. Soc. Mat. Mexicana **24** (1979), 15–26.

100. H. E. Rauch, *On the transcendental moduli of algebraic Riemann surfaces*, Proc. Nat. Acad. Sci. U.S.A. **41** (1955), 42–49.

101. ______, *A transcendental view of the space of algebraic Riemann surfaces*, Bull. Amer. Math. Soc. **71** (1965), 1–39.

102. E. Reich and K. Strebel, *Extremal quasiconformal mappings with given boundary values*, Bull. Amer. Math. Soc. **79** (1973), 488–490.

103. ______, *Extremal quasiconformal mappings with given boundary values*, Contributions to Analysis, Academic Press, New York, 1974, pp. 375–391.

104. H. L. Royden, *Report on the Teichmüller metric*, Proc. Nat. Acad. Sci. U.S.A. **65** (1970), 497–499.

105. ______, *Automorphisms and isometries of Teichmüller space*, Advances in the Theory of Rieman Surfaces, Ann. of Math. Studies, No. 66, 1971, pp. 369–383.

106. M. Schiffer, *A variational method for univalent quasi-conformal mappings*, Duke Math. J. **33** (1966), 395–411.

107. K. Strebel, *Zur Frage der Eindeutigkeit extremaler quasikonformer Abbildungen des Einheitskreises*, Comment. Math. Helv. **36** (1961–62), 306–343; **39** (1964), 77–89.

108. ______, *Über quadratische Differentiale mit geschlossenen Trajektorien und extremale quasikonforme Abbildungen*, Festband zum 70 Geburtstag von Rolf Nevanlinna, Springer, Berlin, 1966, pp. 105–127.

109. ______, *On quadratic differential and extremal quasiconformal mappings*, Proc. Int. Cong. Math. (Vancouver, 1974), pp. 223–227.

110. ______, *On quadratic differentials with closed trajectories on open Riemann surfaces*, Ann. Acad. Sci. Fennicae **2** (1976), 533–551.

111. ______, *On lifts of extremal quasiconformal mappings*, J. d'Analyse Math. **31** (1977), 191–203.

112. ______, *On quasiconformal mappings of open Riemann surfaces*, Comment. Math. Helv. **53** (1978), 301–321.

113. D. Sullivan, *On the ergodic theory of infinity of an arbitrary discrete group of hyperbolic motions*, Riemann Surfaces and Related Topics: Proceedings of the 1978 Stony Brook Conference, Ann. of Math. Studies, No. 97, 1980, pp. 465–496.

114. O. Teichmüller, *Extremale quasikonforme Abbildungen und quadratische Differentiale*, Preuss. Akad. **22** (1939).

115. ______, *Bestimmung der extremalen quasi konformen Abbildungen bei geschlossenen orientierten Riemannschen Flächen*, Preuss Akad. **4** (1943).

116. ______, *Veränderliche Riemannsche Flächen Deutsche Mathematik* **7** (1949), 344–359.

117. W. P. Thurston, *On the geometry and dynamics of diffeomorphisms of surfaces*. I, preprint.

118. ______, *The geometry and topology of three-manifolds*, Lecture Notes, Princeton University.

119. P. Tukia, *Quasiconformal extensions of quasisymmetric mappings compatible with a Fuchsian group* (to appear).

120. A. Weil, *Sur les modules des surfaces de Riemann*, Sém. Bourbaki, 1958.

121. S. Wolpert, *Noncompleteness of the Weil-Petersson metric for Teichmüller space*, Pacific J. Math. **61** (1975), 573–577.

DEPARTMENT OF MATHEMATICS, COLUMBIA UNIVERSITY, NEW YORK, NEW YORK 10027

Proceedings of Symposia in Pure Mathematics
Volume **39** (1983), Part 1

Poincaré and Lie Groups

WILFRIED SCHMID[1]

When I was invited to address this colloquium, the organizers suggested that I talk on the Lie-theoretic aspects of Poincaré's work. I knew of the Poincaré-Birkhoff-Witt theorem, of course, but otherwise was unaware of any contributions that Poincaré might have made to the general theory of Lie groups, as opposed to the theory of discrete subgroups. It thus came as a surprise to me to find that he had written three long papers on the subject, in addition to several short notes. He evidently regarded it as one of the major mathematical developments of his time—the introductions to his papers contain some flowery praise for Lie— but one probably would not do Poincaré an injustice by saying that in this one area, at least, he was not one of the main innovators. Still, his papers are intriguing for the glimpse they give of the early stages of Lie theory. Perhaps this makes a conference on the work of Poincaré an appropriate occasion for some reflections on the origins of the theory of Lie groups.

Sophus Lie (1842–1899) developed his theory of *finite continuous transformation groups*, as he called them, in the years 1874–1893, in a series of papers and three monographs. To Lie, a transformation group is a family of mappings

$$(1a) \qquad\qquad y = f(x, a),$$

where x, the independent variable, ranges over a region in a real or complex Euclidean space; for each fixed a, the identity (1a) describes an invertible map; the collection of parameters a also varies over a region in some $\mathbf{R}^n$ or $\mathbf{C}^n$; and f, as function of both x and a, is real or complex analytic. Most importantly, the family is closed under composition: for two values a, b of the parameter, the composition of the corresponding maps belongs again to the family, i.e.,

$$(1b) \qquad\qquad f(f(x, a), b) = f(x, c),$$

with

$$(1c) \qquad\qquad c = \varphi(a, b)$$

depending analytically on a and b, but not on x. It must be noted that these identities are only required to hold locally; in present-day terminology, (1a–c) define the germ of an analytic group action.

Reprinted from Bulletin Amer. Math. Soc. (N.S.) 6 (1982), 175–186.

1980 *Mathematics Subject Classification.* Primary 22E15.

[1]Supported in part by NSF grant MCS-79-13190.

Differentiating (1a) with respect to the coordinates a_i of the parameter, Lie constructs vector fields

$$(2) \qquad X_i(F) = \sum_j \xi_{ij}(x) \frac{\partial F}{\partial x_j}$$

(Lie's notation), which he calls the *infinitesimal transformation* of the family. That is how he pictures them and, on occasion, calculates with them. The foundations of Lie's theory are embodied is his *three fundamental theorems*. The first consists of a differential equation, involving the X_i, which is equivalent to the group property—an infinitesimal version of (1b). This turns out to be more delicate than one might expect since Lie, initially at least, does not insist on the existence of an identity transformation, or of inverse transformations, within the family (1). According to the second fundamental theorem, the linear span of the X_i is closed under the Lie bracket,

$$(3) \qquad [X_i, X_j] = \sum c_{ijk} X_k.$$

Conversely, any Lie algebra (Hermann Weyl's terminology!) of vector fields generates a group in Lie's sense. His arguments are those that one would use today: the commutators in the Lie algebra correspond infinitesimally to commutators in the group, which leads to the identities (3). On the other hand, the one parameter groups generated by a collection of vector fields $X_1, \ldots, X_n$ fit together as a family, (locally) closed under composition, precisely when the X_i span a Lie algebra.

The third fundamental theorem, finally, states that any set of structural constants $\{c_{ijk}\}$, subject to the obvious necessary conditions, arises from some Lie algebra of vector fields, and hence determines a transformation group. In other words, every (finite dimensional, real or complex) Lie algebra can be realized as a Lie algebra of vector fields. Lie proves the theorem by producing a Lie algebra of functions, with respect to the Poisson bracket, which he obtains as solutions of a system of differential equations. Some years earlier, Lie had published a shorter argument: the given structure constants $\{c_{ijk}\}$ determine vector fields

$$(4) \qquad X_i = \sum c_{jik} x_j \frac{\partial}{\partial x_k},$$

and these form a Lie algebra, provided the c_{ijk} satisfy the appropriate conditions. What amounts to the same, Lie constructs the adjoint group of the group whose existence he wants to establish. If the group in question has a center of positive dimension, it is not locally isomorphic to its adjoint group, and this argument breaks down—a possibility which Lie overlooked at the time.

In Lie's development of the theory, the idea of a *group action* is of primary interest, and the group itself is relegated to a supporting role. However, one can easily recover the group itself in Lie's framework: the composition rule for the parameter (1c), $c = \varphi(a, b)$, in which a may be viewed as the variable and b as the parameter, or vice versa, is a transformation group in the sense of Lie, a group which acts (locally) simply transitively. Lie calls it the first or second parameter group, depending on whether a or b is regarded as the variable. Now one would say that the group acts on itself by left and right translation. To Lie, with the algebraic notion of a group so very far in the background, it was not obvious that the two actions commute; in fact, he credits Engel with this observation.

Friedrich Engel (1861–1941) had been a student of Felix Klein in Leipzig, and in 1884 was sent by Klein to his friend Lie in Norway, where Engel wrote his *Habilitationsschrift*. Two years later, when Lie succeeded Klein in Leipzig, Engel accompanied him. Although Engel eventually became quite active on his own, at first he seems to have limited himself mainly to functioning as Lie's sounding board and selfless secretary—an arrangement with present-day parallels. The foreword to the first volume of *Theorie der Transformationsgruppen* describes Engel's contributions as primarily linguistic, but nonetheless valuable because Lie, in his own words, did not "master any of the major languages completely". Later, in his introduction to the third volume, Lie's credits to Engel become more generous. One may well suspect that Engel influenced his teacher to a greater extent than the latter's acknowledgements suggest: loose definitions and careless mistakes occur frequently in Lie's papers before 1884, but not thereafter.

To put my brief account of Lie's three fundamental theorems into perspective, I ought to remark that the foundations of the theory of continuous groups represent only a small part of his work. Lie saw his theory as a powerful tool, with far-reaching applications to the integration theory of differential equations and to the most basic problems of geometry. He pursued these applications tirelessly, in numerous publications.

Around 1980, Friedrich Schur[2] (1856–1932) published two papers, in which he presented an alternate approach to the foundations of Lie's theory. His point of departure is the observation, first made by Lie, that there is a canonical choice of parameters a in (1a), namely the one for which the straight lines $t \to ta$ correspond to one parameter subgroups. As one would say now, Schur parametrizes a neighborhood of the identity in the group by a neighborhood of the origin in the Lie algebra, via the exponential map. In terms of such canonical coordinates, the compositon rule (1c) also assumes a canonical form: φ can be expressed as a convergent power series, whose coefficients depend polynomially on the structure constants c_{ijk}, but which are otherwise universal—the Campbell-Hausdorff formula in disguise. Schur's power series makes sense and converges near the origin whenever the c_{ijk} satisfy the obvious conditions, i.e., skew symmetry and the Jacobi identity. In particular, this gives a new proof of Lie's third fundamental theorem, one that is much more direct and, incidentally, almost simultaneous with Lie's.

A similar procedure works for any (locally) transitive transformation group. After a linear coordinate change, some of the canonical coordinates become canonical coordinates for the isotopy subgroup at a given point, and the others coordinates for the space on which the group acts. Schur's

---

[2]Not related to Issai Schur, at least not directly.

arguments do not use the analytic nature of the transformation group; two continuous derivatives are enough. It follows that any transitive, C^2 transformation group can be made analytic by means of a suitable coordinate change, a fact which had previously been asserted by Lie, without proof, and without a specific bound on the number of derivatives. This, of course, is the origin of Hilbert's fifth problem.

It is instructive to compare Schur's mathematical style to that of Lie. Schur had been a student of Weierstraß in Berlin, and was strongly influenced by Weierstraß' insistence on rigor and logical completeness. One can almost sense his discomfort with Lie's intuitive reasoning: not once does he refer to the "infinitesimal transformations", although he uses the same letters as Lie for their coefficient functions. One of Schur's papers begins with a detailed proof of the differentiability of solutions of differential equations, as functions of the initial conditions. Elsewhere he carefully estimates the radius of convergence of a power series in several variables. To Lie, such arguments must have appeared overly complicated, and even pedantic. As Engel reports in Schur's obituary, Lie and Schur had very different ideas of what was easy and what was not.

Also around 1890, Wilhelm Killing (1847–1923) wrote a series of five papers in which he established, or came close to establishing, many of the basic structure theorems about complex Lie algebras: the existence of a Levi decomposition of Lie algebras that coincide with their own derived algebras, criteria for semisimplicity—Killing, in fact, coined the term "semisimple"—and most remarkably, the classification of simple Lie algebras. Killing had been led to the classification problem by geometric considerations, to a large extent independently of Lie's work,[3] but in these five papers he generally follows Lie's terminology and notation. Killing's most important tool is the notion of a *root*, i.e., root of the characteristic equation

$$(5) \qquad\qquad \det(\operatorname{ad} X - \omega) = 0,$$

which Killing writes in terms of coordinates and structure constants, of course. Here X is an element of the Lie algebra $\mathfrak{g}$, and $\operatorname{ad} X$ the infinitesimal inner automorphism corresponding to X,

$$(6) \qquad\qquad \operatorname{ad} X(Y) = [X, Y].$$

Lie had already considered the equation (5) when he proved that every $X \in \mathfrak{g}$ lies in a two-dimensional subalgebra, but it was Killing who first studied the root pattern and recognized it as the key to the structure of a Lie algebra.

Now, ninety years later, one can only marvel at Killing's work, especially his list of the exceptional simple Lie algebras, their dimensions and root systems, all discovered during the infancy of the subject. The exposition is flawed, however, by serious gaps and errors, and is often obscure. No wonder his contemporaries remained skeptical, until Élie Cartan, in his thesis, put the results on a solid footing.

---

[3]Cf. Hawkins' article on the origin of Killing's work.

Killing's severest critic was Lie, perhaps not only for mathematical reasons, but also because—hypersensitive, as always in his priority disputes—he felt slighted by Killing's references to his own work. Lie made a habit of reviewing the work of others on what he called, with proprietary undertones, "my theory of groups". One of the pleasures of going back to the early papers in Lie theory is to read these astute, but acerbic reviews. About certain sections of one of Killing's papers, Lie writes " . . . the correct theorems in them are due to Lie, the false ones due to Killing", and in a sweeping damnation of several papers by Killing, " . . . (they) contain not so many results that are correct and new. Proved, correct and new are even fewer". In spite of such harsh language, Lie does acknowledge the great value of Killing's results on the structure of Lie algebras. Among other targets of Lie's criticism, Schur and Maurer are reprimanded for not following the notation which Lie has so carefully chosen, which makes it difficult to see what is really new in their writings. Not even Felix Klein, his friend in earlier days, is spared. So much has been written by Klein's students and friends about the relationship between Lie and Klein, says Lie, that he feels compelled to set the record straight: "I am not a student of Klein, nor is Klein a student of mine, although the latter might come closer to the truth". He then goes on to berate Klein for various offenses.

Before turning to Poincaré, I should mention two short articles of J. E. Campbell (1862–1924), written in 1897, about products of exponentials of non-commuting operators. The opening paragraph of the second neatly describes their point of view: "If x and y are operators which obey the ordinary laws of algebra, we know that $e^y e^x = e^{y+x}$. I propose to investigate the corresponding theorem when the operators obey the distributive and associative laws, but not the commutative". The idea of exponentiating a vector field, or "infinitesimal transformation", to a "finite transformation" already appears in the work of Lie, who considers expressions like

$$f + X(f) + \frac{1}{2!} X^2(f) + \cdots,$$

but does not use the exponential formalism for this purpose. As Campbell observes, the product of exponentials of two "small" vector fields X, Y is itself the exponential of a vector field Z:

$$(7) \qquad\qquad e^X e^Y = e^Z;$$

this follows from the proof of Lie's second fundamental theorem. By laborious calculations he then derives a formula for Z, in terms of X, Y, repeated bracket operations and certain universal coefficients. Although he mentions the word "convergence", his version of the identity (7) has no analytic content. He cites Schur's paper because the same universal coefficients occur there, but remains silent about the close connection between (7) and Schur's proof of the third fundamental theorem.

Except for a short note on the groups of units in hypercomplex systems,[4] Poincaré's essay "Sur les hypothèses fondamentales de la Géométrie" (1887)

---

[4] I.e., finite-dimensional associative algebras over **R**.

is his first publication to mention the theory of continuous transformation groups. After Lobachevsky's description of hyperbolic geometry, the problem of characterizing physical space by suitable axioms had become one of the grand themes of nineteenth century mathematics. Poincaré approaches this problem, in the case of two-dimensional space, with the observation that Euclidean, hyperbolic and elliptic geometry have one important feature in common: their groups of motion act transitively, with one-dimensional isotropy groups. Using Lie's infinitesimal methods, he classifies the two-dimensional homogeneous spaces of three-dimensional groups, up to local equivalence. It is then a relatively simple matter to distinguish among the several possible cases by various geometric properties. The essay is elegantly written, but as Lie points out, with uncharacteristically gentle words, Poincaré seems unaware of earlier investigations of a similar nature, in particular Lie's own classification of three-dimensional (local) group actions on the plane.

Perhaps it is not purely coincidental that Poincaré returned to the theory of Lie groups only after Lie's death, with a Comptes rendus announcement (1899) that outlines a proof of the third fundamental theorem. The details follow a few months later, in the form of a paper dedicated to Sir George Gabriel Stokes, on the occasion of his eightieth birthday. In the interval, Poincaré must have learned of Schur's proof and Campbell's notes: he cites both, remarks that his own results, which overlap theirs to a considerable extent, are not as original as he had thought, and expresses the hope that his arguments contain enough new ideas to merit publication.

The paper begins with a discussion of the exponential formalism for vector fields. Every "infinitesimal transformation" X in a continuous transformation group exponentiates to a one parameter subgroup

$$t \to e^{tX},$$

and these one parameter subgroups generate the group. Campbell's identity (7) thus makes it possible to reconstruct the group law from the bracket operation on the Lie algebra. The third fundamental theorem follows, provided Campbell's formal series is known to converge. This, Poincaré points out, is also the basic mechanism of Schur's proof.

To give concrete meaning to the identity (7), Poincaré in effect introduces the universal enveloping algebra $U(\mathfrak{g})$ of a Lie algebra $\mathfrak{g}$: it consists of all "symbolic polynomials", i.e., formal noncommutative polynomials in the generators of $\mathfrak{g}$, on which he imposes the identifications forced by the equalities

$$(8) \qquad\qquad XY - YX - [X, Y] = 0,$$

with $X, Y \in \mathfrak{g}$. Because of (8), a homogeneous nth degree polynomial is equivalent, as symbolic polynomial, to a *symmetric* homogeneous nth degree polynomial, plus a polynomial of lower degree. This procedure, repeated inductively, makes every symbolic polynomial equivalent to a symmetric polynomial. Less obviously, the symmetric representative is unique—that is the main substance of the Poincaré-Birkhoff-Witt theorem: in present-day terminology, the "symmetrization map"

$$S(\mathfrak{g}) \to U(\mathfrak{g})$$

from the symmetric algebra of $\mathfrak{g}$ to $U(\mathfrak{g})$ defines a linear isomorphism.[5]

Poincaré's proof of the uniqueness of the symmetric representative is complicated and leaves much unsaid. As an illustration of its main idea, let us consider a symmetric polynomial P, of degree three, which can be made equivalent to zero without raising its degree. Expressed in terms of a basis $\{X_1, \ldots, X_n\}$ of $\mathfrak{g}$, P takes the form

$$
\begin{aligned}
P = {} & \sum a_{ijk}(X_iX_j - X_jX_i - [X_i, X_j])X_k \\
& + \sum b_{ijk}X_i(X_jX_k - X_kX_j - [X_j, X_k]) \\
& + \sum c_{ij}(X_iX_j - X_jX_i - [X_i, X_j]).
\end{aligned}
\tag{9}
$$

Since P is symmetric, so are its homogeneous components P_3, P_2, P_1. On the other hand, the leading component

$$P_3 = \sum a_{ijk}(X_iX_jX_k - X_jX_iX_k) + \sum b_{ijk}(X_iX_jX_k - X_iX_kX_j)$$

vanishes when it is symmetrized; hence $P_3 = 0$. This can happen only if the six terms

$$
\begin{aligned}
&(X_iX_j - X_jX_i - [X_i, X_j])X_k, && -X_i(X_jX_k - X_kX_j - [X_j, X_k]), \\
&(X_jX_k - X_kX_j - [X_j, X_k])X_i, && -X_j(X_kX_i - X_iX_k - [X_k, X_i]), \\
&(X_kX_i - X_iX_k - [X_k, X_i])X_j, && -X_k(X_iX_j - X_jX_i - [X_i, X_j])
\end{aligned}
$$

contribute equally to the first two sums in (9). Because of the Jacobi identity, these add up to

$$
\begin{aligned}
& X_i[X_j, X_k] - [X_j, X_k]X_i - [X_i, [X_j, X_k]] \\
& + X_j[X_k, X_i] - [X_k, X_i]X_j - [X_j, [X_k, X_i]] \\
& + X_k[X_i, X_j] - [X_i, X_j]X_k - [X_k, [X_i, X_j]],
\end{aligned}
$$

which is quadratic, equivalent to zero, and thus can be absorbed by the third sum in (9). In other words, P is equivalent to zero already as a second degree polynomial. With considerable effort, Poincaré carries out the analogous argument for symmetric polynomials of arbitrary degree: if such a polynomial is equivalent to zero, its degree can be reduced by one; the theorem follows by induction.

The universal enveloping algebra and Poincaré's description of its structure were forgotten for almost forty years. In 1937, Garrett Birkhoff and Ernst Witt rediscovered Poincaré's result independently, in its most general version, needless to say—for possibly infinite-dimensional Lie algebras, over fields of

---

[5]Over ground fields of nonzero characteristic the theorem must be stated slightly differently.

arbitrary characteristic. Apparently it was Cartan-Eilenberg, in their book on homological algebra, who first affixed Poincaré's name to the theorem.

To Poincaré, Campbell's formula is an identity in the universal enveloping algebra. For $X, Y \in \mathfrak{g}$, the product $(X^m/m!)(Y^n/n!)$ can be expressed uniquely as a sum of homogeneous, symmetric polynomials in the generators of $\mathfrak{g}$, of degree $k \leqslant m + n$,

$$\frac{X^m}{m!} \frac{Y^n}{n!} = \sum_{k \leqslant m+n} Z_{m,n,k}.$$

Hence, in a formal sense,

$$(10) \qquad e^X e^Y = \sum_k Z_k,$$

where

$$(11) \qquad Z_k = \sum_{m,n} Z_{m,n,k}$$

is a formal power series in the variables X and Y, with values in the space of symmetric, homogeneous, kth degree polynomials. The first term

$$(12) \qquad Z_1 = \sum_{m,n} Z_{m,n,1}$$

plays a special role; for algebraic reasons, it must be a sum of repeated brackets in X and Y. Poincaré's version of Campbell's formula has three different aspects:

(a) the product (10) is an exponential series, which means that $Z_k = Z_1^k/k!$, for $k = 0, 1, 2, \ldots$;

(b) the series (12) converges for small X and Y;

(c) the series (12) can be written down explicitly.

For the purpose of proving Lie's third fundamental theorem, (c) is irrelevant, and Poincaré pays little attention to this problem, although the answer follows easily from his methods.

The main tool of his proof of (a) and (b) is a differentiated version of Campbell's formula; in symbolic notation,

$$(13) \qquad e^X e^{\delta Y} = e^{X+\delta X}, \quad \text{with } \delta Y = \frac{1 - e^{-\operatorname{ad} X}}{\operatorname{ad} X} \delta X$$

(cf. (6)). Poincaré gives two separate derivations of the identity (13), one straightforward, by direct calculation with power series, the second a rather curious argument: If $\mathfrak{g}$ is known to be the Lie algebra of a group, Campbell's formula and its infinitesimal analogue (13) follows from Lie's second fundamental theorem. The auxiliary Lie algebra $\tilde{\mathfrak{g}}$, with generators $\{X, X_1, \ldots, X_n\}$ and relations

$$(14) \qquad \begin{array}{ll} \text{(a)} & [X, X_i] = \sum c_{ij} X_j, \\[2mm] \text{(b)} & [X_i, X_j] = 0, \end{array}$$

is center-free for a generic choice of the constants c_{ij}, and hence is the Lie algebra of a linear group, by Lie's early proof of the third fundamental theorem. In $\tilde{\mathfrak{g}}$, then, the identity (13) holds—even if the c_{ij} fail to be generic, as can be shown by a simple degeneration argument. To get the same statement in $\mathfrak{g}$, Poincaré specializes $X_1, \ldots, X_n$ to a basis of $\mathfrak{g}$ and chooses the constants c_{ij} so that (14a) remains an equality. The second set of relations (14b) may be violated in $\mathfrak{g}$, of course, but the commutators which are not specified by (14a) only have a second order effect on Campbell's formula, and hence disappear from the differentiated formula (13)!

To interpret (13) as an analytic identity, Poincaré draws on the residue calculus. If Φ is a polynomial,

$$
\begin{aligned}
\Phi(\operatorname{ad} X) &= \frac{1}{2\pi i} \int (\xi - \operatorname{ad} X)^{-1} \Phi(\xi)\, d\xi \\
&= -\frac{1}{2\pi i} \int \operatorname{cf}(\operatorname{ad} X - \xi) F(\xi)^{-1} \Phi(\xi)\, d\xi,
\end{aligned}
$$

(15)

with $F(\xi) = \det(\operatorname{ad} X - \xi)$ and $\operatorname{cf}(\ldots) =$ cofactor matrix of $(\ldots)$; the integration extends over a circle, centered at the origin, and large enough to enclose the roots $\xi_1, \ldots, \xi_N$ of F. The same formula applies to any holomorphic function Φ, whose Taylor series

$$
\Phi(\xi) = a_0 + a_1 \xi + a_2 \xi^2 + \cdots
$$

has radius of convergence greater than

$$
R_X = \max(|\xi_1|, \ldots, |\xi_N|),
$$

because Φ can then be approximated by polynomials on a suitable circle of integration. In this situation, the formal sum

$$
\Phi(\operatorname{ad} X) = a_0 1 + a_1 \operatorname{ad} X + a_2(\operatorname{ad} X)^2 + \cdots
$$

converges and equals the integral (15). If Φ has no zeroes on the closed disc of radius R_X, the reciprocal series $\Phi^{-1}(\operatorname{ad} X)$ also converges, to the inverse of the operator $\Phi(\operatorname{ad} X)$,

$$
\Phi(\operatorname{ad} X)^{-1} = \Phi^{-1}(\operatorname{ad} X). \tag{16}
$$

In particular, both (13) and the inverted form of that identity,

$$
\delta X = \frac{\operatorname{ad} X}{1 - e^{-\operatorname{ad} X}} \delta Y, \tag{17}
$$

have a definite analytic meaning, the former for all X, the latter whenever the eigenvalues of $\operatorname{ad} X$ lie inside the disc of radius 2π.

As the final step of the proof, Poincaré recovers Campbell's formula from its differentiated version by integration. Because of (13, 17), the relation

$$
e^X e^{tY} = e^{Z(t)}
$$

is equivalent to the differential equation

$$
Z'(t) = \frac{\operatorname{ad} Z(t)}{1 - e^{-\operatorname{ad} Z(t)}} Y, \tag{18}
$$

with initial condition $Z(0) = X$. Any formal solution of (18) converges near $t = 0$, and hence on the interval $[0, 1]$ if only X and Y are small enough; $Z = Z(1)$ satisfies Campbell's formula (7) both formally and analytically.

I want to emphasize precisely what Poincaré proves. The identity (7) is first of all a formal identity, in the universal enveloping algebra of a specific Lie algebra $\mathfrak{g}$, which determines Z as a series in X, Y, and their repeated brackets; the Poincaré-Birkhoff-Witt theorem ensures the uniqueness of Z. Secondly, this formal series converges near $X = Y = 0$, to an analytic function

$$(19) \qquad\qquad Z = \varphi(X, Y).$$

Both statements involve the Lie algebra structure, but make no reference to a group with Lie algebra $\mathfrak{g}$. Finally, the function (19) defines a simply transitive group in Lie's sense—the group property follows easily from the associativity of the multiplication in $U(\mathfrak{g})$.

The residue formula plays a more prominent role in Poincaré's arguments than my summary might suggest. The choice of method is characteristic: to Poincaré, the third fundamental theorem amounts to an application of the residue calculus, to Lie, it is the solution of a partial differential equation, and to Schur, under the influence of Weierstraß, an explicit power series in several variables!

As for the later history of the Campbell-Hausdorff formula, Baker published a new proof in 1905, very much in the spirit of Campbell's original proof, but with a more elaborate formalism taking the place of lengthy calculations. Baker saw the connection between the formula and Schur's proof of the third fundamental theorem, but he does not mention Poincaré's work on the subject. Hausdorff, a year later, said everything about the Campbell-Hausdorff formula that needs to be said. His precision and generality satisfy even Bourbaki—who faults Campbell, Poincaré and Baker for being vague. Like Campbell and Baker, Hausdorff focusses on the algebraic aspects of the formula, but he also establishes the convergence of the formal series by quoting Poincaré's argument. He recounts the previous history fully and accurately; in particular, he characterizes Poincaré's version as the specialization of his own universal identity to the enveloping algebra of a specific Lie algebra.

In 1901 and 1908 Poincaré published his last two papers on Lie groups: long, rambling discussions of such topics as the exponential and logarithm maps, the adjoint group of Lie algebra, and the relationship between a group and its adjoint group. It was Lie who first introduced the adjoint group and gave it its name. The statement that the adjoint group is the homomorphic image of any group with the same Lie algebra amounted to little more than a tautology for Lie, since he defined the notion of homomorphism purely in terms of the structure constants. Poincaré, on the other hand, displays the adjoint homomorphism as a map, with algebraic properties, by writing down formulas like

$$(20) \qquad e^{-X}Ye^{X} = -\frac{1}{2\pi i}\int e^{-\xi}F(\xi)^{-1}\,\mathrm{cf}(\mathrm{ad}\,X - \xi)Y\,d\xi.$$

It must be borne in mind that Poincaré adopts Lie's definition of continuous group; the natural domain of the adjoint homomorphism is the *parameter group*: the set of symbols $\{e^{X} | X \in \mathfrak{G}\}$, endowed with the (local) composition rule (19).

The linear transformations $Y \mapsto e^{-X}Ye^{X}$, which generate the adjoint group, visibly preserve the Lie algebra structure of $\mathfrak{g}$. From this observation Poincaré deduces certain results of Killing about the root space decomposition. For example, if Y_1, $Y_2 \in \mathfrak{g}$ correspond to roots ω_1, ω_2 of X, the bracket $[Y_1, Y_2]$ belongs to the root $\omega_1 + \omega_2$, or vanishes if $\omega_1 + \omega_2$ fails to be a root —not a deep fact, but less evident before the algebraic notion of the adjoint homomorphism became common currency.

As a complex analyst, Poincaré is tempted to approach various global questions from the point of view of analytic continuation. The formula (20) exhibits the adjoint homomorphism as a holomorphic map; if $\mathfrak{g}$ is center-free, this map has an inverse near $X = 0$ (the logarithm map of the adjoint group, in effect), which can be analytically continued. Similarly, the composition rule (19) is defined initially near $X = Y = 0$, but extends to a multiple-valued map: the identity $e^{Z} = e^{X}e^{Y}$ can be solved locally for Z in terms of X and Y, provided exp has maximal rank at Z; φ is well behaved near such points, and ramifies at others. The roots of Killing's equation (5), finally, are multiple-valued functions on the Lie algebra. Poincaré studies the analytic continuations of these three types of functions in great detail, which quickly leads him to consider the center of the group—a vexing problem at a time when the notion of a global Lie group had not been defined.

The discussion of the center and of the multiple-valued composition rule make Poincaré's papers on Lie groups frustrating for a present-day reader. Bourbaki, in his historical notes, dismisses them as hastily written; he criticizes Poincaré for asserting in some places that the exponential map is surjective, and elsewhere giving counterexamples. However, the apparent contradiction can be resolved. When Lie talks of a "group with structure constants c_{ijk}", he means a transformation group; he refers to the underlying (local) abstract group as the parameter group. Poincaré accepts this convention in principle, but in practice considers only the parameter group, the adjoint group, and sometimes also other linear realizations. Unlike Lie, he regards the parameter group as a global object, which is canonically attached to the Lie algebra; indeed, the parameter group really *is* the Lie algebra, equipped with the multiple-valued composition rule (19). The parameter group plays the role that one assigns now to the universal covering group: it maps homomorphically to any (transformation) group with the same Lie algebra—just another way of saying that it is the underlying abstract group of such a transformation group. In this setting, homomorphisms may have singularities and need not be globally defined. Thus it can happen that the exponential map of a linear group fails to be surjective, even though its underlying abstract group, or parameter group, has a surjective exponential map as a matter of definition.

One of Poincaré's examples is especially revealing. He considers two infinitesimal rotations X, Y in the group $SO(3)$, about axes l_X, l_Y in general position, with X normalized so that e^{X} represents a full rotation, through an angle 2π. In $SO(3)$, e^{X} equals the identity, but not in the parameter group. The product $e^{X}e^{Y}e^{-X}$ is a rotation about the axis $e^{X}l_Y = l_Y$, through the same angle as e^{Y}, hence

$$e^{X}e^{Y}e^{-X} = e^{Y}.$$

On the other hand, the two rotations $e^Y e^X e^{-Y}$ and e^X have unequal axes in general, which means that

$$e^Y e^X e^{-Y} \neq e^X,$$

as elements of the parameter group. A paradox, but not a contradiction: the group laws are only required to hold locally, after all.

A satisfactory explanation became possible when the concepts of global Lie group and universal covering group were introduced. Both of these had to await the definition of a manifold, which appeared implicitly, at least, in Hermann Weyl's *Die Idee der Riemannschen Fläche* in 1913, a year after Poincaré's death.

REFERENCES

H. F. Baker, *Alternants and continuous groups*, Proc. London Math. Soc. 3 (1905), 24–47.

G. Birkhoff, *Representability of Lie algebras and Lie groups by matrices*, Ann. of Math. (2) **38** (1937), 526–532.

N. Bourbaki, *Groupes et algèbres de Lie*, Chapitres 2 et 3, Hermann, Paris, 1972.

J. E. Campbell, *On a law of combination of operators bearing on the theory of continuous transformation groups*. I, II, Proc. London Math. Soc. (1) **28** (1897), 381–390; **29** (1898), 14–32.

F. Engel, *Wilhelm Killing*, Jahresber. der Deutsch. Math. Verein. **39** (1930), 140–154.

______, *Friedrich Schur*, ibid. **45** (1935), 1–31.

F. Hausdorff, *Die symbolische Exponentialformel in der Gruppentheorie*, Berichte der köngl. Sächsischen Gesellschaft der Wiss. zu Leipzig (Math.-Phys. Klasse) **63** (1906), 19–48.

T. Hawkins, *Non-Euclidean geometry and Weierstrassian mathematics: the background to Killing's work on Lie algebras*, Historia Math. 7 (1980), 289–342.

W. Killing, *Die Zusammensetzung der stetigen endlichen Transformationsgruppen*. I–IV, Math. Ann. (1888), 252–290; **33** (1889), 1–48; **34** (1889), 57–122; **36** (1890), 161–189.

______, *Bestimmung der größten Untergruppen von endlichen Transformationsgruppen*, Math. Ann. **36** (1890), 239–254.

S. Lie, *Theorie der Transformationsgruppen*. I–III, unter Mitwirkung von F. Engel, Teubner, Leipzig, 1888–1893.

______, *Gesammelte Abhandlungen*. I–VII, Teubner, Leipzig, 1934-1960.

H. Poincaré, *Sur les nombres complexes*, Comptes rendus de l'Acad. des Sciences, Paris **99** (1884), 740–742.

______, *Sur les hypothèses fondamentales de la Géométrie*, Bulletin de la Société math. de France **15** (1887), 203–216.

______, *Sur les groupes continus*, Comptes rendus de l'Acad. des Sciences **128** (1899), 1065–1069.

______, *Sur les groupes continus*, Cambridge Philos. Trans. **18** (1899), 220–255.

______, *Quelques remarques sur les groupes continus*, Rend. Circ. Mat. Palermo **15** (1901), 321–368.

______, *Nouvelles remarques sur les groupes continus*, ibid. **25** (1908), 81–113.

______, *Analyse de travaux scientifiques de Henri Poincaré faite par lui-même*, Acta Math. **38** (1921), 2–135.

E. Witt, *Treue Darstellung Liescher Ringe*, J. Reine Angew. Math. **177** (1937), 152–160.

DEPARTMENT OF MATHEMATICS, HARVARD UNIVERSITY, CAMBRIDGE, MASSACHUSETTS 02138

Proceedings of Symposia in Pure Mathematics
Volume **39** (1983), Part 1

Discrete Conformal Groups and Measurable Dynamics

DENNIS SULLIVAN

Motivated by his study of 2nd order differential equations $a(z)w'' + b(z)w' + c(z)w = 0$ Poincaré (1882) unveiled the vast subject of *discrete* subgroups of conformal transformations, $\{z \to (az + b)/(cz + d)\}$, their associated Riemann surfaces, and the intricacies of the limit set—Cantor sets, nowhere differentiable curves, etc.

In this paper we discuss an interplay between discrete conformal groups acting on any dimensional ball B^{d+1} and measurable dynamics. The development is closely related to ideas considered by Poincaré, for example,

(i) the *Poincaré series* $\sum_{\gamma \in \Gamma} |\gamma' x_0|^s$ where $|\gamma' x|$ is the linear distortion of the Euclidean metric by the conformal transformation γ and x_0 lies in the interior of B^{d+1}.

(ii) The interpretation of interior B^{d+1} with its group of conformal transformations as the *Poincaré model of non-Euclidean or hyperbolic geometry*, $\mathbf{H}^{d+1}$.

The dynamics we consider take place on the d-sphere $= \partial B^{d+1} =$ the visual sphere at ∞ for $\mathbf{H}^{d+1}$. There are basically three parts to the discussion.

Part I. The ergodic properties of the geodesic flow on the associated hyperbolic manifold $\mathbf{H}^{d+1}/\Gamma$ such as (i) the excursion pattern of random geodesics into the cuspidal ends of finite volume noncompact hyperbolic manifolds.

(ii) the ergodicity of the geodesic flow relative to conformal measures and the divergence of the Poincaré series at the *critical exponent* $s = \delta(\Gamma)$.

Part II. The ergodic properties of Γ acting on the tangent spaces to the Poincaré recurrent part of S^d.

Part III. An interrelation between the critical exponent $\delta(\Gamma)$ and other quantities such as (i) the *Hausdorff dimension* of the limit set $\Lambda \subset S^d$, (ii) the "square root" of the *lowest eigenvalue* of the Laplacian acting on $L^2(\mathbf{H}^{d+1}/\Gamma)$, (iii) the *entropy* of the geodesic flow, (iv) general Riemannian manifolds.

We will state theorems and give ideas and references for the proofs. The discussion of Part I works for the usual Lebesgue measure on S^d as well as for any "conformal measure" μ on S^d, that is a finite positive measure μ which satisfies $\gamma^* \mu = |\gamma'|^\delta \mu$, $\gamma \in \Gamma$. These conformal measures are also important in Part III. They sometimes turn out to be Hausdorff measures, or they can be viewed as the boundary values at ∞ of positive eigenfunctions of the Laplacian with the lowest possible eigenvalue, or they determine invariant measures for the geodesic flow with an interesting value for the entropy.

Reprinted from Bulletin Amer. Math. Soc. (N.S.) 6 (1982), 57–73.

1980 *Mathematics Subject Classification.* Primary 32H20, 58F11, 58F17; Secondary 58F18, 10F05.

The discussion of Part II is restricted to Lebesgue measure or to conformal measures which share certain spatial distribution properties with Lebesgue measure. One consequence of the theory in Part II is that relative to Lebesgue measure (or any of these good conformal measures) any two ergodic components of the action on S^d of discrete groups of conformal transformations are *orbit equivalent*. Another consequence is a rather complete extension and analysis of the Mostow rigidity and the Ahlfors-Bers [**B**] deformation theory for infinite volume groups. These results are used by Thurston [**T**] in his existence theory for hyperbolic structures on topological 3-manifolds.

PART I

1. Finite volume groups and diophantine approximation. We begin by recalling the classical ergodicity results for discrete groups Γ of motions of $\mathbf{H}^{d+1}$ with finite volume fundamental domains. To these classical results we add a new result about the excursion pattern of geodesics into the cuspidal ends of a noncompact finite volume hyperbolic manifold.

The discrete group Γ acts on the geodesic lines in $\mathbf{H}^{d+1}$ and each orbit of lines determines one geodesic in the quotient, $V = \mathbf{H}^{d+1}/\Gamma$. These geodesics fill the unit tangent bundle $T(V)$ of V giving a 1-dimensional foliation whose leaves are the orbits of the *geodesic flow*. Thus in particular orbits of the geodesic flow for the quotient $\mathbf{H}^{d+1}/\Gamma$ are in the canonical 1-1 correspondence with Γ orbits of unequal ordered pairs of points on the sphere at infinity S^d of $\mathbf{H}^{d+1}$.

The geodesics may be grouped into asymptotic classes as $t \to \infty$ and as $t \to -\infty$. These classes are the leaves of two transversal foliations which intersect in the 1-dimensional geodesic foliation.

In the middle thirties, Eberhard Hopf used these asymptotic foliations to prove the geodesic flow is ergodic if $\mathbf{H}^{d+1}/\Gamma$ has finite volume. The proof also makes use of Poincaré's famous result that a finite invariant measure forces almost all geodesics to be recurrent. These foliations have been used up to the present to prove the geodesic flow is mixing, structurally stable, Bernoulli, etc. (Hedlund, Anosov, Smale, Ornstein, . . .).

Returning to our discussion, one knows [**T**, p. 8.20] that a finite volume hyperbolic manifold $V^{d+1} = \mathbf{H}^{d+1}/\Gamma$ consists of a compact part to which cuspidal ends homeomorphic to (d-torus) $\times [0, \infty)$ are attached. By the ergodicity of the geodesic flow, the random geodesic wanders continually further and further out each of these ends and returns infinitely often.

We can measure these excursions by the function dist $v(t)$, the distance from the point $v(t)$ in V achieved after time t to a fixed compact starting region of initial vectors $\{v\}$. It is easy to see that the event $\{v: \text{dist } v(t) > T\}$ has volume comparable to e^{-dT} using the volume-preserving character of the geodesic flow and the exponential decay of the size of the torus in the cuspidal end. An elementary use of the Borel-Cantelli lemma shows then for each $\varepsilon > 0$ and for almost all starting directions v the inequality

$$\text{dist } v(t) < (1/d + \varepsilon)\log(t)$$

is eventually true. One is led to conjecture $(1/d)\log t$ is the right upper bound function.

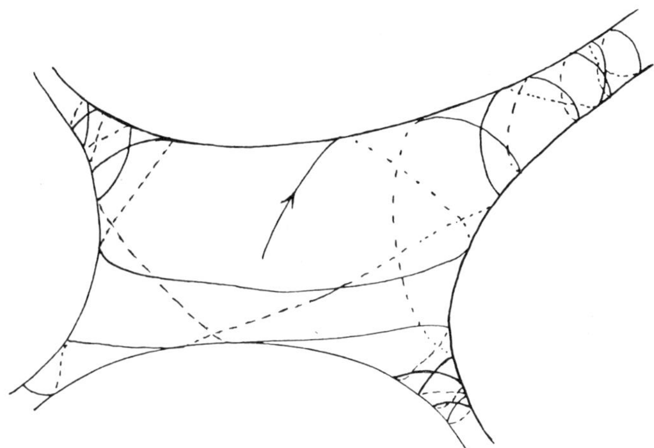

THEOREM 1. *If Γ is any cofinite volume, but not cocompact, discrete group of non-euclidean motions in $\mathbf{H}^{d+1}$ then, for almost all starting directions v in $\mathbf{H}^{d+1}/\Gamma$,*

$$\limsup_{t \to \infty} \frac{\operatorname{dist} v(t)}{\log t} = \frac{1}{d}.$$

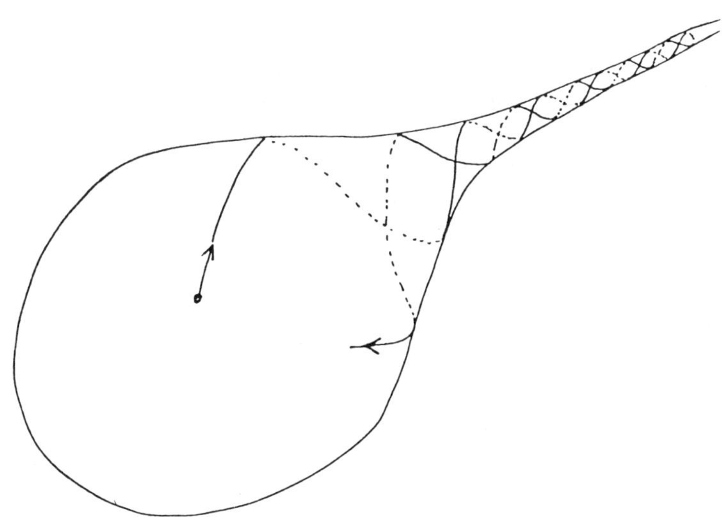

FIGURE 1

This theorem is proved in [S(i)] using a proposition about disjoint spheres resting on a Euclidean hyperplane implying an associated sequence of sets in the hyperplane are sufficiently independent in the sense of probability to have the full Borel-Cantelli lemma—divergence of the sum of measures implies the infinitely often event has positive measure.

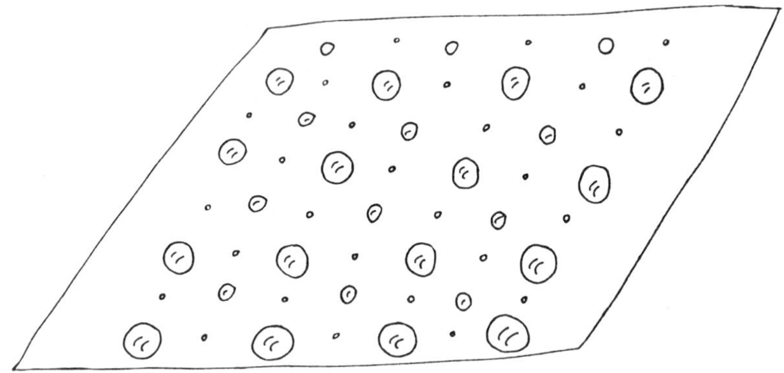

FIGURE 2

To use this proposition there must be enough spheres of each size. In the proof, this abundance of spheres comes from the *mixing* property of the geodesic flow.

The problem of geodesic excursion turns out to be equivalent (an oral communication of David Kahzdan) to the problem of approximating a general point on the sphere at infinity by the orbit of a parabolic cusp. Thus the proof above yields a discrete group—ergodic proof of Khintchine's metric theory of diophantine approximation. For this application of Theorem 1 we take Γ to be $PSl(2, Z)$.

If we consider the analogous approximation discussion for complex numbers and imaginary quadratic fields $Q(\sqrt{-d}\,)$ and let Γ run through the Bianchi groups $PSl(2, \vartheta\sqrt{-d}\,)$ we obtain just as easily a Khintchine theorem in this context. The statement proved in [S(i)] follows.

Let $0 < a(x) \leqslant 1$ be a function so that the value up to bounded ratio only depends on the argument up to bounded ratio. For the positive integer d denote by ϑ the ring of integers in $Q\sqrt{-d}\,$.

THEOREM 2. *For almost all complex numbers z there are infinitely many pairs* $p, \ q \in \vartheta \times \vartheta \ $ *so that* $\ \mathrm{ideal}(p, q) = \vartheta \ $ *and* $\ |z - p/q| < a(|q|)/|q|^2 \ $ *iff* $\int^{\infty}(a(x)^2/x)\,dx = \infty.$

For the Gaussian field, LeVeque (1952) proved a variant of this along the lines of the Khintchine real result.

2. Groups with infinite volume fundamental domains. As suggested already, E. Hopf's foliation proof of ergodicity of the geodesic flow is based on the Poincaré recurrence of the geodesic flow which follows in the finite volume case from the Poincaré recurrence theorem. In fact, E. Hopf proved (1939) the geodesic flow on an infinite volume hyperbolic manifold was either

(i) completely nonrecurrent. Namely, the random geodesic eventually leaves very compact set forever, or

(ii) the geodesic flow is completely recurrent and ergodic.

Hopf asked what geometric conditions might separate these two cases. In [S(ii)] we studied the ergodicity question independently of Hopf's 1939 paper. We found the same dichotomy for the geodesic flow, namely (i) completely nonrecurrent of (ii) ergodic by showing this dichotomy agrees with a dichotomy which exists for any Riemannian manifold [S(ii), §2]. Namely, random motion on a Riemannian manifold is either transient or recurrent.

In classical function theory this dichotomy for a Riemannian surface can be expressed in terms of the existence or nonexistence of a Green's function, a positive function with the standard Green's singularity at a point and harmonic outside the point.

Poincaré himself (1906) analyzed this dichotomy using the fuchsian group representation of a Riemann surface and reformulated the condition in terms of the convergence or divergence of a series (now called the Poincaré series),

$$\sum_{\gamma \in \Gamma} |\gamma'(z_0)| < \infty \text{ or not.}$$

Here Γ is a discrete group of conformal transformations of $\{z: |z| < 1\}$, $|z_0| < 1$, and $|\gamma'(z)|$ is the linear distortion of the Euclidean metric.

The Poincaré dichotomy in terms of the series makes sense for any discrete group of conformal transformations of the Euclidean ball B^{d+1}. The dichotomy becomes

$$\sum_{\gamma \in \Gamma} |\gamma'(x_0)|^d < \infty \text{ or not.}$$

THEOREM 3. *The dichotomy, $\Sigma_\Gamma |\gamma'(x_0)|^d < \infty$ or not, is equivalent to the Hopf dichotomy, the complete nonrecurrence or the ergodicity of the geodesic flow for $\mathbf{H}^{d+1}/\Gamma$.*

The proof in [S(ii)] makes use of random walks and was motivated by a proposition in Garnett [G] about the measurable functions on a compact foliated manifold harmonic in each leaf being constant on almost all leaves.

Thus if the series $\Sigma_{\gamma \in \Gamma} |\gamma'(x_0)|^d$ converges we have a completely noncurrent geodesic flow relative to Lebesgue measure on $\mathbf{H}^{d+1}/\Gamma$. One may try to restore ergodicity by replacing Lebesgue measure by a different geometrically meaningful measure.

To this purpose consider again the Poincaré series $\Sigma_{\gamma \in \Gamma} |\gamma'(x_0)|^s$ with an exponent s. One defines the *critical exponent* $\delta(\Gamma)$ so that the Poincaré series converges for $s > \delta(\Gamma)$ and diverges for $s < \delta(\Gamma)$. In [S(iii)] which generalized Patterson [P] it was shown there is a probability measure μ on the limit set $\Lambda(\Gamma) \subset S^d$ so that $\gamma^*\mu = |\gamma'|^\delta\mu$, $\gamma \in \Gamma$, where $\delta = \delta(\Gamma)$. Moreover, $\delta(\Gamma)$ is the minimum power for which such a measure exists. [S(iii), § 2].

Just as Lebesgue measure is associated to an invariant measure for the geodesic flow, such *conformal measures* μ determine invariant measures for the geodesic flow [**S(iii)**, §4]. Basically, the formula $\gamma^*\mu = |\gamma'|^\delta\mu$ and the formula $|\gamma x - \gamma y|^2 = |\gamma' x|\,|\gamma' y|\,|x - y|^2$ implies $\mu \times \mu/|x - y|^{2\delta}$ is a Γ-invariant measure on $S^d \times S^d$. This invariant measure is combined with arc length to yield a natural invariant measure for the geodesic flow of $\mathbf{H}^{d+1}/\Gamma$.

Theorem 3 above has the following generalization.

THEOREM 4. *The geodesic flow is ergodic relative to a conformal measure μ of dimension δ iff the Poincaré series $\sum_{\gamma \in \Gamma}|\gamma' x_0|^\delta$ diverges. In this divergence case the conformal measure μ satisfying $\gamma^*\mu = |\gamma'|^\delta\mu$, $\gamma \in \Gamma$, is unique.*

It is interesting for what follows to describe the evolution of the proof of Theorem 4. We have mentioned the proof of Theorem 3 using a random walk along the leaves of the asymptotic foliation in [**S(ii)**, §2] motivated by [**G**]. The idea behind that proof was simply that a random path in hyperbolic space hits a definite point at ∞. Thus there is a Γ-invariant map of bi-infinite random paths to geodesics by looking at the endpoints. Recurrence of Brownian motion implies the shift map on bi-infinite paths of $\mathbf{H}^{d+1}/\Gamma$ is ergodic and thus so also is its quotient, the geodesic flow.

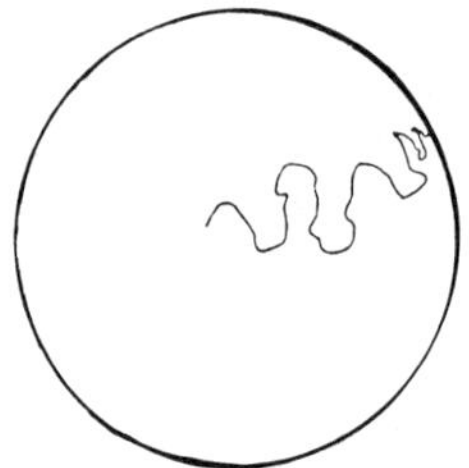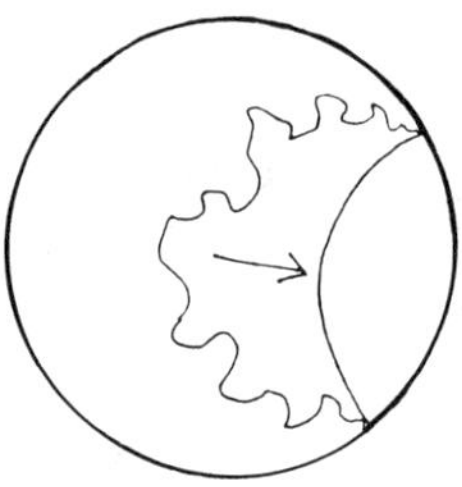

This proof of Theorem 3 was described in [**S(iii)**, §7] together with an extension to those cases of Theorem 4 where $\delta > d/2$. For the extension a conditioned Markov process with transition probabilities

$$e^{-\lambda t}\phi(y)p_t(x, y)/\phi(x)$$

was considered. Here $\lambda = \delta(\delta - d)$, ϕ is a positive λ-eigenfunction of the Laplacian with boundary values at infinity the conformal measure μ, and $p_t(x, y)$ is the transition probability for the usual Brownian motion on $\mathbf{H}^{d+1}/\Gamma$. (See Part III for more discussion of these ideas.)

This Markov process proof does not work for $\delta < d/2$. Recently, a more elementary proof of the case $\delta = d$ was found (related to joint work [**AS**] with Jon Aaronson) which succeeds just as well for all δ. First, E. Hopf's proof that there is a dichotomy between complete nonrecurrence and ergodicity works for any *conformal measure μ* [**S(iii)**, §4]. Second, recurrence can be checked by an abstract Borel-Cantelli lemma analogous to the one used in §1. This step requires an elementary geometric inequality which says the probability that a geodesic starting in a ball B returns to B at a later time t and at an even time $t + s$ is at most a universal constant times the product of the respective probabilities the geodesic returns respectively at times t and time s.

This inequality for the measure μ uses the transformation formula $\gamma^*\mu = |\gamma'|^\delta\mu$ to see that the μ shadow of a ball γB as viewed from B is on the order of $e^{-\delta x}$ where x is the hyperbolic distance from B to γB.

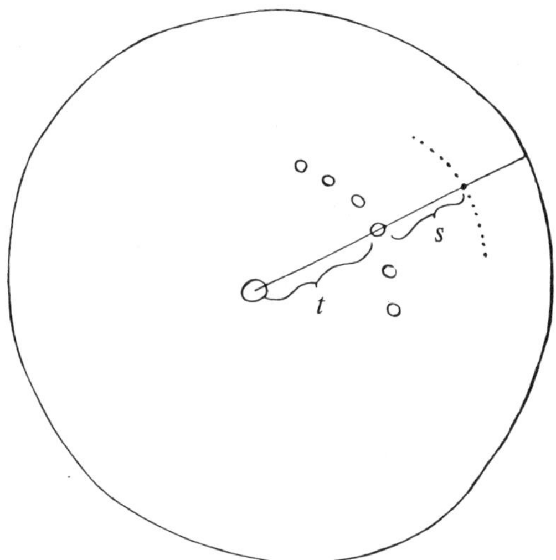

This Markov type inequality has another application in the context of Lebesgue measure, to yield a sharper form of ergodicity when the series $\Sigma|\gamma'x_0|^d$ diverges. This sharper ergodicity is *rational ergodicity* as explained in [A]. It implies in particular that the sojourn times in B up to time T have a pattern of divergence which is just that of the diverging Poincaré series [AS]. Up to now it is a *problem* to derive the rational ergodicity for conformal measures of dimension δ and thus the relation between sojourn times up to time T and the rate of divergence of $\Sigma|\gamma'x_0|^\delta$.

3. There is a relation between this ergodicity of geodesics or equivalently ergodicity of Γ acting on pairs in S^d and a purely measurable form of Mostow rigidity. Suppose Φ is a Borel map: $S^d \to S^d$ conjugating one action of a discrete conformal group Γ_1 to that of another Γ_2. Suppose Φ is nonsingular with respect to conformal measures μ_1 and μ_2 of dimension δ. That is, $\gamma^*\mu_1 = |\gamma'|^\delta\mu_1$ for $\gamma \in \Gamma_1$, $\gamma^*\mu_2 = |\gamma'|^\delta\mu_2$ for $\gamma \in \Gamma_2$ and $\mu_1(A) > 0$ iff $\mu_2(\Phi A) > 0$. Then

THEOREM 5. *If the Poincaré series for* Γ_1 *diverges at* δ, *the map* Φ *agrees* mod μ_1 *with a conformal conjugacy between* Γ_1 *and* Γ_2.

PROOF. Since the Poincaré series for Γ_1 diverges at δ, the measure $\mu_1 \times \mu_1/|x - y|^{2\delta}$ is ergodic for Γ_1 acting on $S^d \times S^d$ (Theorem 4). Since Φ is nonsingular, $\Phi \times \Phi$ is a measurable conjugacy between the two product actions on $S^d \times S^d$. Thus Γ_2 acts ergodically relative to $\mu_2 \times \mu_2/|x - y|^{2\delta}$ and $\Phi \times \Phi$ carries $\mu_1 \times \mu_1/|x - y|^{2\delta}$ to an invariant measure for Γ_2 which must be a constant times $\mu_2 \times \mu_2/|x - y|^{2\delta}$. It follows immediately for μ_1 almost all pairs x, y that $|\Phi x - \Phi y|/|x - y|$ is a product of a function of x

and one of y. Thus Φ preserves cross ratios

$$|x - y| \cdot |z - w|/|x - w| \cdot |y - z| \quad \text{for } \mu_1 \text{ almost all four-tuples.}$$

Renormalizing Φ so that for a generic w, w and $\Phi(w)$ are at infinity and working in R^d, the new Φ preserves $|x - y|/|y - z|$ for μ_1 almost all triples. Cyclically permuting x, y, z shows Φ now takes μ_1 almost all triangles to similar triangles. Renormalizing further by a similarity Φ preserves one triangle. Now it follows Φ preserves distances for almost all pairs. Thus Φ agrees mod μ_1 with an isometry. Returning to the original situation we have a conformal conjugacy between Γ_1 and Γ_2. This proves Theorem 5.

Application of Part I. For a compact Riemann surface S of genus g there is a $6g - 6$ parameter (Teichmüller) space of discrete groups Γ acting on $\mathbf{H}^2$ so that $\mathbf{H}^2/\Gamma$ is homeomorphic to S. All these groups act ergodically on $S^1 \times S^1$ so Theorem 6 implies

COROLLARY 1. *The $6g - 6$ parameter space injects into the abstract ergodic theory of actions of $\Gamma = \pi_1 S$ on the circle. If two of these $\pi_1 S$ actions on S^1 are measurably conjugate (preserving sets of positive measure) the corresponding Riemann surfaces are conformally equivalent.*

Note. In this compact surface case a very different proof by Marina Ratner [**Ra**] has shown a surprising related result—*if two classical horocycle flows are measure-theoretically conjugate the underlying Riemann surfaces are homeomorphic, even conformally isomorphic.*

For general geometrically finite groups Theorem 5 says the *conformal type* is *determined by* the *critical exponent* (which is also *Hausdorff dimension* of the limit set, see Part III) and the *measure-theoretic conjugacy class of the* Γ *action on the limit set relative to the unique conformal measure.*

A more refined analysis by Tukia [**Tu**] has shown that in the geometrically finite case without cusps the critical exponent is even determined by the measure-theoretical conjugacy of the action of Γ on S^d relative to the conformal measure (which in this case is the Hausdorff measure [**S(iii)**, §3]) at least when the conjugacy is continuous.

PART II

Now we discuss the ergodic properties of the action of a discrete conformal group on the tangent spaces to S^d. Say that $\lambda > 0$ belongs to the *ratio set* of a group action with a quasi-invariant measure μ if for each set B of positive measure and each $\varepsilon > 0$ there are *infinitely many groups elements* $\gamma \in \Gamma$ so that γ carries a positive part of B back into B with Radon-Nikodym derivative $D\gamma(x)$ within ε of λ. The ratio set is independent of the choice of μ in the measure class of μ.

The concept of ratio set is only interesting in the *recurrent part* of the action. The *recurrent part* is by definition the complement of the *wandering part*. The *wandering part* is by definition the unique maximal subset (mod null sets) where the action has a fundamental set—one which cuts each orbit in one point.

THEOREM 6. *For the recurrent part in S^d of the action of a discrete conformal group (relative to Lebesgue measure) the ratio set consists of all positive real numbers.*

There are two ingredients in the proof. The first one is an abstract proposition in ergodic theory due to Krieger and K. Schmidt.

PROPOSITION. *For any recurrent group action the number 1 belongs to the ratio set.*

The proof of the proposition may be found in Klaus Schmidt [**Sc**].

This proposition has a strong consequence for any discrete conformal group on its recurrent part relative to Lebesgue measure where of course $D\gamma(x) = |\gamma'(x)|^d$, or relative to any conformal measure μ which by definition satisfies $D\gamma(x) = |\gamma'x|^\delta$.

Namely, working in R^d instead of S^d, we can come back to any recurrent set of positive measure B in R^d by an infinite sequence of elements γ_1, $\gamma_2, \ldots$ which carry the annular regions near the isometric circles from B back into B.

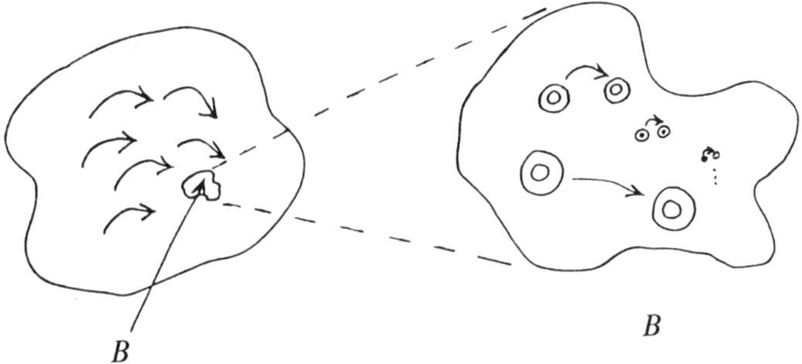

To see this we apply the proposition and recall that a conformal transformation of R^d is the composition of an inversion followed by a Euclidean isometry.

Now $|\gamma'x|$ is just $(a/r)^2$ (coming for the inversion factor) where a is the radius of the isometric sphere ($=$ inversion sphere) and r is the distance of x from the center. So to see λ is in the ratio set consider the annulus between the isometric sphere and the sphere where $(a/r)^2 = \lambda$. These annuli have a definite shape and the diameters are arbitrarily small because the group is discrete and we have infinitely many elements γ so that $|B \cap \gamma^{-1}B| > 0$.

We can apply Lebesgue density to find a small enough size for which such annuli intersecting $B' \subset B$ are filled by points of B to a very high percentage. We can be sure most of these points are carried back to B if at least one is carried to B' since the derivative only varies in a fixed proportion on these annuli. By this argumentation we find λ approximately anywhere we look. For more details see [**S(ii)**, §1].

Now we state a theorem about the action of Γ on tangent spaces to S^d.

THEOREM 7. *In the recurrent part of the action of a discrete conformal group relative to Lebesgue measure the action on tangent spaces to S^d is irreducible—there is no measurable field of Γ invariant subspaces.*

The proof of Theorem 7 begins with the analysis of Theorem 6. Now however instead of moving concentrically from the isometric sphere to find the desired scale distortion of the derivative we move around the isometric sphere to find enough twisting to destroy the invariance of any measurable field of subspaces.

In fact, we look at a set of positive measure where the field is nearly parallel. Then we recur in that set with γ's having small isometric spheres so Lebesgue density applies. The Euclidean isometry factor preserves the parallel property and the inversion destroys it. Thus γ the product cannot preserve it. For more details see [**S(ii)**], §1, §7].

REMARK. The properties of Lebesgue measure used in the above proofs are

(i) the transformation law $\gamma^*\mu = |\gamma'|^d\mu$.

(ii) Lebesgue measure on R^d is not unfairly concentrated near any hypersphere or near any ray.

Now any conformal measure satisfies (i) (with a different power) by definition so this part of the proof is valid. Sometimes one can verify the gemetric property (ii) for particular conformal measures. In those cases Theorems 6 and 7 are then also valid for the recurrent part of a conformal measure.

THEOREM. *For any finitely generated discrete conformal group on S^2 we have that any nonatomic conformal measure on the limit set (assumed to be a proper subset of S^2) is recurrent.*

(This is proved by the same Ahlfors-Borel series argument which shows the finiteness of cusps [**S(vi)**, Part I]. One has in fact $\Sigma_\Gamma|\gamma'z|^N < \infty$ for only finitely many orbits in the limit set. So any nonatomic conformal measure on the limit set is recurrent.)

Now for certain limits of geometrically finite groups—the hyperbolic half cylinder of [**S(v)**]—one knows enough about the canonical conformal measure to verify (ii). Thus for these conformal measures Theorems 6 and 7 are valid.

Applications of Part II. (1) Theorem 7 implies a generalization of the (quasi-conformal $\to$ conformal) part of the Mostow rigidity theory.

COROLLARY 2. *Suppose $\phi\colon S^d \to S^d$ is a quasi-conformal homeomorphism ($d \geqslant 2$) conjugating one discrete conformal group Γ to another Γ' ($\gamma \in \Gamma$ iff $\phi\gamma\phi^{-1} \in \Gamma'$). Further suppose the derivative of ϕ is a similarity transformation at almost all points of the nonrecurrent part of the action of Γ. Then ϕ is a conformal transformation.*

SKETCH OF PROOF. If not, the conformal distortion of ϕ on the recurrent part is expressed (in particular) by a Γ invariant measurable family of proper subspaces. Such a family is impossible by Theorem 7. Thus at almost all points the derivative of ϕ is a similarity transformation. Thus by a standard fact in quasi-conformal mappings ϕ is actually conformal. For more details see [**S(ii)**, §7].

(2) Say two groups actions are *orbit equivalent* relative to quasi-invariant measures if (after discarding invariant sets of zero measure) there is a Borel bijection preserving orbits and sets of positive measure.

Also recall any recurrent group action relative to a quasi-invariant measure can be decomposed into a measurable family of ergodic actions. The base of this family is a separable Lebesgue space consisting of a continuous part (either vacuous or isomorphic to $(0, 1)$) and a countable set of atoms.

So the ergodic decompositions plus the wandering part can be labeled by an element in

$$\{0, 1/2, 1\} \times \{0, 1\} \times \{0, 1, 2, \ldots, \omega\}$$

which we call the *signature* of the action. The $\{0, 1/2, 1\}$ factor tells whether the wandering part has zero, partial, or full measure, the second factor tells whether $(0, 1)$ is present or not, the third factor counts the atoms.

THEOREM 8. (i) *For any discrete group Γ of conformal transformations of S^d there is countable partition of $S^d = X_1 \cup X_2 \cup \cdots$ and elements $\gamma_1, \gamma_2, \ldots$ in Γ so that*

$$\psi = \gamma_1/X_1 \cup \gamma_2/X_2 \cup \cdots$$

is a measurable bijection of S^d whose orbits are precisely the orbits of Γ (modulo the null sets of any quasi-invariant measure).

(ii) *Any two discrete groups of conformal transformations (possibly in different dimensions) with the same signatures relative to Lebesgue measure are orbit equivalent.*

SKETCH OF PROOF. (i) Given ξ in S^d choose an upper half space model R_+^{d+1} of H^{d+1} with ξ at ∞. A fixed reference orbit in H^{d+1} now determines a discrete set in R_+^{d+1}. If we move to ξ' on the same Γ orbit as ξ, the discrete set is only changed by *similarity*. We can view this situation as saying the *orbit in S^d of a discrete conformal group* has the natural structure of a horizontally and vertically discrete set in R_+^{d+1} well defined up to similarity.

This structure implies the orbits of Γ on S^d can be assigned a Z-ordering. An abstract formal proof goes as follows. This structure implies (i) the equivalence relation defined by Γ orbits is *amenable*, bounded measurable functions of pairs of points in Γ orbits can be averaged in a Γ-invariant way (use the discreteness property above and the fact that the group of similarities is amenable to form the average). (ii) Now a new general result of Connes, Feldman, Ornstein, and Weiss [**C et al**] says the orbits can be Z-ordered.

(ii) Kreiger [**K**] has classified actions of Z up to orbit equivalence by the patterns of volume distortion as expressed by the Radon-Nikodym derivative. We use only a part of this which says that if the ratio set contains all values as indicated by Theorem 6, then the orbit equivalence class is determined uniquely (and it is called the unique III_1).

Thus since any discrete conformal group is orbit equivalent to an action of Z by (i) *it is also orbit equivalent to an integral of III_1's over its ergodic decomposition* (on the recurrent part relative to Lebesgue measure).

PART III

Now we discuss the interrelationships between *critical exponent, Hausdorff dimension, the spectrum of the hyperbolic Laplacian, entropy and general Riemannian manifolds*.

Poincaré in his long paper on Kleinian groups (1883) observed the following—if Γ is a discrete conformal group acting on S^2 obtained by a general perturbation of a group Γ_0 whose limit set was a round circle, then the limit set of the perturbed Γ could be a very wiggly curve without a tangent at a dense set of points.

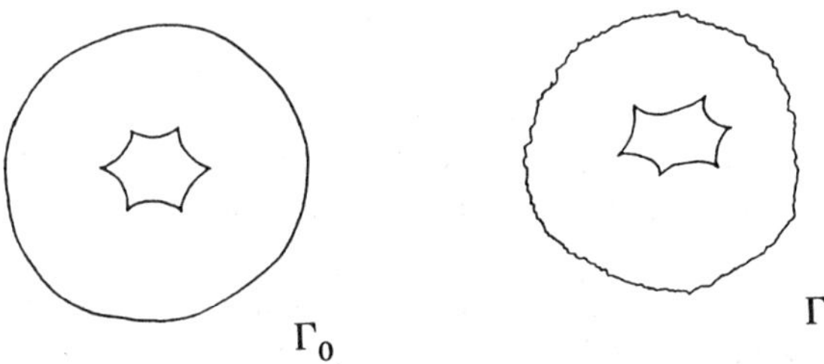

Rufus Bowen (1978) proved first that such wiggly limit curves were nonrectifiable and then that their Hausdorff dimension was greater than one. The nonrectifiability can be seen using Theorem 5 and the Riemann mapping theorem.

Namely, by classical function theory the Riemann map between the round disk and the domain of discontinuity is absolutely continuous on the boundary *if the curve is rectifiable* (Bowen's first step). By Theorem 5 with respect to 1-dimensional Hausdorff measure the boundary map would be part of a conformal conjugacy between Γ_0 and Γ. This argument gives a different proof of a general form of Bowen's nonrectifiability theorem.

THEOREM 9. *Let Γ be a discrete conformal group preserving a simply connected Jordan domain. Suppose for the corresponding group Γ_0 in the disk the Poincaré series diverges at $s = 1$. Then the boundary of the Jordan domain is either nonrectifiable or a round circle.*

The second part of Bowen's Hausdorff dimension proof was to construct, using Markov partitions on the wiggly curve, a finite measure and a real number δ so that the amount of mass in the disk of radius r about any point on the curve is comparable to r^δ. It follows from the definition that δ is the Hausdorff dimension and the Hausdorff measure in dimension δ is finite and positive. (Bowen then concludes the argument by the reasoning—if the Jordan curve is not a circle it is nonrectifiable and thus $\delta > 1$.)

In [S(iii)], this situation concerning Hausdorff measure was extended to all geometrically finite group without cusps (there called *convex cocompact groups*). The new ingredient difference from Bowen's discussion is a construction generalizing one in Patterson's paper [P] which produces a finite conformal measure μ on the limit set of any discrete conformal group Γ. Here $\gamma^*\mu = |\gamma'|^\delta\mu$ for $\gamma \in \Gamma$ and δ is the critical exponent of the Poincaré series ($\Sigma_\Gamma|\gamma'x_0|^s$ converges for $s > \delta$ and diverges for $s < \delta$).

Then rather simple and general arguments [**S(iii)**, §2, §3, §7] show

THEOREM 10. *For a convex-compact group* Γ *in any dimension (that is geometrically finite without cusps) the Hausdorff dimension of the limit set is the critical exponent* δ. *The Hausdorff* δ-*measure* μ *is finite and positive. In fact* μ *(disk of radius r centered on a limit point) is comparable to* r^{δ}. *The group* Γ *acts ergodically on* $\mu \times \mu$ *and the Poincaré series diverges at* δ.

If p belongs to B^{d+1}, *let* $\Phi(p)$ *be the* δ *Hausdorff measure of the limit set in the visual metric on rays from p. Then* Φ *is* Γ-*invariant, it is an eigenfunction of the Laplacian with eigenvalue* $\delta(\delta - d)$. *If* $\delta > d/2$, Φ *is square integrable in* $\mathbf{H}^{d+1}/\Gamma$.

Tukia [**Tu**] has used this theorem to show that if two such actions on the limit set are continuously and absolutely continuously conjugate with respect to Hausdorff measure, the groups are conformally conjugate. (Note the comparison with Theorem 5.)

One consequence is the following analogue of Bowen's Hausdorff dimension result for discrete conformal groups in the 3-sphere. Start with a compact hyperbolic 3-manifold V which contains a totally geodesic embedded surface (e.g. use the units of the quadratic form $x_0^2 + x_1^2 + x_2^2 - \sqrt{2}\, x_3^2$ and set $x_0 = 0$).

Consider the universal cover of V as a hyperplane $\mathbf{H}^3$ contained in $\mathbf{H}^4$. Now following Thurston's ideas for $\mathbf{H}^2 \subset \mathbf{H}^3$ bend the hyperplane $\mathbf{H}^3 \subset \mathbf{H}^4$ equivariantly along the inverse image of the geodesic surface in V. This bends the group and constructs a quasi-conformal deformation of $\Gamma = \pi_1 V^3$ thought of as a discrete group acting on $\mathbf{H}^4$.

The limit set is a topological two-sphere from which infinitely many hemispheres have protruded (the surface analogue of Thurston's "Mickey Mouse" example [**T**]).

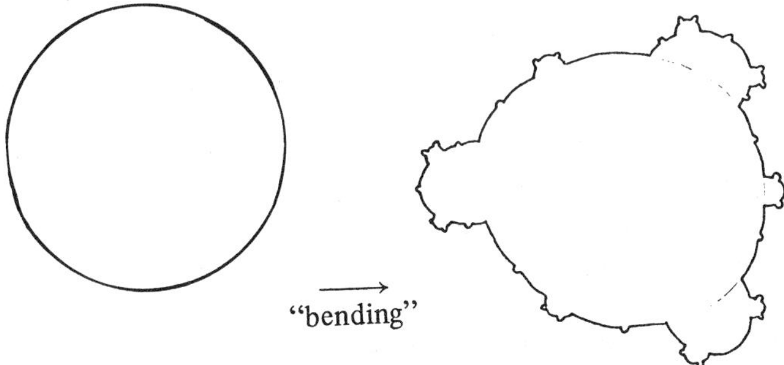

The topological 2-sphere is the boundary of a quasi-conformal 3-ball by construction. Its Hausdorff dimension must be strictly greater than two using Theorems 5 and 10, and the structure of sets of finite Hausdorff 2-measure (Federer).

We have tried to develop the results of Theorem 10 for more general groups and the attempt is mostly successful for general geometrically finite groups with cusps. Some parts of the discussion work for general groups and these even have interpretations for general Riemannian manifolds.

As an indication of this development, we list some results for discrete conformal groups, the Poincaré series $\Sigma|\gamma'x_0|^s$, the critical exponent $\delta(\Gamma)$, conformal measures, entropy, etc.

(1) As mentioned above for $\delta = \delta(\Gamma)$ there is always a finite positive conformal measure in the limit set of dimension δ. Namely, μ satisfies $\gamma^*\mu = |\gamma'|^\delta\mu$, $\gamma \in \Gamma$. This was proved for discrete groups in $\mathbf{H}^2$ by Patterson [**P**] and for groups in $\mathbf{H}^{d+1}$ in [**S(iii)**, §1].

(2) For studying the comparison of μ (ball of radius r) and r^δ it is useful to introduce the Γ-invariant positive eigenfunction of the Laplacian related to μ.

If ξ belongs to $S^d = \partial B^{d+1}$, let y_ξ denote the function on B^{d+1} with a pole at ξ corresponding by stereographic projection to the vertical height y in the upper half space. Since y^δ is a $\delta(\delta - d)$ eigenfunction of the hyperbolic Laplacian Δ any convex combination of the y_ξ^δ is also.

The μ convex combination of the y_ξ^δ is a positive $\delta(\delta - d)$ eigenfunction ϕ_μ which is Γ-invariant when μ is a conformal measure, $\gamma^*\mu = |\gamma'|^\delta\mu$. See [**P**] and [**S(iii)**, §7]. The eigenfunction ϕ_μ can be used to study the ratio μ (disk of radius r)$/r^\delta$. The growth of ϕ_μ along a geodesic in $\mathbf{H}^{d+1}/\Gamma$ is always an upper bound. For certain groups including geometrically finite groups ϕ_μ provides also a lower bound, see [**S(iv)**, §6].

(3) From a Γ-invariant eigenfunction ϕ of the hyperbolic Laplacian Δ one can construct for each $p \in B^{d+1}$ a conformal measure $\mu(\phi)$ at infinity by taking a weak limit as $R \to \infty$ of the normalized measure proportional to

$$\phi \circ (\text{measure on sphere of radius } R \text{ centered at } p).$$

If $\delta \geqslant d/2$, and ϕ is ϕ_μ, one reconstructs μ from ϕ_μ in this way. If $\delta < d/2$ this limit measure is not μ but like a Fourier transform of μ. (This is an instance of the duality that appears in the unitary representation theory of the Moebius group—familiar for the first cases $PSl(2, R)$ and $PSl(2, C)$.)

(4) If Γ is a geometrically finite group the conformal measure is unique and the canonical positive eigenfunction ϕ_μ belongs to $L^2(\mathbf{H}^{d+1}/\Gamma)$ iff $\delta > d/2$. If G denotes the group of conformal transformations of B^{d+1} then of course ϕ_μ also belongs to $L^2(G/\Gamma)$. Because ϕ_μ is an L^2 eigenfunction of the hyperbolic Laplacian Δ the G orbit of ϕ_μ generates an irreducible unitary of G in $L^2(G/\Gamma)$. This irreducible component is the member of the complementary series labeled by δ in $(d/2, d]$. Recalling Theorem 10 we see a *geometric interpretation of the complementary series by their realization in $L^2(G/\Gamma)$ for discrete groups Γ with the parameter being just the Hausdorff dimension of the limit set.*

In [**S(v)**, §7] it was shown there is a continuous family of quasifuchsian groups where δ varied from 1 to 2. *The representations corresponding to ϕ_μ thus cover the entire complementary series as the group Γ varies in this deformation.*

(5) For geometrically finite groups with $\delta \leqslant d/2$ the positive eigenfunction ϕ_μ belongs to L^2 of a part of $\mathbf{H}^{d+1}/\Gamma$ [**S(iv)**]. This part is the *convex core* defined by forming first the convex hull of the limit set in $\mathbf{H}^{d+1}$, then a unit neighborhood of this convex hull, and then the quotient by Γ. It is a *problem*

whether there are any other groups besides geometrically finite groups where ϕ_μ belongs to L^2 of $\mathbf{H}^{d+1}/\Gamma$, $\delta > d/2$ (or to L^2 of the convex core $\mathbf{H}^{d+1}/\Gamma$, $\delta \leqslant d/2$).

Entropy. (6) The property mentioned in (5), the square integrability of the eigenfunction ϕ_μ on the convex core of $\mathbf{H}^{d+1}/\Gamma$, implies the associated invariant measure for the geodesic flow on $\mathbf{H}^{d+1}/\Gamma$ has finite total mass [**S(iv)**].

In the case of finite mass we can define the measure-theoretic entropy of the geodesic flow. For geometrically finite groups one finds [**S(iv)**] this entropy is just the critical exponent $\delta(\Gamma)$.

If there are no cusps one can define a topological entropy for the invariant set of geodesics with both endpoints in the limit set. The topological entropy is again the critical exponent δ [**S(iv)**, §9].

In the geometrically finite case not only is the invariant measure for the geodesic flow of finite total mass, it is also mixing and even Bernoulli [**R**]. This example of a geometrically defined Bernoulli flow provides a new twist on the Anosov-Pesin foliation picture producing this random phenomenon. In these examples the relevant measure class conditioned to the leaves is not the smooth measure (of dimension d) but a Hausdorff type measure of dimension δ.

Hausdorff dimension and measure. (7) When as in (6) the associated measure for the geodesic flow has finite total mass (thus for all geometrically finite groups [**S(iv)**, §5]) the critical exponent is the Hausdorff dimension of the radial limit set Λ_r of Part I [**S(ii)**, Theorem 25]. This number then equals the Hausdorff dimension of the entire limit set for geometrically finite groups [**S(iv)**, §2] because the difference between the radial limit set and the entire limit set is only a countable set in this case.

A corollary *for general discrete groups in* $\mathbf{H}^2$ (or general Riemann surfaces) is that the critical exponent is always the Hausdorff dimension of the radial limit set Λ_r. This follows because any group in $\mathbf{H}^2$ is a union of geometrically finite groups and δ satisfies a simple semicontinuity property for limits [**S(ii)**, §2].

(8) We have seen in Theorem 10 that the Hausdorff δ-measure of a geometrically finite group without cusps is finite and positive (and it equals the unique conformal measure). This is sometimes true when there are cusps.

If all cusps have rank $\leqslant \delta$, the unique conformal measure is again the Hausdorff δ-measure. If all cusps have rank $\geqslant \delta$ the canonical conformal measure may be described metrically by a Hausdorff *packing* construction dual to the customary *covering* definition [**S(iv)**, §7].

If δ strictly separates the ranks of two cusps then it is a *problem* to give a metrical description of the unique conformal measure on the limit set of a geometrically finite group [**S(iv)**].

Riemannian manifolds. (9) The appearance of positive square integrable eigenfunctions above, their uniqueness coming from ergodicity, etc. prompts a more general discussion.

First of all for any complete Riemannian manifold M one can define a real number $\lambda_0 \leqslant 0$ by a sequence of equivalent conditions.

(i) The L^2-spectrum of the Laplacian on M is precisely contained in $(-\infty, \lambda_0]$.

(ii) The infimum of the set of λ so that for $x, y \in M \int_1^\infty e^{-\lambda t} p_t(x, y)\, dt$ converges. Here $p_t(x, y)$ is the heat kernel expressing the probability that a random path starting at x finds itself at y after time t.

(iii) The minimum of the set of λ so that there is a *positive* eigenfunction of Δ.

(10) Any positive eigenfunction ϕ_0 for the smallest eigenvalue λ_0 is called a *ground state or basic eigenfunction*. There are several general results about these.

(i) Any square-integrable eigenfunction of the hyperbolic Laplacian Δ which is *positive* must be a *ground state* and belong to the eigenvalue λ_0.

(ii) Any *square-integrable* eigenfunction of Δ which belongs to λ_0 must be *positive* (or negative) and so define a ground state. (A corollary is that no two such can be orthogonal. Thus the zeroth eigenvalue λ_0 has L^2-multiplicity at most one.)

(iii) If there is a square-integrable ground state then any other ground state (not necessarily square integrable) must be a multiple of the first one. (The proof here and in (v) below uses the modified Markov process mentioned in Part I and [**S(ii)**, §7].)

(iv) More generally if $\int_1^\infty e^{-\lambda_0 t} p_t(x, y)\, dt$ diverges all ground states are multiples of any given one (not necessarily square integrable).

(11) We specialize the general Riemannian manifold to a general hyperbolic manifold $\mathbf{H}^{d+1}/\Gamma$ and connect the zeroth eigenvalue and ground states with critical exponents and Γ-invariant conformal measures. As we mentioned above conformal measures of dimension δ determine $\delta(\delta - d)$ positive eigenfunctions. Furthermore, positive eigenfunctions belonging to $\lambda \leqslant 0$ determine conformal densities of dimension $\delta = d/2 + (\lambda + d^2/4)^{1/2}$ (because $\delta(\delta - d) = \lambda$ and $\delta \geqslant d/2$).

Thus for any hyperbolic manifold $\mathbf{H}^{d+1}/\Gamma$, λ_0 is $\geqslant -d^2/4$ and equality occurs iff the critical exponent $\delta(\Gamma)$ is less than or equal to $d/2$. *Otherwise, where $\delta \geqslant d/2$, λ_0 is $\delta(\delta - d)$.* For $d = 1$, most if not all of this was proved by Elstrodt and Patterson.

The basic eigenfunctions or ground states correspond to the ϕ_μ discussed above. The existence of a conformal measure for any discrete conformal group in case $\delta \geqslant d/2$ is generalized to a statement valid for any complete Riemannian manifold—*the existence of ground states.*

For geometrically finite conformal groups we have a square-integrable ground state if $\delta > d/2$. The uniqueness then following from ergodicity is thus generalized to one for complete Riemannian manifolds—*the uniqueness of square-integrable ground states.* For proofs see [**S(vii)**].

Concluding remarks. We have seen one real number $\delta(\Gamma)$ defined for any discrete conformal group Γ acting on B^{d+1} appearing in other guises for certain groups—the Hausdorff dimension of the limit set, the entropy of the geodesic flow, the "square root" of the lowest eigenvalue λ_0 of the Laplacian $d/2 + (\lambda_0 + d^2/4)^{1/2}$.

For geometrically finite groups all these concepts mesh nicely, and these hyperbolic manifolds of infinite Riemannian volume behave after *renormalization* by the canonical conformal measure just like finite volume manifolds vis a vis ergodicity of the geodesic flow, square-integrable ground states, entropy and dimension, etc.

BIBLIOGRAPHY

[A] Lars Ahlfors, (i) *Finitely generated Kleinian groups*, Amer. J. Math. **86** (1964), 413–429.

(ii) *Some remarks on Kleinian groups*, Tulane Conf. Conformal mappings (1965).

(iii) *Moebius transformations in several dimensions*, Minnesota Book Center, Minneapolis, Minn., 1981.

[Aa] Jon Aaronson, *Pointwise behaviour of infinite measure preserving transformations*, Israel J. Math. Math. **32** (1979).

[As] Jon Aaronson and Dennis Sullivan, *Rational ergodicity of the geodesic flow on infinite volume hyperbolic manifolds* (manuscript).

[B] Lipmann Bers, (i) *Spaces of Kleinian groups*, Several Complex Variables, I, (Maryland, 1970).

(ii) See also these proceedings.

[B] Rufus Bowen, *Hausdorff dimension of quasi-circles*, Inst. Hautes Études Sci. Publ. Math.

[C et al] Alain Connes, Jack Feldman and Benjamin Weiss, *Amenable equivalence relations are hyperfinite*, Inst. Hautes Études Sci. Publ. Math. preprint, 1980.

[G] Lucy Garnett, *Functions and measures harmonic along the leaves of a foliation*, Ph.D. Thesis, Dartmouth College, 1981; Inst. Hautes Études Sci. Publ. Math. preprint, June 1980.

[K] Wolfgang Krieger, *On ergodic flows and isomorphisms of factors*, Math. Ann. **223** (1976), 19–70.

[L] William LeVeque, *Continued fractions for the Gaussian field*, Indag. Math. (1952).

[P] S. J. Patterson, *The limit set of a Fuchsian group*, Acta Math. **B6** (1976), 241–273.

[R] Dan Rudolph, manuscript (Maryland).

[Rn] Marina Ratner, *Rigidity of the horocycle flow*, preprint, Berkeley, 1980.

[S] Dennis Sullivan, (i) *Disjoint spheres, approximation by imaginary quadratic numbers and the logarithm law for geodesics*, Acta Math. (to appear).

(ii) *On the ergodic theory at infinity of an arbitrary discrete group of hyperbolic motions*, Ann. of Math. Studies, No. 97, Princeton Univ. Press, Princeton, N.J., 1981, pp. 465–496.

(iii) *The density at infinity of a discrete group of hyperbolic motions*, Inst. Hautes Études Sci. Publ. Math. **50** (1979).

(iv) *Entropy, Hausdorff measures old and new, and limit sets of geometrically finite Kleinian groups*, Acta Math. (to appear).

(v) *Growth of positive harmonic functions and Kleinian group limit sets of planar measure zero and Hausdorff dimension two*, N. Kuiper 60th Birthday Volume (Holland), Springer-Verlag, 1981.

(vi) *On the finiteness of cusps*, Acta Math. (to appear).

(vii) *The exponential size of a Riemannian manifold relative to heat flow*, Acta Math. (submitted).

[Sc] Klaus Schmidt, *Cocycles of ergodic group actions*, Warwick Notes, 1976.

[T] Bill Thurston, (i) *Geometry and topology of three-manifolds*, preprint, Princeton Univ., 1978; to be published by Princeton Univ. Press, 1982.

(ii) *Hyperbolic structures on 3-manifolds, I. Deformation of acylindrical manifolds*, Ann. of Math. (to appear).

[Tu] Pekka Tukia, *Rigidity of Kleinian groups and dimensionality of the limit set*, preprint, Univ. of Helsinki, 1981.

INSTITUTE DES HAUTE ÉTUDES SCIENTIFIQUE, BURES SUR YVETTE, FRANCE

Section 4

SEVERAL COMPLEX VARIABLES

Proceedings of Symposia in Pure Mathematics
Volume 39 (1983), Part 1

Strictly Pseudoconvex Domains in $\mathbf{C}^n$

MICHAEL BEALS, CHARLES FEFFERMAN AND ROBERT GROSSMAN

CONTENTS

Reprinted from Bulletin Amer. Math. Soc. (N.S.) 8 (1983), 125–322.

1980 *Mathematics Subject Classification*. Primary 32F15.

0. Foreword by Fefferman. The goal of this article is to acquaint readers with analysis and geometry on smooth domains in $\mathbf{C}^n$. For domains with the simplest geometry (strictly pseudoconvex) a wealth of deep results came to light during the 1970s, and we restrict attention to this case. The state of our knowledge of more general (weakly pseudoconvex) smooth domains is much more primitive, although some outstanding results are known, notably on the Cauchy-Riemann equations (Kohn [**40**]) and the Poincaré metric (Cheng-Yau [**10**]). A natural problem is to extend the results presented here to more general domains.

One of our main themes is the close connection between the analysis and local geometry of domains. To understand the picture requires a lot of elementary background in geometry and differential equations. For completeness we have included a long exposition (Chapters 2–5) of the relevant background. Chapter 1 gives a brief introduction and Chapter 6 a detailed introduction to domains in $\mathbf{C}^n$, and finally Chapters 7–12 present the main results.

This paper grew out of a course I gave at Princeton during 1979–80, with notes taken by Beals and extended by Grossman. The contributions of both Beals and Grossman are pervasive, but responsibility for any errors lies with me. Since the course was given, relevant important results were discovered, especially by Kuranishi and by Lee and Melrose. We have incorporated here statements of their results. I spoke on some of the material in this paper at the recent symposium in honor of Poincaré; much of the modern work is rooted in his seminal ideas. The reader should see the excellent brief survey article by Wells [66], also presented at the Poincaré Symposium and overlapping strongly with our long exposition.

Special thanks are due Perry DiVerita, Lauri Hein, Maureen Kirkham, Annette Roselli, and Bonnie Tompsen who cheerfully typed the manuscript despite the pressure of a deadline long past due.

In studying several complex variables, I profited greatly from the deep insights of my colleagues, and it is a pleasure to thank them here. I am especially grateful to S. Bell, J. Faran, J. J. Kohn, J. Moser, L. Nirenberg, D. H. Phong, E. M. Stein, S. Webster, and S. T. Yau. Without a vast amount of work by Beals and Grossman, this paper could never have been completed. Without forceful prodding by F. Browder, it would not have been finished in the twentieth century.

CHAPTER 1. BRIEF INTRODUCTION

We shall study analogues in $\mathbf{C}^n$ of familiar ideas in one complex variable. The topics are as follows:

Cauchy-Riemann equations. A basic problem in several complex variables is to solve the inhomogeneous equations $\bar{\partial}u = \alpha$ with good bounds. Here u is a function of n complex variables $z_k = x_k + iy_k$, and $\bar{\partial}u$ stands for the n functions $\partial u/\partial \bar{z}_k = \frac{1}{2}(\partial u/\partial x_k + i\partial u/\partial y_k)$. The problem is important because, for instance, it allows us to patch local results into global theorems. To illustrate, let us try to find an analytic function $F(z)$ on a domain D which blows up only at a single boundary point p. As a first step, we find such a function F_0 defined only in a small neighborhood U of p. Next, take a smooth cutoff function ϕ supported in U and equal to one near p, and set $\alpha = \bar{\partial}(\phi F_0)$. Now α is globally defined on $\bar{D}$ and has no singularities anywhere, since $\phi = 1$ and F_0 is analytic near p. If we can solve the $\bar{\partial}$-equations with good bounds, then we can find a nice function u which is singular nowhere and satisfies $\bar{\partial}u = \alpha$. Therefore $F = \phi F_0 - u$ will be singular exactly at p, and will be analytic in D since $\bar{\partial}F = \bar{\partial}(\phi F_0) - \bar{\partial}u = \alpha - \alpha = 0$.

The $\bar{\partial}$ equations are overdetermined—there are n equations for one function— so they can be solved only when α satisfies consistency conditions. Also, the solution u is obviously not unique. If $\bar{\partial}u = \alpha$ and F is any analytic function on D, then also $\bar{\partial}(u + F) = \alpha$. So it is natural to try to solve $\bar{\partial}u = \alpha$ in D with the extra condition that u be orthogonal to the subspace $H(D)$ of analytic functions $\subseteq L^2(D)$. This is called *Kohn's solution* of $\bar{\partial}u = \alpha$; it minimizes the L^2-norm among all solutions.

There is also a family of Cauchy-Riemann equations for analytic funtions on the boundary ∂D of a domain in $\mathbf{C}^n$ ($n > 1$). Imagine that F is a function defined in all of $\mathbf{C}^n$, but whose values are known to us only on ∂D. We can easily calculate the derivatives of F in directions tangent to ∂D, but we do not know the normal derivative $\partial F/\partial n$. Now suppose that F is analytic in a neighborhood of ∂D, so that we have the n Cauchy-Riemann equations $\partial F/\partial \bar{z}_k = 0$ on ∂D. We can solve one of these equations for the missing derivative $\partial F/\partial n$, and then substitute the result into the remaining $(n-1)$ Cauchy-Riemann equations. Thus we obtain a system of $(n-1)$ partial differential equations for the restriction of an analytic functions F to ∂D. One writes $\bar{\partial}_b F = 0$, and again we are interested in the inhomogeneous equation $\bar{\partial}_b F = \alpha$ on ∂D.

Cauchy integral formula. The idea of solving $\bar{\partial} u = \alpha$ with u orthogonal to $H(D)$ (the analytic functions) suggests that we study the orthogonal projection π: $L^2(D) \to H(D)$. One shows easily that π is given by an integral kernel, $\pi f(z) = \int_D K(z, w) f(w)\, dw$, where K is called the *Bergman kernel*. We shall explore the relation between the Bergman kernel and the geometry of the domain. The analogue of the Bergman kernel for the $\bar{\partial}_b$-problem is the *Szegö kernel* $K(z, w)$ which realizes the projection

$$(1) \qquad \pi f(z) = \int_{\partial D} K(z, w) f(w)\, dw$$

from $L^2(\partial D)$ to the subspace $H^2(\partial D) = \{$Boundary values of analytic functions in $D\}$. For the unit disc in $\mathbf{C}^1$, the right-hand side of (1) is $(1/2\pi i)\oint_{|w|=1} f(w)/(z-w)\, dw$. So the Szegö kernel is the analogue of the Cauchy integral formula for domains in $\mathbf{C}^n$.

Dirichlet problem. Just as analytic functions are closely related to harmonic functions in one complex variable, so the problems $\bar{\partial} u = \alpha$, $\bar{\partial}_b u = \alpha$ are intimately related to certain second-order equations called $\square$, $\square_b$. To see how these arise, let us try to solve $\bar{\partial} u = \alpha$ in D with u orthogonal to analytic functions. A natural way to produce functions orthogonal to everything analytic is to start with w in the domain of the adjoint operator $\bar{\partial}^*$ and set $u = \bar{\partial}^* w$. If F is analytic, then $\langle u, F \rangle = \langle \bar{\partial}^* w, F \rangle = \langle w, \bar{\partial} F \rangle = 0$. The $\bar{\partial}$-equation therefore takes the form

$$(2) \qquad \bar{\partial}\bar{\partial}^* w = \alpha,$$

$$(3) \qquad w \in \mathrm{Domain}(\bar{\partial}^*).$$

The global condition $u \perp H(D)$ has now been replaced by (3), which is a boundary condition for the second-order differential equation (2). Since (2), (3) come from $\bar{\partial} u = \alpha$, we can hope for solutions only when α satisfies a consistency condition, which we write in the form $\bar{\partial}_1 \alpha = 0$. (Explicitly, $\partial u/\partial \bar{z}_k = \alpha_k$ can be solved only when $\partial \alpha_j/\partial \bar{z}_k - \partial \alpha_k/\partial \bar{z}_k = 0$.) For general α, possibly not satisfying the consistency condition, we replace (2), (3) by the boundary-value problem

$$(4) \qquad \left(\bar{\partial}\bar{\partial}^* + \bar{\partial}_1^*\bar{\partial}_1\right)w = \alpha,$$

$$(5) \qquad w \in \mathrm{Domain}(\bar{\partial}^*), \qquad \bar{\partial}_1 w \in \mathrm{Domain}(\bar{\partial}_1^*).$$

This is called the $\bar{\partial}$-Neumann problem. It can be solved for general α and it reduces to (2), (3) when $\bar{\partial}_1 \alpha = 0$. One finds that the second-order operator (4) is basically the Laplacian, but the boundary conditions (5) are more degenerate than Dirichlet or Neumann conditions and require deep analysis. The analogous construction for the $\bar{\partial}_b$-equation leads to a second-order equation $\Box_b w = \alpha$ on ∂D. Here ∂D is a manifold without boundary, but $\Box_b$ is not elliptic.

Riemann mapping theorem. Given two domains D_1, $D_2 \subseteq \mathbf{C}^n$, we want to know whether there is an analytic mapping Φ which carries D_1 one-to-one and onto D_2. (Φ is called *biholomorphic*.) In more than one complex variable, the answer is almost always "no". For example, a ball is not biholomorphic to an ellipsoid. This leads to *biholomorphic geometry*, the study of those concepts on domains and their boundaries which are preserved under biholomorphic maps.

The most interesting objects in the biholomorphic geometry of a smooth boundary ∂D are local invariants attached to points of ∂D, and a family of distinguished curves in ∂D called *chains*. The chains on the unit sphere S are the circles which arise by intersecting S with a complex line. On more general boundaries, the chains are solution curves of a system of second-order ordinary differential equations, much like geodesics on a hypersurface in $\mathbf{R}^n$.

The local invariants are not trivial to write down, but we can see at once that they exist. It is enough simply to count dimensions: To Nth order about p, a boundary ∂P looks like $\{\operatorname{Re} z_1 = f_N(\operatorname{Im} z_1, \operatorname{Re} z_2, \operatorname{Im} z_3, \ldots, \operatorname{Re} z_n, \operatorname{Im} z_n)\}$ for an Nth degree polynomial f_N. Thus ∂D is described to Nth order by a single Nth degree polynomial in $(2n - 1)$ variables. On the other hand, a biholomorphic map is described to Nth order by n polynomials of degree N in only n variables. For large N, one checks that the space of all possible boundaries has much higher dimension than the space of all possible biholomorphic maps.

It follows at once that many biholomorphic invariants may be attacked to the Taylor expansion of ∂D about p. In particular, a domain D can be biholomorphic to the unit ball only if its boundary satisfies a system of nonlinear partial differential equations. We shall write down these equations explicitly.

Schwartz reflection principle. A domain with real-analytic boundary may be written locally as $D = \{r(z, \bar{z}) < 0\}$ where $r(z, \bar{w})$ is a convergent power series in the independent variables z, $\bar{w}$. (For instance, if D is the ellipsoid $\Sigma_j 2\lambda_j \operatorname{Re}(z_j^2) + |z_j|^2 < 1$, then we may take $r(z, \bar{w}) = \Sigma_j(\lambda_j z_j^2 + \lambda_j \bar{w}_j^2 + z_j \bar{w}_j)$.) The power series $r(z, \bar{w})$ is determined by D up to multiplication by a nonvanishing factor, so the variety $V_w = \{z \in \mathbf{C}^n \mid r(z, \bar{w}) = 0\}$ is associated to D and w, independently of the choice of defining function r. In one complex variable, V_w is a point, and $w \to V_w$ is a conjugate-analytic reflection across ∂D; thus we recover the usual Schwartz reflection principle. What we obtain in higher dimensions is much stronger. For, each V_w is a codimension-one variety, and the family $\mathcal{S} = \{V_w \mid w \in \mathbf{C}^n\}$ is a biholomorphic invariant of the domain D. To see the power of this idea, suppose we try to classify the biholomorphic self-maps Φ of an ellipsoid. From the explicit $r(z, \bar{w})$ above, we see at once that $\mathcal{S}$ is a family of quadric hypersurfaces, and Φ has to carry each quadric in $\mathcal{S}$ to another quadric in $\mathcal{S}$. This is a severe restriction on Φ; for most ellipsoids Φ has to be linear.

Poincaré metric. On a domain $D \subseteq \mathbf{C}^n$ we look for a metric of constant negative curvature which degenerates at ∂D. Such a metric (with constant negative Ricci curvature) is given by

$$ds^2 = \sum_{j,k} \frac{\partial^2}{\partial z_j \partial \bar{z}_k} \log \frac{1}{u} \, dz_j \, \overline{dz_k} \, ,$$

where u is a solution of the complex Monge-Ampère equation

$$(6) \qquad \det\left(\frac{\partial^2}{\partial z_j \partial \bar{z}_k} \log \frac{1}{u} \right) = u^{-(n+1)} \quad \text{in } D,$$

$$u = 0 \quad \text{at } \partial D.$$

For the unit disc in $\mathbf{C}^1$, $u(z) = 1 - |z|^2$, and ds^2 is the familiar Poincaré metric $ds^2 = |dz|^2/(1 - |z|^2)^2$. For suitable smooth domains $D \subseteq \mathbf{C}^n$, equation (6) has a unique solution u, which is known to be smooth in the interior. Near the boundary, u has an asymptotic expansion $u \sim \psi \Sigma_{k=0}^{\infty} \phi_k (\psi^{n+1} \log \psi)^k$, where ϕ_k are smooth functions on $\bar{D}$ and $\psi(z) =$ distance from z to ∂D. In particular, $u \in C^{n+2-\varepsilon}(\bar{D})$. The functions ϕ_k carry a lot of information on the local biholomorphic geometry of ∂D, and appear in the asymptotic expansion of the Bergman kernel.

Throughout this article, we restrict attention to strictly pseudoconvex domains in $\mathbf{C}^n$. This is by far the simplest class to study, and it includes many interesting examples. We recall the elementary definitions: A domain D is strictly pseudo-convex if its Levi form is strictly positive definite at every boundary point. The Levi form of $D = \{z \in \mathbf{C}^n \mid r(z) < 0\}$ with $r \in \mathbf{C}^\infty$, $r' \neq 0$ on ∂D is defined as the restriction of the quadratic form

$$(\xi_k) \to \sum_{j,k} \frac{\partial^2 r}{\partial z_j \partial \bar{z}_k} (p) \xi_j \overline{\xi_k}$$

to the subspace $\{(\xi_k) \in \mathbf{C}^n \mid \Sigma_k (\partial r/\partial z_k)(p) \cdot \xi_k = 0\}$. It is defined up to constant multiples, independently of the choice of r. Unless the Levi form is at least semidefinite at every boundary point, every analytic function on D continues analytically into a fixed $D^+ \supseteq D$.

Our emphasis will be on the "big picture" and the interrelationships of the different ideas. A main theme is the analogy between domains $D \subseteq \mathbf{C}^n$ and curved Riemannian manifolds M:

	Riemannian M	Domains D
Simplest Case	$\mathbf{R}^n$ with Euclidean metric	unit ball
Analytic Problems	Laplace equation	$\partial, \bar{\partial}_b, \Box, \Box_b$
Geometric Invariants	curvature	Chern-Moser invariants

In both settings, the strategy is to get a good understanding of the simplest case, then attack the general case either by approximating a curved M by a flat $\mathbf{R}^n$, or by approximating ∂D by a sphere. The analogy goes further than shown here, but now it is time to present the background information. In Chapter 6 we return to give a more detailed summary of the several complex variables that is our main goal.

CHAPTER 2. MECHANICS

1. Newton's equations and canonical transformations. This chapter contains an introduction to those parts of mechanics which will be needed later in the discussion of pseudodifferential operators (ψDOs) and Fourier integral operators (FIOs). We begin by considering the dynamics of interacting particles with N degrees of freedom, with masses m_i, and whose positions are described by vectors with components q_i. If the particles interact to form a conservative system, then the force exerted on the ith particle can be expressed as $-\partial V(q)/\partial q_i$, for some potential function $V(q)$. The dynamics are determined by Newton's equations

$$(1) \qquad m_i \frac{d^2 q_i}{dt^2} = -\frac{\partial V(q)}{\partial q_i}, \qquad i = 1,\ldots,N.$$

We can expose some of the symmetries hidden in this equation, by introducing the conjugate variables

$$p_i = m_i \dot{q}_i, \qquad i = 1,\ldots,N,$$

and the Hamiltonian

$$H(q, p) = T(p) + V(q),$$

where $T(p)$ is the kinetic energy

$$T(p) = \frac{1}{2} \sum_{i=1}^{N} \frac{p_i^2}{m_i}.$$

In the new variables $(q_1,\ldots,q_N,\, p_1,\ldots,p_N)$ Newton's equations (1) become the Hamilton equations

$$(2) \qquad \dot{q}_i = \frac{\partial H}{\partial p_i}, \qquad \dot{p}_i = \frac{-\partial H}{\partial q_i}, \qquad i = 1,\ldots,N.$$

EXAMPLE. The Kepler problem is concerned with a particle of mass m confined to a plane and moving in a central field with potential energy $V(r) = -k/r$. In polar coordinates (r, θ) the kinetic energy of the particle is

$$T = \frac{m}{2}\left(\dot{r}^2 + r^2\dot{\theta}^2\right).$$

Recall that the angular momentum of the particle is $mr^2\dot{\theta}$. This suggests the coordinate transformation (the Legendre transformation)

$$(3) \qquad (r, \theta, \dot{r}, \dot{\theta}) \to (r, \theta, p_r, p_\theta), \qquad p_r = m\dot{r}, \qquad p_\theta = mr^2\dot{\theta};$$

the kinetic energy is now

$$T = \frac{1}{2m}\left(p_r^2 + \frac{p_\theta^2}{r^2}\right),$$

and the Hamiltonian becomes

$$(4) \qquad H = T + V = \left(p_r^2 + \frac{p_\theta^2}{r^2} \right) - \frac{1}{r}.$$

(Assume that $m = \frac{1}{2}$ and the gravitational constant $k = 1$.) We shall see below that Hamilton's equations (2) still hold in terms of the new p's, q's and H. Since the variable θ does not appear in the Hamiltonian (such a variable is called cyclic), Hamilton's equations (2) imply $0 = -\partial H/\partial\theta = \dot{p}_\theta$, i.e. the angular momentum $p_\theta = l$ is conserved. We can now integrate (3) to find

$$(5) \qquad \theta = 2l \int \frac{dt}{r^2}.$$

Applying (2) again gives

$$\frac{dH}{dt} = \Sigma \left(\frac{\partial H}{\partial q_i} \dot{q}_i + \frac{\partial H}{\partial p_i} \dot{p}_i \right) = 0,$$

and so the total energy $H = E$ is also conserved; hence from (3) and (4), $\dot{r}^2/4 + \omega(r) = E$, or

$$(6) \qquad \int dt = \int \frac{dr}{2\sqrt{E - \omega(r)}},$$

where $\omega(r) = l^2/r^2 - 1/r$ (the effective potential energy). Together the equations (5) and (6) provide a complete description of the path of the particle.

Note that the first step in solving the problem is to change from rectangular to polar coordinates

$$\Phi\colon (x, y) \to (r, \theta).$$

In rectangular coordinates the Hamiltonian involves all four variables $(x, y, \dot{x}, \dot{y})$. Since none of the variables is conserved, we are not able explicitly to integrate the system in these coordinates. This suggests that we look for maps $\Phi^\#$ of *phase space* $\mathbf{R}^{2n} = \{(q_1,\ldots,q_n, p_1,\ldots,p_n)\}$ which preserves the form of Hamilton's equation. We will see in example (2) below that the coordinate change Φ induces a map $\Phi^\#$ taking the Hamiltonian in the $(x, y, \dot{x}, \dot{y})$ variables into the Hamiltonian (4). This is why we are justified in using the coordinates $(r, \theta, p_r, p_\theta)$ to solve the problem.

We begin by defining the Poisson bracket $\{F, G\}$ of two functions $F(q, p)$, $G(q, p)$ on phase space

$$(7) \qquad \{F, G\} \equiv \sum_{k=1}^{n} \left(\frac{\partial F}{\partial p_k} \frac{\partial G}{\partial q_k} - \frac{\partial F}{\partial q_k} \frac{\partial G}{\partial p_k} \right).$$

Consider an integral curve $t \to z(t) = (q(t), p(t))$ of Hamilton's equations (2) and the restriction of a function $F(q, p)$ on phase space to this curve.

PROPOSITION 1.

$$(*) \qquad \frac{dF}{dt}(z(t)) = -\{H, F\}(z(t)).$$

PROOF.

$$\frac{dF}{dt}(q(t), p(t)) = \sum \left(\frac{\partial F}{\partial q_k}\dot{q}_k + \frac{\partial F}{\partial p_k}\dot{p}_k \right) = \left(\frac{\partial F}{\partial q_k}\frac{\partial H}{\partial p_k} + \frac{\partial F}{\partial p_k}\left(\frac{-\partial H}{\partial q_k}\right)\right). \quad \square$$

We define a *canonical transformation* $\Phi: \mathbf{R}^{2n} \to \mathbf{R}^{2n}$ to be a map of phase space which preserves the Poisson brackets, i.e.

$$\{F, G\} \circ \Phi = \{F \circ \Phi, G \circ \Phi\}.$$

If we let F be one of the variables q_i, p_i, we see immediately from Proposition 1 that equation (*) characterizes Hamiltonian paths, so that canonical transformations preserve the form of Hamilton's equations.

EXAMPLE 1.

$$(8) \qquad \Phi: (q, p) \to (Q, P) \equiv (-p, q).$$

Since a canonical transformation $\Phi: (q, p) \to (Q(q, p), P(q,p))$ preserves Poisson brackets, we find after comparing the Poisson brackets $\{P, Q\}$ and $\{p, q\}$ that

$$(9) \qquad \{P_j, P_k\} = \{Q_j, Q_k\} = 0, \qquad \{P_j, Q_k\} = \delta_{jk}.$$

Conversely, we see by checking the definition that if (9) holds then the transformation is canonical. Equation (9) is a system of quadratic equations for the elements of the Jacobian matrix Φ'; in fact, these equations are just a restatement of the matrix equation

$$(10) \qquad (\Phi')^t J(\Phi') = J,$$

where $J = \left(\begin{smallmatrix} 0 & I \\ -I & 0 \end{smallmatrix}\right)$. A matrix A is called symplectic if $A^t J A = J$. We have

PROPOSITION 2. Φ *is canonical* $\Leftrightarrow$ Φ' *is symplectic at every point.*

A simple calculation shows that a symplectic transformation preserves volume and orientation.

We will also need a third means of characterizing canonical transformations. This involves the *symplectic form*

$$(11) \qquad \omega = \sum_{k=1}^{n} dp_k \wedge dq_k.$$

A straightforward calculation shows

PROPOSITION 3. Φ *is canonical* $\Leftrightarrow$ $\Phi^*\omega = \omega$.

EXAMPLE 2. A map $\Phi: q \to Q(q)$ of just the position variables extends to give a canonical map

$$(12) \qquad \Phi^{\#}: (q, p) \to (Q, P) \equiv \left(\Phi(q), \left[(\Phi'(q))^t\right]^{-1} p\right).$$

Indeed, it suffices to check that

$$(13) \qquad \sum p_k dq_k = \sum P_k dQ_k,$$

since the exterior derivative of this equation gives $\Sigma dp_k \wedge dq_k = \Sigma dP_k \wedge dQ_k$ as required. We have

$$\sum P_k dQ_k = \langle P, dQ \rangle = \langle P, \Phi'(q)dq \rangle = \langle (\Phi'(q))'P, dq \rangle.$$

Therefore if we put $p = (\Phi'(q))'P$ or $P = [(\Phi'(q))']^{-1}p$, (13) holds and (12) becomes a canonical transformation. Note that $\Phi^\#$ sends the Hamiltonian

$$H = \frac{1}{2} \sum_j \frac{p_j^2}{2m_j} + V(q)$$

into the Hamiltonian

$$H = \frac{1}{2} \sum_{i,j} g^{ij}(Q)P_i P_j + \tilde{V}(Q),$$

where $\Sigma_j m_j \dot{q}_j^2 = \Sigma_{i,j} g_{ij}(Q)\dot{Q}_i\dot{Q}_j$, $g^{ij} = (g_{ij})^{-1}$ and $\tilde{V}(Q) = V(q)$. We can now finish our discussion of the Kepler problem. The change of variables $\Phi: (x, y) \rightarrow (r, \theta)$ into polar coordinates induces a canonical transformation $\Phi^\#$ carrying the old Hamiltonian into the Hamiltonian (4). We are therefore justified in using the coordinates $(r, p_r, \theta, p_\theta)$ when calculating the orbit of the particle.

EXAMPLE 3. We end this section by giving one final example of a canonical transformation. This is a transformation used by Sundmann in a regularization of the three-body problem. We begin by considering two bodies in $\mathbf{R}^l$ separated by a distance q and put $p = \dot{q}$. The Hamiltonian for gravitational attraction is

$$H = p^2 - 1/q,$$

and we know that H remains constant as p, q evolve by Hamilton's equations. Thus, $H = E$ where E is the energy. For small q, this yields $\dot{q}^2 \sim 1/q$ and so $\dot{q} \sim q^{-1/2}$ or $q^{3/2} \sim t$. This gives

$$q \sim t^{2/3}, \qquad p \sim t^{-1/3}.$$

The map

$$(14) \qquad \Phi: (q, p) \rightarrow (Q, P) = (-p^2 q, 1/p)$$

satisfies $dP \wedge dQ = (-1/p^2)dp \wedge -p^2 dq = dp \wedge dq$ and is therefore canonical. In these new coordinates, $P \sim t^{1/3}$, $Q \sim 1$ as we approach a collision at time $t = 0$.

The treatment of the three-body problem is simplified by using a change of clock. Suppose first that $H = 0$ on the path we are interested in. If we use a new Hamiltonian $H \rightarrow FH$, Hamilton's equations become

$$\frac{dq_i}{dt} = \frac{\partial H}{\partial p_i} \cdot F, \qquad \frac{dp_i}{dt} = \frac{-\partial H}{\partial q_i} \cdot F,$$

and we are led to the change of variables $t \rightarrow \tau$ so that $d\tau = dt/F$. In the general case, we have $H - E = 0$ along a given Hamiltonian path and so we can make a change of clock by using a new Hamiltonian

$$(15) \qquad H \rightarrow F(H - E).$$

After these preliminaries we now turn to the three-body problem itself. We assume that two of the bodies are near collision, while the third body is well separated from the other two. Let q_1, q_2, q_3 measure the displacement between the two nearby bodies and put $p_i = \dot{q}_i$, $i = 1, 2, 3$. The Hamiltonian is

$$(16) \qquad H = p_1^2 + p_2^2 + p_3^2 - \frac{1}{\left(q_1^2 + q_2^2 + q_3^2\right)^{1/2}} + \text{Junk}.$$

Along a given Hamiltonian path H is conserved, $H = E$. We introduce the canonical transformation

$$(17) \qquad \Phi\colon (q, p) \to (Q, P) = \left(L(p)q,\, p/\|p\|^2\right),$$

where $L(p)$ is the linear transformation $L(p) = [(\partial P/\partial p)^t]^{-1} = \|p\|^2 R_p$, and R_p denotes reflection through the plane normal to p. This satisfies $\|Q\| = \|p\|^2\|q\|$ and sends the Hamiltonian (16) into

$$H = \|P\|^{-2} - \|P\|^{-2}\|Q\|^{-1} - E + \text{Junk}.$$

After a change of clock (15) with $F = \|P\|^2$, this becomes

$$(18) \qquad \tilde{H} = 1 - \|Q\|^{-1} + \|P\|^2(\text{Junk} - E).$$

Along our path we have $\tilde{H} = 0$, while near collisions $\|p\| \gg 1$. Thus $\|P\| \ll 1$ and $\|Q\|^{-1} \approx 1$ near the collision, and therefore $\tilde{H}$ has a nice nonsingular solution. We can conclude that the solution of Newton's equations for the three-body problem continues in a natural way past simple collisions. This calculation will reappear in Chapter 11.

2. Generating functions. In the last section we saw that there are at least three ways to decide whether a transformation is canonical or not. On the other hand it is not so clear how to construct canonical transformations. One means of doing this is to use generating functions. Given a canonical transformation

$$\Phi\colon (q, p) \to (Q, P),$$

consider the graph of Φ

$$\Gamma = \{(q, p, Q, P)\colon \Phi(q, p) = (Q, P)\}.$$

By Proposition 3 we know that on Γ

$$d\left(\sum q_i \, dp_i + \sum P_i \, dQ_i\right) = -\left(\sum dp_i \wedge dq_i - \sum dP_i \wedge dQ_i\right) = 0,$$

and we can conclude that locally there exists a function S on Γ such that

$$\sum (q_i \, dp_i + P_i \, dQ_i) = dS.$$

Assume now that Φ is a small perturbation of the identity. This means that on Γ instead of the coordinates (q, p) we can use (p, Q). We have $dS(p, Q) = \Sigma((\partial S/\partial p_i)dp_i + (\partial S/\partial Q_i)dQ_i)$ and therefore

$$q_i = \frac{\partial S(p, Q)}{\partial p_i}, \qquad P_i = \frac{\partial S(p, Q)}{\partial Q_i}.$$

The function S is called a *generating function* for the canonical transformation.

EXAMPLE. The identity map $\Phi: (q, p) \to (Q, P) = (q, p)$ arises from the generating function $S(p, Q) = \Sigma p_i Q_i$.

In fact each of the steps taking us from Φ to S can be reversed, giving us the

PROPOSITION 4. *If $S(p, Q)$ is a small perturbation of $\Sigma p_i Q_i$, then defining*

$$(19) \qquad q_i = \frac{\partial S(p, Q)}{\partial p_i}, \qquad P_i = \frac{\partial S(p, Q)}{\partial Q_i}$$

gives a canonical transformation

$$\Phi: (q, p) \to (Q, P).$$

PROOF. The graph

$$\Gamma = \{(q, p, Q, P): \text{equations (19) hold}\}$$

is a manifold. Because $S(p, Q)$ is a small perturbation of $\Sigma p_i Q_i$, we can by the implicit function theorem use either (q_i, p_i) or (p_i, Q_i) as coordinates. In particular, Γ is the graph of a map $\Phi: (p, q) \to (P, Q)$. By (19) the identity

$$\sum_j P_j dQ_j + \sum q_j dp_j = dS(p, Q)$$

holds on Γ. Therefore

$$\sum_j dP_j \wedge dQ_j - \sum dp_j \wedge dq_j = 0$$

on Γ and Φ is a canonical transformation by Proposition 3. $\quad\square$

3. The Hamilton-Jacobi equation. Consider the transformation of phase space

$$(20) \qquad \Phi_t: (x, \xi) \to (y, \eta)$$

defined by flowing along the integral curves to Hamilton's equations

$$(21) \qquad \frac{dy_j}{dt} = \frac{\partial H}{\partial \eta_j}(y, \eta), \qquad \frac{d\eta_j}{dt} = \frac{-\partial H}{\partial y_j}(y, \eta), \qquad y(0) = x, \eta(0) = \xi,$$

for a time t. This is called the *Hamiltonian flow*.

PROPOSITION 5. *The Hamiltonian flow (20) is canonical.*

PROOF. Note that we make this assertion only for small t, since we are only guaranteed a solution to (21) for small t. We want to show

$$(22) \qquad \Phi^*\omega = \omega.$$

Using the semigroup property of Φ_t: $\Phi^*_{t_0+s}\omega = \Phi^*_{t_0}(\Phi^*_s\omega)$ and differentiating with respect to s gives

$$\frac{d}{dt}(\Phi^*_t\omega)\Big|_{t=t_0} = \Phi^*_{t_0}\frac{d}{ds}(\Phi^*_s\omega)\Big|_{s=0}.$$

We see that we need only check that (22) holds to first order in t at $t = 0$. Since

$$y_j = x_j + t\frac{\partial H}{\partial \xi_j}(x, \xi) + O(t^2), \qquad \eta_j = \xi_j + t\frac{\partial H}{\partial x_j}(x, \xi) + O(t^2),$$

we have

$$\sum d\eta_j \wedge dy_j - \sum d\xi_j \wedge dx_j = \sum_j \left(d\xi_j - t\sum_k \frac{\partial^2 H}{\partial x_j \partial x_k} dx_k - t\sum_k \frac{\partial^2 H}{\partial x_j \partial \xi_k} d\xi_k \right)$$

$$\wedge \left(dx_j + t\sum_k \frac{\partial^2 H}{\partial \xi_j \partial \xi_k} d\xi_k + t\sum_k \frac{\partial^2 H}{\partial \xi_j \partial x_k} dx_k \right) + O(t^2) - \sum d\xi_j \wedge dx_j$$

$$= O(t^2).$$

Thus (22) holds to first order at $t = 0$ and the proposition is proved. $\square$

For small time t the canonical transformation Φ_t of (20) is near the identity and so has a generating function; denote this by $S_t(y, \xi)$. We are going to find a first order nonlinear partial differential equation which the generating function satisfies. We begin by looking at the lower order terms in t of $S_t(y, \xi)$. Equation (19) implies that $dS_t(y, \xi) = \sum x_j d\xi_j + \sum \eta_j dy_j$ and since

$$x_j = y_j - t\frac{\partial H}{\partial \xi_j}(y, \xi) + O(t^2), \qquad \eta_j = \xi_j - t\frac{\partial H}{\partial x_j}(x, \xi) + O(t^2),$$

we have

$$dS_t(y, \xi) = \sum (y_j d\xi_j + \xi_j dy_j) - t\sum \frac{\partial H}{\partial \xi_j}(y, \xi)d\xi_j$$

$$-t\sum \frac{\partial H}{\partial y_j}(y, \xi)dy_j + O(t^2),$$

so that

$$(22) \qquad S_t(y, \xi) = \sum \xi_j y_j - tH(y, \xi) + O(t^2).$$

Before we can begin to derive the nonlinear partial differential equation we need one more preliminary. Consider two successive canonical transformations

$$(23) \qquad (x, \xi) \overset{\Phi}{\to} (y, \eta) \overset{\Psi}{\to} (z, \zeta)$$

with generating functions

$$S(y, \xi) \qquad \text{associated to } \Phi,$$
$$T(z, \eta) \qquad \text{associated to } \Psi$$

and graph

$$\Gamma = \{(x, \xi, y, \eta, z, \zeta): \Phi(x, \xi) = (y, \eta) \text{ and } \Psi(y, \eta) = (z, \zeta)\}.$$

Let $\mathcal{S}(z, \xi)$ be the generating function associated with $\Psi \circ \Phi$.

PROPOSITION 5. $\mathcal{S}(z, \xi) = S(y, \xi) + T(z, \eta) - \sum \eta_j y_j$ whenever $(x, \xi, y, \eta, z, \zeta)$ $\in \Gamma$.

PROOF. Equation (19) implies that on Γ

$$dS(y, \xi) = \sum x_j d\xi_j + \sum \eta_j dy_j, \qquad dT(z, \eta) = \sum y_j d\eta_j + \sum \zeta_j dz_j$$

and so

$$dS(y, \xi) + dT(z, \eta) = \sum x_j d\xi_j + d\left(\sum \eta_j y_j\right) + \sum \zeta_j dz_j$$

or

$$\mathcal{S}(z, \xi) = S(y, \xi) + T(z, \eta) - \sum \eta_j y_j. \quad \square$$

Now assume that the canonical transformations (23) arise as Hamiltonian flows $\Phi = \Phi_{t'}$, $\Psi = \Phi_t$ with $t \ll 1$. Then from (22) we have

$$T(z, \eta) = \sum \eta_j z_j - tH(z, \eta) + O(t^2)$$

and thus by Proposition 5

$$\begin{aligned}
\mathcal{S}(z, \xi) &= S(y, \xi) + \sum \eta_j(z_j - y_j) - tH(z, \eta) + O(t^2) \\
&= S(z, \xi) - \sum \frac{\partial S}{\partial y_j}(z, \xi)(z_j - y_j) + O(t^2) \\
&\quad + \sum \eta_j(z_j - y_j) - tH(z, \eta) + O(t^2) \\
&= S(z, \xi) - tH(z, \eta) + O(t^2),
\end{aligned}$$

since $\eta_j = \partial S / \partial y_j$. Therefore if we put $S_t(z, \xi) = S(z, \xi)$, we have

$$\frac{\partial}{\partial t}\left(S_t(z, \xi)\right) = \lim_{t \to 0} \frac{\mathcal{S}(z, \xi) - S(z, \xi)}{t} = \lim_{t \to 0} - H(z, \eta) = -H\left(z, \frac{S_t(z, \xi)}{\partial z_j}\right).$$

We have proved

THEOREM 1. *If $S_t(z, \xi)$ is the generating function associated with a Hamiltonian flow Φ_t, then $S_t(z, \xi)$ satisfies the first order nonlinear partial differential equation (with parameter ξ)*

$$(24) \qquad \frac{\partial}{\partial t} S_t(z, \xi) + H\left(z, \frac{\partial S_t}{\partial z_j}(z, \xi)\right) = 0.$$

Equation (24) is called the Hamilton-Jacobi equation; it will occur later in our study of the wave equation. To solve it, the theorem tells us that we must simply calculate the integral curves of H and let the solution be the generating function of the resulting flow.

Finally, we mention two general references for this chapter; the book by Arnold [1] and the book by Goldstein [28].

CHAPTER 3. ELLIPTIC PARTIAL DIFFERENTIAL EQUATIONS

In this chapter we will consider the general second-order linear equation

$$(1) \qquad \sum a_{ij}(x) \frac{\partial^2 u}{\partial x_i \partial x_j} + \text{lower order terms} = f,$$

where (a_{ij}) is a positive definite, smoothly varying matrix. We can try to solve (1) by freezing the coefficients at a point x_0 so that it is reduced to the constant coefficient equation

$$\sum a_{ij}(x_0)\frac{\partial^2 u}{\partial x_i \partial x_j} + \text{lower order terms} = f.$$

This approach leads us to the study of singular integral operators (SIOs) and pseudodifferential operators (ψDOs). These are defined in §1 and some of their basic properties are explored in the next three sections. In §5 we prove a regularity theorem about elliptic operators on manifolds.

We will also consider the nonlinear equation

$$(2) \qquad A\left(x, u(x), \partial u/\partial x_i, \partial^2 u/\partial x_i \partial x_j\right) = 0,$$

where $A(x, \omega, \omega_i, \omega_{ij})$ is a smooth function on $\mathbf{R}^n \times \mathbf{R} \times \mathbf{R}^n \times \mathbf{R}^{n^2}$. This defines an elliptic equation if

$$A_{jk} \equiv \left(\partial A/\partial \omega_{jk}\right)$$

is a positive definite matrix. Recall that the Lipschitz space $\Lambda(\alpha)$ consists of those functions f satisfying

$$(3.\text{i}) \qquad \sup_x |f(x+h) - f(x)| \leqslant C|h|^{\alpha},$$

$$(3.\text{ii}) \qquad \sup_x |f(x)| \leqslant C.$$

We assume that $0 < \alpha < 1$. The norm of f is the infimum of the C satisfying (3). In §6 we prove the following regularity theorem from Schauder theory: If u is a solution of (2) and $\partial^2 u/\partial x_i \partial x_j \in \Lambda(\alpha)$, then $u \in C^{\infty}$. Furthermore the norm $\|u\|_{C^k}$ is bounded by a quantity determined only by A and $\|\partial^2 u/\partial x_i \partial x_j\|_{\Lambda(\alpha)}$. The proof requires that we examine the linear equation (1) when the coefficients (a_{ij}) are no longer assumed to be smooth but only in $\Lambda(\alpha)$.

1. SIOs and ψDOs. Consider first the equation

$$(4) \qquad \Delta u = f$$

in $\mathbf{R}^n$. A particular solution of (4) is given by convolving f with the fundamental solution

$$u(x) = c_n \int_{\mathbf{R}^n} \frac{f(x-y)}{|y|^{n-2}}\, dy = c_n \int_{\mathbf{R}^n} \frac{f(y)}{|x-y|^{n-2}}\, dy.$$

Here, as later, c_n denotes a constant depending only on the dimension of the space. We have

THEOREM 1. *If $f \in \Lambda(\alpha)$ and u satisfies (4), then $\partial^2 u/\partial x_i \partial x_j \in \Lambda(\alpha)$.*

To prove this, we must take the second derivatives in x of

$$u(x) = \int_{\mathbf{R}^n} \frac{f(y)}{|x-y|^{n-2}}\, dy.$$

Formally this requires that f be integrated against a kernel which is homogeneous of degree $-n$ and therefore not integrable. This is an example of an SIO.

Consider kernels

$$(5) \qquad\qquad K(x) = \Omega(x)/|x|^n$$

on $\mathbf{R}^n$, where

$$(6)\qquad
\begin{aligned}
&\text{(i)} \quad \Omega(x) \text{ is homogeneous of degree } 0,\\
&\text{(ii)} \quad \Omega(x) \text{ is } C^\infty \text{ away from } 0,\\
&\text{(iii)} \quad \int_{a<|y|<b} K(y)\,dy = 0 \text{ for all } a,\, b.
\end{aligned}$$

By definition an SIO is the operator

$$(7) \qquad f(x) \to K * f(x) \equiv \lim_{\substack{\delta \to 0 \\ m \to \infty}} \int_{\delta<|y|<m} K(y)f(x-y)\,dy.$$

Note that this limit obviously exists when $f \in C_0^\infty$.

EXAMPLE 1. For $f \in L^p(\mathbf{R}^n)$, $1 \leqslant p < \infty$, the Riesz transforms R_j are defined by

$$R_j(f)(x) = \lim_{\varepsilon \to 0} c_n \int_{|y|>\varepsilon} \frac{y_j}{|y|^{n+1}} f(x-y)\,dy,$$

$$(8) \qquad c_n = \Gamma\!\left(\frac{n+1}{2}\right)\bigg/ \pi^{(n+1)/2}, \qquad j = 1,\ldots,n.$$

In other words the R_j are defined by the kernel $K_j(x) = \Omega_j(x)/|x|^n$, with $\Omega_j(x) = c_n x_j/|x|$.

EXAMPLE 2. We claim

$$(9) \qquad\qquad \partial^2 f/\partial x_j \partial x_k = -R_j R_k \Delta f.$$

Indeed, since the Fourier transform of $\partial f/\partial x_j$ is $-2\pi i x_j f(x)$, we have

$$\left(\frac{\partial^2 f}{\partial x_j \partial x_k}\right)^{\!\wedge}(x) = -4\pi^2 x_j x_k \hat{f}(x) = -\left(\frac{i x_j}{|x|}\right)\left(\frac{i x_k}{|x|}\right)\!\left(-4\pi^2 |x|^2\right)\hat{f}(x)$$

$$= -\left(R_j R_k \Delta f\right)^{\!\wedge},$$

verifying (9). Here we have used the fact that

$$(10) \qquad\qquad (R_j f)^{\wedge}(x) = i\frac{x_j}{|x|}\hat{f}(x).$$

Equation (10) can be checked directly or can be deduced by analyzing the effect of rotations on the R_j. Equation (9) shows that if $\Delta u = f$, then

$$(11) \qquad\qquad f \to \partial^2 u/\partial x_i \partial x_j$$

is a composition of two SIOs. In fact it is a linear combination of the identity operator and an SIO; but we won't use that.

In §2 we will prove that SIOs are bounded on $\Lambda(\alpha)$. From Example 2, we see that this proves Theorem 1. In fact SIOs are bounded on L^p, $1 < p < \infty$, but we will not prove that here. An alternate approach to solving elliptic equations is via the Fourier transform. Consider the equation

$$(12) \qquad \sum a_{ij}(x)\frac{\partial^2 u}{\partial x_i \partial x_j} = f.$$

Freezing the coefficients at x_0 gives $\sum a_{ij}(x_0)\partial^2 u/\partial x_i\partial x_j = f$, and then taking Fourier transforms yields $-(\sum a_{ij}(x_0)\xi_i\xi_j)\hat{u}(\xi) = \hat{f}(\xi)$, so that the solution is $u(x) = c_n \int e^{ix\cdot\xi}(-a^{-1}(x_0,\xi))\hat{f}(\xi)\,d\xi$, if we suitably interpret the integral to take into account the zeros of $a(x_0,\xi)$. We might hope that the variable coefficient equation (12) is well approximated by the equation with the coefficients frozen at x_0 so that we would have an approximate solution given by

$$(13) \qquad u(x) \approx \int e^{ix\cdot\xi}(-a^{-1}(x,\xi))\hat{f}(\xi)\,d\xi.$$

In fact when (12) is elliptic, an excellent approximate solution is given by (13), which is an example of a ψDO.

The growth of the function $a^{-1}(x,\xi)$ is important and is captured by considering those functions $p(x,\xi) \in C^\infty(\mathbf{R}^n \times \mathbf{R}^n)$ with the property that, for any multi-indices α and β, there exists a constant $C_{\alpha,\beta}$ such that

$$(14) \qquad \left|D_x^\alpha D_\xi^\beta p(x,\xi)\right| \leqslant C_{\alpha,\beta}(1 + |\xi|)^{m-\rho|\beta|+\delta|\alpha|},$$

for all $(x,\xi) \in \mathbf{R}^n \times \mathbf{R}^n$. The function $p(x,\xi)$ is called an mth order symbol and the class of symbols just defined is denoted $S^m_{\rho,\delta}$. We assume $m, \rho, \delta \in \mathbf{R}$, and that $0 \leqslant \rho, \delta \leqslant 1$. Associated to any symbol $p(x,\xi)$ is a pseudodifferential operator

$$(15) \qquad f(x) \to p(x,D)f(x) \equiv \int e^{ix\cdot\xi}p(x,\xi)\hat{f}(\xi)\,d\xi.$$

EXAMPLE 3. A polynomial in ξ

$$p(x,\xi) = \sum_{|\alpha|\leqslant m} a_\alpha(x)\xi^\alpha$$

is clearly in $S^m_{1,0}$. The Fourier inversion formula shows that $p(x,D)$ defined by (15) is nothing but a differential operator with symbol $p(x,\xi)$. In other words, differential operators are examples of ψDOs.

REMARK. To apply the argument leading to (13), we would need the operators to satisfy $p(x,D)\circ q(x,D) \approx pq(x,D)$ in some well-defined sense so that $p(x,D)\circ p^{-1}(x,D) \approx I$. In §4 we will derive an asymptotic expansion for the composition of ψDOs so that these formulas become precise. It turns out that an elliptic symbol $p(x,\xi) \in S^m_{1,0}$ has a parametrix whose main term is in $S^{-m}_{1,0}$. A parametrix of an operator $p(x,D)$ is an operator $q(x,D)$ satisfying

$$(16) \qquad p(x,D)\circ q(x,D) \sim I, \qquad q(x,D)\circ p(x,D) \sim I.$$

We summarize by listing some of the important properties of these operators:

SIOs: bounded on L^p, $1 < p < \infty$,

bounded on $\Lambda(\alpha)$, $0 < \alpha < 1$;

ψDOs: 0th order operators are bounded on L^2,

there are formulas for compositions and adjoints.

The book by Stein [57] on singular integral operators and the books by Taylor [61] and Trèves [63] on pseudodifferential operators are good references for this chapter.

Finally, we will denote the symbol class $S_{1,0}^m$ by S^m.

2. Boundedness of SIOs on $\Lambda(\alpha)$. Let T be an SIO

$$Tf(x) = \int_{\mathbf{R}^n} K(x - y)f(y)\, dy = \int_{\mathbf{R}^n} K(y)f(x - y)\, dy,$$

and let $f \in \Lambda(\alpha)$.

THEOREM 2. *T is bounded on $\Lambda(\alpha)$.*

PROOF. Without loss of generality we can assume

$$\begin{array}{lll} & \text{(i)} & f \text{ is supported in } |x| \leqslant 2, \\ (17) & \text{(ii)} & |f(x)| \leqslant 1, \\ & \text{(iii)} & |f(x) - f(x')| \leqslant |x - x'|^\alpha. \end{array}$$

With these assumptions it suffices to prove the estimates (3.i) and (3.ii) for $\delta \equiv |x - x'| \leqslant 1/10$. We know by (5) that

$$(18) \qquad |K(x)| \leqslant \frac{c}{|x|^n}, \qquad \left|\frac{\partial K}{\partial x_i}(x)\right| \leqslant \frac{c}{|x|^{n+1}}$$

and (6.iii) implies

$$Tf(x) = \int_{\mathbf{R}^n} K(x - y)f(y)\, dy = \int_{|x-y| \leqslant 100} K(x - y)f(y)\, dy$$

$$= \int_{|x-y| \leqslant 100} K(x - y)[f(y) - f(x)]\, dy;$$

therefore

$$|Tf(x)| \leqslant \int_{|x-y| \leqslant 100} |K(x - y)| \cdot |f(y) - f(x)|\, dy$$

$$\leqslant \int_{|x-y| \leqslant 100} \frac{c}{|x - y|^n} \cdot |x - y|^\alpha\, dy$$

$$\leqslant \int_{|x-y| \leqslant 100} \frac{c}{|x - y|^{n-\alpha}}\, dy < \infty,$$

showing Tf is bounded. We must now prove that $|Tf(x) - Tf(x')| \leqslant c|x - x'|^\alpha$.

We begin by splitting the integral

$$Tf(x) - Tf(x') = \int_{\mathbf{R}^n} K(y)[f(x-y) - f(x'-y)]\, dy$$

$$= \int_{|y| \le 10\delta} K(y)[f(x-y) - f(x'-y)]\, dy$$

$$+ \int_{|y| > 10\delta} K(y)[f(x-y) - f(x'-y)]\, dy$$

$$= I_1 + I_2,$$

and the integral I_1 into

$$I_1 = \int_{|y| \le 10\delta} K(y)f(x-y)\, dy - \int_{|y| \le 10\delta} K(y)f(x'-y)\, dy$$

$$= I_{11} + I_{12}.$$

Using (6.iii) again,

$$I_{11} = \int_{|y| \le 10\delta} K(y)f(x-y)\, dy = \int_{|y| \le 10\delta} K(y)[f(x-y) - f(x)]\, dy$$

so that as before

$$|I_{11}| \le \int_{|y| \le 10\delta} \frac{c}{|y|^n} \cdot |y|^\alpha\, dy = c'\delta^\alpha.$$

Similarly, $|I_{12}| \le c'\delta^\alpha$ and therefore $|I_1| \le 2c'\delta^\alpha$.

Define $K_\delta(y) \equiv K(y)\chi_{10\delta < y < 100}$, where $\chi_{10\delta < y < 100}$ is the characteristic function. Now

$$I_2 = \int_{\mathbf{R}^n} [K_\delta(x-y) - K_\delta(x'-y)] \cdot [f(y) - f(x)]\, dy.$$

For the integrand to be nonzero we must take $|x - y| > 9\delta$ and $|x' - y| > 9\delta$, in which case, by (18) and the mean value theorem we have

$$|K_\delta(x-y) - K_\delta(x'-y)| \le \frac{c\delta}{|x-y|^{n+1}},$$

so that

$$|I_2| \le \int_{|x-y| \ge 9\delta} \frac{c\delta\,|x-y|^\alpha}{|x-y|^{n+1}}\, dy \le c'\delta^\alpha.$$

Therefore $I_1 + I_2 \le c\delta^\alpha$ and the theorem is proved. $\square$

3. Boundedness of ψDOs on L^2. Recall that a symbol $p(x, \xi) \in S^0_{1/2,1/2}$ satisfies estimates of the form

$$(19) \qquad |\partial_x^\alpha \partial_\xi^\beta p(x, \xi)| \le C_{\alpha,\beta}(1 + |\xi|)^{|\alpha|/2 - |\beta|/2}.$$

THEOREM 3. $p(x, \xi) \in S^0_{1/2,1/2} \Rightarrow p(x, D)$ *is bounded on* L^2.

Let Q be a cube in (x, ξ)-space whose sides are parallel to the axes and satisfying

$$\operatorname{diam}_x Q = 1, \qquad \operatorname{diam}_\xi Q = M.$$

LEMMA 1. *It is sufficient to prove Theorem* 3 *for* $p(x, \xi)$ *supported in* Q, *with bound independent of* M.

PROOF OF LEMMA 1. The proof consists of three steps. First we localize to cubes in phase space using a partition of unity. Then we patch together the operators associated with each of these cubes using a lemma due to Cotlar and Stein. Finally we must verify that the estimates occurring in the hypothesis of the Cotlar-Stein lemma are satisfied.

Let $\{Q_\nu\}$ be a partition of phase space into blocks of various sizes, centered at points (x_ν, ξ_ν), satisfying

$$(20) \qquad \text{(i) } \operatorname{diam}_x Q_\nu = 1, \qquad \text{(ii) } \operatorname{diam}_\xi Q_\nu \approx |\xi_\nu| .$$

For example we can partition x-space into cubes of diameter 1 whose sides are parallel to the axes and we can partition ξ-space into similar cubes whose diameters are given by the lengths of the dyadic intervals $[0, 1], [1, 2], [2, 4], \ldots, [2^j, 2^{j+1}], \ldots$. Let $\{\phi_\nu(x, \xi)\}$ be a partition of unity of phase space such that

$$(21) \qquad \begin{array}{l} \text{(i) } \varphi_\nu(x, \xi) = 1 \text{ in a neighborhood of } (x_\nu, \xi_\nu), \\ \text{(ii) } \operatorname{supp} \phi_\nu \subset Q_\nu^*, \end{array}$$

where the double Q_ν^* is obtained from Q_ν by doubling the diameter and keeping the center fixed. We can find such a partition of unity by translating and dilating a fixed cutoff function and then normalizing so that the sum is always one. We may take our partition to satisfy $\Sigma_\nu \chi_{Q_\nu^*} \leqslant C$.

Using the partition of unity we can break up the symbol into pieces

$$(22) \qquad p_\nu(x, \xi) \equiv \phi_\nu(x, \xi) p(x, \xi).$$

Note that $p_\nu(x, \xi)$ is supported in a cube Q_ν^* and that if the center ξ_ν of Q_ν is of the order $|\xi_\nu| \approx M$, then $p_\nu(x, \xi)$ satisfies the estimates

$$(23) \qquad \left| \partial_x^\alpha \partial_\beta^\xi p_\nu(x, \xi) \right| \leqslant C_{\alpha, \beta} M^{|\alpha|/2 - |\beta|/2}.$$

By hypothesis we can assume

$$(24) \qquad p_\nu(x, D) \text{ are uniformly bounded on } L^2.$$

We must show that $p(x, D) = \Sigma p_\nu(x, D)$ is bounded on L^2. This is a consequence of

LEMMA 2 (COTLAR-STEIN). *Let* $\{A_j\}$ *be a sequence of bounded operators on a Hilbert space satisfying*

$$(25) \qquad \begin{array}{ll} \text{(i)} & \|A_j\| \leqslant B, \\ \text{(ii)} & \|A_j^* A_k\| \leqslant C(j - k), \\ \text{(iii)} & \|A_j A_k^*\| \leqslant C(j - k); \end{array}$$

then

$$\left\| \sum A_j \right\| \leq \sum (c(j))^{1/2}.$$

PROOF OF LEMMA 2. Note that

$$\| T^*T \cdots T^*T \| = \| T \|^m,$$

if the product on the left contains m terms. Put $T = \sum_{j=1}^N A_j$ so that

$$\| T \|^m \leq \sum_{j_1, \ldots, j_m = 1}^N \left\| A_{j_1}^* A_{j_2} \cdots A_{j_{m-1}}^* A_{j_m} \right\|.$$

By hypothesis each term on the right satisfies

$$\left\| A_{j_1}^* \cdots A_{j_m} \right\| \leq C(j_1 - j_2) C(j_3 - j_4) \cdots C(j_{m-1} - j_m),$$

$$\left\| A_{j_1}^* \cdots A_{j_m} \right\| \leq BC(j_2 - j_3) C(j_4 - j_5) \cdots C(j_{m-2} - j_{m-1}) B$$

so that

$$\left\| A_{j_1}^* \cdots A_{j_m} \right\| \leq B \left[C(j_1 - j_2) C(j_2 - j_3) \cdots C(j_{m-1} - j_m) \right]^{1/2}.$$

Therefore

$$\| T \|^m \leq NB \left[\sum (c(j))^{1/2} \right]^{m-1},$$

$$\| T \| \leq (NB)^{1/m} \left[\sum (c(j))^{1/2} \right]^{1-(1/m)}.$$

Letting $m \to \infty$ gives $\| T \| \leq \sum (c(j))^{1/2}$. Since the right side is independent of N, we can let $N \to \infty$ and conclude

$$\left\| \sum A_j \right\| \leq \sum (c(j))^{1/2}. \qquad \square$$

COROLLARY. *Instead of* (25), *assume*

$$(25') \qquad \left\| A_j^* A_k \right\| \leq C(j, k), \qquad \left\| A_j A_k^* \right\| \leq C(j, k),$$

with $\sup_j \sum_k \sqrt{C(j, k)} < \infty$. *Then*

$$\left\| \sum A_j \right\| \leq \max_j \sum_k \sqrt{C(j, k)}.$$

PROOF. The same argument with trivial modifications works. $\square$

All that remains is to verify the estimates $(25')$. Now

$$\left\langle p_j(y, D)g, f \right\rangle = \int e^{iy \cdot \eta} p_j(y, \eta) \hat{g}(\eta) \bar{f}(y) \, d\eta \, dy$$

$$= \int e^{i(y-x) \cdot \eta} p_j(y, \eta) g(x) \bar{f}(y) \, dx \, d\eta \, dy.$$

Therefore $p_j(x, D)$ has an adjoint $p_j^*(x, D)$

$$(26) \qquad p_j^*(x, D) f(x) = \int e^{i(x-y) \cdot \eta} \overline{p(y, \eta)} f(y) \, d\eta \, dy,$$

and

$$p_j^* p_k f(x) = \int e^{i(x-y)\cdot\eta} \overline{p_j(y,\eta)} \, e^{i\xi\cdot y} p_k(y,\xi)\hat{f}(\xi)\,d\xi\,d\eta\,dy$$

$$= \int e^{ix\cdot\eta}\left[e^{i(\xi-\eta)\cdot y}\overline{p_j(y,\eta)}\,p_k(y,\xi)\hat{f}(\xi)\,d\xi\,dy\right]d\eta,$$

so that by Plancherel's theorem we need only estimate

$$(27) \qquad \int e^{i(\xi-\eta)\cdot y}\overline{p_j(y,\eta)}\,p_k(y,\xi)\hat{f}(\xi)\,d\xi\,dy.$$

There are three cases.

Case (i). Suppose that the y-support of Q_j^* is distinct from the y-support of Q_k^*. Then the support of $p_j(y,\eta)$ and $p_k(y,\xi)$ are distinct and (27) vanishes.

Case (ii). Consider all those Q_k^* that intersect a fixed Q_j^* and consider the corresponding $p_k(y,\xi)$ and $\overline{p_j(y,\eta)}$. By (21.ii) there are at most $3^{2n}-1$ such $p_k(y,\xi)$. Moreover

$$\left\|p_j^*(y,D)p_k(y,D)\right\|_2 \leqslant \left\|p_j^*(y,D)\right\|_2\cdot\left\|p_k(y,D)\right\|_2,$$

and so the contribution to the sum $\Sigma(c(j,k))^{1/2}$ is bounded.

Case (iii). Assume that $\overline{p_j(y,\eta)}$ and $p_k(y,\xi)$ are supported in blocks which share some y-support and whose centers satisfy $|\eta_j|\approx M$ and $|\xi_k|\approx N$. Now

$$\frac{\left(-\Delta_y\right)^s e^{iy\cdot(\xi-\eta)}}{\left(|\xi-\eta|\right)^{2s}} = e^{iy\cdot(\xi-\eta)},$$

while from (23) we have

$$\left(-\Delta_y\right)^s\left[\overline{p_j(y,\eta)}\,p_k(y,\xi)\right] \leqslant (M+N)^s;$$

therefore after integrating (27) by parts we find

$$\left\|p_j^* p_k f\right\|_2 = \left\|\int e^{i(\xi-\eta)\cdot y}\overline{p_j(y,\eta)}\,p_k(y,\xi)\hat{f}(\xi)\,d\xi\,dy\right\|_2$$

$$= \left\|\int(|\xi-\eta|)^{-2s}e^{i(\xi-\eta)\cdot y}\hat{f}(\xi)\left(-\Delta_y\right)^s\left[\overline{p_j(y,\eta)}\,p_k(y,\xi)\right]dy\,d\xi\right\|_2$$

$$\leqslant C_s(|\eta_j-\xi_k|)^{-2s}\cdot\|f\|_2\cdot(M+N)^s$$

$$\leqslant \frac{C_s}{M^s+N^s}\cdot\|f\|_2.$$

This shows that we can essentially neglect the contribution to the sum $\Sigma(c(j,k))^{1/2}$ arising from these blocks. We have estimated $p_j^*(y,D)p_k(y,D)$; a similar argument yields an estimate for $p_j(y,D)p_k^*(y,D)$.

We have shown that the sum $\Sigma(c(j,k))^{1/2}$ is bounded. By the corollary to Lemma 2, the operator $p(x,D) = \Sigma_j p_j(x,D)$ is bounded on L^2. This finishes the proof of Lemma 1. $\square$

PROOF OF THEOREM 3. By Lemma 1, we can assume $p(x, \xi)$ is supported on a block Q in (x, ξ)-space of diameter $1 \times M$. We will rerun the proof of Lemma 1, this time using a finer localization. Let $\{Q_\nu\}$ be a partition of Q into blocks with

$$\operatorname{diam}_x Q_\nu = M^{-1/2}, \qquad \operatorname{diam}_\xi Q_\nu = M^{1/2},$$

and let $1 = \Sigma \phi_\nu(x, \xi)$ be a partition of unity with

$$\phi_\nu(x, \xi) = 1 \quad \text{near the center of } Q_\nu,$$
$$\operatorname{supp} \phi_\nu \subset Q_\nu^*.$$

As before, put

$$p(x, D) = \Sigma p_\nu(x, D) \quad \text{where } p_\nu(x, \xi) \equiv \phi_\nu(x, \xi) p(x, \xi).$$

We must now verify the hypotheses of Cotlar's lemma. We begin by showing that each of the $p_\nu(x, D)$ is bounded on L^2. Suppose that $p_0(x, \xi) \in C^\infty$ and is supported in $|x| < 10$, $|\xi| < 10$ so that $|p_0(x, D)u(x)| \leqslant C \|u\|_2 \chi_{|x| < 10}$, where χ is the characteristic function. By integrating we get

$$(28) \qquad \|p_0(x, D)u\|_2 \leqslant C \|u\|_2.$$

Let (x_ν, ξ_ν) be the center of Q_ν. Using the change of variables

$$(29) \qquad (x, \xi) \to (y, \eta) \equiv \left(M^{1/2}(x - x_\nu), M^{-1/2}(\xi - \xi_\nu) \right),$$

we can change the support of $p_\nu(x, \xi)$ to $|y| < 10$, $|\eta| < 10$ and apply estimate (28). This together with the fact that $p_\nu \in S^0_{1/2,1/2}$ uniformly in ν imply

$$\|p_\nu(x, D)u\|_2 \leqslant C \|u\|_2,$$

where C is independent of ν.

We must now estimate $\|p_j^*(x, D)p_k(x, D)\|_2$. Using formula (26) for the adjoint we have

$$(30) \qquad p_j^* p_k u(x) = \int e^{i(x-z)\cdot\zeta} \overline{p_j(z, \zeta)} \left[p_k(z, D)u(z) \right] dz d\zeta$$

$$= \int e^{i[(x-z)\cdot\zeta + (z-y)\cdot\xi]} \overline{p_j(z, \zeta)}\, p_k(z, \xi)u(y)\, dy d\xi dz d\zeta.$$

We proceed here just as we did at the end of Lemma 1. Let $(x_j, \xi_j), (x_k, \xi_k)$ be the centers of the cubes Q_j^*, Q_k^* supporting $\overline{p_j}$ and p_k. There are three cases.

Case (i). If $|x_j - x_k| \geqslant CM^{-1/2}$, then $\overline{p_j}p_k = 0$ and $\|p_j^* p_k\|_2 = 0$.

Case (ii). Assume $|x_j - x_k| < CM^{-1/2}$ but $|\xi_j - \xi_k| \geqslant CM^{1/2}$. We rewrite (30) as

$$p_j^* p_k u(x) = \int e^{i[(x-z)\cdot\zeta + z\cdot\xi]} \overline{p_j(z, \zeta)}\, p_k(z, \xi)\hat{u}(\xi)\, d\xi dz d\zeta$$

$$= \int e^{ix\cdot\xi} p_{jk}(x, \xi)\hat{u}(\xi)\, d\xi = p_{jk}(x, D)u(x),$$

where

$$(31) \qquad p_{jk}(x, \xi) = \int e^{i[(x-z)\cdot(\zeta-\xi)]} \overline{p_j(z, \zeta)}\, p_k(z, \xi)\, dz d\zeta.$$

To estimate $p_{jk}(x, \xi)$, we use the identities

$$(32.\text{i}) \qquad -|\zeta - \xi|^{-2}\Delta_z e^{i[(x-z)\cdot(\zeta-\xi)]} = e^{i[(x-z)\cdot(\zeta-\xi)]},$$

$$(32.\text{ii}) \qquad -|x - z|^{-2}\Delta_\zeta e^{i[(x-z)\cdot(\zeta-\xi)]} = e^{i[(x-z)\cdot(\zeta-\xi)]}.$$

To use these, note that $|\zeta - \xi| \neq 0$ in the support of the integrand of (31); and for $|x - x_j| \geq CM^{-1/2}$, we have also that $|x - z| \neq 0$ in the support of the integrand of (31).

If we substitute the expressions (32.i) and (32.ii) for the exponential in (31) and integrate by parts repeatedly, we obtain

$$(33.\text{i}) \qquad p_{jk}(x, \xi) = \int e^{i[(x-z)\cdot(\zeta-\xi)]}\left(-|\zeta - \xi|^{-2}\Delta_z\right)^s\left[\,\overline{p_j(z, \zeta)}\, p_k(z, \xi)\right] dz\, d\zeta,$$

$$(33.\text{ii}) \qquad p_{jk}(x, \xi) = \int e^{i[(x-z)\cdot(\zeta-\xi)]}\left(-|x - z|^{-2}\Delta_\zeta\right)^s\left[\,\overline{p_j(z, \zeta)}\, p_k(z, \xi)\right] dz\, d\zeta.$$

Equation (33.i) holds for all (x, ξ) and (33.ii) holds for $|x - x_j| \geq CM^{-1/2}$.

Since the integrands are supported on the cube Q_j^* of volume ~ 1, and since $|\xi - \zeta| \sim |\xi_j - \xi_k|$ in (33.i) and $|x - z| \sim |x - x_j|$ in (33.ii), our estimates on the derivatives of $p_j(z, \zeta)$, $p_k(z, \xi)$ now yield

$$|p_{jk}(x, \xi)| \leq C_s\left(\frac{M^{1/2}}{|\xi_j - \xi_k|}\right)^{2s} \quad \text{for all } (x, \xi),$$

$$|p_{jk}(x, \xi)| \leq C_s\left(\frac{M^{-1/2}}{|x - x_j|}\right)^{2s} \quad \text{for all } |x - x_j| \geq CM^{-1/2}.$$

Therefore

$$|p_{jk}(x, \xi)| \leq C_s\left(M^{-1/2}|\xi_j - \xi_k| + M^{1/2}|x - x_j|\right)^{-2s} \quad \text{for all } (x, \xi)$$

while a glance at the definition shows that $p_{jk}(x, \xi)$ is supported in the projection Q_k^ξ of Q_k^* onto ξ-space. So we can estimate

$$|p_j^*(x, D)p_k(x, D)u(x)| \leq \left|\int e^{ix\cdot\xi}p_{jk}(x, \xi)\hat{u}(\xi)\, d\xi\right|$$

$$\leq C_s\left(M^{-1/2}|\xi_j - \xi_k| + M^{1/2}|x - x_j|\right)^{-2s}\int_{Q_k^\xi}|\hat{u}(\xi)|\, d\xi,$$

from which we get immediately

$$\|p_j^*(x, D)p_k(x, D)u\|_2 \leq C_s\left(M^{-1/2}|\xi_j - \xi_k|\right)^{-s}\|u\|_2;$$

in other words

$$\|p_j^*(x, D)p_k(x, D)\|_2 \leq C_s\left(M^{-1/2}|\xi_j - \xi_k|\right)^{-s}.$$

Similarly, we find

$$\|p_j(x, D)p_k^*(x, D)\|_2 \leq C_s\left(M^{-1/2}|\xi_j - \xi_k|\right)^{-s}.$$

Case (iii). The terms arising when $|x_j - x_k| < CM^{-1/2}$ and $|\xi_j - \xi_k| < CM^{-1/2}$ contribute only a finite amount to the sum appearing in the Cotlar-Stein lemma and need not be estimated.

We have considered all the cases and can now invoke the Cotler-Stein lemma to conclude that $p(x, D) = \Sigma p_\nu(x, D)$ is bounded on L^2. This proves Theorem 3. $\square$

We have just shown that pseudodifferential operators arising from 0th order symbols are bounded on L^2; on the other hand, in general, mth order symbols do not yield operators that are bounded on L^2. But consider the space

$$H^k(\mathbf{R}^n) = \left\{ u \in L^2(\mathbf{R}^n) \colon D^\alpha u \in L^2(\mathbf{R}^n), \text{ for } |\alpha| \leq k \right\},$$

for k a positive integer. Using Theorem 3 it is not hard to show that $p(x, \xi) \in S^m$ implies

$$p(x, D)\colon H^k(\mathbf{R}^n) \to H^{k-m}(\mathbf{R}^n)$$

is bounded. See [**61**, p. 51], for a proof.

Note that by the Plancherel theorem

$$D^\alpha u \in L^2(\mathbf{R}^n), \qquad |\alpha| \leq k \Leftrightarrow \xi^\alpha \hat{u}(\xi) \in L^2(\mathbf{R}^n), \qquad |\alpha| \leq k,$$

or equivalently,

$$\Leftrightarrow \left(1 + |\xi|\right)^k \hat{u}(\xi) \in L^2(\mathbf{R}^n).$$

Therefore we can generalize the spaces $H^k(\mathbf{R}^n)$ by defining for $s \in \mathbf{R}$,

$$H^s(\mathbf{R}^n) = \left\{ \text{tempered distributions } u\colon \quad \begin{array}{l} \text{(i) } \hat{u} \in L^2_{\text{loc}}(\mathbf{R}^n) \\[4pt] \text{(ii) } (1 + |\xi|)^s \hat{u} \in L^2(\mathbf{R}^n) \end{array} \right\}$$

and

$$\|u\|_{H^s} = \left[\int \left(1 + |\xi|^2\right)^s |\hat{u}(\xi)|^2 \, d\xi \right]^{1/2}.$$

These are the *Sobolev spaces*.

4. Stationary phase and the calculus of ψ DOs. In this section we will have to deal with integrals of the form

$$(34) \qquad F_M(x, \eta) = \int e^{iM(x-y)\cdot(\eta-\xi)} \psi(x, y, \xi, \eta) \, dy d\xi.$$

This integral converges if ψ has compact support. The estimates in Lemma 4 below allow us to pass to the limit and interpret the integral for a general symbol ψ.

We now prove the important composition formula.

THEOREM 4. *If $a(x, \xi) \in S^m$ and $b(x, \xi) \in S^n$, then $a(x, D) \circ b(x, D)$ is a pseudodifferential operator, whose symbol $a \circ b$ has an asymptotic expansion*

$$a \circ b \sim \sum_{0 \leq |\alpha|} \frac{1}{\alpha!} \left(\frac{1}{i} \partial_\xi\right)^\alpha a (\partial_x)^\alpha b.$$

More precisely, we have for each N the estimate

$$(35) \qquad a \circ b - \sum_{0 \leqslant |\alpha| < N} \frac{1}{\alpha!} \left(\frac{1}{i} \partial_\xi \right)^\alpha a (\partial_x)^\alpha b \in S^{m+n-N}.$$

PROOF. Using a localization argument like that in Lemma 1, we can assume that $a(x, \xi)$ and $b(x, \xi)$ are supported in a ring $M \leqslant |\xi| \leqslant 2M$.

With

$$a(x, D)u(x) = \iint e^{i(x-y)\cdot\xi} a(x, \xi) u(y) \, dy d\xi,$$

$$b(y, D)u(y) = \iint e^{i\eta \cdot y} b(y, \eta) \hat{u}(\eta) \, d\eta,$$

the composition becomes

$$a(x, D) \circ b(x, D)u(x) = \int e^{i(x-y)\cdot\xi + y\cdot\eta} a(x, \xi) b(y, \eta) \hat{u}(\eta) \, d\eta dy d\xi$$

$$= \int e^{ix\cdot\eta} (a \circ b)(x, \eta) \hat{u}(\eta) \, d\eta,$$

where we have defined

$$(36) \qquad (a \circ b)(x, \eta) \equiv \int e^{i(x-y)\cdot(\xi-\eta)} a(x, \xi) b(y, \eta) \, dy d\xi.$$

The integral (36) converges when $a(x, \xi)$ and $b(y, \eta)$ have compact support. As we remarked at the beginning of the section, we will be able to drop the assumption of compact support after we prove Lemma 4. According to the asymptotic expansion, the main contributions to the integral (36) occur when $(y, \xi) = (x, \eta)$.

This is an illustration of the principle of stationary phase, which holds more generally for integrals of the form

$$F(t) = \int e^{it\phi(z)} \psi(z) \, dz,$$

where $\phi(z) \in C^\infty(\mathbf{R}^n)$ and $\psi(z) \in C_0^\infty(\mathbf{R}^n)$. We are interested in the asymptotic behavior of $F(t)$ as $t \to \infty$. Using the identity

$$\frac{(\partial/\partial z_j) e^{it\phi(z)}}{it(\partial/\partial z_j)\phi(z)} = e^{it\phi(z)},$$

and integrating by parts k times gives

$$F(t) = \left(\frac{1}{it} \right)^k \int e^{it\phi(z)} \frac{\psi(z)}{\left[(\partial/\partial z_j)\phi(z) \right]^k} \, dz.$$

If $(\partial/\partial z_j)\phi(z) \neq 0$ on the support of ψ, then $F(t) \to 0$ very fast as $|t| \to \infty$. Breaking up $\psi = \Sigma_l \psi_l$ with a partition of unity so that supp ψ_l is small shows that we may neglect the region where $\partial_z\phi(z) \neq 0$. Therefore the only important points are those z such that $\partial_z\phi(z) = 0$. These are the *stationary points*.

We will use this principle in the form

LEMMA 4. *Let*

$$F_M(x, \eta) = \iint e^{iM(x-y)\cdot(\eta-\xi)}\psi(x, y, \xi, \eta)\, dy d\xi,$$

where ψ satisfies

(i) *$\psi \in C_0^\infty$ with support in $\{|x - y| + |\eta - \xi| < 1\}$,*

(ii) *ψ vanishes to order k' at the stationary point $(y, \xi) = (x, \eta)$.*

Then $\partial_{x,\eta}^\alpha F_M = O(M^{-k})$ for $|\alpha| \leq k$ as $M \to \infty$. Here k may be made arbitrarily large by taking k' large.

PROOF. Observe

$$\frac{\left[-\left(\partial_y^2 + \partial_\xi^2\right)\right]^s e^{iM(x-y)\cdot(\eta-\xi)}}{\left[M^2\left(|\eta - \xi|^2 + |x - y|^2\right)\right]^s} = e^{iM(x-y)\cdot(\eta-\xi)};$$

therefore after integrating by parts we get

$$F_M(x, \eta) = \int \frac{e^{iM(x-y)\cdot(\eta-\xi)}}{M^{2s}}\left[-\left(\partial_y^2 + \partial_\xi^2\right)\right]^s\left\{\frac{\psi(x, y, \xi, \eta)}{\left[|\eta - \xi|^2 + |x - y|^2\right]^s}\right\} dy d\xi.$$

If ψ vanishes to high enough order at $(x, \xi) = (y, \eta)$, then

$$\frac{\psi(x, y, \xi, \eta)}{\left[|\eta - \xi|^2 + |x - y|^2\right]^s} \in C^{2s},$$

so that

$$\left|\left(\partial_y^2 + \partial_\xi^2\right)^s\left\{\frac{\psi(x, y, \xi, \eta)}{\left[|\eta - \xi|^2 + |x - y|^2\right]^s}\right\}\right| \leq C_s.$$

Consequently, recalling that ψ is supported in $|y| \leq 1, |\xi| \leq 1$, we find

(37) $$|F_M(x, \eta)| \leq C_s/M^{2s}.$$

We can obtain similar estimates for $\partial_{x,\eta}^\alpha F_M$ by noting $\partial_{x,\eta}^\alpha F_M(x, \eta)$ is a sum of terms

$$\int e^{iM(x-y)\cdot(\xi-\eta)}M^\beta(x - y)^\gamma(\xi - \eta)^\delta \psi_\mu(x, y, \xi, \eta)\, dy d\xi,$$

with $\beta \leq |\alpha|$. These terms are

$$M^\beta \cdot \int e^{iM(x-y)\cdot(\xi-\eta)}\tilde{\psi}(x, y, \xi, \eta)\, dy\, d\xi,$$

with $\tilde{\psi} = (x - y)^\gamma(\xi - \eta)^\delta\psi_\mu$, so (37) applies. Therefore,

$$|\partial_{x,\eta}^\alpha F_M| \leq \frac{C_s'}{M^{2s}} \cdot M^{|\alpha|}.$$

Taking s large enough completes the proof. $\quad\square$

We now return to the proof of the composition formula. Let $\theta \in C_0^\infty$ be a cutoff function

$$\theta(x, \xi, y, \eta) = \begin{cases} 1 & \text{for } |x+y| + M^{-1}|\xi - \eta| \leq 1, \\ 0 & \text{for } |x+y| + M^{-1}|\xi - \eta| \geq 2 \end{cases}$$

and consider the integral

$$I_\theta = \iint e^{i(x-y)\cdot(\xi-\eta)}\theta(x, \xi, y, \eta)a(x, \xi)\,b(y, \eta)\,dy d\xi.$$

This is the integral (36) with the cutoff function θ introduced so that Lemma 4 applies. It will turn out that putting in θ has negligible effect on the integral (36). By Taylor's theorem the product $a \cdot b$ satisfies

$$a(x, \xi)b(y, \eta) \sim \sum_{|\alpha|,|\beta| \geq 0} \left[\frac{1}{\alpha!}\partial_\eta^\alpha a(x, \eta)(\xi - \eta)^\alpha\right]\left[\frac{1}{\beta!}\partial_x^\beta b(x, \eta)(y - x)^\beta\right].$$

If we truncate this sum after sufficiently many terms, the remainder satisfies the hypotheses of Lemma 4 and hence may be neglected. Thus I_θ is essentially

$$(38) \quad \sum_{0 \leq |\alpha|,|\beta| \leq N} \frac{1}{\alpha!\beta!}$$

$$\cdot \left[\int e^{i(x-y)\cdot(\xi-\eta)}\theta(x, \xi, y, \eta)(\xi - \eta)^\alpha(y - x)^\beta\,dy d\xi\right]\left(\partial_\eta^\alpha a\partial_x^\beta b\,\big|_{(x,\eta)}\right).$$

To evaluate this integral we can remove $\theta(x, \xi, y, \eta)$ since a simple stationary phase argument shows that the effect is only $O(M^{-s})$. This leaves us with a finite sum of integrals of the form

$$J = \iint e^{i(x-y)\cdot(\xi-\eta)}(\xi - \eta)^\alpha(y - x)^\beta\,dy d\xi,$$

which we interpret as

$$\lim_{\varepsilon_k,\delta_k \to 0^+} \iint e^{i(x-y)\cdot(\xi-\eta)}(\xi - \eta)^\alpha(y - x)^\beta e^{-\Sigma_k \varepsilon_k(x_k-y_k)^2+\delta_k(\xi_k-\eta_k)^2}\,dy d\xi.$$

It is sufficient to evaluate the translated integral $\iint e^{iy\cdot\xi}y^\beta\xi^\alpha\,dy d\xi$. By analyzing the effect of dilations we see that the integral vanishes unless $\alpha = \beta$. Now the Dirac δ function satisfies

$$\left(\partial_y^k\right)\delta(y) = \int e^{iy\cdot\xi}(i\xi)^k\,dy,$$

therefore

$$\iint e^{iy\cdot\xi}y^k\xi^k\,dy d\xi = \frac{1}{(i)^k}\int y^k\left(\delta^{(k)}(y)\right)\,dy = \frac{1}{(i)^k}k!,$$

and hence $J = k!/(i)^k$. This establishes the correct coefficients in the asymptotic expansion (38) and finishes the proof. $\square$

COROLLARY. *Let U be an open set and $u \in C^\infty(U)$. Then $p(x, D)u(x) \in C^\infty(U)$.*

PROOF. Let $\phi, \psi \in C_0^\infty$ with $\psi = 1$ on supp ϕ. It is enough to show that $p(x, D)u$ is smooth in supp ϕ whenever u is smooth in a neighborhood of supp(ψ). We write $\phi(x)p(x, D)u = \phi(x)p(x, D)(\psi u) + \phi(x)p(x, D)\cdot (1 - \psi(x))u$. The first term on the right is obviously smooth since ψu is smooth and compactly supported. To handle the second term we note that $\phi(x)p(x, D)\cdot (1 - \psi(x))$ is a composition of three pseudodifferential operators whose symbols have no common support. The composition law (Theorem 4) for ψDOs shows that $\phi(x)p(x, D)(1 - \psi(x))$ is a smoothing operator ($\in S^{-\infty}$), so the second term must also be smooth. Therefore $\phi(x)p(x, D)u$ is smooth, and the proof is complete. $\square$

Although pseudodifferential operators, unlike differential operators, do not necessarily decrease the support of functions on which they act, they do according to the corollary decrease the singular support of the functions on which they act, i.e., the complement of the open set on which the function is smooth. Such an operator is called *pseudolocal*.

5. Elliptic regularity. We begin by sketching a proof that elliptic operators have parametrices (Equation 16). A symbol $a(x, \xi) \in S^m$ is called elliptic if

$$c(1 + |\xi|)^m \leqslant |a(x, \xi)| \leqslant c'(1 + |\xi|)^m.$$

Step 1. Put $b_1(x, \xi) = \theta(\xi)/a(x, \xi)$, where $\theta(\xi)$ is a cutoff function that vanishes in a bounded set containing the zeros of $a(x, \xi)$ and is 1 elsewhere. It is easy to check that $b_1(x, \xi)$ satisfies the estimates required to make $b_1 \in S^{-m}$. By Theorem 4,

$$a(x, D) \circ b_1(x, D) = I + e_1(x, D),$$

where I is the identity and $e_1(x, \xi) \in S^{-1}$.

Step 2. Put $b_2(x, \xi) = -\theta(\xi)e_1(x, \xi)/a(x, \xi) \in S^{-m-1}$ so that

$$a(x, D) \circ b_2(x, D) = -e_1(x, D) + e_2(x, D),$$

where $e_2(x, \xi) \in S^{-2}$.

Step k. In general put $b_k(x, \xi) = -\theta(\xi)e_{k-1}(x, \xi)/a(x, \xi) \in S^{-m-k+1}$ so that

$$a(x, D) \circ b_k(x, D) = -e_{k-1}(x, D) + e_k(x, D),$$

where $e_k(x, \xi) \in S^{-k}$

Formally setting $b(x, \xi) \equiv \Sigma_1^\infty b_k(x, \xi)$, we have

$$a(x, D) \circ b(x, D) = I + e(x, D),$$

where $e(x, \xi) \in S^{-\infty} \equiv \cap_m S^m$. We need

LEMMA 5. *Given a formal sum $\Sigma_{k=1}^\infty b_k$, with $b_k(x, \xi) \in S^{m-k+1}$, there exists $b \in S^m$ such that*

$$b - \left(\sum_1^N b_k \right) \in S^{m-N}.$$

SKETCH OF PROOF. Set $b = \Sigma \chi_k(\xi) b_k$, where $b_k \in C^\infty$ and

$$\chi_k(\xi) = \begin{cases} 1, & |\xi| > R_k, \\ 0, & |\xi| < R_k/2. \end{cases}$$

If $R_k \nearrow \infty$ sufficiently rapidly, then $b(x, \xi) \in S^m$ as required. $\square$

We have proved

THEOREM 5. *An elliptic symbol $a \in S^m$ has a parametrix $b \in S^{-m}$ satisfying*

$$a(x, D) \circ b(x, D) = I + e(x, D), \qquad b(x, D) \circ a(x, D) = I + \tilde{e}(x, D),$$

where $e(x, D), \tilde{e}(x, D) \in S^{-\infty} \equiv \cap_m S^m$ (smoothing errors). (See Remark below.)

REMARK. Strictly speaking, we only checked

$$a(x, D) \circ b(x, D) = I + e(x, D)$$

where $e \in S^{-\infty}$. Similarly, we can construct a symbol $\tilde{b}$ so that

$$\tilde{b}(x, D) \circ a(x, D) = I + \tilde{e}(x, D)$$

with $\tilde{e} \in S^{-\infty}$. Modulo errors in $S^{-\infty}$ we have

$$\tilde{b} = \tilde{b} \circ (a \circ b) = (\tilde{b} \circ a) \circ b = b.$$

So b is both a left and right parametrix for a.

We will now discuss briefly a regularity theorem for elliptic systems of equations on compact manifolds. First we examine the invariance of pseudodifferential operators under change of variables. Let $\tau: U_1 \to U_2$ be a diffeomorphism between two open sets in $\mathbf{R}^n$. Fix a symbol $a(x, \xi) \in S^m$ with compact support (with respect to x) in U_1. We want to define a symbol $b(x, \xi)$ in U_2 that arises from $a(x, \xi)$ under the action of τ. For f supported in U_2, define

$$f_\tau(x) \equiv f(\tau^{-1}x);$$

$f_\tau(x)$ is supported in U_1.

THEOREM 6. *There exists a symbol $b(x, \xi) \in S^m$ such that*
(i) $(a(x, D)f_\tau)_{\tau^{-1}}(x) = b(x, D)f(x)$,
(ii) $b(x, \xi) = a(\tau(x), [(\partial\tau/\partial x)^t]^{-1}\xi) + S^{m-1}$.

Here $[(\partial\tau/\partial x)^t]^{-1}$ is the inverse of the transpose of the Jacobian of τ. We will see in the next chapter that this is a special case of a theorem describing the composition of two Fourier integral operators. Although we could give a direct proof here, we will instead note that this theorem is a corollary of Theorem 4 in Chapter 4.

We can now define a pseudodifferential operator on a manifold. First use a partition of unity so that the symbol and the functions on which it operates are supported in a local coordinate neighborhood (U_α, ϕ_α). Then Theorem 6 applies and we see by letting τ be the change of variables $\phi_\beta \circ \varphi_\alpha^{-1}$ that the principal symbol is invariantly defined on the cotangent bundle.

Let $u_1, \ldots, u_p$ and $v_1, \ldots, v_p$ be functions defined on a manifold and consider the matrix equation

$$A \begin{pmatrix} u_1 \\ \vdots \\ u_n \end{pmatrix} = \begin{pmatrix} v_1 \\ \vdots \\ v_n \end{pmatrix},$$

where A is a matrix of pseudodifferential operators (a_{ij}) defined on the manifold and satisfying

(i) $a_{ij} \in S^m$,

(ii) the inverse matrix (a^{ij}) has entries in S^{-m}.

Condition (ii) amounts to saying that $\det(a_{ij})$ is an elliptic symbol of the appropriate order.

THEOREM 7.

(i) ker A and coker A are finite dimensional and composed of smooth functions.

(ii) ran A is closed.

(iii) $Au = v$ has a solution u if $v \perp$ coker A.

(iv) $Au = v$ and $v \in C^\infty \Rightarrow u \in C^\infty$.

SKETCH OF PROOF. Recall that an operator T is called Fredholm if there exists an operator S such that $ST - \mathrm{I}$ and $TS - \mathrm{I}$ are compact. We have the standard

Fact. If T is Fredholm, then

(i) ker T and coker T are finite dimensional.

(ii) ran T is closed.

By the assumptions (i) and (ii) above about A and by Theorem 5, we can find a matrix B of pseudodifferential operators such that $AB - \mathrm{I}$ and $BA - \mathrm{I}$ are smoothing. We claim $AB - \mathrm{I}$ and $BA - \mathrm{I}$ are compact. Indeed both these operators can be represented by a matrix of integral operators with smooth kernels. For the symbols of the operators are in the intersection $\cap_m S^{-m}$. In particular, the kernel is in L^2. Therefore the operator is Hilbert-Schmidt and $AB - \mathrm{I}$ and $BA - \mathrm{I}$ are compact.

This together with the fact above proves the first half of (i) and (ii). Since A has a closed range and since the closure of the range of A is the orthogonal complement of ker $A^* =$ coker A, we get (iii). Now $(BA - \mathrm{I})u \in C^\infty$ and by hypothesis $Au \in C^\infty$. By the corollary to Theorem 4, $BAu \in C^\infty$, therefore $u \in C^\infty$. This gives (iv) and the fact that $u \in$ ker $A \Rightarrow u \in C^\infty$. To prove the rest of (i), just note that a similar argument shows $u \in$ ker $T^* \Rightarrow u \in C^\infty$. $\quad\square$

6. Schauder theory. Consider the nonlinear equation

$$A\left(x, u, \partial u/\partial x_i, \partial^2 u/\partial x_i \partial x_j\right) = 0,$$

where

(39)
$$\quad \text{(i)} \quad A(x, \omega, \omega_i, \omega_{ij}) \text{ is smooth,}$$
$$\quad \text{(ii)} \quad \left(\partial A/\partial \omega_{ij}\right) > 0 \text{ for all } x, \omega, \omega_i, \omega_{ij}.$$

If $u \in C^2$ and $\partial^2 u/\partial x_i \partial x_j \in \Lambda(\alpha)$, we write $u \in C^{2+\alpha}$.

THEOREM 8. *If $u \in C^{2+\alpha}$ is a solution of (39), then $u \in C^\infty$ with $\|u\|_{C^k}$ bounded only by a quantity determined by A and $\|u\|_C^{2+\alpha}$.*

Here

$$\|u\|_{C^k} \equiv \sum_{l=0}^{k} \max_{|\alpha|=l} \|D^\alpha u\|_{C^0}, \qquad \|u\|_{C^{k+\alpha}} \equiv \|u\|_{C^k} + \max_{|\alpha|=l} \|D^\alpha u\|_{\Lambda(\alpha)},$$

where $\|u\|_{C^0(\Omega)} = \sup_{\overline{\Omega}} |u(x)|$. We stress that this theorem asserts neither the existence of a solution nor the existence of a $\|\cdot\|_{C^{2+\alpha}}$ bound on the solution, rather just the regularity of such a solution. In practice finding a $\|\cdot\|_{C^{2+\alpha}}$ bound on the solution is the most difficult part of the analysis.

We motivate the proof by differentiating (39):

$$0 = \frac{\partial}{\partial x_k}\left(A\left(x, u, \frac{\partial u}{\partial x_i}, \frac{\partial^2 u}{\partial x_i \partial x_j} \right) \right)$$

$$= \sum_{ij} A_{\omega_{ij}}(x, u, u_i, u_{ij}) \frac{\partial^3 u}{\partial x_i \partial x_j \partial x_k} + \text{lower order terms}.$$

The first term is the only one involving third order differentiations of u. The coefficients $a_{ij}(x) \equiv A_{\omega_{ij}}(x, u, u_i, u_{ij})$ of this term are not C^∞ but rather only in $\Lambda(\alpha)$. The proof consists of four steps. The first two steps prove results on linear PDE's with rough coefficients, that permit us to conclude that $\partial^3 u/\partial x_i \partial x_j \partial x_k \in \Lambda(\alpha)$. The third step is a bootstrapping argument that allows us to repeat this procedure and reach the desired conclusion. The final step removes one of the initial a priori assumptions.

Step 1. Assume that u is supported in $|x| \leq 3/2$ and is a C^∞ solution of

$$\left[\Delta + \sum b_{ij}(x) \frac{\partial^2}{\partial x_i \partial x_j} \right] u = f, \qquad |x| \leq 2,$$

where

(i) $f, b_{ij} \in C^\infty$,

(ii) $\|f\|_{\Lambda(\alpha)} \leq B$, $\|b_{ij}\|_{\Lambda(\alpha)} \leq \delta$, $\delta > 0$.

Then $\|\partial^2 u/\partial x_i \partial x_j\|_{\Lambda(\alpha)} \leq B'$, where B' depends on B and δ.

PROOF. By (9) we can rewrite (40) using the Riesz transforms as

$$\left[I - \sum b_{ij}(x) R_i R_j \right] \Delta u = f.$$

By Theorem 2, $R_i R_j$ are bounded on $\Lambda(\alpha)$. Now $\sum b_{ij}(x) R_i R_j$ has a small norm as an operator on $\Lambda(\alpha)$; therefore $I - \sum b_{ij}(x) R_i R_j$ is a small perturbation of the identity, and the required estimate follows from Theorem 1.

Note that by assuming

(iii) $\|u\|_{\Lambda(\alpha)} \leq B$, $\|\partial u/\partial x_i\|_{\Lambda(\alpha)} \leq B$,

we can drop the assumption that u be supported in $|x| \leq 3/2$. Indeed let θ be an appropriate cutoff function and consider

$$L(u\theta) = \theta f + \text{remainder}.$$

Assumption (iii) guarantees that remainder $\in \Lambda(\alpha)$ and satisfies the required bounds. $\square$

Step 2. Assume that u is a C^∞ solution of

$$\sum a_{ij}(x)\frac{\partial^2 u}{\partial x_i \partial x_j} = f, \qquad (a_{ij}(x)) > \delta I, \delta > 0,$$

with

(i) $\|u\|_{\Lambda(\alpha)} \leq B$, $\|\partial u/\partial x_i\|_{\Lambda(\alpha)} \leq B$.

Also assume

(ii) $f, a_{ij} \in C^\infty$ in $|x| \leq 2$,

(iii) $\|a_{ij}\|_{\Lambda(\alpha)} \leq B$.

Then $\|\partial^2 u/\partial x_i \partial x_j\|_{\Lambda(\alpha)} \leq B'$, where B' depends only on B and δ.

PROOF. Let $\mathcal{B}$ be any ball of radius $\eta > 0$ such that $\mathcal{B} \subset \{|x| \leq 2\}$. To simplify the notation, assume that the center of $\mathcal{B}$ is 0. We will make an estimate on the inner half $\mathcal{B}_0$ of $\mathcal{B}$. Let $y = 2\eta^{-1}x$, so that $\mathcal{B}_0 = \{|y| \leq 2\}$, and put $\tilde{u}(y) = u(\eta y/2)$, so that $\sum a_{ij}(\eta y/2)\partial^2 \tilde{u}/\partial x_i \partial x_j = \eta^2 f(\eta y/2) = \tilde{f}(y)$. Now on $|y| \leq 2$ we have

$$\|\tilde{f}\|_{\Lambda(\alpha)} \leq \|f\|_{\Lambda(\alpha)}, \qquad \|a_{ij}(\eta y/2) - a_{ij}(0)\|_{\Lambda(\alpha)} \leq B\eta^\alpha.$$

After a linear change of coordinates (so that $(a_{ij}(0)) = I$) we can take η small enough so that the result above applies and conclude

$$\left\|\frac{\partial^2 u}{\partial y_i \partial y_j}\right\|_{\Lambda(\alpha)} \leq B' \quad \text{on } |y| \leq 1,$$

or

$$\left\|\frac{\partial^2 u}{\partial x_i \partial x_j}\right\|_{\Lambda(\alpha)} \leq B'\eta^{-2\alpha} \quad \text{for } x \in \mathcal{B}_0.$$

To finish the proof, cover $\{|x| \leq 1\}$ by these $\mathcal{B}_0$'s. The conclusion then follows since the η are bounded below. $\square$

Step 3. Take $u \in C^\infty$ in $|x| \leq 2$ that satisfies the estimate

$$\|u\|_{C^{2+\alpha}} \leq B$$

and consider the equation

$$(40) \qquad A\left(x, u, \frac{\partial u}{\partial x_i}, \frac{\partial^2 u}{\partial x_i \partial x_j}\right) = 0.$$

Differentiating (40) gives

$$0 = \sum_{ij}\left(A_{\omega_{ij}}\left(x, u, \frac{\partial u}{\partial x_i}, \frac{\partial^2 u}{\partial x_i \partial x_j}\right)\right)\frac{\partial^3 u}{\partial x_i \partial x_j \partial x_k} + \text{lower order terms},$$

where the lower order terms involve derivatives of u of order ≤ 2. Put $G_{ij}(x) = A_{\omega_{ij}}(x, u, \partial u/\partial x_i, \partial^2 u/\partial x_i \partial x_j)$, and rewrite the equation above as

$$\sum_{ij} G_{ij}(x) \frac{\partial^2 u}{\partial x_i \partial x_j} \left(\frac{\partial u}{\partial x_k} \right) = f',$$

where $f' \in \Lambda(\alpha)$, and, moreover, the bounds on f and G_{ij} depend only on the bounds for the original equation (3). By Step 1, $\partial u/\partial x_k \in C^{2+\alpha}$, or $u \in C^{3+\alpha}$ on a slightly smaller ball with the required bounds on its norm. We complete the bootstrap by using the induction on the degree k of differentiation: assume $\|u\|_{C^{k+\alpha}} \leq B$, where B depends on the original $C^{2+\alpha}$ bound. Let D be a differential monomial of order $k - 1$ and apply D to (3)

$$0 = \sum_{ij} A_{\omega_{ij}} \left(x, u, \frac{\partial u}{\partial x_i}, \frac{\partial^2 u}{\partial x_i \partial x_j} \right) \frac{\partial^2}{\partial x_i \partial x_j} (Du) + \text{lower order terms},$$

where the lower terms involve differentiations of u of order $\leq k$. Just as above, we have $\sum_{ij} A_{ij}(x)(\partial^2/\partial x_i \partial x_j)(Du) = f$, with $\|f\|_{\Lambda(\alpha)} \leq \text{const } B$, or $\|Du\|_{C^{2+\alpha}} \leq B'$, giving

$$\|u\|_{C^{k+1+\alpha}} \leq B',$$

as required. $\square$

Step 4. It is not difficult to remove the a priori assumption that $u \in C^{\infty}$ from Step 2. To remove the assumption from Step 3 assume that $u \in C^{2+\alpha}$ is any solution of

$$A \left(x, u, \frac{\partial u}{\partial x_i}, \frac{\partial^2 u}{\partial x_i \partial x_j} \right) = 0.$$

Replace the derivatives in this equation by difference quotients

$$\begin{aligned}
0 &= \frac{A\big(u_{ij}(x + h)\big) - A\big(u_{ij}(x)\big)}{h} \\
&= \frac{1}{h} \int_0^1 \frac{\partial}{\partial t} \left\{ A\big(tu_{ij}(x + h) + (1 - t)u_{ij}(x)\big) \right\} dt \\
&= \sum_{ij} \left\{ \int_0^1 A_{\omega_{ij}}\big(tu_{ij}(x + h) + (1 - t)u_{ij}(x)\big) \, dt \right\} \left(\frac{u_{ij}(x + h) - u_{ij}(x)}{h} \right) \\
&\quad + \text{other terms}.
\end{aligned}$$

This is the only place where we require $(A_{\omega_{ij}}) > 0$; in fact we require only that $(A_{\omega_{ij}}) > 0$ near the solution u. Now apply the theory of elliptic PDE's with rough coefficients to these difference coefficients and conclude that $u \in C^{3+\alpha}$. Iterating this argument removes the a priori assumption for $A_{\omega_{ij}} = 0$; a similar argument works in general. This finishes Step 4 and completes the proof of the Schauder Theorem. $\square$

7. Boundary value problems. So far we have studied elliptic equations locally on $\mathbf{R}^n$ and on manifolds without boundary. We now show how to use pseudodifferential operators to study elliptic equations with boundary conditions. To illustrate the ideas, we shall discuss the problem

$$(41) \qquad\qquad Lu = 0 \quad \text{in } D \subseteq \mathbf{R}^n,$$

$$(42) \qquad\qquad Xu = f \quad \text{on } \partial D,$$

where L is an elliptic second order partial differential operator, and X is a complex vector field.

Our plan is as follows. Set $u_+ = u$ in D; 0 outside D. For (41) to hold, we must have $Lu_+ = g$ on $\mathbf{R}^n$, where g is a distribution on $\mathbf{R}^n$ whose support lies in ∂D. In other words, our solution is given by

$$(43) \qquad\qquad u_+ = L^{-1}g, \quad g \text{ a distribution supported on } \partial D.$$

Note that (41) now holds automatically whenever u_+ is defined by (43). We shall pick g so that (42) holds also. It turns out that finding such a g amounts to solving a pseudodifferential equation on the manifold ∂D. Once g is known, formula (43) gives the solution u to our boundary problem.

To carry out this plan, we first have to study what kind of distributions g arise by applying L to u_+. It is convenient to work in a coordinate system in which ∂D is straightened out. Thus, assume in local coordinates:

$$(44) \qquad\qquad D = \left\{ (x', x_n) \in \mathbf{R}^{n-1} \times \mathbf{R}^1 \,\middle|\, x_n > 0 \right\},$$

$$L = -\sum_{jk} a_{jk}(x)\frac{\partial^2}{\partial x_j \partial x_k} + \cdots, \qquad (a_{jk}) > 0,$$

$$X = \sum_k b_k(x')\frac{\partial}{\partial x_k}.$$

In our coordinates, we have the distribution equations

$$(45) \qquad \frac{\partial}{\partial x_j}u_+ = \left(\frac{\partial}{\partial x_j}u\right)_+ + u(x',0)\delta(x_n)\cdot\delta_{jk},$$

$$(46) \qquad \frac{\partial^2}{\partial x_j \partial x_k}u_+ = \left(\frac{\partial^2 u}{\partial x_j \partial x_k}\right)_+ \quad \text{if } j, k \neq n,$$

$$(47) \qquad \frac{\partial^2}{\partial x_j \partial x_n}u_+ = \left(\frac{\partial^2 u}{\partial x_j \partial x_n}\right)_+ + \frac{\partial u}{\partial x_j}(x',0)\delta(x_n) \quad \text{if } j \neq n,$$

$$(48) \qquad \frac{\partial^2}{\partial x_n^2}u_+ = \left(\frac{\partial^2 u}{\partial x_n^2}\right)_+ + \frac{\partial u}{\partial x_n}(x',0)\delta(x_n) + u(x',0)\delta'(x_n).$$

Here $\delta(t)$ is the Dirac delta, while δ_{jk} is the Kronecker delta. Equation (45) follows by differentiating $u_+ = u\chi_{\{x_n>0\}}$, while successive applications of (45) yield (46)–(48). Now equations (44)–(48) show that

$$Lu_+ = (Lu)_+ + g_0(x')\delta(x_n) + g_1(x')\delta'(x_n),$$

with $g_1 = -a_{nn}(x',0)\cdot u(x',0)$. In our application, we want $(Lu)_+ = 0$, so that

$$(49) \qquad u_+ = L^{-1}\{g_0(x')\delta(x_n) + g_1(x')\delta'(x_n)\} \quad \text{on } R^n.$$

The two unknown functions g_0, g_1 will be determined by the boundary condition (49) and the consistency requirement that u_+ defined by (49) must vanish for $x_n < 0$.

To carry this out, we now have to compute $L^{-1}\{g_0(x')\delta(x_n) + g_1(x')\delta'(x_n)\}$ using the formula for L^{-1} as a pseudodifferential operator. The answer is most easily expressed if in addition to (44), we normalize our coordinate system to satisfy

$$(50) \qquad a_{jn}(x',0) = 0 \quad \text{for } j < n.$$

A coordinate system satisfying only (44) may be transformed locally to satisfy both (44) and (50) by the simple change of variable $(x', x_n) \to (x' + x_n F(x'), x_n)$ for suitable F. With normalizations taken care of, we can now state the result of the calculations with (49).

PROPOSITION 1. *Let D, L, X be as in (44), (50), and define*

$$u_+ = L^{-1}\{g_0(x')\delta(x_n) + g_1(x')\delta'(x_n)\}.$$

If

$$\lim_{x_n \to 0-} u_+(x', x_n) = 0 \quad \text{and} \quad \lim_{x_n \to 0+} Xu_+(x', x_n) = f(x_1),$$

then g_0, g_1 satisfy the pseudodifferential equations

$$(51) \quad g_0 = \tilde{p}(x', D')g_1 \quad \text{with } \tilde{p}(x', \xi') = -a_{nn}^{-1/2}\left(\sum_{j,k<n} a_{jk}\xi_j\xi_k\right)^{1/2} \text{mod } S^0$$

$$(52) \quad p^+(x', D')g_1 = f$$

$$\text{with } p^+(x', \xi') = \frac{b_n}{a_{nn}^{3/2}}\left(\sum_{j,k<n} a_{jk}\xi_j\xi_k\right)^{1/2} + i\sum_{k<n}\frac{b_k}{a_{nn}}\xi_k \text{ mod } S^0.$$

Here b_n and a_{jk} are to be evaluated at $(x',0)$.

The proposition tells us that our boundary-value problem (41), (42) is reduced to the pseudodifferential equation (51) on ∂D. In fact, once g_1 is obtained by solving (52), we can then use (51) to find g_0 and then substitute g_0, g_1 into (49) to obtain our solution u of (41), (42). Let us illustrate this procedure in a simple example. Suppose X has real coefficients and is everywhere transverse to the boundary ($b_n(x',0) \neq 0$). Then the symbol p^+ in (52) is elliptic, since already

$$\text{Re } p^+ = b_n a_{nn}^{-3/2}\left(\sum_{j,k<n} a_{jk}\xi_j\xi_k\right)^{1/2} \text{mod } S^0$$

is elliptic. Theorem 7 now applies to equation (52), so that we can read off as a consequence

Elliptic regularity of the Neumann problem. Equations (41), (42) have a solution for all f orthogonal to a finite-dimensional cokernel. The solution is unique modulo a finite-dimensional kernel. If f is smooth, then u is smooth up to the boundary. The kernel and cokernel both consist of smooth functions.

If X has complex coefficients, then equation (52) becomes quite hard. In general it is not understood. Significant regularity theorems for (52) have been obtained by Egorov [17, 19, 20], Hörmander [36] and Kohn [40]. Egorov and Hörmander imposed hypotheses insuring that the S^0 error in (52) may be regarded as a trivial perturbation. Kohn's work [40] deals with the $\bar{\partial}$-Neumann problem, a system for which the first-order part of p^+ is positive semidefinite, and the zero-order correction term plays an essential role. We shall return later to discuss the $\bar{\partial}$-Neumann problem in its simplest case (strictly pseudoconvex domains).

PROOF OF THE PROPOSITION. We have to compute u_+ from formula (49). Now the main term in the symbol of the operator L^{-1} is $1/a(x, \xi)$ where $a(x, \xi) = \sum_{j,k} a_{jk}(x)\xi_j\xi_k$, so the main term in (49) is the integral

$$(53) \qquad u_+(x', x_n) \approx \int_{R^n} \frac{e^{ix_n\xi_n}e^{ix'\cdot\xi'}}{a(x', x_n, \xi', \xi_n)} \left[\hat{g}_0(\xi') - i\xi_n\hat{g}_1(\xi') \right] d\xi'd\xi_n.$$

For fixed real x', x_n, ξ', the function $\xi_n \to 1/(a(x', x_n, \xi', \xi_n))$ is of course meromorphic with two simple poles $\xi_n = \tau_+(x', x_n, \xi')$ and $\xi_n = \tau_-(x', x_n, \xi')$ lying in the upper and lower half-planes respectively. Therefore, we can evaluate the ξ_n-integral in (53) by residues. The result for $x_n < 0$ is

$$u_+(x', x_n) \approx \int_{R^{n-1}} \frac{e^{ix_n\xi_-}e^{ix'\cdot\xi'}}{(\partial a/\partial\xi_n)(x', x_n, \xi', \xi_-)} \left[\hat{g}_0(\xi') - i\tau_-\hat{g}_1(\xi') \right] d\xi',$$

so that

$$(54) \qquad \lim_{x_n\to 0-} u_+(x', x_n) \approx \int_{R^{n-1}} \frac{e^{ix'\cdot\xi'}}{(\partial a/\partial\xi_n)(x', 0, \xi', \tau_-)} \hat{g}_0(\xi')\, d\xi'$$

$$+ \int_{R^{n-1}} \frac{e^{ix'\cdot\xi'}(-i\tau_-)}{(\partial a/\partial\xi_n)(x', 0, \xi', \tau_-)} \hat{g}_1(\xi')\, d\xi'.$$

The integrals on the right in (54) are pseudodifferential operators applied to g_0, g_1. In particular, one checks easily that τ_- and $(\partial a/\partial\xi_n)(x', x_n, \xi', \tau_-)$ are first-order elliptic symbols for x_n near 0. (Recall that $a \sim a_{nn}\xi_n^2 + \sum_{j,k<n}a_{jk}\xi_j\xi_k$ by (50).)

Now the formulas (53), (54) are only approximate because we used only the principal symbol $1/(a(x, \xi))$ for L^{-1} in place of its full symbol. However, a glance at the recursive procedure for computing L^{-1} shows that the lower-order correction terms in the symbol for L^{-1} also continue analytically in ξ_n to meromorphic functions with poles only at $\tau_\pm$. Therefore a residue calculation like the one we carried out yields the corrected form of formula (54)

$$(55) \qquad \lim_{x_n\to 0-} u_+ = p_0(x', D')g_0 + p_1(x', D')g_1$$

where

$$p_0(x', \xi') = \frac{1}{(\partial a/\partial \xi_n)(x', 0, \xi', \tau_-)} \quad \mathrm{mod}\ S^{-2},$$

$$p_1(x', \xi') = \frac{i\tau_-}{(\partial a/\partial \xi_n)(x', 0, \xi', \tau_-)} \quad \mathrm{mod}\ S^{-1}.$$

Again note that p_0 is elliptic of order -1 while p_1 is elliptic of order 0. If $\lim_{x_n \to 0-} u_+(x', x_n) = 0$, then (55) shows that $p_0(x', D')g_0 = -p_1(x', D')g_1$, so that (since p_0 is elliptic) $g_0 = \tilde{p}(x', D')g_1$ with $\tilde{p} = -(p_1/p_0)\,\mathrm{mod}\ S^0$. A computation with (44), (50) easily gives the formula (51) for $\tilde{p}$.

Next we want to calculate $\lim_{x_n \to 0+} Xu_+(x', x_n)$.

We can save part of the work by recalling that $g_1 = -a_{nn}(x', 0)u(x', 0)$. (See the discussion just before (49).) It follows at once that

$$\lim_{x_n \to 0+} \sum_{k<n} b_k(x') \frac{\partial u_+}{\partial x_k}(x', x_n) = \sum_{k<n} b_k(x') \frac{\partial}{\partial x_k} \left\{ \frac{-g_1(x')}{a_{nn}(x', 0)} \right\}.$$

Comparing this with the definition (44) of X, we see that it remains to find $\lim_{x_n \to 0+}(\partial/\partial x_n)u_+(x', x_n)$. To compute this, we differentiate (53) under the integral sign and again use residues to evaluate the ξ_n-integral. This time since $x_n > 0$, it is $\tau_+(x', x_n, \xi')$ that enters. Again the lower-order correction terms in L^{-1} are not important. The result is

$$\lim_{x_n \to 0+} \frac{\partial}{\partial x_n} u_+(x', x_n) = p_0^\#(x', D')g_0 + p_1^\#(x', D')g_1,$$

where $p_k^\#(x', \xi') = -(-i\tau_+)^{k+1}/(\partial a/\partial \xi_n)(x', 0, \xi', \tau_+)\,\mathrm{mod}$ lower terms, $k = 0, 1$. Substituting (51) to eliminate g_0, we obtain a formula for

$$\lim_{x_n \to 0+}(\partial/\partial x_n)\, u_+(x', x_n)$$

as a pseudodifferential operator applied to g_1. Therefore

$$\lim_{x_n \to 0+} Xu_+(x', x_n) = b_n(x') \lim_{x_n \to 0+} \frac{\partial}{\partial x_n} u_+(x', x_n)$$

$$+ \lim_{x_n \to 0+} \sum_{k<n} b_k(x') \frac{\partial u_+}{\partial x_k}(x', x_n)$$

is expressed as a pseudodifferential operator applied to g_1. Carrying out the details using (50) to simplify formulas, we arrive at formula (52). $\square$

REMARK. If L^{-1} is expressed as a (variable-coefficient) singular integral instead of a pseudodifferential operator, then it is very easy to read off the formula for $u_+(x', x_n)$, $x_n \neq 0$. We have no need of the residue calculations used above. The price we pay for this is that it takes some work to see what happens as $x_n \to 0\pm$, whereas the passage to the limit in the proof given above is immediate.

Finally, note that it makes no difference to replace $(41, 42)$ by the seemingly more general problem

$$Lu = f_1 \quad \text{in } D,$$

$$Xu = f_2 \quad \text{on } \partial D.$$

The reason is that we can extend f_1 to a convenient function $\tilde{f}_1$ on $\mathbf{R}^n$, solve $Lv = \tilde{f}_1$ in $\mathbf{R}^n$, and then observe that $(u - v)$ satisfies

$$L(u - v) = 0 \quad \text{in } D,$$

$$X(u - v) = f_2 - Xv \equiv f \quad \text{on } \partial D.$$

CHAPTER 4. THE WAVE EQUATION

Let $(b_{ij}(x))$ be a positive definite, smoothly varying matrix and let

$$\Box = \frac{\partial^2}{\partial t^2} - \sum_{i,j=1}^{n} b_{ij}(x) \frac{\partial^2}{\partial x_i \partial x_j}$$

denote the wave operator. This chapter is concerned with the linear hyperbolic equation

$$(1) \qquad \Box u = 0, \qquad u\big|_{t=0} = f, \qquad \frac{\partial u}{\partial t}\bigg|_{t=0} = g.$$

In the last chapter we saw how ψDOs provided parametrices for elliptic operators; §1 of this chapter uses techniques from geometrical optics to construct a parametrix for the wave operator. This is an example of a Fourier Integral Operator (FIO). §2 sketches how this parametrix can be used to prove a theorem of Hörmander about the asymptotic distribution of the eigenvalues of selfadjoint elliptic operators with positive principal symbols. The next section defines (local) Fourier integral operators and amplifies the discussion in the third chapter on the principle of stationary phase. Egorov's theorem is proved in §4. This theorem gives a formula for the conjugation of a ψDO by an FIO. The calculus of FIOs is discussed in the next section. §6 completes the proof of Hörmander's theorem on eigenvalues, and the final section contains a few remarks about the global theory of FIOs and FIOs with complex phase. A good general reference for these topics is the book by Trèves [**63**].

1. A parametrix for $\Box$. Our calculations will be a bit simpler if we use

$$g^{ij}(x) = \begin{cases} 1, & i = j = 0, \\ -b_{ij}(x), & 1 \le i, j \le n, \\ 0, & \text{otherwise,} \end{cases}$$

so that

$$\Box = \sum_{i,j=0}^{n} g^{ij}(x) \frac{\partial^2}{\partial x_i \partial x_j};$$

for convenience we will assume that the matrix $(g^{ij}(x))$ is symmetric. Consider first the constant coefficient wave equation $\partial^2/\partial t^2 - \partial^2/\partial x_1^2 - \cdots - \partial^2/\partial x_n^2$ and its plane wave solutions

$$(2) \qquad u_\xi(x,t) = e^{i[x \cdot \xi \pm t|\xi|]}.$$

If the initial datum $f(x) = u(x,0)$ is suitable, then the Fourier decomposition $f(x) = \int_{\mathbf{R}^n} e^{ix \cdot \xi} \hat{f}(\xi)\, d\xi$ expresses f as a "sum" of plane waves $e^{ix \cdot \xi}$. Since $u_\xi(x,0) = e^{ix \cdot \xi}$, we can solve our original boundary value problem by superposition.

Turning now to the boundary value problem (1), we rewrite (2) as

$$e^{i\lambda[x \cdot \xi^0 \pm t|\xi^0|]},$$

where $\xi = \lambda(\xi/|\xi|) = \lambda\xi^0$. This suggests that we try to solve $\Box u = 0$ using a wave

$$(3.\text{i}) \qquad u(x,t) = a(x,t,\lambda,\xi^0)e^{i\lambda S(x,t,\xi^0)},$$

with

$$(3.\text{ii}) \qquad a(x,t,\lambda,\xi^0) = \sum_{k=0}^\infty a_k(x,t,\xi^0)\lambda^{-k}.$$

We require

$$(3.\text{iii}) \qquad a(x,0,\lambda,\xi^0) = 1, \qquad S(x,0,\xi^0) = x \cdot \xi^0,$$

so that (3) reduces to a plane wave at $t = 0$. It turns out that it is reasonable to assume that $a(x,t,\lambda,\xi^0) \in S^0$ so that it satisfies estimates of the form

$$(4) \qquad \left| \partial_x^\alpha \partial_{\lambda\xi^0}^\beta \partial_t^\gamma a \right| \leq C_{\alpha\beta\gamma}\left(1 + |\lambda\xi^0|\right)^{-|\beta|},$$

and to assume that

$$(5) \qquad \begin{aligned} &\text{(i)} \quad S \text{ is a smooth, real-valued function,} \\ &\text{(ii)} \quad \text{for } \xi^0 \neq 0,\ \nabla_x S \text{ is never } 0 \text{ on the support of } a(x,t,\lambda,\xi^0). \end{aligned}$$

Since

$$\frac{\partial}{\partial x_j}\left(ae^{i\lambda S}\right) = \frac{\partial a}{\partial x_j}e^{i\lambda S} + i\lambda\frac{\partial S}{\partial x_j}ae^{i\lambda S},$$

$$\frac{\partial^2}{\partial x_i \partial x_j}\left(ae^{i\lambda S}\right)$$

$$= e^{i\lambda S}\left[\frac{\partial^2 a}{\partial x_i \partial x_j} + i\lambda\left(\frac{\partial a}{\partial x_j}\frac{\partial S}{\partial x_i} + \frac{\partial a}{\partial x_i}\frac{\partial S}{\partial x_j} + \frac{\partial^2 S}{\partial x_i \partial x_j}a \right) - \lambda^2\frac{\partial S}{\partial x_i}\frac{\partial S}{\partial x_j}a \right],$$

we have

$$(6) \quad \Box\left(ae^{i\lambda S}\right) = \sum g^{ij}(x)\frac{\partial^2}{\partial x_i \partial x_j}\left(ae^{i\lambda S}\right)$$

$$= e^{i\lambda S}\sum g^{ij}(x)$$

$$\times \left[\frac{\partial^2 a}{\partial x_i \partial x_j} + 2i\lambda\frac{\partial a}{\partial x_j}\frac{\partial S}{\partial x_i} + i\lambda\frac{\partial^2 S}{\partial x_i \partial x_j}a - \lambda^2\frac{\partial S}{\partial x_i}\frac{\partial S}{\partial x_j}a \right].$$

We will choose $S(x, t, \xi^0)$ and $a_k(x, t, \xi^0)$ so that the right-hand side of (6) is 0. Beginning with the highest-order term, we require $\Sigma^n_{i,j=0} g^{ij}(x)(\partial S/\partial x_i)(\partial S/\partial x_j) = 0$, or in the earlier notation

$$\left(\frac{\partial S}{\partial t}\right)^2 - \sum_{i,j=1}^{n} b_{ij}(x)\frac{\partial S}{\partial x_i}\frac{\partial S}{\partial x_j} = 0,$$

which becomes

$$(7) \qquad \frac{\partial S}{\partial t} = \pm\left(\sum_{i,j=0}^{n} b_{ij}(x)\frac{\partial S}{\partial x_i}\frac{\partial S}{\partial x_j}\right)^{1/2} = H\left(x, \frac{\partial S}{\partial x_i}\right),$$

if we define $H(x, \zeta) = \pm(\Sigma b_{ij}(x)\zeta_i\zeta_j)^{1/2}$. To solve this equation, subject to the initial condition (3.iii), we choose, say, the positive sign and invoke Theorem 1 of Chapter 2. According to Theorem 1, the flow determined by the Hamiltonian $H(x, \zeta)$ yields for small time a family of canonical transformations Φ_t: $(y, \eta) \to (x, \zeta)$ whose generating functions $S_t(x, \eta)$ satisfies (7). Equation (7) is called the eikonal equation.

Having defined $S(x, t)$, we now ask that

$$\sum g^{ij}(x)\left(\frac{\partial^2 a}{\partial x_i \partial x_j} + 2i\lambda\frac{\partial a}{\partial x_j}\frac{\partial S}{\partial x_i} + i\lambda\frac{\partial^2 S}{\partial x_i \partial x_j}a\right) = 0.$$

Collecting terms of first order in λ and using the expansion (3.ii) of $a(x, t, \lambda, \xi^0)$ yields the next requirement

$$\sum\left(2g^{ij}(x)\frac{\partial S}{\partial x_i}\frac{\partial a_0}{\partial x_j} + g^{ij}(x)\frac{\partial^2 S}{\partial x_i \partial x_j}a_0\right) = 0.$$

This is a first-order linear partial differential equation for a_0 called the transport equation. Since at $t = 0$ (and hence for small time), $\Sigma g^{ij}(x)(\partial S/\partial x_i)(\partial/\partial x_j)$ is a nondegenerate vector field, we can solve the transport equation by integrating along the integral curves $\dot{x}_i = \Sigma_j g^{ij}(x)\partial S/\partial x_j$. We use the initial condition $a_0(x, 0, \xi^0) = 1$ so that (3.iii) is satisfied. Note, that except for a time change, the integral curves of $\dot{x}_i = \Sigma_j g^{ij}(x)\partial S/\partial x_j$ are the same as the integral curves determined by the Hamiltonian $H(x, \zeta)$ and agree with the geodesics in the metric in which $\Sigma b_{ij}\partial^2/\partial x_i\partial x_j + \cdots$ is the Laplacian.

In general, if we have found $a_0,\ldots,a_k$, collecting terms in (6) produces the requirement

$$(8) \quad \sum g^{ij}(x)\frac{\partial^2 a_k}{\partial x_i\partial x_j} + 2\sum g^{ij}(x)\frac{\partial S}{\partial x_i}\frac{\partial a_{k+1}}{\partial x_j} + \sum g^{ij}(x)\frac{\partial^2 S}{\partial x_i\partial x_j}a_{k+1} = 0.$$

This is a first-order linear ordinary differential equation for a_{k+1}. That is, (8) has the form $Xa_{k+1} + \phi a_{k+1} = b$ for known functions ϕ, b and a vector field X. It is called the transport equation and is solved just as above. For the initial conditions we can take $a_k(x, 0, \xi^0) = 0$, for $k = 2, 3, \ldots$.

THEOREM 1. *A solution of the wave equation* (1) *is given for small time by*

$$(9) \qquad T^t u \equiv \int_{\mathbf{R}^n} \sum_{\pm} a_{\pm}(x, t, \xi) e^{iS_{\pm}(x,t,\xi)} \hat{f}(\xi)\, d\xi + \varepsilon f$$

$$+ \int_{\mathbf{R}^n} \sum_{\pm} a'_{\pm}(x, t, \xi) e^{iS_{\pm}(x,t,\xi)} \hat{g}(\xi)\, d\xi + \varepsilon' g,$$

where $a_{\pm}$, $a'_{\pm}$ *are symbols satisfying the estimates*

$$\left| \partial_x^\alpha \partial_\xi^\beta \partial_t^\gamma a_{\pm} \right| \le C_{\alpha\beta\gamma} (1 + |\xi|)^{-|\beta|}, \qquad \left| \partial_x^\alpha \partial_\xi^\beta \partial_t^\gamma a'_{\pm} \right| \le C_{\alpha\beta\gamma} (1 + |\xi|)^{-1-|\beta|},$$

S *satisfies the requirements* (5), *and* ε, ε' *are smoothing operators.*

PROOF. Recall that a smoothing operator is a ψDO $\varepsilon(x, D)$ with symbol $\varepsilon(x, \xi) \in S^{-\infty} \equiv \cap_m S^m$. First apply the construction above to the boundary value problem

$$\Box u = 0, \qquad u \big|_{t=0} = f, \qquad \frac{\partial u}{\partial t} \bigg|_{t=0} = 0,$$

by breaking f into plane waves using the Fourier decomposition and letting a_k be the functions determined by the transport equations. To achieve $\partial u / \partial t = 0$, we use linear combinations of our two solutions of $\Box u = 0$, corresponding to $S_{\pm}$ which arise from the two choices of sign in the eikonal equation. By Lemma 5 of Chapter 3, there exists a symbol $a(x, \xi)$ such that formally $a - \sum_{k=0}^\infty a_k = \varepsilon$, where $\varepsilon \in S^{-\infty}$. See the lemma for the precise statement. The proof is completed by repeating this procedure for the boundary value problem

$$\Box u = 0, \qquad u \big|_{t=0} = 0, \qquad \frac{\partial u}{\partial t} \bigg|_{t=0} = g,$$

and adding the results. $\Box$

As a corollary of this theorem we can prove a result about the propagation of singularities for the wave equation $\Box u = 0$. Let Φ_τ be the Hamiltonian flow associated with the generating function $S_\tau(x, \xi)$ in (9).

THEOREM 2. *Assume that the initial data* $u \big|_{t=0}$, $\partial u / \partial t \big|_{t=0}$ *of* $\Box u = 0$ *are supported at* $x = 0$. *Then the singularities of* $T^t u$ *lie on the light cone*

$$\{ x: \exists \eta \text{ with } \Phi_\tau(0, \eta) = (x, \xi) \text{ for some } \xi \}.$$

We recall that the light cone is given by geodesics starting at x.

PROOF. Since up to a smooth error

$$T^t u = \int a(x, t, \eta) e^{iS(x,t,\eta)} \hat{u}(0, \eta)\, d\eta$$

$$+ \int a'(x, t, \eta) e^{iS(x,t,\eta)} \frac{\partial \hat{u}}{\partial t}(0, \eta)\, d\eta,$$

we know by the principle of stationary phase that the singular points of $T^t u$ occur where $\partial S / \partial \eta_j = 0$. In general, if $\Phi_t \colon (y, \eta) \to (x, \xi)$ is the Hamiltonian flow associated with the generating function $S_t(x, \eta)$, then $\partial S / \partial \eta_j = y_j$. So in our case, the singularities can occur only at those points x such that $(0, \eta) \to (x, \xi)$, for some η, ξ. But this is precisely the light cone. $\square$

2. Hörmander's Theorem. The operator $T'u$ in (9) is an example of a Fourier integral operator. We have seen in Theorem 2 how this representation can be used to prove theorems about the propagation of singularities. In this section we will show how it can be used to prove a theorem about eigenvalue asymptotics. We will give a detailed proof of this theorem in §6 after we have developed some of the machinery necessary in order to study FIOs.

We begin with an example. Let M be the n-dimensional torus $S^1 \times \cdots \times S^1$ and $A = -\Delta = -\Sigma_{i=1}^n (\partial^2 / \partial \theta_i^2)$ the Laplacian. The eigenvectors are $e^{im \cdot \theta}$, where $m = (m_1, \ldots, m_n) \in \mathbf{Z}^n$, and the eigenvalues are $|m|^2$. Let

$$(10) \qquad N(\lambda) = \# \{ \text{eigenvalues of } A < \lambda^2 \}.$$

Now the number of integer lattice points $N'(\lambda)$ inside a ball B of radius λ is

$$N'(\lambda) = (\text{volume } B) + O(\lambda^{n-1}).$$

Indeed by approximating the ball from above and below by unions of unit cubes we see that $N'(\lambda) - (\text{volume } B)$ is bounded by the area of the boundary of B. Since volume $B = c_n \lambda^n$ we have

$$N(\lambda) = c_n \lambda^n + O(\lambda^{n-1}).$$

We will see later how to improve the error term.

This is illustrative of the general case. Let M be a compact n-dimensional manifold and $A = a(x, D)$ a mth-order elliptic, selfadjoint pseudodifferential operator whose principal symbol $a_m = \lim_{\lambda \to \infty} \lambda^{-m} a(x, \lambda \xi)$ is positive. Put

$$(11) \qquad N(\lambda) = \# \{ \text{eigenvalues of } A < \lambda \}.$$

THEOREM 3 (HÖRMANDER) [**33**].

$$(12) \qquad N(\lambda) = c_{a,n} \lambda^{n/m} + O(\lambda^{(n-1)/m}),$$

where $c_{a,n} = c_n \text{ measure}\{(x, \xi) \in T^*M : a_m(x, \xi) < 1\}$.

Since $N(\lambda)$ in (10) counts the number of eigenvalues $< \lambda^2$, we see after replacing λ by λ^2 that the estimate above for the torus is a special case of (12). It should be noted that Theorem 3 is part of an extensive literature on eigenvalues going back to H. Weyl, and that the connection with the wave equation is also quite old. See [**33**] for detailed references. To simplify the discussion, we will assume in this section that $a(x, D)$ is first order. We begin by using the eigenfunctions of $a(x, D)$ to solve the "half-wave" equation

$$(13) \qquad \left(\frac{1}{i} \frac{\partial}{\partial t} + A \right) u = 0, \qquad u \big|_{t=0} = f$$

on $M \times [0, T)$. Let $\psi_1, \psi_2, \ldots$ be the eigenfunctions of A and $\lambda_1, \lambda_2, \ldots$ the eigenvalues. If the initial data has an expansion $f = \Sigma c_k \psi_k$, then the solution of

(13) is $u(x, t) = \sum e^{i\lambda_k t} c_k \psi_k$. Formally the operator

$$U^t : f \to u(x, t) = \sum e^{i\lambda_k t} c_k \psi_k$$

is diagonal with entries $e^{i\lambda_k t}$ and its trace is $\sum e^{i\lambda_k t}$.

On the other hand the construction in §1 may be adopted to (13) to yield a parametrix

$$T^t : f \to u(x, t) = \int b(x, t, \eta) e^{iS(x, t, \eta)} \hat{f}(\eta) \, d\eta.$$

All the information about the distribution of the eigenvalues of A is contained in the measure

$$(14) \qquad\qquad \mu = \sum \delta_{\lambda_k},$$

where δ_λ is the delta function. Interpreted as a distribution the Fourier transform of μ is $\hat{\mu} = \sum_k e^{i\lambda_k t}$. We will study the asymptotic distribution of the eigenvalues of A by comparing U^t and T^t and calculating their traces (in an appropriate sense).

To be precise take $\phi \in C_0^\infty(\{|t| < t_0\})$ and consider the operator $f \to \int_{-\infty}^\infty \phi(t) T^t f \, dt$. It turns out that this operator is of trace class; so we may define a distribution F acting on test functions $\phi \in C_0^\infty$ by

$$\langle F, \phi \rangle \equiv \operatorname{Trace} \int_{-\infty}^\infty \phi(t) T^t \, dt.$$

We calculate that $\langle F, \phi \rangle = \langle \hat{\mu}, \phi \rangle$ for ϕ supported in $(-t_0, t_0)$ and hence that $(\theta F)\hat{\ } = \mu * \hat{\theta}$, for a suitable cutoff function $\theta \in C_0^\infty$.

We now take advantage of the Fourier integral operator representation of the solution. Recall that when the integral operator $f(x) \to \int K(x, y) f(y) \, dy$ has a nice kernel $K(x, y)$, we have $\operatorname{Trace} = \int K(x, x) \, dx$. In our case, for $\phi \in C_0^\infty(-t_0, t_0)$,

$$\left(\int \phi(t) T^t \right)(f(x)) = \iiint \phi(t) b(x, t, \eta) e^{i[S(x, t, \eta) - \eta \cdot y]} f(y) \, dt \, d\eta \, dy,$$

so that

$$(15) \qquad \langle F, \phi \rangle = \iiint \phi(t) b(x, t, \eta) e^{i[S(x, t, \eta) - \eta \cdot x]} \, dt \, d\eta \, dx.$$

By the principle of stationary phase, the important contributions are the points (x, η) at which

$$\partial S / \partial \eta_j = x_j, \qquad \partial S / \partial x_j = \eta_j.$$

Note that these are the points which at time t have been carried back to themselves by the Hamiltonian flow associated with the generating function $S(x, t, \eta)$ in (15). After working a bit on (15), we find

$$\int_0^\lambda d[\hat{\theta} * \mu] = c_{a,n} \lambda^n + O(\lambda^{n-1}) \quad \text{as } \lambda \to \infty,$$

so that by applying a Tauberian theorem we get the desired estimate

$$N(\lambda) = \int_0^\lambda d\mu = c_{a,n} \lambda^n + O(\lambda^{n-1}).$$

3. Fourier integral operators and stationary phase. In this section we will have to work with operators of the form

$$Af(y) = \int e^{iS(y,\xi)} a(y,\xi) \hat{f}(\xi)\, d\xi$$

$$= \iint e^{i[S(y,\xi) - x\cdot\xi]} a(y,\xi) f(x)\, dx d\xi,$$

where

$$(16) \qquad
\begin{aligned}
&\text{(i)} \quad a(y,\xi) \in S^m, \\
&\text{(ii)} \quad f(x) \in C_0^\infty(\mathbf{R}^n), \\
&\text{(iii)} \quad T(x,\xi; y) \equiv S(y,\xi) - x\cdot\xi \text{ is smooth, real valued,} \\
&\qquad\quad \text{with } d_{x,\xi} T \neq 0, \text{ for any } (x,\xi).
\end{aligned}$$

Condition (iii) implies that $T(x,\xi; y)$ does not have an extremum with respect to x and ξ, for any value of the parameter y; we will relax this condition shortly. Our first task is to give a precise meaning to the integral (16), which will not be absolutely convergent in general.

Let $\theta \in C_0^\infty(\mathbf{R}^n)$ be a cutoff function that is 1 near the origin and 0 elsewhere and put

$$a_\varepsilon(y,\xi) = a(y,\xi)\theta(\xi/\varepsilon),$$

so that the product $a_\varepsilon f$ has compact support in both x and ξ. If $m + n < 0$, then $Af(y)$ is absolutely convergent. If this is not the case, then let L be a first-order differential operator on $\mathbf{R}^n$ with smooth coefficients satisfying $LT/i = 1$. We can always find such an L locally: a partition of unity then produces a globally defined L. We have $Le^{iT} = e^{iT}$ and therefore integrating by parts formally k times gives

$$A_\varepsilon f(y) \equiv \iint e^{iT(x,\xi; y)} (^t L)^k \big[a_\varepsilon(y,\xi) f(x) \big]\, dx d\xi,$$

where $^t L$ is the transpose of L and maps $^t L\colon S^m \to S^{m-1}$. By taking k large enough, we get $m + n - k < 0$, so that the integral above becomes absolutely convergent uniformly in ε. In this case, $A_\varepsilon f$ has a limit Af as $a_\varepsilon(y,\xi) \to a(y,\xi)$, for $\varepsilon \to \infty$. We take this as the definition of the integral (16). Note that this procedure defines a map

$$C_0^\infty(\mathbf{R}^n) \to C^\infty(\mathbf{R}^n),$$

$$f \to Af,$$

for each fixed $a(y,\xi)$ and $T(x,\xi; y)$, which is a compactly supported distribution of order k. This is because at most k derivatives of f appear in the integral $A_\varepsilon f$. By taking duals we can extend this map to a map from compactly supported distributions to general distributions $\mathcal{E}'(\mathbf{R}^n) \to \mathcal{D}'(\mathbf{R}^n)$. In particular, we have defined Af for

$$(16.\text{ii})' \qquad\qquad f(x) \in C^\infty(\mathbf{R}^n).$$

This is called an *oscillatory integral*.

REMARK. This first order operator L involves dividing by the derivatives of T, which is where we used the hypothesis (16.iii).

We now relax the requirement (16.iii) and assume

$(16.\text{iii})'$
$$T(x, \xi; y) \equiv S(y, \xi) - x \cdot \xi \text{ has a unique}$$
$$\text{nondegenerate critical point } (x_0, \xi_0; y).$$

We repeat that the critical point is with respect to x and ξ; y is just a parameter. In §4 of Chapter 3 we saw how the principle of stationary phase applied to the integral

$$(17) \qquad F(y, \eta) = \iint e^{iM(x-y)\cdot(\xi-\eta)} a(y, \xi) f(x) \, dx d\xi.$$

This is the integral appearing in Lemma 4 of Chapter 3 with $\psi(x, y, \xi, \eta) = a(y, \xi) f(x)$ and with x and y interchanged. The Morse lemma with parameters (which is our next proposition) will allow us to find the main term in an asymptotic expansion of the composition of two integral operators (16) by reducing it to this form.

PROPOSITION 1. *Suppose that $\phi_\lambda(z)$ is smooth and satisfies*

$$\phi_{\lambda_0}(0) = \phi'_{\lambda_0}(0) = 0,$$

$$\phi''_{\lambda_0}(0) = \begin{pmatrix} 1 & & & & & & 0 \\ & \ddots & & & & & \\ & & 1 & & & & \\ & & & -1 & & & \\ & & & & \ddots & & \\ 0 & & & & & & -1 \end{pmatrix}$$

where $z \in \mathbf{R}^n$, $\lambda \in \mathbf{R}^m$. Then near $z = 0$, $\lambda = \lambda_0$, there exists a smooth change of variables $z \to \tilde{z}$, where $\tilde{z}$ also depends smoothly on λ, such that

$$\phi_\lambda(\tilde{z}) = \tilde{z}_1^2 + \cdots + \tilde{z}_k^2 - \tilde{z}_{k+1}^2 - \cdots - \tilde{z}_n^2 + C(\lambda).$$

PROOF. Put $(z_1, \ldots, z_n) = (z_1, z')$ and let $f_\lambda(z')$ be the solution of the equation

$$\frac{\partial}{\partial z_1} \phi_\lambda(z_1, z') = 0,$$

with $z_1 = f_\lambda(z')$. By the implicit function theorem $f_\lambda(z')$ is smooth in λ and z'. After a change of variables $(z_1, z') \to (z_1 - f_\lambda(z'), z')$, we can assume $f_\lambda = 0$, i.e.,

$$\frac{\partial}{\partial z_1} \phi_\lambda(z_1, z') = 0 \quad \text{when } z_1 = 0.$$

Now by Taylor's theorem we can write

$$\phi_\lambda(z_1, z') = \phi_\lambda(0, z') + a_\lambda(z_1, z') z_1^2,$$

where $a_\lambda(z_1, z')$ is smooth and $a_\lambda(0,0) > 0$. Letting $\tilde{z}_1 = (a_\lambda(z_1, z'))^{1/2}z_1$, $\tilde{z}' = z'$ gives

$$\phi_\lambda(\tilde{z}) = \tilde{z}_1^2 + \tilde{\phi}_\lambda(\tilde{z}'),$$

allowing us to finish the proof inductively. $\quad\square$

In fact this proposition allows us to deal with integrals slightly more general than (16); but, before we define these integrals, we must introduce an appropriate class of symbols. Let $S^m(1 \times M \times 1 \times N)$ denote those smooth functions $a(x, \xi, y, \eta)$ which are supported on blocks of Q of dimension

$$\mathrm{diam}_x = 1, \qquad \mathrm{diam}_y = 1, \qquad \mathrm{diam}_\xi = M, \qquad \mathrm{diam}_\eta = N,$$

and which satisfy estimates of the form

$$(18) \qquad \left|\partial_x^\alpha \partial_\xi^\beta \partial_y^\gamma \partial_\eta^\delta a(x, \xi, y, \eta)\right| \leqslant C_{\alpha\beta\gamma\delta} M^{m-|\beta|} N^{-|\delta|},$$

for $(x, \xi, y, \eta) \in Q$. Note that this is a simple generalization of the estimates (23) in §3 of Chapter 3. The symbol class $S^m(1 \times M)$ is defined in the obvious way. Now consider integrals of the form

$$(19) \qquad F(y, \eta) = \iint e^{iT(x,\xi,y,\eta)} a(x, \xi, y, \eta)\, dxd\xi,$$

where

$$(20) \qquad
\begin{array}{ll}
\text{(i)} & a(x, \xi, y, \eta) \in S^m(1 \times M \times 1 \times N), \\
\text{(ii)} & T \text{ is real valued,} \\
\text{(iii)} & T \in S^1(1 \times M \times 1 \times N), \\
\text{(iv)} & T(x, \xi, y, \eta) \text{ is a small perturbation of} \\
& y \cdot \xi + x \cdot \eta - x \cdot \xi \text{ in the topology of } S^1(1 \times M \times 1 \times N).
\end{array}$$

Condition (iv) means that there exist a small constant ε and a large constant E such that

$$\left|\partial_x^\alpha \partial_\xi^\beta \partial_y^\gamma \partial_\eta^\delta (T - (y \cdot \xi + x \cdot \eta - x \cdot \xi))\right| \leqslant \varepsilon M^{1-|\beta|} N^{-|\delta|},$$

for $|\alpha|, |\beta|, |\gamma|, |\delta| \leqslant E$.

PROPOSITION 2. *Assume that $T(x, \xi, y, \eta)$ and $a(x, \xi, y, \eta)$ satisfy (20) and that T has a unique nondegenerate critical point $(x_0(y, \eta), \xi_0(y, \eta), y, \eta)$, with respect to x and ξ; the critical point may depend on the parameters y and η. Then the main term in an asymptotic expansion of the integral (19) is*

$$(21) \qquad (\textit{Jacobian factor})\ a(x_0(y, \eta), \xi_0(y, \eta), y, \eta) e^{iT(x_0,\xi_0,y,\eta)}.$$

PROOF. Putting $\lambda = (y, \eta/M)$, $z = (x, \xi/M)$ and applying Proposition 1 to (19) gives

$$F(y, \eta) = \iint e^{i(x-x_0(y,\eta)) \cdot (\xi-\xi_0(y,\eta))} \tilde{a}(x, \xi, y, \eta)\, dxd\xi.$$

The proof is completed by invoking Lemma 4 in §4 of Chapter 3. $\quad\square$

REMARK. In particular this proposition gives us the main term in an asymptotic expansion of (16) around a critical point satisfying (16.iii)'.

A *Fourier integral operator* is an operator

$$(22) \qquad Af(y) = \int e^{iS(y,\xi)} a(y,\xi) \hat{f}(\xi) \, d\xi$$

$$= \int\int e^{i[S(y,\xi) - x \cdot \xi]} a(y,\xi) f(x) \, dx d\xi$$

where

$$(23) \qquad \begin{aligned} &\text{(i)} \quad a(y,\xi) \in S^m(1 \times M), \\ &\text{(ii)} \quad S(y,\xi) \text{ is real valued and homogeneous of degree 1} \\ &\qquad \text{in } \xi \text{ outside } |\xi| < 1, \\ &\text{(iii)} \quad S(y,\xi) \in S^1(1 \times M), \\ &\text{(iv)} \quad S(y,\xi) - y \cdot \xi \text{ is small in the topology of } S^1. \end{aligned}$$

As before (iv) means that there exists a small constant ε and a large constant E such that

$$(24) \qquad \left| \partial_y^\alpha \partial_\xi^\beta (S(y,\xi) - y \cdot \xi) \right| \leq \varepsilon M^{1-|\beta|},$$

for all $|\alpha|, |\beta| \leq E$. The function $a(y,\xi)$ is called the *amplitude* and $S(y,\xi)$ is called the *phase function*.

EXAMPLE 1. If $a(y,\xi) \in S^m$, then the pseudodifferential operator

$$a(y,D) f(y) = \int e^{iy \cdot \xi} a(y,\xi) \hat{f}(\xi) \, d\xi$$

$$= \int\int e^{i[y \cdot \xi - x \cdot \xi]} a(y,\xi) f(x) \, dx d\xi$$

is clearly a Fourier integral operator.

EXAMPLE 2. Let $\Phi: (x,\xi) \to (y,\eta)$ be a canonical transformation with generating function $S(y,\xi)$ and let $a(y,\xi) \in S^m$ be a symbol. Define the Fourier integral operator U^Φ by

$$(25) \qquad U^\Phi f(y) = \int e^{iS(y,\xi)} a(y,\xi) \hat{f}(\xi) \, d\xi$$

$$= \int\int e^{i[S(y,\xi) - x \cdot \xi]} a(y,\xi) f(x) \, dx d\xi.$$

A short calculation shows that $S(y,\xi)$ satisfies the requirements (23.ii), (23.iii), (23.iv).

EXAMPLE 3. Let $a(y,\xi)$, Φ and $S(y,\xi)$ have the same meaning as in the last example.

The adjoint operator of the Fourier integral operator (25) is given by

$$(26) \qquad (U^\Phi)^* f(x) = \int\int e^{-i[S(y,\xi) - x \cdot \xi]} \bar{a}(y,\xi) f(y) \, dy d\xi.$$

We shall see that $(U^\Phi)^*$ is a Fourier integral operator arising from the canonical transformation Φ^{-1}.

EXAMPLE 4. For fixed t, the parametrix for $\square u = 0$ described in Theorem 1 is the sum of two Fourier integral operators, the first of which is

$$A^t f(y) = \int e^{iS(y,t,\xi)} a(y, t, \xi) \hat{f}(\xi)\, d\xi$$

$$= \iint e^{i[S(y,t,\xi) - x \cdot \xi]} a(y, t, \xi) f(x)\, dx d\xi.$$

According to Theorem 1, $a(y, t, \xi) \in S^0$; since $S_t(y, \xi) = S(y, t, \xi)$ arises as a canonical transformation, it satisfies the necessary requirements by Example 2. The second Fourier integral operator is handled in the same way.

The next proposition clarifies the nature of the requirements (23) of the phase function $S(y, \xi)$.

PROPOSITION 3. *Conditions* (23.ii), (23.iii), (23.iv) *imply that* $S(y, \xi)$ *is the generating function for a canonical transformation* $\Phi \colon (x, \xi) \to (y, \eta)$ *satisfying*

(27)

(i) Φ *is homogeneous of degree* 1 *in* ξ *outside* $|\xi| < 1$,

$\quad$ *i.e., for* $\lambda > 0$, $\Phi \colon (x, \lambda\xi) \to (y, \lambda\eta)$,

(ii)   *for* $|\xi| = 1$, Φ *is a small perturbation of the identity*.

PROOF. This follows from Proposition 4 of Chapter 2. $\square$

Recall that as a corollary to Theorem 4 of Chapter 3 we proved that pseudodifferential operators decrease the singular support of functions on which they act. We conclude this section by extending this discussion to Fourier integral operators. We begin with an important refinement of the notion of singular support; this will also play a role in §8.

If u is a distribution on $\mathbf{R}^n$ and $(x_0, \xi_0) \in \mathbf{R}^n \times \mathbf{R}^n = T^*(\mathbf{R}^n)$ with $\xi_0 \neq 0$, then we call u regular at (x_0, ξ_0) if $\sigma(x, D)u \in C^\infty$ for some symbol $\sigma(x, \xi) \in S^0$ for which $\sigma(x_0, \lambda\xi_0)$ stays bounded away from zero as $\lambda \to \infty$. The *wave-front set* $\mathrm{WF}(u)$ consists of all (x_0, ξ_0) at which u is not regular. This refines the notion of singular support because $x^0 \in \operatorname{sing\,supp}(u)$ iff $(x^0, \xi^0) \in \mathrm{WF}(u)$ for some ξ_0. If $x^0 \in \operatorname{sing\,supp}(u)$, then for a cutoff function ϕ supported near x_0 we have $\phi u \notin C^\infty$; but $\mathrm{WF}(u)$ tells us the directions in which $(\phi u)\hat{}(\xi)$ fails to decrease rapidly. A Fourier integral operator, $Tf(y) = \int a(y, \xi) e^{iS(y,\xi)} \hat{f}(\xi)\, d\xi$, with $S =$ generating function of Φ, may be written as $Tf(y) = \int K(y, x) f(x)\, dx$ for a "Fourier integral distribution" K on $\mathbf{R}^n \times \mathbf{R}^n$.

Later on it will be useful to know the wave-front set of K. By stationary phase we can calculate $(\phi K)\hat{}$, and the answer is $\mathrm{WF}(K) = \{(y, \eta, x, \xi) \mid \Phi(x, \xi) = (y, \eta)\}$. In particular $\mathrm{WF}(Tf) \subset \Phi\mathrm{WF}(f)$, which strengthens and generalizes the corollary to Theorem 4 of Chapter 3.

4. Egorov's theorem. Changing variables $x \rightarrow y$ in the equation $p(x, D)u = f$ yields a new equation $\tilde{p}(y, D)\tilde{u} = \tilde{f}$ which may be easier to solve. Egorov's idea was to consider more generally canonical transformations $\Phi: (x, \xi) \rightarrow (y, \eta)$ of the symbol $p(x, \xi)$ so that in the new variables $\tilde{p} = p \circ \Phi$ and to show that the equations $p(x, D)u = f$ and $\tilde{p}(y, D)\tilde{u} = \tilde{f}$ are still equivalent. For example, if the principal symbol $p(x, \xi)$ is real and has zeros in ξ of at most first order (in this case $p(x, \xi)$ is said to be of real principal type), then locally there is a canonical transformation such that $p(x, \xi) = \eta_1$. Thus equations of real principal type are reduced to

$$\frac{1}{i} \frac{\partial}{\partial x_1} u(x) = f(x).$$

Note that the wave equation is of real principal type.

THEOREM 4 (EGOROV) [**18**]. *Suppose $a(y, \xi) \in S^0$ is elliptic and that Φ is a canonical transformation satisfying*

$$(28) \quad \begin{array}{ll} \text{(i)} & \Phi: (x, \lambda\xi) \rightarrow (y, \lambda\eta) \text{ for } \lambda > 0, \\ \text{(ii)} & \text{for } |\xi| = 1, \Phi \text{ is a small perturbation of the identity.} \end{array}$$

Let U^Φ and $(U^\Phi)^*$ be the FIO associated with $a(y, \xi)$ and Φ by equations (23) and (24) respectively. Let $p(y, \eta) \in S^m$ be a symbol. Then

$$(29) \quad \begin{array}{ll} \text{(i)} & \tilde{p}(x, D) \equiv (U^\Phi)^* \circ p(y, D) \circ (U^\Phi) \text{ satisfies} \\ & \tilde{p}(x, \xi) \in S^m, \text{ and} \\ \text{(ii)} & \text{by choosing } a(x, \xi) \text{ appropriately, we have} \\ & \tilde{p}(x, \xi) = p \circ \Phi(x, \xi) \bmod S^{m-1}. \end{array}$$

PROOF. By adapting the localization argument used in Lemma 1 of Chapter 3, we can assume

$$a(x, \xi) \in S^m(2 \times 2M), \qquad p(y, \eta) \in S^m(1 \times M).$$

From equation (26) and the definition of a pseudodifferential operator, we have

$$(U^\Phi)^* f(x) = \iint \bar{a}(y, \xi) e^{-i[S(y,\xi) - x \cdot \xi]} f(y) \, dy d\xi$$

and

$$p(y, D)g(y) = \iint p(y, \eta) e^{i\eta(y - y')} g(y') \, dy' d\eta;$$

therefore, since

$$U^\Phi f(y) = \int a(y, \xi) e^{iS(y,\xi)} \hat{f}(\xi) \, d\xi,$$

we have

$$(U^\Phi)^* p(y, D) U^\Phi f(x) = \int \bar{a}(y, \xi) e^{i[x \cdot \xi - S(y,\xi)]}$$

$$\cdot p(y, \eta) e^{i\eta(y - y')} a(y', \xi') e^{iS(y',\xi')} \hat{f}(\xi') \, dy d\xi dy' d\eta d\xi',$$

where each of these integrals is interpreted as an oscillatory integral. Define $\tilde{p}(x, \xi')$ by

$$(U^{\Phi})^* p(y, D) U^{\Phi} f(x) = \int \tilde{p}(x, \xi') e^{ix \cdot \xi} \hat{f}(\xi') \, d\xi',$$

so that

$$(30) \qquad \tilde{p}(x, \xi') = \int\!\!\int\!\!\int\!\!\int \bar{a}(y, \xi') a(y', \xi) p(y, \eta) e^{iT} \, dy \, d\xi' \, dy' \, d\eta,$$

where

$$T = x \cdot (\xi' - \xi) - S(y, \xi') + S(y', \xi) + \eta \cdot (y - y').$$

We will complete the proof by applying to equation (30) a stationary phase argument using Proposition 2. Unfortunately, Proposition 2, as it stands, does not apply; indeed, we must replace the symbol class $S^m(1 \times M \times 1 \times N)$ by the symbol class $S^m(1 \times M \times 1 \times N \times 1 \times P)$, which is defined in the obvious way. It is clear that $\bar{a}(y, \xi') a(y', \xi) p(y, \eta) \in S^m$ and that T is real valued. Since by hypothesis Φ is a small perturbation of the identity, $S(y, \xi')$ and $S(y', \xi)$ are small perturbations of $y \cdot \xi'$ and $y' \cdot \xi$, respectively, for these are the generating functions of the identity. This implies that $T \in S^1$ and that T is a small perturbation of

$$T_{\text{lin}} = x \cdot \xi - y \cdot \xi' + y' \cdot \xi - x' \cdot \xi + \eta \cdot y - \eta \cdot y'$$

in the topology of S^1. This completes the verification of the hypotheses of Proposition 2. Now we must locate the stationary points of T.

The equation $\nabla_{(y, \xi', y', \eta)} T = 0$ is equivalent to

$$\frac{\partial S}{\partial y_j}(y, \xi') = \eta_j, \qquad \frac{\partial S}{\partial \xi'_j}(y, \xi') = x_j,$$

$$\frac{\partial S}{\partial y'_j}(y', \xi) = \eta_j, \qquad y_j = y'_j.$$

Using the fact that S is a small perturbation of the generating function of the identity, we get

$$(31.\text{i}) \qquad\qquad y = y', \qquad \xi = \xi',$$

$$(31.\text{ii}) \qquad\qquad \partial S/\partial y_j = \eta_j, \qquad \partial S/\partial \xi_j = x_j.$$

Comparing (31.ii) and Proposition 4 of Chapter 2, we find that

$$(31.\text{iii}) \qquad\qquad \Phi(x, \xi) = (y, \eta),$$

where Φ is the canonical transformation associated with S. We can now evaluate the integral (30) using Proposition 2 and conclude that the principal symbol of $\tilde{p}(x, \xi')$ is

$$(\text{Jacobian factor}) \quad (\bar{a}(y, \xi) a(y, \xi)) p(\Phi(x, \xi)).$$

This proves (29.i). Now choose $a(y, \xi) \in S^0$ so that

$$(\text{Jacobian factor}) \quad (\bar{a}(y, \xi) a(y, \xi)) = 1 \quad \mod S^{m-1},$$

giving

$$\tilde{p}(x, \xi) = p \circ \Phi(x, \xi) \quad \bmod S^{m-1}. \quad \square$$

Note that instead of the hypotheses (28) in Egorov's theorem we could have simply stated the theorem for the Fourier integral operators defined by (22) and (23). Proposition 3 would then have given us a canonical transformation Φ satisfying (28). This was not done since we wish to emphasize that the canonical transformation can be chosen to suit the problem at hand. This is illustrated by the next theorem. Recall that a symbol $p(x, \xi)$ is of *principal type* if its principal symbol p_m satisfies for fixed x

$$(32) \qquad p_m(x, \xi) = 0 \quad \text{and} \quad \xi \neq 0 \Rightarrow \nabla_\xi p_m(x, \xi) \neq 0.$$

THEOREM 5. *If $p(x, \xi)$ is real and of principal type, then locally there exists a canonical transformation Φ: $(x, \xi) \to (y, \eta)$ such that*

$$p(x, \xi) = \eta_1.$$

PROOF. See [**63**, vol. 2, p. 468]. $\square$

5. The composition formula. Fourier integral operators enjoy a calculus just as pseudodifferential operators do. In this section we prove a composition formula for Fourier integral operators corresponding to different phase functions. Consider

$$Au(y) = \int a(y, \xi) e^{iS(y, \xi)} \hat{u}(\xi) \, d\xi,$$

$$Bv(z) = \int b(z, \eta) e^{iT(z, \eta)} \hat{v}(\eta) \, d\eta,$$

where the amplitudes $a(y, \xi) \in S^m(1 \times M)$, $b(z, \eta) \in S^{m'}(1 \times M)$ and the phases $S(y, \xi)$, $T(z, \eta)$ satisfy the requirements (23). By Proposition 3, S and T are the generating functions of canonical transformations Φ and Ψ. Let Γ be the graph

$$\Gamma = \left\{ (x, \xi, y, \eta, z, \zeta): \begin{array}{l} \Phi: (x, \xi) \to (y, \eta) \text{ and} \\ \Psi: (y, \eta) \to (z, \zeta) \end{array} \right\}$$

THEOREM 6. *$B \circ A$ is an FIO of the form*

$$(B \circ Au)(z) = \int c(z, \xi) e^{iS \square T(z, \xi)} \hat{u}(\xi) \, d\xi,$$

where

(33)

(i)   *$S \square T(z, \xi)$ is the generating function for the canonical transformation $\Psi \circ \Phi$.*

(ii)   *$c(z, \xi) \in S^{m+m'}$.*

(iii)   *Mod $S^{m+m'-1}$, $c(z, \xi) = \theta(z, \xi) b(z, \eta) a(y, \xi)$ on Γ, with $\theta(z, \xi)$ an elliptic symbol, depending on S, T but not on a, b.*

PROOF. We will argue formally first. We have

$$Bv(z) = \int\int b(z, \eta)e^{i[T(z,\eta) - \eta \cdot y]}v(y)\, dy\, d\eta$$

and, therefore, putting $v = Au$ gives

$$(34) \quad (B \circ Au)(z) = \int\int\int b(z, \eta)a(y, \xi)e^{i[T(z,\eta) + S(y,\xi) - \eta \cdot y]}\hat{u}(\xi)\, d\xi\, d\eta\, dy.$$

To complete the calculation we will compute

$$F(z, \xi) = \int\int b(z, \eta)a(y, \xi)e^{i[T(z,\eta) + S(y,\xi) - \eta \cdot y]}\, dy\, d\eta$$

using stationary phase. Put $R = T(z, \eta) + S(y, \xi) - \eta \cdot y$. The important points are the critical points, where

$$\partial R/\partial y = 0, \qquad \partial R/\partial \eta = 0,$$

or

$$\eta = \partial S(y, \xi)/\partial y, \qquad y = \partial T(z, \eta)/\partial \eta.$$

By Proposition 4 of Chapter 2, we see

$$\Phi\colon (x, \xi) \to (y, \eta), \qquad \Psi\colon (y, \eta) \to (z, \zeta),$$

showing that the stationary points coincide with the graph Γ. Proposition 5 of Chapter 2 tells us that R is the generating function of $\Psi \circ \Phi$. We can now apply Proposition 2 and conclude that $F(z, \xi)$ is asymptotic to

$$F(z, \xi) \sim \theta(z, \xi)b(z, \eta)a(y, \xi)\big|_{\Gamma}e^{iS \square T(z,\xi)},$$

and this gives

$$(B \circ Au)(z) = \int c(z, \xi)e^{iS \square T(z,\xi)}\hat{u}(\xi)\, d\xi,$$

with the amplitude and the phase satisfying (33).

To justify the application of Proposition 2, we can mimic the proof used to justify the use of stationary phase in Egorov's theorem. We have already tacitly assumed that the symbols have been localized and are elements of $S^m(1 \times M)$. As before, this follows from a modification of the localization argument in Lemma 1 of Chapter 3.

We must now check hypotheses (20) of Proposition 2. Clearly $b(z, \eta)a(y, \xi)$ is a symbol of the proper type and R is real valued. By (23.iii) we know $S, T \in S^1(1 \times M)$; therefore $R = T(z, \eta) + S(y, \xi) - \eta \cdot y \in S^1(1 \times M \times 1 \times M)$. Finally, since $S(y, \xi) - y \cdot \xi$ and $T(z, \eta) - z \cdot \eta$ are small in the topology of S^1, we see that R is a small perturbation of

$$R_{\text{linear}} = z \cdot \eta + y \cdot \xi - \eta \cdot y$$

in the topology of S^1. This verifies all the hypotheses of Proposition 2. All that remains is to check that the symbol $\theta(z, \xi)$ arising from the Jacobian factor is elliptic. This is a consequence of the proof of Proposition 2. $\square$

EXAMPLE 1. Let $T = y \cdot \xi$ so that B is a pseudodifferential operator. Then $B \circ A$ is a Fourier integral operator whose amplitude is multiplied by the symbol $b(y, \xi)$ of the pseudodifferential operator B.

EXAMPLE 2. Let U^{Φ} be the Fourier integral operator defined by the elliptic symbol $a \in S^0$ and the canonical transformation Φ according to (25). Thinking of the pseudodifferential operator $b(x, D)$ as a Fourier integral operator and applying the composition formula twice to

$$(U^{\Phi^{-1}})b(x, D)(U^{\Phi})$$

gives another proof of Egorov's theorem.

6. Proof of Hörmander's theorem. Let M be a compact n-dimensional manifold and $A = a(x, D)$ an mth order elliptic, selfadjoint operator whose principal symbol is positive. Recall that

$$N(\lambda) = \#\{\text{eigenvalues of } A < \lambda\}.$$

In this section we will prove

THEOREM 3 (HÖRMANDER). $N(\lambda) = c_{a,n}\lambda^{n/m} + O(\lambda^{(n-1)/m})$.

PROOF. Note that we will not actually compute the constant $c_{a,n}$. Assume to start with that A is first order. The main idea is to study the half-wave equation

$$(35) \qquad \left(\frac{1}{i}\frac{\partial}{\partial t} + A\right)u = 0, \qquad u\big|_{t=0} = f$$

on $M \times \mathbf{R}^n$.

Step 0. In the spirit of §1 we shall write down a parametrix for (35) locally in $\mathbf{R}^n$. This is the heart of the analysis.

We look for a solution of (35) given by

$$u(t, y) = \int e^{iS(t,y,\xi)}\hat{F}_t(\xi)\,d\xi = U_t F_t \quad \text{with } F_t \text{ to be determined.}$$

The composition law for Fourier integral operators lets us compute $A(y, D_y)U_t$ and yields

$$\left[\frac{1}{i}\frac{\partial}{\partial t} + A(y, D_y)\right]u$$

$$= \int \frac{1}{i}\left\{\frac{\partial S}{\partial t} + A\left(y, \frac{\partial S}{\partial y}\right) + (\text{lower-order symbol})\right\}e^{iS(t,y,\xi)}\hat{F}_t(\xi)\,d\xi$$

$$+ \int e^{iS(t,y,\xi)}\frac{1}{i}\frac{\partial \hat{F}_t}{\partial t}(\xi)\,d\xi.$$

Therefore if $S(t, y, \xi)$ is taken to satisfy the Hamilton-Jacobi equation, we obtain

$$U_t^{-1}\left[\frac{1}{i}\frac{\partial}{\partial t} + A(y, D_y)\right]u = \left[\frac{1}{i}\frac{\partial}{\partial t} + B_t(x, D_x)\right]F_t$$

with $B_t(x, D)$ a zero-order pseudodifferential operator depending smoothly on t.

So $u = U_t F_t$ satisfies (35), provided F_t satisfies

$$(35a) \qquad \left[\frac{1}{i} \frac{\partial}{\partial t} + B_t(x, D_x) \right] F_t = 0, \qquad F \big|_{t=0} = f.$$

However, it is easy to write down a parametrix for (35a). We just set $F_t = b(t, x, D_x) f$ where $b(t, x, \xi)$ is a symbol with the asymptotic expansion $b(t, x, \xi) \sim \sum_{\mu=0}^{\infty} b_\mu(t, x, \xi)$, $b_\mu \in S^{-\mu}$.

Equations (35a) hold modulo smoothing operators if the symbols b_μ satisfy

$$(35b) \qquad \left[\frac{1}{i} \frac{\partial}{\partial t} + B_t(x, \xi) \right] b_\mu = g_\mu(t, x, \xi), \qquad b_\mu \big|_{t=0} = \begin{cases} 1, & \mu = 0, \\ 0, & \mu > 0, \end{cases}$$

where $g_\mu \in S^{-\mu}$ is determined by $b_{\mu'}$ ($\mu' < \mu$). Since $B_t \in S^0$, the integrating factor for the ordinary differential equation (35b) is elliptic in S^0. Thus (35b) can be solved with $b_\mu \in S^{-\mu}$, and we have our parametrix for (35a). The resulting parametrix $u = U_t F_t = U_t b(t, x, D_x) f$ for (35) has the form

$$u(t, y) = \int b(t, y, \xi) e^{iS(t, y, \xi)} \hat{f}(\xi) \, d\xi \quad \text{with } b \in S^0,$$

in view of the composition law for Fourier integral operators.

Step 1. We shall solve (35) on the manifold $M \times [-T, T]$ by patching together the local solutions derived in Step 0. Let $\{U_j\}$ be a cover of M by coordinate patches and let $\{\phi_j\}$ be a partition of unity subordinate to this cover. The construction of Step 0 yields a parametrix T^t for (35) in the following way. We break up the initial data f using the ϕ_j and iet T_j^t be the parametrix constructed in Step 0 for

$$(36) \qquad \left(\frac{1}{i} \frac{\partial}{\partial t} + a(x, D) \right) u = 0, \qquad u \big|_{t=0} = \phi_j f.$$

Put $u_j(x, t) = T_j^t \phi_j f(x)$. Observe that $u_j(x, t)$ is defined only for $x \in U_j$. We glue together the $u_j(x, t)$ by using a second partition of unity $\{\psi_j\}$ satisfying

   (i) $\psi_j = 1$ on a neighborhood of supp ϕ_j,

   (ii) $\mathrm{supp}(\psi_j) \subset U_j$

and by defining $u(x, t)$ by

$$u = \left(\sum_j \psi_j T_j^t \phi_j \right)(f).$$

Since $\mathrm{supp}(\phi_i f) \subseteq \mathrm{supp}(\phi_i)$, we have $\mathrm{sing\,supp}(T_j^t \phi_j f) \subseteq \{\psi_j = 1\}$, for $|t| \ll 1$; this means that the cutoff functions ψ_j introduce only C^∞ errors, and, therefore $(\sum_j \psi_j T_j^t \phi_j)$ is a parametrix for the half-wave equation (35) on $M \times [-T, T]$.

We now show that it is sufficient to prove the theorem for a first-order operator.

PROPOSITION 4. *Let $a(x, \xi) \in S^m$ be an elliptic symbol with positive principal part and with $a(x, D)$ selfadjoint. Then there exists a symbol $b(x, \xi) \in S^1$ satisfying*

   (i) *$b(x, D)$ is selfadjoint,*

   (ii) *$[b(x, D)]^m - a(x, D)$ is a bounded operator,*

   (iii) *$b - a^{1/m} \in S^0$.*

PROOF. Since $a(x, \xi)$ is positive and elliptic, $b_1(x, \xi) \equiv a^{1/m}(x, \xi)$ satisfies

$$b_1^m - a \in S^{m-1},$$

with $b_1(x, \xi) \in S^1$. Now choose $b_0(x, \xi) \in S^0$ so that

$$(b_1 + b_0)^m - a \in S^{m-2}.$$

In general choose $b_{-k}(x, \xi) \in S^{-k}$ recursively so that

$$(b_1 + b_0 + \cdots + b_{-k})^m - a \in S^{m-(k+2)}.$$

By Lemma 5 of Chapter 3, there exists $b(x, \xi) \in S^1$ satisfying

$$b \sim b_1 + b_0 + b_{-1} + \cdots, \qquad [b(x, D)]^m - a = \text{(bounded operator)}.$$

One checks that $b(x, D)$ is selfadjoint if $a(x, D)$ is. $\square$

This means that the eigenvalues of $[b(x, D)]^m$ and $a(x, D)$ differ by a bounded amount; therefore, the asymptotic behavior of the eigenvalues of these operators is the same. Note that even if we had started out with (35) a partial differential equation, this procedure would still lead to a pseudodifferential equation. In summary, instead of studying the eigenvalues of $a(x, D)$ by solving (35) for higher-order operators, we can solve (36) for $a(x, D)$ a first-order operator.

Step 2. Let $T^t: f \to u(\cdot, t)$ be the parametrix for (36) described above. To simplify the notation, we have denoted the initial data $\phi_i f$ by f. The operator maps f to the solution u and has the form

$$(37) \qquad T^t f(x) = \int b(x, t, \eta) e^{i[S(x,t,\eta) - y \cdot \eta]} f(y) \, dy d\eta + \varepsilon_t f,$$

where ε_t is a smoothing operator $b(x, t, \eta) \in S^0$, and S_t is the generating function for the Hamiltonian flow $\Phi_t: (y, \eta) \to (x, \xi)$, determined by the Hamiltonian $H(x, \xi) = a(x, \xi)$, $a = $ symbol of A.

Let $\phi(t) \in C_0^\infty(-T, T)$ and consider the operator

$$T_\phi = \int_{-T}^{T} \phi(t) T^t \, dt.$$

We claim

$$(\mathrm{i}) \qquad T_\phi f(y) = \int_M K(y, x) f(x) \, dx, \text{ with } K \in C^\infty(M \times M),$$

$$(38) \qquad (\mathrm{ii}) \qquad \text{Trace } T_\phi = \int_M K(y, y) \, dy,$$

$$(\mathrm{iii}) \qquad |\text{Trace } T_\phi| \leqslant c \|\phi\|_{C^k}, \text{ for } k \text{ large},$$

$$(\mathrm{iv}) \qquad \text{the map } F: \phi \to \text{Trace } T_\phi \text{ is a distribution}.$$

We begin with the verifiction of (i). From (37) we have

$$T_\phi f(x) = \int \phi(t) b(x, t, \eta) e^{i[S(x,t,\eta) - y \cdot \eta]} f(y) \, dy d\eta dt$$

$$+ \int \phi(t)(\varepsilon_t f) \, dt.$$

The kernel of the second integral is obviously smooth; the kernel of the first integral is

$$K(x, y) = \iint \phi(t)b(x, t, \eta)e^{i[S(x,t,\eta)-y\cdot\eta]}\,d\eta\,dt.$$

In polar coordinates $\eta = \rho\eta^0 = |\eta|\,\eta^0$, this becomes

$$(39) \qquad K(x, y) = \iiint \phi(t)b(x, t, \rho, \eta^0)e^{i\rho[S(x,t,\eta^0)-y\cdot\eta^0]}\rho^{n-1}\,d\rho\,d\eta^0\,dt.$$

By changing the smoothing operator $\varepsilon_t f$, we can assume that b is supported in $\rho > 1$. Theorem 1 of Chapter 2 states that the generating function S_t satisfies the Hamilton-Jacobi equation

$$\frac{\partial S_t}{\partial t}(x, \eta^0) = H(x, \xi) = a_1(x, \xi),$$

where $\Phi_t\colon (y, \eta) \to (x, \xi)$. By hypothesis the principal symbol $a_1(x, \xi)$ of $a(x, \xi)$ is positive; therefore,

$$(40) \qquad \frac{\partial}{\partial t}\left[S_t - \eta^0\cdot y\right] > 0.$$

This shows that the exponential in (39) has no stationary points; we can conclude by stationary phase that the kernel $K(x, y)$ is smooth. Since we are on a compact manifold, we have $K(x, y) \in C^\infty(M \times M)$. This proves (38.i).

The implications (i) $\Rightarrow$ (ii) and (iii) $\Rightarrow$ (iv) follow immediately from the definitions. Thus it remains to verify (iii). However, by going carefully through the proof of (i) we can see that $|K(y, x)| \leqslant C\|\phi\|_{C^k}$ for k large. (Essentially, k is determined by the number of integrations by parts needed to apply stationary phase to (39).) Now (iii) follows at once from (ii).

Let λ_k and ψ_k be the eigenvalues and eigenfunctions of $A = a(x, D)$. Put $\mu = \Sigma_k\delta_{\lambda_k}$. We have just shown that $F\colon \phi \to \text{Trace }T_\phi$ is a well-defined distribution. We claim

$$(41) \qquad \langle F, \phi\rangle = \langle \hat\mu, \phi\rangle \quad \text{for } \phi \in C_0^\infty(-T, T),$$

where $\hat\mu$ is the Fourier transform of the distribution μ. Indeed if the initial data has an expansion $f = \Sigma c_k\psi_k$, then $u(x, t) = \Sigma e^{it\lambda_k}c_k\psi_k$, and we see that

$$T_\phi f = \sum_k\left[\int\phi(t)e^{it\lambda_k}\,dt\right]c_k\phi_k;$$

i.e., T_ϕ is diagonal with entries $\int\phi(t)e^{it\lambda_k}\,dt$. Therefore

$$\text{Trace }T_\phi = \sum_k\left[\int\phi(t)e^{it\lambda_k}\,dt\right] = \sum_k\hat\phi(\lambda_k) = \langle\mu, \hat\phi\rangle,$$

as desired.

Step 3. For $\phi \in C_0^\infty(-T, T)$ we claim

$$(42) \qquad \langle F, \phi\rangle = \left\langle \sum_{k=0}^\infty \frac{c_k}{t^{n-k}} + h(t), \phi\right\rangle,$$

where $h(t) \in C^\infty$ and c_k are constants which are defined below. Indeed, by integrating (39) we find

$$\langle F, \phi \rangle = \text{Trace } T_\phi$$

$$= \iiiint \phi(t) b(y, t, \rho, \eta^0) e^{i\rho[S(y,t,\eta^0) - \eta^0 \cdot y]} \rho^{n-1} \, d\rho \, d\eta^0 \, dt \, dy.$$

We can evaluate this by substituting the expansion

$$b(y, t, \rho, \eta^0) = \sum_k b_k(y, t, \eta^0) \rho^{-k}$$

and using the identity

$$\int_{0+}^{\infty} \rho^{n-1-k} e^{iz\rho} \, d\rho = \frac{\text{const}}{z^{n-k}},$$

with $z = S_t(y, \eta^0) - y \cdot \eta^0$. Here $1/z^\mu$ is the distribution defined by

$$\frac{1}{z^\mu} = \lim_{\varepsilon \to 0^+} \frac{1}{(z + i\varepsilon)^\mu}.$$

Now $\partial S/\partial t = a$ and by hypothesis the principal symbol a_1 is positive; also recall $S_t(y, \eta^0) = y \cdot \eta^0$, when $t = 0$; therefore

$$z \sim te(y, \eta^0),$$

where $e(y, \eta^0)$ is smooth and nonvanishing. This leaves

$$(43) \qquad \langle F, \phi \rangle = \int \phi(t) \left\{ \sum_{k=0}^{\infty} \frac{\tilde{c}_k(t)}{t^{n-k}} \right\} dt,$$

with

$$\tilde{c}_k(t) = \int \frac{b_k(y, t, \eta^0)}{\left((S_t(y, \eta^0) - y \cdot \eta^0)/t \right)^{n-k}} \, dy \, d\eta^0$$

smooth in t. This proves (42).

Now we can calculate the Fourier transform of $1/t^\mu$ with the table

Function	Fourier transform
$\text{sign } \lambda$	$c(1/t)$
1	δ_0
$\chi_{[0, \infty)}$	$c\left(\dfrac{1}{t} + \delta_0 \right) = \lim_{\varepsilon \to 0^+} \dfrac{c}{t + i\varepsilon}$
$\lambda^{n-1} \chi_{[0, \infty)}$	$\lim_{\varepsilon \to 0^+} \dfrac{c}{(t + i\varepsilon)^n}$

We have

$$(44) \qquad \langle F, \phi \rangle = \int \phi(t) \left\{ \frac{c_0}{t^n} + (\text{less singular terms}) \right\} dt$$

for $\phi \in C_0^\infty(-T, T)$.

Step 4. We shall complete the proof using a Tauberian theorem. Equation (42) determines the distribution

$$(45) \qquad F_\theta = \theta\hat{\mu}$$

for $\theta \in C_0^\infty(-T, T)$. We may pick θ so that

$$(46) \qquad \begin{array}{ll} \text{(i)} & \hat{\theta} \geqslant 0 \text{ and even,} \\[4pt] \text{(ii)} & \hat{\theta}(\lambda) \geqslant \tfrac{1}{10} \text{ on } |\lambda| \leqslant 1, \\[4pt] \text{(iii)} & \int \hat{\theta}(\lambda)\, d\lambda = 1. \end{array}$$

Now from (44) and (45) and the table above,

$$(46a) \qquad \hat{F}_\theta(\lambda) = \mu * \hat{\theta}(\lambda) = c_n \lambda^{n-1}\chi_{[0,\infty)}(\lambda) + O(\lambda^{n-2}), \qquad \lambda \to \infty,$$

so

$$(47) \qquad (\mu * \hat{\theta})[0, \lambda] = c\lambda^n + O(\lambda^{n-1}), \qquad \lambda \to \infty.$$

We claim

$$(48) \qquad \mu(\lambda - 1, \lambda) = O(\lambda^{n-1}), \qquad \lambda \to \infty.$$

Indeed

$$\hat{F}_\theta(\lambda) = \int \hat{\theta}(\nu)\, d\mu(\lambda - \nu) \leqslant c\lambda^{n-1}$$

by (46a), so that (46.i), (46.ii) imply

$$\frac{1}{10}\mu(\lambda - 1, \lambda) \leqslant \int_{|\nu| \leqslant 1} \hat{\theta}(\nu)\, d\mu(\lambda - \nu) \leqslant c\lambda^{n-1},$$

which is (48).

We want to estimate

$$\mu[0, \lambda] = \int_{-\infty}^{\infty} \chi_{[0,\lambda]}(\nu)\, d\mu(\nu)$$

$$= \int_{-\infty}^{\infty} \chi_{[0,\lambda]} * \hat{\theta}(\nu)\, d\mu(\nu) + \int_{-\infty}^{\infty} \left[\chi_{[0,\lambda]} - \chi_{[0,\lambda]} * \hat{\theta}\right](\nu)\, d\mu(\nu)$$

$$= I + II.$$

The first integral is estimated using (47)

$$I = c\lambda^n + O(\lambda^{n-1}), \qquad \lambda \to \infty;$$

for the second we have using (48)

$$II \leqslant \int_{-\infty}^{\infty} \frac{c_N\, d\mu(\nu)}{(1 + |\lambda - \nu|)^N} = \sum_{\text{integers } k} \int_{|\nu - k| \leqslant 1/2} \frac{c_N\, d\mu(\nu)}{(1 + |\lambda - \nu|)^N}$$

$$\leqslant \sum_{\text{integers } k} \frac{c_N}{(1 + |\lambda - k|)^N} \cdot O(|k|^{n-1} + 1)$$

$$= O(\lambda^{n-1}), \qquad \lambda \to \infty,$$

if we take N large enough. This gives

$$N(\lambda) = \mu[0, \lambda] = c\lambda^n + O(\lambda^{n-1}), \qquad \lambda \to \infty,$$

and finishes the proof of Hörmander's theorem. $\square$

If $A = (-\Delta)^{1/2}$ and $M =$ the n-sphere, then according to Hörmander's theorem

$$N(\lambda) = c_n\lambda^n + O(\lambda^{n-1}).$$

A direct count using spherical harmonics shows that the error term cannot be improved. On the other hand consider $A = -\Delta$ acting on the n-torus. We know from the discussion following equation (10) that

$$N(\lambda) = c_n\lambda^n + O(\lambda^{n-1}),$$

where $N(\lambda) = \{$eigenvalues of $A < \lambda^2\}$. Since eigenvalues are square roots of integers, and since there are approximately $\sim \lambda$ square roots of integers between λ and $\lambda + 100$, we see that $N(\lambda)$ is a step function with $\sim \lambda$ jumps in $[\lambda, \lambda + 100]$ and an increment of $\sim \lambda^{n-1}$ in that interval. Therefore at least one jump of $N(\lambda)$ exceeds λ^{n-2}; this means that we cannot have an error better than

$$N(\lambda) = c_n\lambda^n + O(\lambda^{n-2}).$$

In fact, we have the following theorem which is due to Van der Corput, Hlawka, Herz and Hardy-Littlewood; see [**30**].

THEOREM 7. $N(\lambda) = c_n\lambda^n + O(\lambda^{n-1-a})$, where $a = a(n) \to 1$, as $n \to \infty$.

PROOF. Recall that with $f \in \mathcal{S}(\mathbf{R}^n)$, the Poisson summation formula states

$$(49) \qquad \sum_{k \in \mathbf{Z}^m} f(k) = c_n \sum_{k \in \mathbf{Z}^m} \hat{f}(k).$$

We will need this formula later; for now fix an approximation to the identity $\phi \in C_0^\infty$ satisfying

$$(i) \qquad \phi \geqslant 0,$$

$$(ii) \qquad \phi(x) \text{ supported in } |x| \leqslant 1,$$

$$(iii) \qquad \int \phi(x)\, dx = 1,$$

and define

$$(50.\text{i}) \qquad \phi_\zeta(x) = \zeta^{-n}\phi(x/\zeta),$$

$$(50.\text{ii}) \qquad N(\lambda, \zeta) = \sum_{k \in \mathbf{Z}^n} \chi_{B(\lambda)} * \phi_\zeta(k),$$

where $\chi_{B(\lambda)}$ is the characteristic function of the ball $\{x \in \mathbf{R}^n : |x| \leqslant \lambda\}$. Note that if $|k| < \lambda - 10\zeta$ or $|k| > \lambda + 10\zeta$, then $\chi_{B(\lambda)} * \phi_\zeta = \chi_{B(\lambda)}$ and so

$$(51) \qquad N(\lambda - 10\zeta, \zeta) \leqslant N(\lambda) \leqslant N(\lambda + 10\zeta, \zeta).$$

By (49) and (50)

$$(52) \qquad N(\lambda, \zeta) = \sum_{k \in \mathbf{Z}^m} \hat{\chi}_{B(\lambda)}(k)\hat{\phi}_\zeta(k) = \lambda^n \sum_{k \in \mathbf{Z}^m} \hat{\chi}(k\lambda)\hat{\phi}_\zeta(k),$$

where $\chi = \chi_{B(1)}$ and $k = 0$ is the main term. We need the following estimates

$$|\hat{\chi}(k\lambda)| \leqslant \frac{c_n}{(|k|\lambda)^{(n+1)/2}}, \qquad |\hat{\phi}(\zeta)| \leqslant \frac{c_m}{(1+|\zeta|)^m}.$$

The second follows from the fact that $\phi \in \mathcal{S}$ and the first follows from writing

$$\int_{\mathbf{R}^n} e^{i\lambda \cdot x} \chi(x)\, dx = \int_{|x| \leqslant 1} e^{i\lambda \cdot x}\, dx = \int_{-1}^{1} e^{i|\lambda||x_1|} \left(1 - x_1^2\right)^{\text{power}} dx_1.$$

To evaluate the integral we expand $(1 - x_1^2)^{\text{power}}$ in powers of $1 - x$ near $x = 1$ and in powers of $1 + x$ near $x = -1$. This gives asymptotic formulas for the contribution near $x = \pm 1$. After the usual argument with integration by parts and stationary phase, we see that apart from the contributions near $x = \pm 1$ the integral decreases rapidly as $\lambda \to \infty$. For details, see [30]. With these estimates (52) becomes

$$N(\lambda, \zeta) = \omega_n \lambda^n + O\left(\lambda^n \left\{ \sum_{\substack{k \neq 0 \\ k \in \mathbf{Z}^m}} \left[\frac{c_n}{(|k|\lambda)^{(n+1)/2}} \frac{c_m}{(1+|k|\zeta)^m} \right] \right\} \right),$$

where $\omega_n = \text{volume}\{|x| \leqslant 1\}$.

Next we estimate the term in braces $\{\ \}$. The main contribution occurs when $|k| < \zeta^{-1}$ and so the error is

$$\lambda^n \sum_{\substack{0 \neq k \in \mathbf{Z}^m \\ |k| \leqslant \zeta^{-1}}} \frac{c_n}{(|k|\lambda)^{(n+1)/2}} \approx \lambda^{(n-1)/2} \sum_{\substack{0 \neq k \in \mathbf{Z}^m \\ |k| \leqslant \zeta^{-1}}} \frac{1}{(1+|k|)^{(n+1)/2}}$$

$$\approx \lambda^{(n-1)/2} \int_{|z| < \zeta^{-1}} \frac{dz}{(1+|z|)^{(n+1)/2}}$$

$$\approx \lambda^{(n-1)/2} \zeta^{-((n-1)/2)}.$$

So

$$N(\lambda, \zeta) = \omega_n \lambda^n + O\left(\lambda^{(n-1)/2} \zeta^{-((n-1)/2)}\right).$$

Substituting $\lambda \pm 10\zeta$ in place of λ here and combining with (51), we find that

$$N(\lambda) = \omega_n \lambda^n + O\left(\lambda^{n-1}\zeta + \lambda^{(n-2)/2}\zeta^{-((n-1)/2)}\right).$$

Finally if we pick ζ to minimize the remainder, we get

$$N(\lambda) = \omega_n \lambda^n + O\left(\lambda^{n-2+2/(n+1)}\right). \qquad \square$$

7. Introduction to global Fourier integral operators. For short time $|t| \ll 1$ we solved the wave equation by

$$T^t u(y) = \int a(y, \xi) e^{i[S_t(y,\xi) - \xi \cdot x]} u(x)\, dx d\xi,$$

where S_t is the generating function of a Hamiltonian flow Φ_t. However, for long times t, Φ_t will not have a generating function. The starting point of the global theory of Fourier integral operators (Hörmander [**34**]; see also Maslov [**45**]) is that even for long times t we can solve the wave equation by an operator

$$(53) \qquad T^t u(y) = \int a(y, x, \zeta) e^{i\phi_t(y,x,\zeta)} u(x) \, dx d\zeta,$$

where ϕ_t is homogeneous in ζ of degree 1, and smooth in x, y, $\zeta/|\zeta|$.

The idea is that the t-axis is covered by overlapping small intervals $\{I_\alpha\}$; in each I_α we can represent T^t in the form (53) with suitable a, ϕ_t. However, if we pass from I_α to an overlapping interval I_β then the form of the representation (53) may change completely. Thus, the amplitude a, the phase function ϕ, and even the dimension of the space of dummy variables ζ may be different for I_α and I_β. For small I_α near $t = 0$ we can take $\phi_t(y, x, \zeta) = S_t(y, \zeta) - \zeta \cdot x$, but for most of the t-intervals I_β this will not be possible.

To carry out this idea, a main issue is clearly to decide when two integrals of the form (53) define approximately the same operator. Since $Tu(y) = \int K(y, x)u(x) \, dx$ with

$$(54) \qquad K(y, x) = \int a(y, x, \zeta) e^{i\phi(y,x,\zeta)} \, d\zeta,$$

it is enough to classify distributions of the form (54). So we consider locally in $\mathbf{R}^n$ two distributions

$$(55) \qquad K(z) = \int a(z, \tau) e^{i\phi(z,\tau)} \, d\tau,$$

$$(56) \qquad K'(z) = \int a'(z, \tau') e^{i\phi'(z,\tau')} \, d\tau'.$$

Our assumptions on a and ϕ are as follows:

— $\quad \phi(z, \tau)$ is defined on $\mathbf{R}^n \times \mathbf{R}^N$, real valued and homogeneous of degree 1 in τ;
— $\quad a(z, \tau)$ is supported in a small conic neighborhood in (z, τ)-space;
— $\quad a \in S^m$, that is $|\partial_\tau^\alpha \partial_z^\beta a| \leqslant C_{\alpha\beta}(1 + |\tau|)^{m-|\alpha|}$;
— $\quad \phi$ is nondegenerate in the sense that $d_{x,\tau}\phi(x, \tau) \neq 0$ and $d_{x,\tau}(\partial\phi/\partial\tau_1), \ldots, d_{x,\tau}(\partial\phi/\partial\tau_N)$ are linearly independent on the set $C_\phi = \{(x, \tau) \in \mathbf{R}^n \times \mathbf{R}^N : d_\tau\phi(x, \tau) = 0\}$.

(The last condition vastly simplifies analysis of the critical points of ϕ. It turns out that the wave equation for long time can be solved using nondegenerate phase functions.)

Assume a' and ϕ' are defined on $\mathbf{R}^n \times \mathbf{R}^{N'}$ and satisfy analogous conditions; in particular $a' \in S^{m'}$.

(57)   *Problem.* Decide when the distributions K and K' are approximately the same.

The first step in the classification is to determine the wave-front set of K. Stationary phase suggests that the important contributions to the integral (55) arise at $C_\phi = \{(z, \tau) \mid (\partial\phi/\partial\tau)(z, \tau) = 0\}$. Furthermore, one computes that $(z, \tau) \in C_\phi$ contributes $\iota_\phi(z, \tau) \equiv (z, (\partial\phi/\partial z)(z, \tau)) = (z, \xi)$ to WF(K). Thus, WF(K) $= \iota_\phi(C_\phi) \equiv \Lambda_\phi$.

The nondegeneracy conditions on ϕ guarantee that Λ_ϕ is a manifold, homogeneous in the ξ-variable (i.e., $(z, \xi) \in \Lambda_\phi$ implies $(z, \delta\xi) \in \Lambda_\phi$, $\delta > 0$). Also Λ_ϕ has a very important special property: The symplectic form $\omega = \Sigma_k d\xi_k \wedge dz_k$ vanishes when restricted to Λ_ϕ. One says that Λ_ϕ is *Lagrangian*.

EXAMPLE. Suppose $z = (x, y)$, $\phi(z, \tau) = S(y, \tau) - \tau \cdot x$, where S is the generating function of a canonical transformation Φ. This is of course precisely the phase function considered before. Then $\Lambda_\phi = \{(x, \xi, y, -\eta) \mid \Phi(x, \xi) = (y, \eta)\}$. The vanishing of the symplectic form on Λ_ϕ just means that $\Sigma_k d\xi_k \wedge dx_k - \Sigma_k d\eta_k \wedge dy_k = 0$, i.e., Φ is canonical.

Now the classification problem (57) is answered by the following result [**34**].

THEOREM 8. *Let $\phi(z, \tau)$ and $\phi'(z, \tau')$ be phase functions defined on $\mathbf{R}^n \times \mathbf{R}^N$ and $\mathbf{R}^n \times \mathbf{R}^M$ respectively as above. Suppose $\Lambda_\phi = \Lambda_{\phi'}$. Then given any symbol $a \in S^m$, there is a symbol $a' \in S^{m'}$ with $m' = m + \frac{1}{2}(N - M)$ so that the Fourier integral distributions* (55), (56) *differ only by a C^∞ error. Moreover, a' can be computed modulo lower-order errors by the formula $(a \circ \iota_\phi^{-1}) \approx \theta \cdot (a' \circ \iota_{\phi'}^{-1})$ on Λ_ϕ. Here, θ is an elliptic symbol depending on ϕ, ϕ' but not on a.*

To prove this, one simply calculates the asymptotic behavior of the Fourier transforms of K, K' at infinity using stationary phase.

The above theorem lets us define a class of Fourier integral distributions associated to a Lagrangian manifold Λ, since locally one can easily find nondegenerate phase functions ϕ so that $\Lambda = \Lambda_\phi$. The choice of ϕ is irrelevent by Theorem 8, and one can then patch together local Fourier integral distributions into global ones by pseudodifferential partitions of unity. By calculating explicitly θ in Theorem 8, one can define intrinsically the principal symbol of a Fourier integral distribution as a section of a suitable bundle.

This means that we can associate Fourier integral operators to canonical transformations, even when there is no generating function. As expected, taking the product of two Fourier integral operators corresponds to composing the canonical transformations from which they arise. In particular, taking $\Phi_t =$ Hamiltonian flow as in the short-time case, we can show that the (half)-wave equation is solved globally in time by Fourier integral operators arising from the canonical transformation Φ_t.

These ideas have interesting applications. For instance, they can be used to relate the eigenvalues of the Laplacian to the lengths of closed geodesics on a manifold. (See Duistermaat and Guillemin [**15**].) We shall not pursue this here. We have meant these fragmentary remarks only as a lead-in to the following explanation, contributed by D. H. Phong, of Fourier integral operators with complex phase functions. These will be needed for the Bergman and Szegö kernels in Chapter 12.

8. Introduction to Fourier integral operators with complex phase. The Bergman and Szegö projections can be represented as Fourier integral operators, but with complex phase functions. This can be easily seen in the model case of the Mizohata operator

$$(57a) \qquad\qquad P = \partial/\partial t + t\,|D_x|\,,$$

which can be viewed as an analogue of the $\bar{\partial}_b$ equation and will play an important role in the sequel. The null space N of P is nontrivial, but the orthogonal projection S onto N together with a right inverse Q with values in $N^\perp$ can be contructed as follows. Taking Fourier transforms with respect to x transforms P into $d/dt + t\,|\xi|$, which vanishes over $\phi_0(t, \xi) = e^{-(1/2)t^2|\xi|}/|\xi|^{1/4}$. On the other hand $Pf = \mu$ implies

$$(d/dt + t\,|\xi|)\tilde{f}(t, \xi) = \tilde{\mu}(t, \xi)$$

and thus

$$\tilde{f}(t, \xi) = \int_0^t e^{-(1/2)|\xi|(t^2 - s^2)}\tilde{\mu}(s, \xi)\, ds + K\phi_0(t, \xi).$$

The condition $\langle\, \tilde{f}(t, \xi), \phi_0(\cdot\,, \xi)\rangle_{L^2(\mathbf{R})} = 0$ yields

$$K(\xi) = -|\xi|^{1/4}\int_{-\infty}^{\infty} e^{-|\xi|\alpha^2}\int_0^\alpha e^{(1/2)s^2|\xi|}\tilde{\mu}(s, \xi)\, ds d\alpha.$$

From this we obtain the following formulae for S and Q:

$$(Su)(t, x) = \int e^{i\langle x, \xi\rangle}\big\langle \tilde{\mu}(\cdot\,, \xi), \phi_0(\cdot\,, \xi)\big\rangle_{L^2(\mathbf{R})}\phi_0(t, \xi)\, d\xi$$

$$= \iint e^{i\langle x, \xi\rangle}e^{-(1/2)(s^2 + t^2)|\xi|}\tilde{\mu}(s, \xi)\frac{1}{|\xi|^{1/2}}\, ds d\xi$$

$$(58) \qquad\qquad = \iiint e^{i\langle x - y, \xi\rangle}e^{-(1/2)(s^2 + t^2)|\xi|}\frac{1}{|\xi|^{1/2}}u(s, y)\, dy ds d\xi;$$

$$(Qu)(t, x) = \int e^{i\langle x, \xi\rangle}\tilde{f}(t, \xi)\, d\xi$$

$$= \iint e^{i\langle x, \xi\rangle}e^{it\tau}\sigma(t, \tau, \xi)\tilde{\mu}(\tau, \xi)\, d\tau d\xi,$$

where

$$(59) \qquad \sigma(t, \tau, \xi) = \left(\frac{|\xi|}{\pi}\right)^{1/2}\int_{-\infty}^{\infty} e^{-|\xi|\alpha^2}\int_\alpha^t e^{-(1/2)|\xi|(t^2 - s^2)}e^{-i(t - s)\tau}\, ds d\alpha.$$

A direct calculation shows that

$$(60) \qquad\qquad PQ = \mathrm{I}, \qquad QP = \mathrm{I} - S,$$

which reduces existence and regularity questions for P to the study of Q and S. Now Q and S are pseudodifferential operators of class $S_{1/2,1/2}$, and sharp estimates for Q can be obtained by imbedding Q in more restrictive classes (see Nagel-Stein [**49**], in which the formula (59) first appears); however, one can also view S as a Fourier integral operator with a classical symbol, but a complex phase function

$$(61) \qquad \phi(x, t; y, s; \xi) = \langle x - y, \xi \rangle + i(t^2 + s^2)|\xi|/2.$$

We now turn to a study of such integrals, following Melin-Sjöstrand [**46**]. The discussion will be local stressing only the main differences with the real case. Once more the key tool will be the method of stationary phase, suitably extended to complex functions.

Consider then the integral

$$I(x, \lambda) = \int e^{i\lambda\phi(x,y)}\mu(y, \lambda)\, dy, \qquad x \in \mathbf{R}^n, y \in \mathbf{R}^N,$$

where $\operatorname{Im}\phi \geq 0$, and $|D_x^\alpha D_\lambda^\beta \mu(y, \lambda)| \leq C_{\alpha\beta}\lambda^{m-|\beta|}$. The main contributions to the integral will come from the critical points of ϕ, which will be assumed to be nondegenerate. Thus assume $\mu(y, \lambda)$ is supported in a small neighborhood of 0, $d_y\phi(0,0) = 0$, and $d_{yy}^2\phi(0,0) \neq 0$. Unlike the real case, the set $C_{\phi,\mathbf{R}} = \{(x, y) \in \mathbf{R}^n \times \mathbf{R}^N; d_y\phi(x, y) = 0\}$ will not in general be a C^∞ manifold, nor if it is will it have the proper dimension. If ϕ is analytic we may view $C_{\phi,\mathbf{R}}$ as the set of real points of $C_\phi = \{(x, y) \in \mathbf{C}^n \times \mathbf{C}^N; \partial_y\phi(x, y) = 0\}$ which is C^∞. When ϕ is merely C^∞ we are naturally led to consider almost analytic extensions of ϕ.

Thus let ω, Ω be open sets in $\mathbf{R}^n$ and $\mathbf{C}^n$ respectively, with $\Omega \cap \mathbf{R}^n = \omega$, and let f be a C^∞ function on Ω. f is said to be almost analytic if $\bar{\partial}f$ vanishes to infinite order on ω. Any C^∞ function on ω admits an infinite number of almost analytic extensions which are all equivalent in the sense that the difference between any two vanishes to infinite order at the real points. We shall not distinguish between equivalent functions. If $\mu \in S^m$ is a symbol on a cone, then the extension can be chosen to satisfy analogous estimates. An almost analytic manifold $\Lambda \subset \Omega$ is a C^∞ manifold defined locally near real points by almost analytic functions $f_1(x) = \cdots = f_k(x) = 0$, with $\partial f_1(x), \ldots, \partial f_k(x)$ linearly independent over $\mathbf{C}$. Two almost analytic manifolds Λ_1 and Λ_2 will be said to be equivalent if $d(z, \Lambda_1) \leq C_N|\operatorname{Im} z|^N$ for $z \in \Lambda_2$ (here $d(z, \Lambda_1)$ denotes the distance from z to Λ_1). This is easily seen to be an equivalence relation which will allow us to identify manifolds in the same class. Almost analytic functions on Λ will be restrictions to Λ of almost analytic functions on Ω.

Returning now to the integral $I(x, \lambda)$, observe that the set

$$(62) \qquad C_{\tilde{\phi}} = \left\{(x, y) \in \mathbf{C}^n \times \mathbf{C}^N; \partial_y\tilde{\phi}(x, y) = 0\right\},$$

where $\tilde{\phi}$ denotes an almost analytic extension of ϕ, is an almost analytic manifold. The points of $C_{\tilde{\phi}}$ will still be called critical points of ϕ, despite the fact that this notion does not coincide with the usual one when x or y is complex. Near 0, parametrize $C_{\tilde{\phi}}$ by $x \to (x, Y(x))$. Then

THEOREM 9 (METHOD OF STATIONARY PHASE). *Locally near 0, we have*

$$(63) \qquad \text{(a)} \quad \text{Im}\,\tilde{\phi}(x, Y(x)) \geq C\,|\,\text{Im}\,Y(x)\,|^2.$$

(b) *Let $\tilde{\mu}(y, \lambda)$ be an almost analytic extension of μ with good estimates. Then*

$$(64) \qquad I(x, \lambda) \sim \lambda^{-N/2} e^{i\lambda\tilde{\phi}(x,Y(x))} \sum_{k=0}^{\infty} \lambda^{-k}(R_k(x, D)\tilde{\mu})(Y(x), \lambda).$$

Here $R_k(x, D)$ are differential operators of order $\leq 2k$, with

$$(65) \qquad R_0(x, D) = (2\pi)^{N/2}\Big[\det\big(i^{-1}\text{Hess}_{yy}\,\tilde{\phi}(x, Y(x))\big)\Big]^{-1/2}.$$

(The branch of the square root is chosen so that it equals 1 at 1; this can be achieved since the eigenvalues of $i^{-1}\text{Hess}_{yy}\,\phi(0,0)$ do not lie on the negative real axis.)

PROOF OF (a). Let $A = \text{Hess}_{yy}\,\tilde{\phi}(x, Y(x))$. The almost analyticity of $\tilde{\phi}$ and Taylor's formula imply

$$(66) \qquad \tilde{\phi}(x, y) = \tilde{\phi}(x, Y(x)) + \tfrac{1}{2}\big\langle A(y - Y(x)), y - Y(x)\big\rangle$$
$$+ O\big(|\,\text{Im}\,Y(x)\,|^3 + |\,y - Y(x)\,|^3\big).$$

Since $\text{Im}\,\phi(x, y) \geq 0$ for real values of the argument, it suffices to exhibit y in the real ball centered at $\text{Re}\,Y(x)$ and of radius $\text{Im}\,Y(x)$ such that

$$(67) \qquad \text{Im}\big\langle A(y - Y(x)), (y - Y(x))\big\rangle \leq -C\,|\,\text{Im}\,Y(x)\,|^2.$$

Writing $y = Y(x) = i[\text{Im}\,Y(x) + ik]$, $k \in \mathbf{R}^N$, (67) reduces to finding k such that $|\,k\,| < |\,\text{Im}\,Y(x)\,|$ and

$$\text{Im}\big\langle A(\text{Im}\,Y(x) + ik), \text{Im}\,Y(x) + ik\big\rangle \leq C\,|\,\text{Im}\,Y(x)\,|^2.$$

Now for $\varepsilon > 0$ small enough the equation

$$(68) \qquad (I + Q^t)A(I + Q) = A + i\varepsilon I$$

admits a solution $Q(\varepsilon)$ since it does for $\varepsilon = 0$ and the differential of the mapping $Q \to (I + Q^t)A(I + Q)$ from the space of complex $N \times N$ matrices to the space of complex symmetric matrices admits a right inverse when A is invertible. Obviously $Q(\varepsilon) \to 0$ as $\varepsilon \to 0$. Letting $T(\varepsilon) = (\text{Im}\,Q)(I + \text{Re}\,Q)^{-1}$ and multiplying (68) by $(I + \text{Re}\,Q)^{-1}$ yields

$$(I + iT^t)A(I + iT) = (I + \text{Re}\,Q^t)^{-1}(A + i\varepsilon I)(I + \text{Re}\,Q)^{-1}.$$

Set $k = T(\varepsilon)(\text{Im}\,Y(x))$. In view of the fact that $\text{Im}\,A \geq 0$ we have

$$\text{Im}\big\langle A(\text{Im}\,Y(x) + ik), \text{Im}\,Y(x) + ik\big\rangle \geq \varepsilon\,|\,(I + \text{Re}\,Q^t)^{-1}k\,|^2 \geq C\,|\,\text{Im}\,Y(x)\,|^2,$$

as was to be shown. $\square$

PROOF OF (b). We begin by choosing for each x an appropriate system of coordinates for $\mathbf{C}^N$ near $Y(x)$. According to Taylor's formula

$$(69) \quad \tilde{\phi}(x, y) = \tilde{\phi}(x, Y(x)) + l(x, y) + \tfrac{1}{2}\big\langle R(x, y)(y - Y(x)), y - Y(x)\big\rangle.$$

Here the presence of the "linear term" $l(x, y)$ is caused by the fact that $Y(x)$ is not a critical point of ϕ in the usual sense when $Y(x)$ is complex. The terms l and R satisfy the estimates

$$(70) \qquad |\bar{\partial}_y R(x, y)| + |l(x, y)| \leq C_K(|\operatorname{Im} y|^K + |\operatorname{Im} Y(x)|^K).$$

Let P be a matrix with $P'R(0,0)P = i\mathrm{I}$. Now the map $Q \to iQ'Q$ from $\mathrm{GL}(N, \mathbf{C})$ into the space of symmetric matrices is analytic with surjective differential at $Q = P^{-1}$, and thus for (x, y) small we may find $Q(x, y)$ analytic in (x, y) such that

$$iQ'(x, y)Q(x, y) = R(x, y), \qquad Q(0,0) = P^{-1}$$

and

$$|\bar{\partial}_x Q(x, y)| \leq C_K(|\operatorname{Im} y|^K + |\operatorname{Im} Y(x)|^K).$$

For each x define now new coordinates in $\mathbf{C}^N$ by

$$y \to \zeta(x, y) = Q(x, y)(y - Y(x)).$$

Then (69) becomes

$$(71) \qquad \tilde{\phi}(x, y) = \tilde{\phi}(x, Y(x)) + i\langle \zeta, \zeta \rangle/2 + l(x, y).$$

Set $\zeta = \xi + i\eta$. Observe that at $x = 0$, $Y(0) = 0$, $|\xi|^2 - |\eta|^2 = \operatorname{Im} \phi(0, y) \geq 0$ for $y \in \mathbf{R}^N$, so that in the new coordinates $\mathbf{R}^N$ is given by an equation $\eta = g(\xi, x)$. We write $y(\zeta)$ for the inverse map of $y \to \zeta(x, y)$ and introduce the following chains $\Gamma_{x,t}$ in $\mathbf{C}^N$:

$$\zeta_t(\xi) = \xi + itg(\xi, x), \qquad 0 \leq t \leq 1, \xi \in \mathbf{R}^N,$$

$$\Gamma_{x,t} : \xi \to y(\zeta_t(\xi)).$$

We now show that

$$(72) \quad \int_{\mathbf{R}^N} e^{i\lambda\phi(x,y)}\mu(y, \lambda)\, dy - \int_{\Gamma_{x,0}} e^{i\lambda\tilde{\phi}(x,y)}\tilde{\mu}(y, \lambda)\, dy_1 \wedge \cdots \wedge dy_N = O(\lambda^{-K})$$

for any K. Indeed by Stokes' formula the left-hand side of (72) can be estimated by

$$\left| \int_{\substack{\cup \Gamma_{x,t} \\ 0 \leq t \leq 1}} \bar{\partial}_y \left(e^{i\lambda\tilde{\phi}(x,y)}\tilde{\mu}(y, \lambda) \right) dy_1 \wedge \cdots \wedge dy_N \right|$$

$$\leq C \sup_{\substack{\xi \,\mathrm{near}\, 0 \\ x \,\mathrm{near}\, 0 \\ 0 \leq t \leq 1}} \left| e^{i\lambda\tilde{\phi}(x,y(\zeta_t(\xi)))} \left[i\lambda\tilde{\mu}(y(\zeta_t), \lambda)\bar{\partial}_y\tilde{\phi}(x, y(\zeta_t)) + \bar{\partial}_y\tilde{\mu}(y(\zeta_t), \lambda) \right] \right|,$$

and thus (72) is a consequence of the almost analyticity of $\tilde{u}$, $\tilde{\phi}$, and the following inequalities:

$$(73) \qquad (1 - t)(|\operatorname{Im} Y(x)| + |\xi|) \geq C|\operatorname{Im} y(\zeta_t)|,$$

$$(74) \qquad |\operatorname{Im} \tilde{\phi}(x, y(\tilde{\zeta}_t))| \geq C(1 - t)(|\operatorname{Im} Y(x)|^2 + |\xi|^2).$$

To establish (73), write

$$|\operatorname{Im} y(\zeta_t(\xi))| = |\operatorname{Im} y(\zeta_t(\xi)) - \operatorname{Im} y(\zeta_1(\xi))| \leqslant C(1-t)|g(\xi, x)|$$
$$\leqslant C(1-t)(|g(0, x)| + |\xi|)$$

and observe that $|g(0, x)| \sim |\operatorname{Im} Y(x)|$; as for (74) it follows from (63), (73), the identity

$$\operatorname{Im} \phi\big(x, y(\tilde{\zeta}_t)\big) = t^2 \operatorname{Im} \phi\big(x, y(\tilde{\zeta}_1)\big) + (1 - t^2)\big[\operatorname{Im} \phi(x, Y(x)) + \tfrac{1}{2}|\xi|^2\big]$$
$$+ (1 - t^2)\operatorname{Im} l\big(x, y(\tilde{\zeta}_1)\big) + \big\{\operatorname{Im}\big(l\big(x, y(\tilde{\zeta}_t)\big) - l\big(x, y(\tilde{\zeta}_1)\big)\big)\big\}$$

(which is a corollary of (71)), and from the estimates (70) which hold for l and its derivatives.

We are thus reduced to the study of the asymptotic expansion of the integral

$$(75)\quad \int_{\Gamma_{x,0}} e^{i\lambda\tilde{\phi}(x,y)}\tilde{\mu}(y, \lambda)\, dy_1 \wedge \cdots \wedge dy_N$$

$$= \int_{\mathbf{R}^N} e^{i\lambda(\tilde{\phi}(x,Y(x)) + i|\xi|^2/2 + l(x,y(\xi)))}\tilde{\mu}(y(\xi), \lambda)\left(\det\left(\frac{\partial y}{\partial \xi}\right)\right) d\xi$$

$$= e^{i\lambda\tilde{\phi}(x,Y(x))}\int e^{-\lambda|\xi|^2/2}\tilde{\mu}(y(\xi), \lambda)\left(\det\left(\frac{\partial y}{\partial \xi}\right)\right) d\xi$$

$$+ \int\int_0^1 e^{i\lambda(\tilde{\phi}(x,Y(x)) + i|\xi|^2/2 + sl(x,y(\xi)))}l(x, y(\xi))\tilde{\mu}(y(\xi), \lambda)\left(\det\left(\frac{\partial y}{\partial \xi}\right)\right) d\xi\, ds$$

which becomes

$$(76)\quad \int_{\Gamma_{x,0}} e^{i\lambda\tilde{\phi}(x,y)}\tilde{\mu}(y, \lambda)\, dy_1 \wedge \cdots \wedge dy_N$$

$$\sim e^{i\lambda\tilde{\phi}(x,Y(x))}\int e^{-\lambda|\xi|^2/2}\tilde{\mu}(y(\xi), \lambda)\left(\det\left(\frac{\partial y}{\partial \xi}\right)\right) d\xi,$$

since the integrand in the expression between brackets in (75) is bounded by

$$C\lambda^{(\text{fixed power})}\big(|\operatorname{Im} Y(x)|^K + |\xi|^K\big)\big(e^{-c\lambda(|\operatorname{Im} Y(x)|^2 + |\xi|^2)}\big)$$

in view of (63), (70), and (73).

The phase in (76) being real, we can now proceed as in the classical case by applying Parseval's formula and expanding $(e^{-\lambda|\xi|^2/2})^{\hat{}} = \lambda^{-(N/2)}e^{-|\xi|^2/2\lambda}$ in a Taylor series. The identity (65) follows from an inspection of the coefficients. The proof of Theorem 2 is complete. $\quad\square$

We now consider Fourier integral distributions. Let $\phi(x, \theta)$ be a C^∞ function on $\mathbf{R}^n \times \mathbf{R}^N$ satisfying the following conditions:

(i) $\phi(x, \lambda\theta) = \lambda\phi(x, \theta), \lambda \in \mathbf{R}^+$;

(ii) $d_{x,\theta}\phi(x, \theta) \neq 0$;

(iii) $d_{x,\theta}(\partial\phi/\partial\theta_1), \ldots, d_{x,\theta}(\partial\phi/\partial\theta_N)$ are linearly independent over $\mathbf{C}$ on the set

$$C_{\phi,\mathbf{R}} = \{(x, \theta) \in \mathbf{R}^n \times \mathbf{R}^N \setminus 0; d_\theta\phi(x, \theta) = 0\};$$

(iv) $\operatorname{Im} \phi(x, \theta) \geqslant 0$;

and let $a(x, \theta)$ be a classical symbol of order m. The distribution

$$\langle I(a, \phi), u \rangle = \iint e^{i\phi(x,\theta)} a(x, \theta) u(x)\, dx\, d\theta, \qquad u \in C_0^\infty(\mathbf{R}^n),$$

can then be defined as an oscillatory integral, with wave front set contained in

$$\mathrm{WF}(I(a, \phi)) \subset \{(x, d_x\phi(x, \theta)); (x, \theta) \in C_{\phi,\mathbf{R}}\} = \Lambda_{\phi,\mathbf{R}}$$

(observe that $(d_x\phi)(x, \theta)$ is real when $(x, \theta) \in C_{\phi;\mathbf{R}}$). We next determine to what extent the class of distributions $I(a, \phi)$ for various symbols a depends on the phase function ϕ. The method of stationary phase outlined above suggests considering not the sets $C_{\phi,\mathbf{R}}$ and $\Lambda_{\phi,\mathbf{R}}$, but rather the corresponding almost analytic manifolds defined by almost analytic extensions of ϕ. Thus let $\tilde{\phi}$ be an almost analytic extension of ϕ and set

$$C_{\tilde{\phi}} = \{(\tilde{x}, \tilde{\theta}) \in \mathbf{C}^n \times (\mathbf{C}^N \setminus 0); d_{\tilde{\theta}}\tilde{\phi}(\tilde{x}, \tilde{\theta}) = 0\},$$

$$\Lambda_{\tilde{\phi}} = \text{Image of } C_{\tilde{\phi}} \text{ under the mapping } T_{\tilde{\phi}}: (x, \theta) \to (\tilde{x}, d_{\tilde{x}}\tilde{\phi}(\tilde{x}, \tilde{\theta})).$$

Then $C_{\tilde{\phi}}$ and $\Lambda_{\tilde{\phi}}$ are almost analytic manifolds, $C_{\tilde{\phi},\mathbf{R}} = C_{\tilde{\phi}} \cap (\mathbf{R}^n \times \mathbf{R}^N \setminus 0)$, $T_{\tilde{\phi}}(C_{\tilde{\phi},\mathbf{R}}) = \Lambda_{\tilde{\phi}} \cap (\mathbf{R}^n \times (\mathbf{R}^N \setminus 0))$, and $\Lambda_{\tilde{\phi}}$ is a positive conic Lagrangian manifold in the sense that near a given real point (x_0, ξ_0), there always exist real symplectic coordinates (x, ξ) with almost analytic extensions $(\tilde{x}, \tilde{\xi})$ so that Λ is given by the equations

$$(77) \qquad\qquad \tilde{x} = \partial_{\tilde{\xi}} S(\tilde{\xi})$$

for some homogeneous almost analytic function $S(\tilde{\xi})$ with positive imaginary part. It follows that in any other system of symplectic coordinates (y, η) with the projection $(\tilde{y}, \tilde{\eta}) \in \Lambda_{\tilde{\phi}} \to \tilde{y} \in \mathbf{C}^n$ having surjective differential $\Lambda_{\tilde{\phi}}$ can be represented in the same way as in (77) with a suitable choice of generating functions S.

THEOREM 10 (INVARIANCE UNDER CHANGE OF PHASE FUNCTIONS). *Let $\phi(x, \theta)$ and $\psi(x, \omega)$ be phase functions defined on $\mathbf{R}^n \times (\mathbf{R}^N \setminus 0)$ and $\mathbf{R}^n \times (\mathbf{R}^M \setminus 0)$ respectively. Assume that $d_\theta\phi(x_0, \theta_0) = 0$, $d_\omega\psi(x_0, \omega_0) = 0$ and $\Lambda_\phi \sim \Lambda_\psi$ near (x_0, ξ_0) $= (x_0, d_x\phi(x_0, \theta_0)) = (x_0, d_x\psi(x_0, \omega_0))$. Then any distribution $I(a, \phi)$ for $a \in S^m(\mathbf{R}^n \times (\mathbf{R}^N \setminus 0))$ with conic support in a small conic neighborhood of (x_0, θ_0) can also be represented as a distribution $I(b, \psi)$ for a suitable symbol b in $S^{m+(1/2)(N-M)}(\mathbf{R}^n \times \mathbf{R}^M \setminus 0)$. (Here we identify distributions which differ by C^∞ functions in a neighborhood of x_0.) If a is classical, then b can be chosen to be classical also.*

PROOF OF THEOREM 10. With appropriate coordinates on $\mathbf{R}^n$, we may represent Λ_ϕ near (x_0, ξ_0) by

$$\Lambda_\phi = \{(\tilde{x}, \tilde{\xi}) \in \mathbf{C}^n \times (\mathbf{C}^n \setminus 0); \tilde{x} = \tilde{x}(\tilde{\xi})\},$$

with $\tilde{x}(\tilde{\xi})$ almost analytic and positively homogeneous of degree 0. It is then easily seen that the critical points of the function $(\tilde{x}, \tilde{\theta}) \to \tilde{\phi}(\tilde{x}, \tilde{\theta}) - \langle \tilde{x}, \tilde{\xi} \rangle$ are exactly the pull backs of $(\tilde{x}(\tilde{\xi}), \tilde{\xi})$ under the map $T_{\tilde{\phi}}$, which we shall denote by

$(\tilde{x}(\tilde{\xi}), \tilde{\theta}(\tilde{\xi}))$. Now the Fourier transform of $I(a, \phi)$ at η is given by

$$(I(a, \phi))^{\wedge}(\eta) = \lambda^N \iint e^{i\lambda(\phi(x,\theta) - \langle x, \xi \rangle)} a(x, \lambda\theta) \, dx d\theta,$$

where we have written $\eta = \lambda\xi$, $|\xi| = 1$, $\lambda \in \mathbf{R}^+$. The integral over the region away from the critical points of $\phi(x, \theta) - \langle x, \xi \rangle$ is of order $C_N \lambda^{-N}$, and thus Theorem 9 applied to the integral near the critical point yields

$$(I(a, \phi))^{\wedge}(\eta) \sim \lambda^{N - (n/2)} e^{i\lambda(\tilde{\phi}(\tilde{x}(\xi), \tilde{\theta}(\xi)) - \langle \tilde{x}(\xi), \xi \rangle)}$$

$$\times \sum_{k=0}^{\infty} \lambda^{-k} R_k(\xi, x, D_x, D_\theta)(\tilde{a}(\tilde{x}(\xi), \lambda\tilde{\theta}(\xi))),$$

where $\tilde{a}$ denotes an almost analytic extension of a. However $\tilde{\phi}(\tilde{x}, \tilde{\theta})$ is $O(|\operatorname{Im} \tilde{x}(\xi)|^K + |\operatorname{Im} \tilde{\theta}(\xi)|^K)$ since $\tilde{\phi}$ is homogeneous and $\tilde{\theta}$ is a critical point of $\tilde{\phi}$, while the arguments leading to (63) still yield

$$\operatorname{Im}\{\tilde{\phi}(\tilde{x}(\xi), \tilde{\theta}(\xi)) - \langle \tilde{x}(\tilde{\xi}), \xi \rangle\} \geq C(|\operatorname{Im} \tilde{x}(\xi)|^2 + |\operatorname{Im} \tilde{\theta}(\xi)|^2).$$

It thus follows that

$$(I(a, \phi))^{\wedge}(\eta) \sim \lambda^{N - (n/2)} e^{i\lambda\langle \tilde{x}(\xi), \xi \rangle} \sum_{k=0}^{\infty} \lambda^{-k} R_k(\xi, x, D_x, D_\theta)(\tilde{a}(\tilde{x}(\xi), \tilde{\theta}(\xi))).$$

Observing that the phase now no longer depends on ϕ and only on Λ_ϕ, we see that $(I(b, \phi))^{\wedge}(\eta)$ is given by the same kind of expression with the same phase. The proof of the theorem can be completed by showing that given any symbol $\alpha(x, \eta)$ we may choose $a(x, \theta)$ so that

$$\sum_{k=0}^{\infty} |\eta|^{-k} R_k(\eta/|\eta|, x, D_x, D_\theta)(\tilde{a}(\tilde{x}(\eta/|\eta|), \tilde{\theta}(\eta))) \sim \alpha(x, \eta).$$

This can be done by successive approximations since $R_0(\eta/|\eta|, x, D_x, D_\theta)$ is $\neq 0$ in view of (65). $\quad\square$

With Theorem 10 it is now possible to associate to a given closed conic positive almost analytic Lagrangian manifold Λ a class of distributions $I^m(\mathbf{R}^n, \Lambda)$. A distribution A will be said to be in $I^m(\mathbf{R}^n, \Lambda)$ if $\operatorname{WF}(A) \subseteq \Lambda_{\mathbf{R}} (\equiv \Lambda \cap \mathbf{R}^n)$, and for each $(x_0, \xi_0) \in \Lambda$ we have $(x_0, \xi_0) \notin \operatorname{WF}(A - I(a, \phi))$ for some phase function ϕ with $\Lambda = \Lambda_\phi$ near (x_0, ξ_0), and some classical symbol a of order $m + \frac{1}{4}(n - 2N)$. Theorem 10 then simply says that $I^m(\mathbf{R}^n, \Lambda)$ and $I^m(\mathbf{R}^n, \Lambda^\dagger)$ are identical modulo C^∞ functions when Λ and $\Lambda^\dagger$ are equivalent. The converse is also true when the symbols involved are required to be classical and not merely in $S^{m + (1/4)(n - 2N)}$.

Fourier integral operators can next be defined through their distribution kernels. Let X and Y be open sets in $\mathbf{R}^{n_x}$ and $\mathbf{R}^{n_y}$ respectively. Given a submanifold C of $T^*(X \times Y)$, we shall denote by C' the manifold $\{((x, \xi), (y, \eta)) \in T^*(X \times Y); ((x, \xi), (y, -\eta)) \in C\}$. A positive canonical relation C is a submanifold of $(T^*(X \times Y) \setminus 0)^{\sim}$ such that $C_{\mathbf{R}} \equiv C \cap T^*(X \times Y) \subseteq (T^*(X) \setminus 0) \times (T^*(Y) \setminus 0)$ is a closed set, and C' is a positive conic almost analytic Lagrangian manifold. A Fourier integral operator A associated to a positive canonical relation C is an operator with distribution kernel K_A in $I^m(X \times Y; C')$. Such operators can be extended as operators from compactly supported distributions on Y to compactly supported distributions on X by neglecting regularizing kernels. Now let A_1 and A_2 be Fourier integral operators with kernels $K_{A_1} \in I^{m_1}(X \times Y, C_1')$, $K_{A_2} \in I^{m_2}(Y \times Z, C_2')$, where X, Y, Z are open sets in $\mathbf{R}^{n_x}, \mathbf{R}^{n_y}, \mathbf{R}^{n_z}$, and C_1, C_2 are positive canonical relations; then the composition $A_1 \circ A_2 \colon \mathcal{E}'(Z) \to \mathcal{E}'(X)$ is a well-defined operator with singularities determined by

$$\mathrm{WF}\left(K_{A_1 \circ A_2}\right) \subseteq \mathrm{WF}\left(K_{A_1}\right) \circ \mathrm{WF}\left(K_{A_2}\right) \subseteq C_{1,\mathbf{R}} \circ C_{2,\mathbf{R}}.$$

To insure that $A_1 \circ A_2$ also be a Fourier integral operator, however, we will need some addition conditions. If

$$\Delta = T^*(X) \times \left(\mathrm{diag}\, T^*(Y) \times T^*(Y)\right) \times T^*(Z)$$

and $\tilde{\Delta}$ is its almost analytic complexification, we have

THEOREM 11 (COMPOSITION OF FOURIER INTEGRAL OPERATORS). *Assume that $C_1 \times C_2$ and $\tilde{\Delta}$ intersect transversally at $(C_{1,\mathbf{R}} \times C_{2,\mathbf{R}}) \cap \Delta$ and that the projection $C_{1,\mathbf{R}} \times C_{2,\mathbf{R}} \to (T^*(X) \setminus 0) \times (T^*(Z) \setminus 0)$ is one-to-one and proper. Then the mapping*

$$(C_1 \times C_2) \cap \tilde{\Delta} \to (T^*(X, Z) \setminus 0)^{\tilde{}}, \qquad (x, \xi, y, \eta; y, \eta, z, \zeta) \to (x, \xi, z, \zeta)$$

is a local diffeomorphism near real points whose image is a conic positive canonical relation which will be denoted by $C_1 \circ C_2$ and satisfies $(C_1 \circ C_2)_\mathbf{R} = C_{1,\mathbf{R}} \circ C_{2,\mathbf{R}}$. Furthermore, the operator $A_1 \circ A_2$ is a Fourier integral operator associated to $C_1 \circ C_2$ of order $m_1 + m_2$.

PROOF OF THEOREM 11. We may represent $K_{A_1 \circ A_2}$ as

$$(78) \quad K_{A_1 \circ A_2}(x, z) = \iiint e^{i(\phi_1(x, y, \theta) + \phi_2(y, z, \sigma))} a_1(x, y, \theta) a_2(y, z, \sigma)\, dy\, d\theta\, d\sigma,$$

where ϕ_1, ϕ_2, a_1, a_2 are phase functions and symbols defining A_1 and A_2. Partial integration with respect to (x, θ) and (z, σ) respectively show that the integrals over the regions $|\sigma| < \varepsilon |\theta|$ and $|\theta| < \varepsilon |\sigma|$ are of class C^∞ in (x, z) so that we may restrict our attention to the region where $|\theta| \sim |\sigma|$. There however the integral in (78) may be rewritten as

$$(79) \quad \iiint e^{i\phi(x_1, \theta, \sigma, \omega)} a(x, z; \theta, \sigma, \omega)\, d\theta\, d\sigma\, d\omega,$$

where (θ, σ, ω) are the new frequency variables, and

$$\phi(x, z; \theta, \sigma, \omega) = \phi_1\left(x, \omega\left(|\theta|^2 + |\sigma|^2\right)^{-1/2}, \theta\right) + \phi_2\left(\omega\left(|\theta|^2 + |\sigma|^2\right)^{-1/2}, z, \sigma\right)$$

$$a(x, z; \theta, \sigma, \omega) = a_1\left(x, \omega\left(|\theta|^2 + |\sigma|^2\right)^{-1/2}, \theta\right) a_2\left(\omega\left(|\theta|^2 + |\sigma|^2\right)^{-1/2}, z, \sigma\right) .$$

$$\times \left(|\theta|^2 + |\sigma|^2\right)^{-n_y/2}$$

are phase functions and symbols satisfying good estimates. That (79) is a Fourier integral distribution associated to $C_1 \circ C_2$ can now be verified by calculating the critical points of ϕ, using the transversality conditions, and applying Theorem 10.
$\square$

EXAMPLES. Returning to the operator (58) with the phase function (61) we readily see that C_ϕ is given by $x - y + i(t^2 + s^2)\xi/(2|\xi|) = 0$, and thus the canonical relation C corresponding to ϕ is the manifold of points $\{(x, \xi), (t, \tau); (y, \eta), (s, \rho)\}$ satisfying

$$\eta = \xi, \qquad\qquad\qquad \tau = it|\xi|,$$
$$y = x + i(t^2 + s^2)\xi/(2|\xi|), \quad \rho = -is|\xi|.$$

Let $V = \{(x, \xi, t, \tau) \in T^*(\mathbf{R}^{n+1}) \setminus 0; t = \tau = 0\}$ and $\tilde{V} = \{(x, \xi, t, \tau) \in (T^*(\mathbf{R}^{n+1}) \setminus 0)^\sim; i\tau + t|\xi| = 0\}$ be respectively the real and complex characteristic varieties of P; then it is not difficult to see that C is the only positive canonical relation which contains diag V and is contained in $\tilde{V} \times \tilde{V}$.

Once the model case P is understood we can construct parametrices and projections for systems of equations with noninvolutive characteristics by bringing them back to P by a canonical transformation. That such a canonical relation does exist has been established in the analytic case by Sato-Kawai-Kashiwara [55] and in the C^∞ case by Duistermaat-Sjöstrand [16] and Bontet de Monvel [6]. Applying a canonical transformation ϕ corresponds to conjugating with a Fourier integral operator F whose canonical relation is just the graph of ϕ. The transversality conditions of Theorem 11 are then satisfied, and the theorem implies that FSF^* is a Fourier integral operator with complex phase associated to the canonical relation $\mathcal{C} = \phi \circ C \circ \phi^{-1}$.

The case of $\bar{\partial}_b$ on a strongly pseudoconvex boundary $b\Omega \subset \mathbf{C}^n$ is one such example. The above procedure leads then to a representation for the Szegö projection S as a Fourier integral operator whose canonical relation $\mathcal{C}$ is easily determined in terms of the characteristic variety of $\bar{\partial}_b$. Then Theorem 10 allows us to change phases in the formula for S. A defining phase function for $\mathcal{C}$ is calculated to be

$$(x, y, \tau) \to \tau\psi(x, y), \qquad \tau \in \mathbf{R}^+,$$

where $\psi(x, y)$ is an almost analytic extension of the defining function r of $b\Omega$ (i.e., $\psi(x, x) = r(x)/i$, $\psi(x, y) = -\overline{\psi(y, x)}$, $\bar{\partial}_x\psi, \partial_x\psi$ vanish of infinite order for $y = x$), and thus K_S can be written as

$$\int_0^\infty e^{i\tau\psi(x,y)} a(x, y, \tau) \, d\tau.$$

Since $a(x, y, \tau)$ is a classical symbol in τ, Hadamard's principal value formulas immediately yield the well-known asymptotic expansion for the Szegö kernel. We return to these matters in Chapter 12.

CHAPTER 5. ELEMENTARY DIFFERENTIAL GEOMETRY AND THE HEAT EQUATION

The first three sections of this chapter concern the geometry of Riemannian and complex manifolds. Much of this is background material for the later chapters. The fourth and fifth sections introduce the Laplacian $\Box = dd^* + d^*d$ on a Riemannian n-manifold M associated to the exterior derivative d and its adjoint d^* and then relate it to the cohomology of the manifold and to the Euler characteristic $\chi(M)$. In fact we prove

$$
(1) \qquad \begin{aligned}
&\text{Trace}(e^{-t\Box_{\text{even}}}) - \text{Trace}(e^{-t\Box_{\text{odd}}}) \\
&= \dim(\ker \Box_{\text{even}}) - \dim(\ker \Box_{\text{odd}}) = \chi(M),
\end{aligned}
$$

where $\Box_{\text{even}}$ ($\Box_{\text{odd}}$) is the Laplacian restricted to forms of even (odd) degree.

§6 is concerned with the scalar heat equation (i.e. the heat operator acts on functions rather than forms)

$$
(2) \qquad \partial u/\partial t - \Delta_x u = 0, \qquad (x, t) \in M \times [0, \infty), \qquad u|_{t=0} = f.
$$

The solution has the form $u(x, t) = \int_M K_t(x, y) f(y) d\text{vol}(y)$, and we shall derive the asymptotics of the kernel $K_t(x, y)$. In particular, we shall see that

$$
(3) \qquad K_t(x, x) \sim \sum_{l \geq 0} t^{-n/2+l} \gamma_l(x) \quad \text{for small } t,
$$

where the γ_l are polynomials in the curvature tensor and its covariant derivatives.

Since the operator $T_t: f \to u(\,\cdot\,, t)$ is formally $e^{-t\Delta}$, we have from (3)

$$
(4) \qquad \text{Trace}(e^{-t\Delta}) \sim \sum_{l \geq 0} t^{-n/2+l} \int_M \gamma_l(x) d\text{vol}(x).
$$

On the other hand, diagonalizing Δ by its eigenfunctions ψ_k and eigenvalues λ_k, we see that

$$
(5) \qquad \text{Trace}(e^{-t\Delta}) = \sum_k e^{-t\lambda_k}.
$$

So there is a connection between integrals of "curvatures" $\gamma_l(x)$ and the eigenvalues of the Laplacian.

It is a simple matter to extend these results to the Laplacian on forms. Then by comparing (1) and (4) we obtain (in principle) a formula for the Euler characteristic in terms of curvature, namely

$$
(6) \qquad \chi(M) = \int_M P(R_{pqrs/\sigma}) d\text{vol},
$$

where P is a polynomial in the indeterminates $R_{pqrs/\sigma}$.

The remainder of the chapter contains the invariant theory necessary to discover the form of $P(\cdot)$. The classical invariant theory of Weyl is sufficient to show that the polynomials corresponding to the heat kernel $K_{t,\Box_i}(x, x)$ on forms of degree i are composed of terms of the form

$$
\text{Trace}\left[R_{abcd/\sigma} \otimes \cdots \otimes R_{ijkl/\tau} \right].
$$

This is discussed in §8. §9 contains a theorem of Gilkey in invariant theory, which is powerful enough to say exactly which polynomial P enters into (6). It turns out that P involves only the curvature and none of its covariant derivatives. P is called the Pfaffian, and (6) is the Chern-Gauss-Bonnet theorem.

1. Riemannian manifolds. Gauss considered the geometry of a surface in, for example, $\mathbf{R}^3$. Locally we we can write the surface as $z = u(x, y)$. In general we can translate and rotate the coordinates so that this takes the form

$$z = \lambda x^2 + \mu y^2 + O\big(|(x, y)|^3\big).$$

The product $K = \lambda\mu$ is called the Gaussian curvature.

EXAMPLE. Consider the sphere of radius A and center $(0, 0, +A)$ in $\mathbf{R}^3$,

$$x^2 + y^2 + (z - A)^2 = A^2.$$

We have

$$z = A\left(1 - \frac{x^2 + y^2}{A^2}\right)^{1/2} - A = \frac{-1}{A}x^2 - \frac{1}{A}y^2 + O\big(|(x, y)|^3\big),$$

so that $K = (-1/A)\cdot(-1/A) = 1/A^2$.

This definition of Gaussian curvature depends prima facie on the embedding of the surface in $\mathbf{R}^3$. Gauss discovered that in fact K depends only on the intrinsic geometry of the surface, not on the embedding. There are many different ways of defining the curvature of a manifold. In this section we give two: the first is very geometric but not obviously intrinsic; the second is obviously intrinsic but not as geometric.

Let

$$(7) \qquad\qquad x^i = f^i(u^1,\dots,u^m), \qquad i = 1,\dots,n,$$

be the components of a parametrized surface $\Sigma \subset \mathbf{R}^n$ and consider a parametrized curve γ, $u^i = g^i(t)$ on the surface. Suppose we have a vector field

$$t \to v(t) \in \mathbf{R}^n$$

along γ such that the vector $v(t)$ lies in the tangent space of Σ at $\gamma(t)$; i.e.

$$v(t) \in \operatorname{span}\{\Sigma_1,\dots,\Sigma_m\} \quad \text{at } \gamma(t),$$

where Σ_i are the coordinate tangent vectors $\Sigma_i \equiv (\partial f^1/\partial u^i,\dots,\partial f^n/\partial u^i)$. The vector field $v(t)$ is said to be *parallel in the sense of Levi-Civita* if the derivative dv/dt along the curve is always normal to the surface, i.e.

$$\frac{dv}{dt}\cdot\Sigma_i = 0, \qquad i = 1,\dots,m.$$

In this case two vector $v(t)$ and $v(t')$ are said to be obtained from one another by *parallel transport*.

EXAMPLE. Vectors in the hypersurface $\Sigma = \mathbf{R}^{n-1} \in \mathbf{R}^n$ that are parallel in the sense of Levi-Civita are parallel in the ordinary sense.

Suppose that X and Y are vector fields on Σ. Using parallel transport we will define a vector field $\nabla_X Y$, called the covariant derivative of Y with respect to X. If $p \in \Sigma$, let $\gamma(t)$ be the integral curve of X through $p = \gamma(0)$ (i.e. $\dot{\gamma}(t) = X(\gamma(t))$) and let Y_t be the parallel transport of $Y(\gamma(t))$ to p. Then

$$(8) \qquad (\nabla_X Y)(p) \equiv \lim_{t \to 0} \frac{Y_t - Y_0}{t}$$

satisfies

$$(9) \qquad
\begin{aligned}
&\text{(i)} \quad && \nabla_X(\alpha Y_1 + \beta Y_2) = \alpha \nabla_X Y_1 + \beta \nabla_X Y_2, \\
&\text{(ii)} \quad && \nabla_{\alpha X_1 + \beta X_2} Y = \alpha \nabla_{X_1} Y + \beta \nabla_{X_2} Y, \\
&\text{(iii)} \quad && \nabla_{fX} Y = f \nabla_X Y, \\
&\text{(iv)} \quad && \nabla_X(fY) = X(f)Y + f \nabla_X Y, \\
&\text{(v)} \quad && \nabla_X Y - \nabla_Y X = [X, Y] \text{ (the commutator)}, \\
&\text{(vi)} \quad && Z\langle X, Y \rangle = \langle \nabla_Z X, Y \rangle + \langle X, \nabla_Z Y \rangle,
\end{aligned}$$

where α, β are numbers, f is a function, and $\langle \cdot, \cdot \rangle$ is the inner product on $\mathbf{R}^n$. We call

$$T(X, Y) = \nabla_X Y - \nabla_Y X - [X, Y]$$

the torsion; (v) says that the covariant derivative (8) has no torsion. A covariant derivative satisfying (vi) is called *Riemannian*.

These properties can be used intrinsically to define a covariant derivative on a general Riemannian manifold M; this approach is due to Kozul. First we recall a few facts. Let M be a manifold. If $u^1, \ldots, u^m$ are a fixed set of local coordinates in a coordinate neighborhood $U \subset M$, then a vector field X on U can be expressed uniquely as

$$X = \sum_{k=1}^{m} x^k U_k,$$

where x^k are real-valued functions on U and $U_k = \partial / \partial u^k$. If we evaluate this at a point,

$$X(p) = \sum_{k=1}^{n} x^k(p) U_k,$$

we get a tangent vector; the set $T_p(M)$ of all such tangent vectors is called the tangent space at $p \in M$; a metric is the assignment of a positive-definite (or at least nondegenerate) inner product $\langle \cdot, \cdot \rangle_p$ on each vector space $T_p(M)$ in a manner that varies smoothly with respect to p. If $X = \Sigma x^k U_k$, $Y = \Sigma y^k U_k$ are two vector fields, then for appropriate functions g_{ij} on U we can define

$$(10) \qquad \langle X(p), Y(p) \rangle_p = \sum_{i,j=1}^{n} g_{ij}(p) x^i(p) y^j(p),$$

so that g_{ij} gives a metric on $U \subset M$. We abbreviate this as $ds^2 = \Sigma g_{ij}(p) du^i du^j$. If $t \to u(t) = (u^i(t))_{i=1,\ldots,n}$ is a path in the coordinate neighborhood, then the *length* of the path is defined as

$$L = \int ds = \int \sqrt{\sum_{ij} g_{ij}(u(t)) \frac{du^i}{dt} \frac{du^j}{dt}} \, dt.$$

A surface imbedded in $\mathbf{R}^m$ has an obvious induced metric $g_{ij} = \Sigma_l(\partial f^l/\partial u_i)(\partial f^l/\partial u_j)$, which gives the right formula for lengths of curves. The pair $(M, \langle \cdot, \cdot \rangle)$ is called a Riemannian manifold. It is convenient at this point to switch notation and let X_i denote the coordinate vector fields U_i.

By these remarks the covariant derivative is completely determined by the n^3 functions Γ_{ij}^k on U defined by

$$(11) \qquad \nabla_{X_i} X_j = \sum_{k=1}^n \Gamma_{ij}^k X_k.$$

PROPOSITION 1. *Given a Riemannian manifold* $(M, \langle \cdot, \cdot \rangle)$ *we can uniquely define the* Γ_{ij}^k *so that the covariant derivative* (11) *satisfies the axioms* (9).

PROOF. Define

$$(12) \qquad g_{jk} = \langle X_j, X_k \rangle$$

and require

$$\frac{\partial}{\partial x^i}(g_{jk}) = X_i(g_{jk}) = \langle \nabla_{X_i} X_j, X_k \rangle + \langle X_j, \nabla_{X_i} X_k \rangle,$$

so that (9.vi) holds. Now we cyclically permute i, j, k to get 3 equations in the 3 unknowns

$$\langle \nabla_{X_i} X_j, X_k \rangle, \quad \langle \nabla_{X_j} X_k, X_i \rangle, \quad \langle \nabla_{X_k} X_i, X_j \rangle;$$

note there are 3 rather than 6 unknowns since $\nabla_{X_i} X_j = \nabla_{X_j} X_i$ by (9.v). If we solve this system we get

$$(13) \qquad \langle \nabla_{X_i} X_j, X_k \rangle = \frac{1}{2}\left(\frac{\partial g_{jk}}{\partial x^i} + \frac{\partial g_{ik}}{\partial x^j} - \frac{\partial g_{ij}}{\partial x^k} \right).$$

But by (10) and (11), $\langle \nabla_{X_i} X_j, X_k \rangle = \Sigma_{l=1}^n \Gamma_{ij}^l g_{lk}$; therefore

$$(14) \qquad \Gamma_{ij}^l = \sum_{k=1}^n \frac{1}{2} g^{kl}\left(\frac{\partial g_{jk}}{\partial x^i} + \frac{\partial g_{ik}}{\partial x^j} - \frac{\partial g_{ij}}{\partial x^k} \right),$$

where g^{kl} is the inverse of the matrix g_{lk}. Now extend the definition (10) to expressions of the form

$$\nabla_{(\Sigma_i r^i X_i)}\left(\sum_j s^j X_j \right) \qquad \text{for functions } r^i, s^j$$

using the properties (9). It is easy to check explicitly that ∇ satisfies all the required axioms. $\square$

Equations (13) and (14) are known as the first and second Christoffel identities and the connection Γ_{ij}^k is called the *Riemannian connection*.

EXAMPLE 1. Let $M = \mathbf{R}^n$ with the flat metric $\langle \cdot, \cdot \rangle$ (the standard inner product). For $R = \Sigma r^i X_i$, $S = \Sigma s^i X_i$, define

$$(15.\text{i}) \qquad \nabla_R S = \sum_{i=1}^{n} (Rs^i) X_i,$$

where $Rs^i = \langle (r_1,\ldots,r_n), \nabla s^i \rangle$.

EXAMPLE 2. Let $\Sigma \subset \mathbf{R}^n$ be a hypersurface and define

$$(15.\text{ii}) \qquad \nabla^{\Sigma}_{R(p)} S = \text{projection of } \left(\nabla_R S \text{ defined by } (15.\text{i}) \right)$$
$$\text{onto the tangent space } T_p(\Sigma).$$

EXAMPLE 3. Let $\Sigma = \{(x_1, x_2) \in \mathbf{R}^2 \,|\, x_2 > 0\}$ with $ds^2 = (dx_1^2 + dx_2^2)/x_2^2$ (Poincaré half-plane). We leave it as an exercise to compute Γ^k_{ij}.

Let γ be a curve in M with a tangent vector field T. A vector field Y along γ is called *parallel in the sense of Kozul* if

$$\nabla_T Y = 0 \quad \text{on } \gamma.$$

EXAMPLE. We return to the parametrized surface $\Sigma \subset \mathbf{R}^m$ containing the curve γ,

$$\Sigma: x^i = f^i(u^1,\ldots,u^n), \qquad i = 1,\ldots,m,$$
$$\gamma: u^i = g^i(t), \qquad i = 1,\ldots,n,$$

and the vector field V with components $(v^1(t),\ldots,v^n(t))$ defined on γ and lying in the tangent space of Σ; γ has a field of tangent vectors $T = (dx^1/dt,\ldots,dx^n/dt)$. The vector field V is parallel along $\gamma \Leftrightarrow \nabla_T V = 0$ along $\gamma \Leftrightarrow$

$$(16) \qquad \frac{dv^k}{dt} + \sum_{i,j=1}^{n} \Gamma^k_{ij} \frac{dx^i}{dt} v^j = 0, \qquad k = 1,\ldots,n, \text{ along } \gamma,$$

by (9) and (11). It is not hard to see from this that V is parallel in the sense of Levi-Civita along γ.

Recall that the Lie bracket $[X, Y]$ of two vector fields about p_0 has the following geometric interpretation. Fix a small $t > 0$ and flow along the integral curve of X through p_0 for t units of time to reach p_1; then flow along the integral curve through p_1 of Y for t units of time to reach p_2; then flow along the integral curve through p_2 of $(-X)$ to reach p_3; finally flow along the integral curve through p_3 of $(-Y)$ to reach $\gamma_0(t)$. Define $\gamma(t) = \gamma_0(\sqrt{t})$. Then

$$[X, Y]_p = \frac{d\gamma}{dt}\bigg|_0.$$

This means that by flowing along the integral curve through $\gamma_0(t)$ of $-[X, Y]_p$ for t^2 units of time, the procedure above gives a closed loop up to second order in t. If we parallel transport a tangent vector Z around this loop, then Z is taken to

$$Z + t^2\left(\nabla_X \nabla_Y Z - \nabla_Y \nabla_X Z - \nabla_{[X,Y]} Z\right) + O(t^3).$$

This suggests that we define the *curvature* $R(X, Y)Z$ of the covariant derivative ∇ to be the vector field

$$(17) \qquad R(X, Y)Z = \nabla_X \nabla_Y Z - \nabla_Y \nabla_X Z - \nabla_{[X,Y]} Z,$$

for vector fields X, Y, Z.

PROPOSITION 2. $R(X, Y)Z$ *is a tensor, i.e.*

(i) *the value of $R(X, Y)Z$ at a point $p \in M$, depends only on the values X_p, Y_p, Z_p of the vector fields at this point,*

(ii) *the map*

$$(X_p, Y_p, Z_p) \to R(X_p, Y_p)Z_p$$

is trilinear.

PROOF. Property (ii) is obvious. To verify (i), replace X by fX; then

$$\nabla_X \nabla_Y Z \to f \nabla_X \nabla_Y Z, \qquad -\nabla_Y \nabla_X Z \to -(Yf)\nabla_X Z - f \nabla_Y \nabla_X Z,$$

$$-\nabla_{[X,Y]} Z \to -f \nabla_{[X,Y]} Z + (Yf)\nabla_X Z.$$

Adding these we get

$$R(fX, Y)Z = fR(X, Y)Z,$$

and similarly for $R(X, fY)Z$ and $R(X, Y)(fZ)$. Therefore if $X = \sum_i x^i U_i$, $Y = \sum_j y^j U_j$, $Z = \sum_k z^k U_k$, we have

$$R(X, Y)Z = \sum R(x^i U_i, y^j U_j)(z^k U_k) = \sum x^i y^j z^k R(U_i, U_j)U_k$$

and so

$$(R(X, Y)Z)_p = \sum x^i(p) y^j(p) z^k(p)\big(R(U_i, U_j)U_k\big)_p.$$

That is, $(R(X, Y)Z)_p$ depends only on the values of the vector fields X, Y, Z at the point p and not on the values at nearby points. $\square$

We switch notation again and denote the standard coordinate vector fields by X_i. The *Riemann-Christoffel tensor* is defined by

$$(18) \qquad R(X_i, X_j)X_k = \sum_{l=1}^{n} R^l_{ijk} X_l.$$

Since $[X_i, X_j] = 0$ for all i, j,

$$R(X_i, X_j)X_k = \nabla_{X_i} \nabla_{X_j} X_k - \nabla_{X_j} \nabla_{X_i} X_k,$$

and we see that the curvature $R(\cdot, \cdot)\cdot$ measures the extent to which the second covariant derivatives $\nabla_{X_i} \nabla_{X_j}$ fail to be symmetric. Using (8) and (11) it is easy, but tedious, to calculate

$$(19) \qquad R^l_{ijk} = \frac{\partial \Gamma^l_{ik}}{\partial x^j} - \frac{\partial \Gamma^l_{ij}}{\partial x^k} + \sum_m \big(\Gamma^m_{ik}\Gamma^l_{mj} - \Gamma^m_{ij}\Gamma^l_{mk}\big).$$

The *Ricci tensor* $\mathrm{Ric}(X, Y)$ is defined by

$$(20.\mathrm{i}) \qquad \mathrm{Trace}(Z \to R(Z, X)Y),$$

so its components are given by

$$(20.ii) \qquad \mathrm{Ric}_{ik} = \sum_{j=1}^{n} R^{j}_{ijk}.$$

We define the *scalar curvature* to be the trace of Ric_{ik}, that is

$$(21) \qquad R = \sum_{ij} g^{ij} R_{ij},$$

where as always (g^{ij}) is the inverse of (g_{ij}). Finally if we use the inner product to define $g_{im} = \langle X_i, X_m \rangle$, we can define the *Riemann curvature tensor*

$$(22) \qquad R_{ijkl} = \sum_{m=1}^{n} g_{im} R^{m}_{jkl}.$$

EXAMPLE. We return to the sphere $\Sigma \subset \mathbf{R}^3$ of radius A defined by

$$Z = \frac{-1}{A} x^2 - \frac{1}{A} y^2 + O\left(\left|(x, y)\right|^3\right)$$

and its Gaussian curvature $1/A^2$. If we parametrize the sphere by

$$x^1 = A \sin u^1 \cos u^2, \qquad x^2 = A \sin u^1 \sin u^2, \qquad x^3 = A \cos u^1,$$

then the metric g_{ij} is given by (15.iii)

$$g_{11} = A^2, \qquad g_{12} = g_{21} = 0, \qquad g_{22} = A^2 \sin^2 u^1,$$

and after a long calculation using (14), (19), (20) and (21), we find $R = 1/A^2$. We have finally attached an intrinsic significance to the Gaussian curvature, independent of the embedding.

Before leaving the covariant derivative, we show how it can be extended to arbitrary tensors. Suppose we are given an arbitrary tensor, say a 100-tensor

$$A : (X_1, \ldots, X_{100}) \to \text{numbers}.$$

We want the *covariant derivative* of A at p to be a 101-tensor $A(X_1, \ldots, X_{100}, Y)$. First let $\gamma(t)$ be the integral curve through $p = \gamma(0)$ of Y and let X_i' be the parallel transport to $\gamma(t)$ of the tangent vector X_i at p. Define

$$(23.i) \quad \nabla A(X_1, \ldots, X_{100}, Y) = \lim_{t \to 0} \frac{1}{t} \left[A_{\gamma(t)}(X_1', \ldots, X_{100}') - A_p(X_1, \ldots, X_{100}) \right].$$

Using an argument like that in Proposition 2, we can check that A is indeed a 101-tensor; it has components

$$(\nabla A)_{i_1 \cdots i_{100} j} = \frac{\partial A_{i_1 \cdots i_{100}}}{\partial x^j} - \sum_{\mu=1}^{100} \sum_{l=1}^{n} \Gamma^{l}_{i_\mu j} A_{i_1 \cdots i_{\mu-1} l i_{\mu+1} \cdots i_{100}}.$$

We now characterize those paths that locally minimize the distance D between two points on a Riemannian manifold (M, g_{ij}). Let $(x^1, \ldots, x^n)$ be local coordinates about the two points and let

$$x^i = r^i(t), \qquad 0 \leqslant t \leqslant 1,$$

be a path connecting them. We define

$$(24.\text{i}) \qquad D(r) = \int_0^1 L\big(r(t), \dot{r}(t)\big)\, dt$$

with

$$(24.\text{ii}) \qquad L(x, \dot{x}) = \left(\sum_{i,j=1}^{n} g_{ij}(x)\dot{x}^i\dot{x}^j \right)^{1/2}.$$

Note that this is the usual definition of length on $\mathbf{R}^n$ with the flat metric. The paths that locally minimize the distance between the two points will be stationary points of the functional (24). To find these, let

$$x^i = s^i(t), \qquad 0 \leqslant t \leqslant 1,$$
$$s^i(0) = s^i(1) = 0$$

be a smooth path and consider the variation r_σ of r given by

$$x^i = r^i(t) + \sigma s^i(t)$$

and its length

$$D(r_\sigma) = \int_0^1 L\big(r_\sigma(t), \dot{r}_\sigma(t)\big)\, dt.$$

By integrating by parts we get

$$\frac{\partial D(r_\sigma)}{\partial \sigma} = \int_0^1 \sum_{j=1}^{n} \left[\frac{\partial L}{\partial x^j} s^j + \frac{\partial L}{\partial \dot{x}^j} \dot{s}^j \right] dt$$

$$= \int_0^1 \sum_{j=1}^{n} \left[\frac{\partial L}{\partial x^j} - \frac{d}{dt}\left(\frac{\partial L}{\partial \dot{x}^j} \right) \right] s^j\, dt.$$

At a stationary point $\partial D(r_\sigma)/\partial \sigma = 0$ for all variations $x^i = s^i(t)$; therefore we get the *Euler-Lagrange equations*

$$(25) \qquad \frac{\partial L}{\partial x^j} = \frac{d}{dt}\left(\frac{\partial L}{\partial \dot{x}^j} \right).$$

Substituting (24.ii) yields

$$\frac{1}{2} L^{-1} \sum_{p,q=1}^{n} \frac{\partial g_{pq}}{\partial x^j} \dot{x}^p\dot{x}^q = \frac{d}{dt}\left(\frac{1}{2} L^{-1} 2 \sum_{q=1}^{n} g_{jq}\dot{x}^q \right).$$

When the curve is parametrized by arclength, L is constant and this reduces to

$$\sum_{p,q=1}^{n} \frac{\partial g_{pq}}{\partial x^j} \dot{x}^p\dot{x}^q = 2 \sum_{r=1}^{n} \frac{\partial g_{jq}}{\partial x^r} \dot{x}^r\dot{x}^q + \sum_{q=1}^{n} g_{jq}\ddot{x}^q;$$

using the expressions (14) for Γ_{ij}^l in terms of the derivatives of the metric gives

$$(26) \qquad \ddot{x}^l + \sum_{i,j=1}^{n} \Gamma_{ij}^l \dot{x}^i\dot{x}^j = 0.$$

A curve $\gamma = \{x^i = r^i(t)\}$ satisfying (26) is called a *geodesic*. If we let the vector field V in (16) be the tangent field T of the curve γ, we see that (26) holds $\Leftrightarrow \nabla_T T = 0$ along γ.

We end this section by defining a coordinate system around $p \in M$ in which geodesics through p are straight lines through the origin. Let X be a tangent vector in $T_p(M)$. As long as $|X|$ is small, we can find a geodesic $\gamma = \{x^i = r^i(t)\}$ as a solution of (26) satisfying

$$\gamma(0) = p, \qquad \frac{d\gamma}{dt}\bigg|_{t=0} = X.$$

The map

$$(27) \qquad \exp: T_p(M) \to M, \qquad X \to \gamma(1)$$

is a diffeomorphism between a neighborhood of 0 in $T_p(M)$ and a neighborhood of p in M. These local coordinates in M are called *normal coordinates*. The manifold M is called *complete* if exp is defined for all points p and all vectors X. This means that the geodesics continue for all time.

The following observation will be useful later.

PROPOSITION 3. *In a normal coordinate system, the coefficients in the Taylor expansion of $g_{ij}(x)$ about the origin are polynomials in the curvature R_{ijkl} and its covariant derivatives at* 0.

PROOF. At the origin in a normal coordinate system, one has $g_{ij}(0) = \delta_{ij}$, $(\partial g_{ij}/\partial x_k)(0) = 0$, so the derivatives of order 0 and 1 are already known. A long computation with (19), (14) and the definition of normal coordinates shows that

$$R_{ijkl,\mu_1 \cdots \mu_s}\big|_0 = (\text{const})\left(\frac{\partial^{s+2} g_{ik}}{\partial x_j \partial x_l \partial x_{\mu_1} \cdots \partial x_{\mu_s}}\right)\bigg|_0$$
$$+ \left(\begin{array}{l}\text{polynomial in lower order} \\ \text{derivatives of the } g_{ab}\end{array}\right).$$

From this it is clear by induction that all derivatives of g_{ik} at 0 may be expressed as polynomials in the $R_{ijkl,\mu_1 \cdots \mu_s}$. $\square$

2. Complex manifolds. We begin with some preliminaries about complex vector spaces. Suppose that we are given a real vector space V and an automorphism $J: V \to V$ satisfying $J^2 = -\text{id}$. Then V becomes a complex vector space if we define $\sqrt{-1}\, v = J(v)$, $v \in V$. Conversely given a complex vector space, we can naturally define a pair (V, J) in the obvious way. We will be concerned below with the case in which V is the tangent space of a manifold at a point.

Let V^* denote the dual space of V consisting of all real-valued linear functions on V; then $V^* \otimes \mathbf{C}$ is the space of all complex-valued $\mathbf{R}$-linear functions on V. When V is the tangent space, $V^* \otimes \mathbf{C}$ becomes the space of complex-valued

1-forms. We say

(31)
$$\phi \in V^* \otimes \mathbf{C} \text{ is of type } (1,0) \quad \text{if } \phi(JZ) = \sqrt{-1}\, Z,\, Z \in V.$$
$$\phi \in V^* \otimes \mathbf{C} \text{ is of type } (0,1) \quad \text{if } \phi(JZ) = -\sqrt{-1}\, Z,\, Z \in V.$$

It is easy to check that

(32)
$$V^* \otimes \mathbf{C} = V_{10} \oplus V_{01},$$

where V_{10} (V_{01}) is the space of all $(1,0)$-forms ($(0,1)$-forms). Whenever a vector space V splits, the exterior powers $\Lambda^r V$ inherit the splitting. In our case the splitting (32) induces a splitting

(33)
$$\Lambda^r(V^* \otimes \mathbf{C}) = \bigoplus_{p+q=r} \Lambda^p(V_{10}) \wedge \Lambda^q(V_{01}).$$

An element in $\Lambda^p(V_{10}) \wedge \Lambda^q(V_{01})$ is called a (p,q)-*covector*. Finally we can dualize this construction and define (p,q)-*multivectors*.

Let $U \subset \mathbf{C}^n$ be an open set and $f\colon U \to \mathbf{C}^n$ a C^1 map. We say

(34)
$$f \text{ is holomorphic} \Leftrightarrow (df)_z \colon \mathbf{C}^n \to \mathbf{C}^n \text{ is complex linear, } z \in U$$
$$\Leftrightarrow J(df)_z = (df)_z J.$$

A *complex manifold* M is a manifold with an atlas of coordinate charts $\{(U_\alpha, \phi_\alpha)\colon U_\alpha \to \mathbf{C}^n)\}$ such that the coordinate changes $\phi_\beta \circ \phi_\alpha^{-1}$ are holomorphic, when defined. Now for every point $p \in \mathbf{C}^n$, multiplication by $\sqrt{-1}$ maps a tangent vector at p into a tangent vector at p. By (34) the coordinate changes $\phi_\beta \circ \phi_\alpha^{-1}$ preserve the notion of multiplication by $\sqrt{-1}$; therefore we get a map

(35)
$$J\colon T_p(M) \to T_p(M);$$

in fact, it is not hard to check that J is well defined globally. It is called the *almost complex structure*. If $(z_1 = x_1 + \sqrt{-1}\, y_1, \ldots, z_n = x_n + \sqrt{-1}\, y_n)$ are local coordinates for M and $(\bar{z}_1 = x_1 - \sqrt{-1}\, y_1, \ldots, \bar{z}_n = x_n - \sqrt{-1}\, y_n)$ are the conjugate variables, then

$$\frac{\partial}{\partial x_1}, \frac{\partial}{\partial y_1}, \ldots, \frac{\partial}{\partial x_n}, \frac{\partial}{\partial y_n} \text{ is a basis of } T_p(M),$$

$$\frac{\partial}{\partial z_1}, \frac{\partial}{\partial \bar{z}_1}, \ldots, \frac{\partial}{\partial z_n}, \frac{\partial}{\partial \bar{z}_n} \text{ is a basis of } T_p(M),$$

with

$$J\colon \frac{\partial}{\partial x_i} \to \frac{\partial}{\partial y_i}, \qquad \frac{\partial}{\partial y_i} \to \frac{-\partial}{\partial x_i},$$

$$J\colon \frac{\partial}{\partial z_i} \to \frac{\partial}{\partial z_i}, \qquad \frac{\partial}{\partial \bar{z}_i} \to \frac{-\partial}{\partial \bar{z}_i};$$

the $\partial/\partial z_i$ are $(1,0)$-vectors, while the $\partial/\partial \bar{z}_i$ are $(0,1)$-vectors; finally a (p,q)-form may be written locally as

(36)
$$\sum \phi_{i_1 \cdots i_p j_1 \cdots j_q}\, dz_{i_1} \wedge \cdots \wedge dz_{i_p} \wedge d\bar{z}_{j_1} \wedge \cdots \wedge d\bar{z}_{j_q}.$$

We will denote the space of all C^∞ (p,q)-forms on M by $\Lambda^{p,q}(M)$.

Consider a Riemannian metric $g(\cdot, \cdot)$ on a complex manifold M satisfying

$$(37) \qquad g(J_p X, J_p Y) = g(X, Y),$$

for all $p \in M$, and $X, Y \in T_p(M)$. If we set $\omega(X, Y) = g(X, JY)$, then

$$(38) \qquad h(X, Y) = g(X, Y) + i\omega(X, Y)$$

is a complex-valued sesquilinear form called the *Hermitian metric*, which may be extended to the complexification

$$T_p^c(M) \equiv T_p(M) \otimes \mathbf{C}$$

of the tangent space. Since

$$\omega(X, Y) = g(X, JY) = g(JY, X) = g(J^2 Y, JX) = -\omega(Y, X),$$

$\omega(\cdot, \cdot)$ is an exterior 2-form. It is called the *Kähler form*. If $(z_1, \ldots, z_n)$ are local complex coordinates, then

$$Z_\alpha \equiv \frac{\partial}{\partial z_\alpha} \qquad (\text{type}(1, 0)),$$

$$\bar{Z}_\alpha \equiv Z_{\bar\alpha} \equiv \frac{\partial}{\partial \bar z_\alpha} \qquad (\text{type}(0, 1)), \alpha = 1, \ldots, n,$$

are a basis for the complex tangent spaces $T_p^c(M)$; we abbreviate this to

$$Z_i, \qquad i = 1, \ldots, n, \bar 1, \ldots, \bar n.$$

If we set

$$h_{ij} = h(z_i, z_j),$$

then by (37), $h_{\alpha\beta} = h_{\bar\alpha\bar\beta} = 0$, while $h_{\alpha\bar\beta} = \overline{h_{\beta\bar\alpha}}$. For this reason the metric (38) is often written

$$(41) \qquad ds^2 = \sum_{\alpha,\beta=1}^{n} h_{\alpha\bar\beta}\, dz_\alpha\, d\bar z_\beta.$$

Corresponding to ds^2 we have the Kähler form $\omega = \sum_{\alpha\beta} h_{\alpha\bar\beta}\, dz_\alpha \wedge d\bar z_\beta$. This is a globally defined form of type $(1, 1)$. For the rest of this section we will use the convention

$$\begin{aligned}
&\text{Roman indices } i, j, k, &&1, \ldots, n, \bar 1, \ldots, \bar n,\\
&\text{Greek indices } \alpha, \beta, \gamma, &&1, \ldots, n,\\
&\bar{\bar{j}} = j.
\end{aligned}$$

Suppose now that M is equipped with a connection allowing us to parallel transport vectors Z. In general parallel transport does not preserve complex structure; that is

$$\text{parallel transport of } JZ \neq J \, (\text{parallel transport of } Z).$$

But if the Riemannian connection ∇ induced by the metric $g(\cdot, \cdot)$ according to Proposition 1 satisfies

$$(42) \qquad \nabla_X(JY) = J(\nabla_X Y),$$

for all vector fields X, Y, then parallel transport does preserve the complex structure. In this case the Hermitian metric defined by (38) is called *Kähler*.

PROPOSITION 4. *The following are equivalent*:

(i) $\nabla_X(JY) = J(\nabla_X Y)$;

(ii) $d\omega = 0$;

(iii)   *there exists a smooth real-valued function* $u(z)$ *such that*

$$(43) \qquad h_{\alpha\bar{\beta}} = \frac{\partial^2 u(z)}{\partial z_\alpha \partial \bar{z}_\beta}.$$

PROOF. See [**38**, vol. 2, Chapter 9].

Note that (43) implies

$$\frac{\partial h_{\alpha\bar{\beta}}}{\partial z_{\bar{\gamma}}} = \frac{\partial h_{\alpha\bar{\gamma}}}{\partial z_{\bar{\beta}}}.$$

EXAMPLE 1. If $M = \mathbf{C}^n$, then

$$g = \sum_{k=1}^{n} |dz_k|^2 = \sum_{k=1}^{n} (dx_k)^2 + (dy_k)^2$$

and

$$h_{\alpha\bar{\beta}}(z) = \frac{1}{2} \frac{\partial^2}{\partial z_\alpha \partial \bar{z}_\beta} |z|^2,$$

so that $\mathbf{C}^n$ is Kähler.

EXAMPLE 2. If $M = \mathbf{CP}^n$ (complex projective n-space), then

$$h_{\alpha\bar{\beta}}(z) = \frac{1}{2} \frac{\partial^2}{\partial z_\alpha \partial \bar{z}_\beta} \log |z|^2$$

induces a metric called the Fubini-Study metric; therefore $\mathbf{CP}^n$ is also Kähler.

In later chapters we shall study carefully certain Kähler metrics on domains in $\mathbf{C}^n$.

If $h_{\alpha\bar{\beta}}$ is a Kähler metric on M, then using the defining property (42), we find

$$(44) \qquad \begin{array}{ll} \nabla_{Z_\alpha}\bar{Z}_\beta = 0, & \nabla_{Z_\alpha}Z_\beta \text{ is of type } (1,0), \\[2mm] \nabla_{\bar{Z}_\alpha}Z_\beta = 0, & \nabla_{\bar{Z}_\alpha}\bar{Z}_\beta \text{ is of type } (0,1), \end{array}$$

when Z_α, $\bar{Z}_\alpha$ are the coordinate vector fields defined by (39). Because of these formulas, equations involving a Kähler metric are often simpler then their counterparts for a general Hermitian metric.

We illustrate this by calculating the components Γ_{ij}^k of the Riemannian connection of a Kähler manifold M with metric $h_{\alpha\bar{\beta}}$. The real part of the Hermitian metric $h_{\alpha\bar{\beta}}$ is a Riemannian metric (compare equation (38)) and by Proposition 1, it is associated with the Riemannian connection ∇ such that

$$\nabla_{Z_i}Z_j = \Gamma_{ij}^k Z_k.$$

Note that the formulas (44) and (9) imply

$$\Gamma^\gamma_{\alpha\beta} = \Gamma^\gamma_{\beta\alpha}, \qquad \Gamma^{\bar\gamma}_{\alpha\bar\beta} = \Gamma^{\bar\gamma}_{\bar\beta\alpha}, \qquad \text{other } \Gamma^k_{ij} = 0.$$

Now

$$\langle \nabla_{Z_\alpha} Z_\beta, \overline{Z}_\gamma \rangle = \sum_\delta \Gamma^\delta_{\alpha\beta} h_{\delta\bar\gamma}.$$

On the other hand by (9.vi) and (44), we have

$$\langle \nabla_{Z_\alpha} Z_\beta, \overline{Z}_\gamma \rangle = Z_\alpha \langle Z_\beta, \overline{Z}_\gamma \rangle = \frac{\partial h_{\beta\bar\gamma}}{\partial z_\alpha}.$$

Comparing these last two equations we see

$$(45.\mathrm{i}) \qquad \qquad \Gamma^\gamma_{\alpha\beta} = \sum_\delta h^{\gamma\bar\delta} \frac{\partial h_{\alpha\bar\delta}}{\partial z_\beta},$$

where $h^{\alpha\bar\beta}$ is the inverse of the matrix $h_{\alpha\bar\beta}$. Similarly we find

$$(45.\mathrm{ii}) \qquad \qquad \Gamma^{\bar\gamma}_{\alpha\bar\beta} = \sum_\delta h^{\bar\gamma\delta} \frac{\partial h_{\bar\alpha\delta}}{\partial \bar z_\beta}.$$

Note how much simpler this expression is compared to expression (14) for the Riemannian case.

The formulas for the Riemannian curvature and Ricci curvature are also much simpler. Instead of the formulas (44), we use the second-order version

$$\nabla_{Z_\alpha} \nabla_{Z_\beta} Z_\gamma = \nabla_{Z_\beta} \nabla_{Z_\alpha} Z_\gamma \qquad \text{(take the inner product of both sides}$$
$$\text{with } \overline{Z}_\delta \text{ and apply } (45.\mathrm{i})),$$

$$(46) \quad \begin{aligned} &\nabla_{\bar Z_\alpha} \nabla_{\bar Z_\beta} Z_\gamma = 0, \\[4pt] &\nabla_{\bar Z_\alpha} \nabla_{Z_\beta} Z_\gamma = \nabla_{\bar Z_\alpha}\!\left(\sum_\delta \Gamma^\delta_{\beta\gamma} Z_\delta \right) = \sum_\delta \Gamma^\delta_{\beta\gamma} Z_\delta, \\[4pt] &\nabla_{Z_\beta} \nabla_{\bar Z_\alpha} Z_\gamma = 0; \end{aligned}$$

it turns out that

$$(47) \qquad \qquad R_{\alpha\bar\beta\gamma\delta} = \sum_\varepsilon \left(\frac{\partial}{\partial \bar z_\beta} \Gamma^\varepsilon_{\alpha\gamma} \right) h_{\varepsilon\bar\delta}.$$

When the metric is expressed as

$$h_{\alpha\bar\beta}(z) = \frac{\partial^2}{\partial z_\alpha \partial \bar z_\beta} u(z),$$

we can rewrite this as

$$(48) \qquad \qquad R_{\alpha\bar\beta\gamma\delta} = u_{\alpha\bar\beta\gamma\delta} - \sum_{\sigma,\tau} u^{\sigma\bar\tau} u_{\alpha\gamma\bar\tau} u_{\bar\beta\bar\delta\sigma},$$

where $u^{\sigma\bar\tau}$ is the inverse of the matrix $u_{\sigma\bar\tau}$ and

$$u_{\bar\alpha\bar\beta\gamma} \equiv \frac{\partial^3 u}{\partial \bar z_\alpha \partial \bar z_\beta \partial z_\gamma}.$$

Finally using the identity

$$\frac{d}{dx} \log \det(A_{\alpha\bar\beta}) = \mathrm{Trace}\left(\frac{dA}{dx} A^{-1} \right),$$

we find

$$(49) \qquad \mathrm{Ric}_{\alpha\bar\beta} = c\frac{\partial^2}{\partial z_\alpha \partial \bar z_\beta}(\log \det h_{\alpha\bar\beta}).$$

We end this section by discussing the exterior differentiation of (p, q)-forms. For $f \in \Lambda^{0,0}(M)$, define

$$(50.\mathrm{i}) \quad \partial f = \sum_{\alpha=1}^{n} \frac{\partial f}{\partial z_\alpha} dz_\alpha \in \Lambda^{1,0}(M), \qquad \bar\partial f = \sum_{\alpha=1}^{n} \frac{\partial f}{\partial \bar z_\alpha} d\bar z_\alpha \in \Lambda^{0,1}(M)$$

and extend these in the standard way:

$$(50.\mathrm{ii}) \qquad \begin{aligned} \partial(f\, dz_I \wedge d\bar z_J) &= (\partial f) \wedge dz_I \wedge d\bar z_J \in \Lambda^{p+1,q}(M), \\ \bar\partial(f\, dz_I \wedge d\bar z_J) &= (\bar\partial f) \wedge dz_I \wedge d\bar z_J \in \Lambda^{p,q+1}(M), \end{aligned}$$

for I, J multi-indices with $|I| = p$, $|J| = q$. Finally extend $\partial, \bar\partial$ to all of $\Lambda^{p,q}(M)$ by linearity.

3. Bundles over Riemannian manifolds. In this section we give two additional definitions of connection, including the definition of a connection in a principal fibre bundle, a concept which we will need in Chapter 10.

Let ∇ be a connection on a Riemannian n-manifold $(M, \langle \cdot, \cdot \rangle)$ and let $U \subset M$ be an open set. We call the vector fields $X_1,\ldots,X_n$ an orthonormal frame on U if

$$(51) \qquad \begin{aligned} &(\mathrm{i}) \quad X_1(p),\ldots,X_n(p) \text{ is a basis of } T_p(M), \text{ all } p \in U, \\ &(\mathrm{ii}) \quad \langle X_i(p), X_j(p) \rangle_p = \delta_{ij}, \text{ all } p \in U. \end{aligned}$$

Note that we do not necessarily assume that the X_i are coordinate vector fields. In particular, they need not commute. Let ω_i be the dual 1-forms defined by

$$(52.\mathrm{i}) \qquad \omega_i(X_j) = \delta_{ij}$$

and define the 1-forms ω_{ij} by

$$(52.\mathrm{ii}) \qquad \nabla_Y X_i = \sum_{j=1}^{n} \omega_{ij}(Y) X_j.$$

The 1-forms ω_i, ω_{ij} are called the *connection forms*.

Let $R(\cdot, \cdot)\cdot$ be the curvature of ∇ (equation (17)) and define the 2-forms Ω_{ij} by

$$R(Y, Z)X_i = \sum_{j=1}^{n} \Omega_{ij}(Y, Z) X_j,$$

i.e. $\Omega_{ij}(Y, Z) = \omega_i([\nabla_Y\nabla_Z - \nabla_Z\nabla_Y - \nabla_{[Z,Y]}]X_j)$. It is easy to check that

$$(53) \qquad \Omega_{ij} = d\omega_{ij} + \sum_{k=1}^{n} \omega_{ij} \wedge \omega_{kj}.$$

The 2-forms Ω_{ij} are called the *curvature forms*.

We can reverse this process: given $(M, \langle \cdot, \cdot \rangle)$ and an orthonormal frame $\{X_i\}$, the *Cartan connection* ∇ is defined by (52.ii) and its curvature by (53). We define the torsion $T(\cdot, \cdot)$ of ∇ by

$$(54.\text{i}) \qquad T(Y, Z) = \nabla_Y Z - \nabla_Z Y - [Y, Z]$$

and the torsion 2-forms by

$$(54.\text{ii}) \qquad T(Y, Z) = \sum_{i=1}^{n} T_i(Y, Z)X_i.$$

PROPOSITION 5. (i) $d\omega_i = -\sum_{j=1}^{n} \omega_{ij} \wedge \omega_j + T_i$.
(ii) $d\omega_{ij} = -\sum_{k=1}^{n} \omega_{ik} \wedge \omega_{kj} + \Omega_{ij}$.

PROOF. We only prove (i); (ii) follows by a similar argument. Before we begin note that for a 1-form ω and for a 2-form $\omega \wedge \eta$, we have

$$(55.\text{i}) \qquad d\omega(Y, Z) = Y\omega(Z) - Z\omega(Y) - \omega([Y, Z]),$$

$$(55.\text{ii}) \qquad (\omega \wedge \eta)(Y, Z) = \omega(Y)\eta(Z) - \eta(Y)\omega(Z).$$

Now from the definition (54) of the torsion,

$$\sum_i T_i(Y, Z)X_i = \nabla_Y Z - \nabla_Z Y - [Y, Z]$$

$$= \nabla_Y\left(\sum_j \omega_j(Z)X_j\right) - \nabla_Z\left(\sum_j \omega_j(Y)X_j\right) - \sum_j \omega_j([Y, Z])X_j$$

$$= \sum_j \{Y\omega_j(Z) - Z\omega_j(Y) - \omega_j[Y, Z]\}X_j$$

$$+ \sum_j \{\omega_j(Z)\omega_{ij}(Y) - \omega_j(Y)\omega_{ij}(Z)\}X_j.$$

Therefore comparing components and using (55.ii) we have

$$T_i(Y, Z) - \left(\sum_j \omega_{ij} \wedge \omega_j\right)(Y, Z) = Y\omega_i(Z) - Z\omega_i(Y) - \omega_i[Y, Z],$$

which by (55.i) becomes

$$d\omega_i(Y, Z) = \left(-\sum_j \omega_{ij} \wedge \omega_j\right)(Y, Z) - T_i(Y, Z).$$

REMARK. Equations (i) and (ii) are called the *first and second structural equations*.

There is a closely related construction in which G is a Lie group and $X_1, \ldots, X_n$ are left-invariant vector fields on G which span the tangent space $\tilde{G} = T_{\text{id}}(G)$. We can identify $\tilde{G}$ with the Lie algebra of G. We define the structure constants c_{ij}^k of $\tilde{G}$ by

$$(56) \qquad \left[X_i, X_j\right] = \sum_{k=1}^{n} c_{ij}^k X_k;$$

they are clearly antisymmetric in i and j. Let $\omega_1, \ldots, \omega_n$ be the dual 1-forms to $X_1, \ldots, X_n$ (equation 52.i). Now

$$d\omega_i(X_j, X_k) = -\omega_i\left(\left[X_j, X_k\right]\right) = -c_{jk}^i,$$

$$-\frac{1}{2}\left(\sum_{r,s} c_{rs}^i \omega_r \wedge \omega_s\right)(X_j, X_k) = c_{jk}^i$$

and so,

$$(57) \qquad d\omega_i + \sum c_{jk}^i \omega_j \wedge \omega_k = 0,$$

which is the structure equation for the Lie group G.

If we let

$$TM = \bigcup_{p \in M} T_p(M)$$

and let $\pi\colon TM \to M$ be the natural projection, then a vector field may be thought of as a map $s\colon U \subset M \to TM$ satisfying $\pi S = \mathrm{id}$ (i.e. a *section*). We can think of an orthonormal frame $\{X_1, \ldots, X_n\}$ on U in a similar manner. Let

$$E = \left\{ \begin{array}{c} (m; e_1, \ldots, e_n)\colon e_1, \ldots, e_n \text{ is an oriented orthonormal} \\ \text{basis of } T_m(M) \end{array} \right\}$$

and let $\pi\colon E \to M$ be the natural projection. An orthonormal frame $X_1, \ldots, X_n$ over $U \subset M$ is just a section of $E \to M$.

We now describe some of the structure of $E \to M$. For $g \in \mathrm{SO}(n)$, let

$$R_g\colon E \to E,$$

$$b = (m; e_1, \ldots, e_n) \mapsto bg = \left(m; \sum g_{i1}e_i, \ldots, \sum g_{in}e_i\right)$$

then $(bg_1)g_2 = b(g_1 g_2)$, i.e. this is a right action of $\mathrm{SO}(n)$ on E. For a fixed $b_0 \in E$, define

$$f_{b_0}\colon \mathrm{SO}(n) \to E, \qquad g \mapsto bg.$$

We call

$$E_m \equiv \pi^{-1}(m) \subset E$$

the (vertical) fibre. The map f_{b_0} is a diffeomorphism onto the fibre E_m; that is, after choosing a frame b_0, we can identify $\mathrm{SO}(n)$ and the fibre. The derivative of this map is

$$f'_{b_0}\colon T(\mathrm{SO}(n)) \to T(E)$$

in which the image consists of vectors X satisfying

$$\pi_*(X) = 0;$$

these are called vertical vectors. We let

$$V_b = \left\{ X \in T_b(E) \colon \pi_* X = 0 \right\}.$$

These vectors are tangent to the fibre. We can identify $T(\mathrm{SO}(n))$ with the Lie algebra $\mathrm{so}(n)$ of $\mathrm{SO}(n)$. A basis for $\mathrm{SO}(n)$ is given by the matrices e_{ij} containing a 1 in the (i, j)-slot, a -1 in the (j, i)-slot, and 0 elsewhere. Then $X_{ij} \equiv f'_{b_0}(e_{ij})$ are vertical vector fields on E, which are in fact independent of the b_0 used to identify the fibre with $\mathrm{SO}(n)$. Now the X_{ij} do not give a framing of E; indeed, we must somehow choose n additional vector fields $X_1, \ldots, X_n$ to frame E. These are called horizontal vector fields. Unlike the construction of the X_{ij}, these cannot be defined in a canonical manner. Once they are chosen, set

$$H_b = \mathrm{span}\{ X_1(b), \ldots, X_n(b) \};$$

then we expect

(i) $T_b(E) = V_b \oplus H_b$,
(ii) $V_b \cap H_b = \{0\}$,
(iii) $H_{bg} = (R_g)_* H_b$.

The assignment $b \mapsto H_b \subset T_b(E)$ is called a connection.

If Y is a vector field on $U \subset M$, then the assignment $b \to H_b$ gives a *unique* horizontal vector field $\overline{Y}$ on $\overline{U} \equiv \pi^{-1}(U)$ satisfying

$$Y_b \in H_b, \quad \pi_*\big(\overline{X}_b\big) = X_{\pi(b)} \quad \text{for all } b \in \overline{U};$$

it is called the *horizontal lift* of Y. Now if γ is a curve on M with a tangent vector field T, then T lifts to a horizontal vector field $\overline{T}$ (after extending T to some neighborhood of U) and the integral curves of $\overline{T}$ define a curve $\overline{\gamma}$ in E called the lift of γ. A point $b(t) = (m(t); e_1(t), \ldots, e_n(t))$ on $\overline{\gamma}$ can be thought of as the parallel transport of the frame $(e_1(0), \ldots, e_n(0))$ at $m = m(0)$ to $m(t)$. This is why the map $b \mapsto H_b$ is called a connection.

In Chapter 10 we will need this construction in greater generality. We define a *principal fibre bundle* with structure group G as a smooth map of manifolds

$$(58) \qquad\qquad\qquad \pi \colon P \to M$$

together with an action of G on P satisfying

(i) $R \colon G \times P \to P, (g, p) \mapsto R_g(p) = pg$, with $pg = p$ for all $p \Leftrightarrow g = \mathrm{id}$.
(ii) $M = P/B$.
(iii) If $\{U_\alpha\}$ is a cover of M, then P is locally a product $U \times G$, i.e. there are maps $\psi_U \colon \pi^{-1}(U) \to U \times G, p \mapsto (s_U(p), t_U(p))$ which make the diagram

$$
\begin{array}{ccc}
\pi^{-1}(U) & \overset{\psi_u}{\to} & U \times G \\
& \searrow{\scriptstyle \pi} & \downarrow{\scriptstyle \mathrm{proj}} \\
& & U
\end{array}
$$

commute and satisfy

$$(59) \qquad\qquad s_U(pg) = (s_U(p))g \quad \text{for all } p \in \pi^{-1}(U).$$

EXAMPLE. We have just shown that the frame bundle $E \to M$ described above is a principal fiber bundle. Let

$$(60) \qquad V_p = \left\{ X \in T_p(P) : \pi_*(X) = 0 \right\}$$

be the *vertical space*. A *connection* is an assignment $p \mapsto H_p \subset T_p(P)$ of *horizontal spaces* satisfying

$$(61) \qquad \begin{aligned} &\text{(i) } H_p \oplus V_p = T_p(P),\ H_p \cap V_p = \{0\}, \\ &\text{(ii) } H_{pg} = (R_g)_* H_p. \end{aligned}$$

This is almost our final definition of connection. We have had several, including

Coordinates	vector field X_i	$\nabla_{X_i} X_j = \Sigma \Gamma_{ij}^k X_k$ Γ_{ij}^k Christoffel symbols
Frames	orthonormal frame X_i	$\nabla_X X_i = \Sigma \omega_{ij}(X) X_j$ ω_{ij} connection forms
Bundles	section of $P \to E$	$T_p(P) = H_p \oplus V_p$ H_p horizontal space

Observe that on the principal bundle E a connection induces a natural $\mathfrak{I}$-valued one-form, i.e. a natural map from tangent vectors $X \in T_p(E)$ into $\mathfrak{I}$, the Lie algebra of G. In fact, the connection splits X uniquely into horizontal and vertical components, $X = X_h + X_v$ with X_v in the tangent space of the fibre. Since the tangent space of a fibre is naturally identified with the Lie algebra $\mathfrak{I}$, the linear map $X \to X_v \in \mathfrak{I}$ is a Lie-algebra-valued 1-form, which we call ω. Note that the horizontal subspace of $T_p(E)$ is the kernel of ω.

We close the discussion with a simple example showing how connections on frame bundles relate to the construction of canonical forms on Lie groups (equation (57)). The example is as follows.

In $\mathbf{R}^n$ with the standard metric $ds^2 = \Sigma_k dx_k^2$, fix an origin O and an orthonormal frame (e_α). Any Euclidean motion of $\mathbf{R}^n$ carries O to a new origin and (e_α) to a new frame; the new origin and frame uniquely specify the transformation. Therefore the bundle of orthonormal frames on $\mathbf{R}^n$ may be identified with the group E of Euclidean motions. The fibres in E are the cosets of $SO(n) \subseteq E$, while $\mathbf{R}^n \simeq E/SO(n)$.

On the Lie algebra level, we have $\mathfrak{E} \simeq so(n) \oplus \mathbf{R}^n$ as a direct sum of vector spaces, so we have a basis $X_{ij} \in so(n)$, $X_i \in \mathbf{R}^n$ for $\mathfrak{E}$. The X's may be thought of as left-invariant vector fields on E. In particular, the X_{ij} span the tangent space of a fibre, while the X_i may be considered horizontal. This gives rise to a notion of parallel transport in E which agrees with the obvious parallel transport of frames in $\mathbf{R}^n$. Passing from X_{ij}, X_i to the dual 1-forms ω_{ij}, ω_i, we can check that the structural equations (57) for the Lie group E take the form

$$(62) \qquad d\omega_i = -\sum_j \omega_{ij} \wedge \omega_j, \qquad d\omega_{ij} = -\sum_k \omega_{ij} \wedge \omega_{kj}.$$

Comparing with Proposition 5, we see that our flat connection on $\mathbf{R}^n$ has zero curvature and torsion.

This simple example can be used to construct the Levi-Civita connection and curvature on a Riemannian manifold M with general (nonflat) metric. For, any connection in the bundle E_M of orthonormal frames induces the 1-forms ω_i, ω_{ij} on E_M, but we no longer have (62). Rather, the best we can do is to introduce the error terms T_i, Ω_{ij} as in Proposition 5, and try to pick the connection to make them as "close to zero" as possible. Computation shows that the T_i vanish for exactly one connection, which we will call the Levi-Civita connection. Then the Ω_{ij} with respect to the Levi-Civita connection are the natural obstructions to the flatness of M or of the bundle E_M. We call Ω_{ij} the curvature forms on M.

All of this seems unnecessarily abstract when applied to familiar elementary Riemannian geometry. However, using different groups in place of the Euclidean motions and $\mathrm{SO}(n)$, we shall have a framework for defining curvatures at boundary points of strictly pseudoconvex domains in $\mathbf{C}^n$. This will be carried out in Chapter 11.

4. The Laplace operator. The Laplace operator Δ on $\mathbf{R}^n$ satisfies

$$\int_{\mathbf{R}^n} \langle \nabla u, \nabla v \rangle \, d\mathrm{vol} = -\int_{\mathbf{R}^n} v\Delta u \, d\mathrm{vol},$$

for u, v compactly supported C^2 functions. On a compact Riemannian manifold M with metric tensor g_{ij} the left-hand side still makes sense; therefore, we define Δ on M by

$$\int_M \left(\sum_{i,j=1}^n g_{ij} \frac{\partial u}{\partial x^i} \frac{\partial u}{\partial x^j} \right) d\mathrm{vol} = -\int_M v\Delta u \, d\mathrm{vol},$$

where

$$d\mathrm{vol} = \sqrt{g}\, dx^1 \cdots dx^n, \qquad \sqrt{g} = \left(\det(g_{ij})\right)^{1/2}$$

is the volume form. A short calculation shows

$$(63) \qquad \Delta u = \sum_{i,j=1}^n \frac{1}{\sqrt{g}} \frac{\partial}{\partial x^i}\left(\sqrt{g}\, g^{ij} \frac{\partial}{\partial x^j} \right) u,$$

where $g^{ij}g_{ij} = \delta_{ij}$.

Throughout the remainder of this chapter we let $(M, \langle \cdot, \cdot \rangle)$ be a compact Riemannian n-manifold and let $\Lambda^p(M)$ denote the space of C^∞ p-forms on M. We motivate the definition of the adjoint d^* and the Laplacian $\square$ by considering the map

$$d^*: \Lambda^1(\mathbf{R}^n) \to \Lambda^0(\mathbf{R}^n), \qquad \omega \mapsto \mathrm{div}(\omega^\#),$$

where $\mathrm{div}(\cdot)$ is the divergence of a vector field and $\omega^\#$ is the dual vector field associated to the 1-forms ω by the Euclidean structure. Using some identities from vector calculus, we find

$$\Delta u = d^*du, \qquad \langle (du)^*, v \rangle = \langle u, d^*(v^\#) \rangle,$$

where $\langle \cdot, \cdot \rangle$ is the standard inner product on $\mathbf{R}^n$ and $u, v \in \Lambda^0(M)$. We now give the formal definition. Recall that if V has an inner product, then $\Lambda^p V$ inherits an inner product. In particular we can define an inner product

$$\langle \alpha, \beta \rangle_m, \qquad \alpha, \beta \in \Lambda^p(M),$$

for each $m \in M$ and by integration an inner product on $\Lambda^p(M)$

$$(64) \qquad \langle \alpha, \beta \rangle = \int_M \langle \alpha, \beta \rangle_m \, d\operatorname{vol}(m).$$

Then $\Lambda^p(M)$ becomes a pre-Hilbert space with respect to the inner product (64); we let $L_p^2(M)$ denote its completion and let

$$(65) \qquad d^*\colon L_p^2(M) \to L_p^2(M)$$

be the Hilbert space adjoint of d. Finally we define the Laplace-Beltrami operator

$$(66) \qquad \square = dd^* + d^*d.$$

Harmonic forms (functions) are those annihilated by $\square(\Delta)$.

Let M be a compact manifold. The cohomology of the sequence

$$\Lambda^0(M) \overset{d}{\to} \Lambda^1(M) \overset{d}{\to} \cdots \overset{d}{\to} \Lambda^p(M) \overset{d}{\to} \cdots$$

is isomorphic by DeRham's theorem to the cohomology of the manifold

$$H^p(M) \approx \ker(d\colon \Lambda^p \to \Lambda^{p+1})/d(\Lambda^{p-1}).$$

Let us examine this sequence in more detail. Given $\omega \in \Lambda^p(M)$ with $d\omega = 0$, consider the equation

$$(67.\text{i}) \qquad d\lambda = \omega, \qquad \lambda \in \Lambda^{p-1}(M).$$

Note that λ is not unique. Indeed, since $d^2 = 0$, $\lambda + d\mu$, for any $\mu \in \Lambda^{p-2}(M)$ is also a solution. Among all such solutions λ of this equation, we can single out one by requiring $\lambda \perp \operatorname{Ran} d$ (with respect to the inner product (63)); since $0 = \langle \lambda, d\mu \rangle = \langle d^*\lambda, \mu \rangle$, for $\mu \in \Lambda^{p-2}(M)$, this is equivalent to

$$(67.\text{ii}) \qquad d^*\lambda = 0.$$

Applying $\square$ to the solution λ of the system (67) yields

$$\square\lambda = (d^*d + dd^*)\lambda = d^*\omega + d(0) = d^*\omega.$$

In other words, the system (67) leads naturally to the equation

$$(68) \qquad \square\lambda = \alpha, \qquad \lambda, \alpha \in \Lambda^p(M).$$

If we introduce local coordinates we can check that $\square$ is a second-order elliptic operator in the sense that it satisfies hypotheses (i) and (ii) of Theorem 7 in Chapter 3. We can also check that it is self adjoint.

THEOREM 1. (i) $L_p^2(M) = \operatorname{Ker} \square \oplus \operatorname{Ran} \square$.

(ii) $\{$*harmonic p-forms*$\} \subseteq \Lambda^p(M)$.

(iii) $H^p(M) \approx \{$*harmonic p-forms*$\}$.

(iv) $\dim H^p(M) < \infty$.

PROOF. As we remarked, Theorem 7 of Chapter 3 applies. Therefore $L_p^2(M) =$ Coker $\square \oplus$ Ran $\square$, with $\dim(\text{Coker } \square) < \infty$ and $\dim(\text{Ker } \square) < \infty$; moreover Ker $\square$ consists of smooth functions. This gives (ii). Since $\square$ is selfadjoint, Coker $\square =$ Ker $\square^* =$ Ker $\square$ and we get (i).

Since $\square = (d + d^*)(d + d^*)$, $\square\omega = 0$ implies $(d + d^*)\omega = 0$. Therefore $0 = \langle(d + d^*)\omega, d\omega\rangle = \|d\omega\|^2 + \langle\omega, dd\omega\rangle$ and $0 = \langle(d + d^*)\omega, d^*\omega\rangle = \|d^*\omega\|^2 + \langle dd\omega, \omega\rangle$, so that $d\omega = 0$ and $d^*\omega = 0$; in particular, harmonic forms are closed. Therefore the map

$$j: \{\text{harmonic } p\text{-forms}\} \to H^p(M), \qquad \omega \mapsto \omega + \text{Ran } d$$

is well defined. To show that j is injective, take $\omega \in$ Ran $d \cap$ Ker $\square$. Then $\omega = d\lambda$, for some λ with $\square(d\lambda) = 0$; that is, $dd^*d\lambda = 0$. Now $0 = \langle dd^*d\lambda, d\lambda\rangle = \|d^*d\lambda\|^2$, so $d^*d\lambda = 0$. Similarly $0 = \langle d^*d\lambda, \lambda\rangle = \|d\lambda\|^2$, so $d\lambda = 0$. This gives $\omega = 0$ and j is injective as desired. We now show j is surjective. By (i) any closed form β may be written as $\beta = \omega + \square\alpha$ with $\square\omega = 0$. Now $d\omega = 0$ and $d\beta = 0$, so $0 = d\square\alpha = d(dd^* + d^*d)\alpha = dd^*d\alpha$. Therefore $0 = \langle dd^*d\alpha, d\alpha\rangle = \|d^*d\alpha\|^2$, which implies $\square\alpha = d(d^*\alpha) \in$ Range d. So $\beta = \omega + d(\text{something})$, with ω harmonic, and j is surjective. The proof of (iii) is complete. Part (iv) is an immediate consequence of the finiteness of $\{\text{harmonic } p\text{-forms}\}$ and (iii). $\square$

COROLLARY. *The Euler characteristic*

$$\chi(M) = \sum_{i=0}^{n} (-1)^i \dim H^i(M)$$

is finite.

5. The Euler characteristic. In this section we show how we can express the Euler characteristic $\chi(M)$ in terms of the action of $\square$ on the space of forms. Let

$$\Lambda_{\text{even}}(M) = \bigoplus_{p \text{ even}} \Lambda^p(M), \qquad \Lambda_{\text{odd}}(M) = \bigoplus_{p \text{ odd}} \Lambda^p(M).$$

Consider the operator

$$L = d + d^*: \Lambda_{\text{even}}(M) \to \Lambda_{\text{odd}}(M).$$

Since

$$(69) \qquad L^*L = \square\big|_{\Lambda_{\text{even}}(M)}, \qquad LL^* = \square\big|_{\Lambda_{\text{odd}}(M)},$$

by Theorem 1 we get

$$\dim \text{Ker}(L^*L) = \sum_{p \text{ even}} \dim H^p(M),$$

$$\dim \text{Ker}(LL^*) = \sum_{p \text{ odd}} \dim H^p(M),$$

and so

$$(70) \quad \dim \text{Ker}(L^*L) - \dim \text{Ker}(LL^*) = \sum_{p=1}^{n} (-1)^p \dim H^p(M) = \chi(M).$$

If $0 \leqslant \lambda_1 \leqslant \lambda_2 < \cdots$ are the eigenvalues, listed according to their multiplicities, of $\square_{\text{even}}$ acting on $\lambda_{\text{even}}(M)$, then

$$(71.\text{i}) \qquad \text{Trace}(e^{-t\square_{\text{even}}}) = \sum e^{-t\lambda_j}$$

and similarly

$$(71.\text{ii}) \qquad \text{Trace}(e^{-t\square_{\text{odd}}}) = \sum e^{-t\mu_j},$$

where $0 \leqslant \mu_1 \leqslant \mu_2 \leqslant \cdots$ are the eigenvalues of $\square_{\text{odd}}$. The relation between the λ's and the μ's is given by

LEMMA 1. *Assume that L^*L has eigenvalues λ_k, that LL^* has eigenvalues μ_k and that the spectra of both operators are discrete. Then if those λ_k and μ_k which equal 0 are deleted, the list of the λ_k's and the μ_k's coincide.*

PROOF. Let $\lambda \neq 0$ be an eigenvalue of L^*L; then

$$L^*L\phi = \lambda\phi,$$

for some nonzero ϕ. This implies $L\phi \neq 0$ and

$$LL^*(L\phi) = \lambda(L\phi);$$

i.e. λ is an eigenvalue of LL^* as required. The same argument shows that the multiplicity of λ for LL^* is at least as great as for L^*L. Since the argument is symmetric, we are done. $\square$

Because the Laplacian acting on a compact manifold has a discrete spectrum, we can apply the lemma with $L = d + d^*$. In this case the number of 0 eigenvalues of L^*L (respectively LL^*) is equal to $\dim(\text{Ker } \square_{\text{even}})$ (respectively $\dim(\text{Ker } \square_{\text{odd}})$). If we take note of equations (69) and (70), this gives

PROPOSITION 6.

$$\text{Trace}(e^{-t\square_{\text{even}}}) - \text{Trace}(e^{-t\square_{\text{odd}}}) = \dim(\text{Ker } \square_{\text{even}}) - \dim(\text{Ker } \square_{\text{odd}}) = \chi(M).$$

6. Asymptotics of the heat kernel. To simplify the notation, we work primarily with functions and the Laplacian Δ, rather than forms and the Laplacian $\square$. In this section we will construct an asymptotic approximation to the fundamental solution of the heat equation

$$(72) \qquad (\partial/\partial t - \Delta_x)u(t, x, y) = 0, \qquad u(t, x, y)|_{t=0} = \delta_y(x)$$

on $M \times (0, \infty)$ (the *heat kernel*).

We begin by considering $M = \mathbf{R}^n$ and

$$u(t, x, y) = (4\pi t)^{-n/2} e^{-|x-y|^2/4t}.$$

Setting $y = 0$ and $a = -n/2$, we find

$$(73) \qquad \Delta u = (4\pi t)^a e^{-|x|^2/4t}\left\{ \frac{|\nabla|x|^2|^2}{16t^2} - \frac{\Delta|x|^2}{4t} \right\},$$

$$\frac{\partial u}{\partial t} = (4\pi t)^a e^{-|x|^2/4t}\left\{ \frac{|x|^2}{4t^2} - \frac{a}{t} \right\},$$

and since $|\nabla|x|^2|^2 = 4|x|^2$, $\Delta|x|^2 = 2n$, we see that u is a solution of (72) with $y = 0$ and that

$$K_t(x, y) = u(t, x, y)$$

is the required heat kernel. Note that this is an exact expression for the heat kernel and not an asymptotic one.

This suggests that for a general Riemannian manifold $(M, \langle \cdot, \cdot \rangle)$ we look for a solution of the heat equation with an asymptotic expansion of the form

$$(74) \qquad u(t, x, y) \sim (4\pi t)^{-n/2} e^{-\mathrm{dist}^2(x, y)/4t} \left\{ \sum_{i \geq 0} \phi_i(x, y) t^i \right\}, \qquad t \to 0^+.$$

Here $\mathrm{dist}(x, y)$ is the *geodesic distance* induced by the metric $\langle \cdot, \cdot \rangle$, i.e. the infimum of the length of all geodesics connecting x and y. This expansion is due to Minakshisundaram and Pleijel [47].

We proceed now to derive (74). Let y^i be a system of normal coordinates about the point x (equation (27)). Recall that in this coordinate system geodesics $t \to y^i(t)$ passing through x are simply straight lines; therefore the geodesic distance $\mathrm{dist}(x, y)$ is simply the length of the line connecting x and y. Put $r = \mathrm{dist}(x, y)$ (we use this notation throughout the remainder of the section) and consider a function Ψ on M that depends only upon the geodesic distance r. A short calculation (see [3, p. 134]) shows that

$$\Delta(\Psi) = \frac{d^2\Psi}{dr^2} + \frac{n-1}{r}\frac{d\Psi}{dr} + \frac{d\log\sqrt{g}}{dr}\frac{d\Psi}{dr}.$$

This in turn gives

$$(75) \quad \Delta(\Phi\Psi) = \Phi\left(\frac{d^2\Psi}{dr^2} + \frac{n-1}{r}\frac{d\Psi}{dr} + \frac{d\log\sqrt{g}}{dr}\frac{d\Psi}{dr} \right) + 2\frac{d\Phi}{dr}\frac{d\Psi}{dr} + \Psi\Delta\Phi,$$

for Φ an arbitrary function.

Put

$$u_N(t, x, y) = (4\pi t)^{-n/2} e^{-r^2/4t} \left\{ \phi_0 + \phi_1 t + \cdots + \phi_N t^N \right\},$$

$$\Psi = (4\pi t)^{-n/2} e^{-r^2/4t}, \qquad \Phi = \left\{ \phi_0 + \phi_1 t + \cdots + \phi_N t^N \right\}$$

and substitute this into (75):

$$\Delta(\Phi\Psi) = \Psi\left\{ \left(\frac{r^2}{4t^2} - \frac{1}{2t} - \frac{n-1}{2t} - \frac{d\log\sqrt{g}}{dr}\frac{r}{2t} \right) \right.$$

$$\left. -\frac{r}{t}\left(\frac{d\phi_0}{dr} + t\frac{d\phi_1}{dr} + \cdots + t^N\frac{d\phi_N}{dr} \right) + \left(\Delta\phi_0 + t\Delta\phi_1 + \cdots + t^N\Delta\phi_N \right) \right\};$$

this implies

$$(76) \quad \left(\frac{\partial}{\partial t} - \Delta_x \right) u_N(t, x, y)$$

$$= \Psi \sum_{i=0}^{N} \left\{ \left(\frac{r^2}{4t^2} + \frac{i}{2} - \frac{n}{2t} - \frac{r^2}{4t^2} + \frac{1}{2t} + \frac{n-1}{2t} + \frac{d \log \sqrt{g}}{dr} \frac{r}{2t} \right) \right.$$

$$\left. \cdot t^i \phi_1 + t^{i-1} r \frac{d\phi_i}{dr} - t^i \Delta_x \phi_i \right\}.$$

To make this 0, we set the coefficient of Ψt^{i-1} equal to 0; i.e. we require $\phi_1 \equiv 0$,

$$(77) \quad r \frac{d\phi_i}{dr} + \frac{r}{2} \frac{d \log \sqrt{g}}{dr} \phi_i + i\phi_i = \Delta \phi_{i-1}, \qquad i = 0, 1, \ldots, N.$$

We claim that there are C^∞ solutions to these equations. Indeed, if we write (77) as

$$r \frac{d}{dr} \left(r^i g^{1/4} \phi_i \right) = r^i g^{1/4} \Delta_x \phi_i,$$

then we get such solutions by setting

$$\phi_0(x, y) = \left(\frac{g(x)}{g(y)} \right)^{1/4}$$

and

$$\phi_i(x, y) = \frac{\phi_0(x, y)}{r^i(x, y)} \int_x^y \frac{r^{i-1}(x, z) \Delta_z \phi_{i-1}(x, z)}{\phi_0(x, z)} dr(x, z), \qquad i = 1, \ldots, N,$$

where $r(x, y) = \mathrm{dist}(x, y)$. We can check by induction on i that ϕ_i is smooth. This is clear for ϕ_0. If ϕ_{i-1} is smooth, then in a normal coordinate system about x, $\phi_i(x, y)$ has the form

$$\Phi(y) = \frac{1}{|y|^i} \int_0^{|y|} r^{i-1} g\left(r \frac{y}{|y|} \right) dr = \int_0^1 t^{i-1} g(ty) \, dt \quad \text{with } g \in C^\infty.$$

Evidently $\Phi \in C^\infty$ also, so ϕ_i is smooth. Note that with the ϕ_i defined this way,

$$(78) \quad (\partial/\partial t - \Delta_x) u_N(t, x, y) = (4\pi t)^{-n/2} e^{-r^2/4t} \Delta \phi_N t^N.$$

This completes the construction of the asymptotic solution (74).

Now put

$$(79.\mathrm{i}) \quad K_t^N(x, y) = \theta(\mathrm{dist}(x, y), t) u_N(t, x, y),$$

where $\theta(\cdot, \cdot) \in C_0^\infty$ is a cutoff function with

$$(79.\mathrm{ii}) \quad \theta(x, t) = \begin{cases} 1 & \text{near the origin,} \\ 0 & \text{outside a small neighborhood} \\ & \text{of the origin.} \end{cases}$$

We define a *parametrix* $K(t, x, y)$ for the (scalar) heat operator by the properties

$$
\begin{array}{ll}
\text{(i)} & K(t, x, y) \in C^\infty(M \times M \times (0, \infty)), \\
\text{(ii)} & (\partial/\partial t - \Delta_x)K(t, x, y) \text{ and its derivatives up to high order} \\
& \text{tend to zero uniformly as } t \to 0^+, \\
\text{(iii)} & \lim_{t \to 0^+} K(t, \cdot, y) = \delta_y(\cdot).
\end{array}
$$
(80)

A parametrix $K^p(t, x, y)$ for the heat operator $\partial/\partial t - \square_p$ acting on $\Lambda^p(M)$ is defined similarly.

THEOREM 2. (i) *For large N, a parametrix for the scalar heat operator is given by*

$$
K_t^N(x, y) = \theta(r, t)(4\pi t)^{-n/2} e^{-r^2/4t}\{\phi_0(x, y) + t\phi_1(x, y) + \cdots + t^N\phi_N(x, y)\}
$$

where $r = \operatorname{dist}(x, y)$.

(ii) *On the diagonal $x = y$, the $\phi_i(x, y)$ are real-valued polynomial functions of the derivatives $\partial^\alpha g_{ij}$ of the metric tensor in normal coordinates about x.*

(iii) *There is a similar expansion for a parametrix*

$$
K^{p,N}(x, y) = \theta(r, t)(4\pi t)^{-n/2} e^{-r^2/4t}\{\phi_0^p(x, y) + \cdots + \phi_N^p(x, y)t^N\}.
$$

Here $\phi_i(x, y)$ is an endomorphism from $\Lambda^p T^*M|_y$ into $\Lambda^p T^*M|_x$ with $T^*M = $ cotangent bundle, and ϕ_i is smooth in $(x, y) \in M \times M$. In local coordinates, this makes ϕ_i a matrix of smooth functions of x, y.

(iv) *On the diagonal $x = y$ in a local coordinate system, the entries of the matrix ϕ_i^p are polynomial functions of the derivatives $\partial^\alpha g_{ij}$ of the components of the metric tensor.*

In order to use this we need the following lemma about the regularity of solutions to the heat equation

$$
(81) \qquad\qquad (\partial/\partial t - \Delta)u = g, \qquad u|_{t=0} = 0
$$

on $M \times (0, \infty)$.

LEMMA 2. *If the datum g satisfies*
  (i) *$\Delta^N g \in L^2$, for some large N,*
  (ii) *g vanishes to order N^1 in t at $t = 0$;*
then the solution u satisfies
  (iii) *$\Delta^P u \in L^2$, for some large P, depending upon N, N^1,*
  (iv) *u vanishes to order P^1, depending upon N, N^1.*

PROOF. By expanding the data and solution in the eigenfunctions of Δ, we can reduce the lemma to familiar questions about ordinary differential equations in t. $\square$

The lemma tells us that if the data is very smooth and vanishes to high order, then the same is true of the solution. Consequently, our parametrix (which is annihilated to high order by the heat operator) must agree to high order with the exact solution of the heat equation.

PROOF OF THEOREM 2. We prove only the statements about the scalar heat operator; the corresponding statements for the heat operator $\partial/\partial t - \square_p$ follow by a similar argument.

PROOF OF (i). We must check that $K_t^N(x, y)$ satisfies the properties (80.i), (80.ii), (80.iii) of a parametrix. Property (80.i) is obvious; (80.ii) follows at once from the formal calculations used to construct the $\phi_i(x, y)$; and (80.iii) follows from a short calculation which we omit.

PROOF OF (ii). We work in normal coordinates about $x = 0$. Say that $f(y)$ is *polynomially determined* if all derivatives $\partial^\beta f(0)$ are polynomials in $\partial^\alpha g_{ij}(0)$. Since $g_{ij}(0) = \delta_{ij}$, manipulations with Taylor series show that (g^{ij}) and $g^{\pm 1/2}$ are polynomially determined. Thus $\phi_0(0, y)$ is polynomially determined. Again by manipulating Taylor series, we can check by induction on i that $\phi_i(0, y)$ is polynomially determined. In particular, $\phi_i(0,0)$ is a polynomial in $\partial^\alpha g_{ij}(0)$, and this is the assertion of (ii). Note that we used only $g_{ij}(0) = \delta_{ij}$, not the full force of the normal coordinate system. This will be needed in Gilkey's invariant theory (Theorem 9). $\square$

The properties (80) that define the parametrix of the heat operator are useful for many purposes, but very often a stronger notion is needed. A *fundamental solution* $H(t, x, y)$ of the scalar heat equation is defined by

$$(81) \qquad
\begin{aligned}
&\text{(i)} \quad H(t, x, y) \text{ is } C^1 \text{ in } t \text{ and } C^2 \\
&\qquad\;\; \text{in } x \text{ and } y \; (t > 0), \\
&\text{(ii)} \quad (\partial/\partial t - \Delta_x)H(t, x, y) = 0, \\
&\text{(iii)} \quad \lim_{t \to 0^+} H(t, \,\cdot\,, y) = \delta_y(\cdot).
\end{aligned}$$

There is an analogous definition for the fundamental solution $H^p(t, x, y)$ of the heat operator $\partial/\partial t - \square_p$ acting on $\Lambda^p(M)$. Again we note the following as a consequence of Lemma 2: If $H(t, x, y)$ is the fundamental solution and $K(t, x, y)$ is a parametrix for the heat equation, then $H(t, x, y) - K(t, x, y)$ has many derivatives on $[0, \infty) \times M \times M$ and vanishes to high order at $t \to 0^+$. So K is an excellent approximation to H.

THEOREM 3. (i) *Let f_i be an orthonormal basis of eigenfunctions on M and let λ_i be the corresponding eigenvalues. Then the fundamental solution of the scalar heat equation may be written*

$$H_t(x, y) = \sum e^{-\lambda_i t} f_i(x) \,\overline{f_i(y)}\,.$$

In particular

$$\sum e^{-\lambda_i t} = \int_M H_t(x, x)\, d\mathrm{vol}$$

is well defined.

(ii) *As* $t \to 0^+$, $H_t(x, y) \sim K_t(x, y)$.

(iii) $H_t^p(x, y) = \sum e^{-\lambda_i t} f_i^p(x) \otimes f_i^p(y)$, $H_t^p(x, y) \sim K_t^p(x, y)$, *as* $t \to 0^+$, *where f_i^p are the normalized eigenfunctions of $\square_p$ on M with eigenvalues λ_i.*

PROOF. The proof is straightforward. See [3] for details. $\square$

7. The Chern-Gauss-Bonnet theorem. In this section we pull together the last several sections and show what the asymptotics of the heat kernel have to do with the Euler characteristic. We begin by reviewing what we know. By Theorem 3 the heat operator $\partial/\partial t - \square_p$ has a fundamental solution

$$(82) \quad H_t^p(x, y)$$

$$\sim \theta(r, t)(4\pi t)^{-n/2} e^{-r^2/4t}\{\phi_0^p(x, y) + t\phi_1^p(x, y) + t^2\phi_2^p(x, y) + \cdots\},$$

where θ is the cutoff function (79.ii) and $r = \mathrm{dist}(x, y)$. By Theorem 2 the coefficients of the ϕ_i^p on the diagonal $x = y$ are polynomials in the derivatives $\partial^\alpha g_{ij}$ of the components of the metric tensor g_{ij}. Assume now that we introduce normal coordinates; then by Proposition 3 the $\partial^\alpha g_{ij}$ are polynomials in the covariant derivatives $R_{pqrs/\sigma}$ of the Riemann curvature tensor R_{pqrs} and, hence, so are the ϕ_i^p.

On the other hand note that because $H_t^p(x, y)$ is a fundamental solution of $\partial/\partial t - \square_p$ it satisfies

$$(83) \qquad \mathrm{Trace}(e^{-t\square_p}) = \int_M H_t^p(x, x)\, d\mathrm{vol}(x).$$

Now by Proposition 7,

$$(84) \qquad \chi(M) = \mathrm{Trace}(e^{-t\square_{\mathrm{even}}}) - \mathrm{Trace}(e^{-t\square_{\mathrm{odd}}}).$$

After comparing (82), (83) and (84), we see that

$$(85.\mathrm{i}) \qquad \chi(M) = \int_M P(R_{pqrs/\sigma})\, d\mathrm{vol},$$

where P is some polynomial in the $R_{pqrs/\sigma}$.

The rest of this section is a heuristic description of how we can use invariant theory to discover the form of $P(\cdot)$. Later, after we have developed the necessary invariant theory, we will be able to show that $P(\cdot)$ is the Pfaffian

$$(85.\mathrm{ii}) \quad \mathrm{Pff}(R_{ijkl})\, dx^1 \wedge \cdots \wedge dx^n = c \sum \mathrm{sgn}(i)\, \mathrm{sgn}(j) R_{i_1 i_2 j_1 j_2}\, dx_{j_1} \wedge dx_{j_2}$$

$$\wedge \cdots \wedge R_{i_{2k-1} i_{2k} j_{2k-1} j_{2k}}\, dx_{j_{2k-1}} \wedge dx_{j_{2k}},$$

where $\mathrm{sgn}(i) = \mathrm{sgn}(i_1, i_2, \ldots, i_{2k-1}, i_{2k})$. The relation (85) for the Euler characteristic is the Chern-Gauss-Bonnet theorem.

Consider the parametrix for the scalar heat equation

$$(86) \qquad K_t(x, x) = \sum_{i \geqslant 0} (4\pi t)^{-n/2} \phi_i(x, x) t^i.$$

Let $(M, \langle \cdot, \cdot \rangle)$ be a Riemannian manifold and let M' be the same space but equipped with the metric

$$\langle \cdot, \cdot \rangle' = \lambda \langle \cdot, \cdot \rangle,$$

obtained by dilating $\langle \cdot, \cdot \rangle$. Let $\{e_i\}$ be an orthonormal frame on M. Then $\{e_i' = \lambda^{-1/2}e_i\}$ is an orthonormal frame on M' and $\{e_i^{*'} = \lambda^{1/2}e_i^*\}$ is the corresponding coframe. We have

$$R'_{ijkl} = \langle R'(e_i', e_j')e_k', e_l' \rangle = \lambda^{-1}\langle R(e_i, e_j)e_k, e_l \rangle = \lambda^{-1}R_{ijkl}$$

and similarly

$$(87) \qquad R'_{ijkl/\sigma_1\cdots\sigma_m} = \lambda^{-m/2-1}R_{ijkl/\sigma_1\cdots\sigma_m},$$

$$(88) \qquad \phi_l'(x, y) = \lambda^{-l}\phi_i(x, y).$$

Now $\phi_l(x, y)$ is a polynomial in the $R_{ijkl/\sigma_1\cdots\sigma_m}$; that is, a sum of monomials of the form

$$(89) \qquad R_{ijkl/\sigma_1\cdots\sigma_m} \otimes \cdots \otimes R_{i'j'k'l'/\sigma_1'\cdots\sigma_m'} \qquad (p \text{ terms}).$$

In order for $\phi_l(x, y)$ to transform correctly after a scale change

$$e_i \mapsto e_i' = \lambda^{-1/2}e_i,$$

we see, by comparing (87) and (88), that each monomial term (88) that occurs in $\phi_l(x, y)$ must satisfy

$$(90) \qquad \sum_{\text{factors } \sigma} 2 + |\sigma| = 2l,$$

where the sum is over the p factors $R_{ijkl/\sigma_1\cdots\sigma_m}$, and where $|\sigma|$ denotes the length of σ, i.e. the number of indices.

Where $l = 0$, this formula tells us that $\phi_0(x, x)$ must be a constant, say c_0; (86) becomes

$$K_t(x, x) = (4\pi t)^{-n/2}\{C_0 + \phi_1(x, x)t + \cdots\}.$$

Now, orthogonal transformations preserve normal coordinates and hence the invariants $\phi_l(x, x)$ are defined by $O(n)$-invariant polynomials in $R_{ijkl/\sigma}$. In the next section we prove the following

Fact. The $O(n)$-invariants ϕ_l must be a sum of terms of the form

$$\text{Trace}\left(R_{ijkl/\sigma} \otimes \cdots \otimes R_{i'j'k'l'/\sigma'}\right).$$

When $l = 1$, this fact together with the weight formula (90) implies that each monomial contains just one factor R_{ijkl}. Therefore

$$\phi_1 = (\text{const})R,$$

where $R = \sum g^{ik}g^{jl}R_{ijkl}$ is the scalar curvature (21). The parametrix now takes the form

$$K_t(x, x) = (4\pi t)^{-n/2}\{C_0 + C_1 Rt + \cdots\}.$$

When $l = 2$, we need to consider $\Sigma(2 + |\sigma|) = 4$. Each monomial must involve one factor with $|\sigma| = 2$ or two factors with $|\sigma| = 0$. It turns out that only the term $(\text{const})\Delta R$ arises from $R_{ijkl/\sigma_1\sigma_2}$ and only R^2, $\|R_{ijkl}\|^2$ and $\|\text{Ric}\|^2$ (see (20),

(21)) arise from R_{ijkl}, $R_{i'j'k'l'}$. This means that

$$\phi_2 = C_2 R^2 + C_3 \|\mathrm{Ric}\|^2 + C_4 \|R_{ijkl}\|^2 + C_5 \Delta R$$

and

$$K_t(x, x)$$

$$= (4\pi t)^{-n/2}\left\{ C_0 + C_1 Rt + \left[C_2 R^2 + C_3 \|\mathrm{Ric}\|^2 + C_4 \|R_{ijkl}\|^2 C_5 \Delta R \right] t^2 + \cdots \right\}.$$

We can determine the constants C_i by making explicit comptuations for simple manifolds. Note that although the C_i do not depend on the particular manifold, they do depend on the dimension of the manifold.

This type of information is all we can determine from a single heat kernel $K_t(x, x)$. On the other hand, for the alternating sum of heat kernels associated with the operator

$$e^{-t\Box}{}_{\text{even}} - e^{-t\Box}{}_{\text{odd}}$$

some remarkable cancellations occur. It turns out that the resulting polynomial depends only upon the Riemann curvature tensor R_{ijkl} and not on its covariant derivatives $R_{ijkl/\sigma}$. When n is odd, this polynomial is identically 0; when n is even the polynomial is the Pfaffian, which we now define.

Let A_{ij} be a matrix on $\mathbf{R}^{2k}$ with $A_{ij} = -A_{ji}$. It can be shown that $\det(A_{ij})$ is the square of a polynomial. The Pfaffian $\mathrm{Pf}(A_{ij})$ associated with A_{ij} is defined by

$$\det(A_{ij}) = \left[\mathrm{Pf}(A_{ij})\right]^2.$$

When A_{ij} is a two-tensor on $\mathbf{R}^{2k}$, we define the associated Pfaffian by

$$\mathrm{Pf}(A_{ij}) = C_n \sum \mathrm{sgn}(i_1 \cdots i_{2k}) A_{i_1 i_2} A_{i_3 i_4} \cdots A_{i_{2k-1} i_{2k}},$$

where $\mathrm{sgn}(\cdot)$ is the sign of the permuation if all the i_j are different and 0 otherwise. Finally we let

$$R_{ij} = \sum_{k,l} R_{ijkl}\, dx^k \wedge dx^l.$$

Since 2-forms commute and $R_{ij} = -R_{ji}$, $\mathrm{Pf}(R_{ij})$ is a well-defined n-form, and we define the Pfaffian of the Riemann curvature tensor by

$$(91.\mathrm{i}) \qquad \mathrm{Pf}(R_{ij}) = \mathrm{Pff}(R_{ijkl})\, dx^1 \wedge \cdots \wedge dx^n,$$

or, equivalently

$$(91.\mathrm{ii}) \quad \mathrm{Pff}(R_{ijkl})\, dx^1 \wedge \cdots \wedge dx^n = c \sum \mathrm{sgn}(i)\, \mathrm{sgn}(j) R_{i_1 i_2 j_1 j_2}\, dx^{j_1} \wedge dx^{j_2}$$

$$\wedge \cdots \wedge R_{i_{2k-1} i_{2k} j_{2k-1} j_{2k}}\, dx^{j_{2k-1}} \wedge dx^{j_{2k}},$$

where $\mathrm{sgn}(i) = \mathrm{sgn}(i_1 \cdots i_{2k})$.

8. Hermann Weyl's invariant theory. First we prove an easy theorem about the action of $\mathrm{SO}(n)$ on n-vectors $v^1, \ldots, v^N$. The problem is to find those polynomials in the components of the n-vectors that are invariant under the action of $\mathrm{SO}(n)$.

There are two obvious examples:

$$\text{(i) the inner product } \langle v^j, v^k \rangle = v_1^j v_1^k + \cdots + v_n^j v_n^k;$$

(92)
$$\text{(ii) the bracket factor } \left[v^{j_1}, \ldots, v^{j_n} \right] = \det \begin{pmatrix} v_1^{j_1} & \cdots & v_1^{j_n} \\ \vdots & & \vdots \\ v_n^{j_1} & \cdots & v_n^{j_n} \end{pmatrix}.$$

In fact we have

THEOREM 4. *Examples* (i) *and* (ii) *generate the invariant polynomials.*

Next we consider the more complicated problem of determining the $SO(n)$-invariant polynomials in the components of a collection of tensors of fixed rank, $\{A_{ij\cdots p}\}$.

THEOREM 5. (i) *We can write down an infinite list of polynomials generating the invariant polynomials as a vector space.*

(ii) *Finitely many terms from this list generate all the invariant polynomials as a ring.*

The third problem we will consider also concerns the set of tensors of fixed rank, $\{A_{ij\cdots p}\}$. The problem is to find all $SO(n)$-invariant linear maps

$$l \colon \{A_{ij\cdots p}\} \to \mathbf{C}.$$

There are two obvious examples:

(93)
   (i)   the map formed by pairing indices and contracting to scalars;

   (ii)   the map formed by first tensoring $A_{ij\cdots p}$ with an alternating n-tensor Ω to get $A \otimes \Omega$ and then pairing indices and contracting to scalars,

and a theorem.

THEOREM 6. *Linear combinations of examples* (i) *and* (ii) *are the only possible* $SO(n)$-*invariant linear maps.*

PROOF OF THEOREM 4. We will derive this theorem as a consequence of Theorem 6. Let $P(v^1, \ldots, v^N)$ be an $SO(n)$-invariant polynomial, i.e.,

(94)
   (i) $P(gv^1, \ldots, gv^N) = P(v^1, \ldots, v^N)$ for all $g \in SO(n)$,

   (ii) P is a polynomial in the components $v_1^i, \ldots, v_n^i$ of the n-vectors v^i.

We may assume that P is homogeneous of degree m_k in the vector v^k, that is

(95)
$$P(v^1, \ldots, \lambda v^k, \ldots, v^N) = \lambda^{m_k} P(v^1, \ldots, v^k, \ldots, v^N).$$

Consider a map ϕ

$$\phi \colon (v^1, \ldots, v^N) \to v^1 \otimes \cdots \otimes v^1 \otimes \cdots \otimes v^N \otimes \cdots \otimes v^N,$$

where the vector v^i is repeated m_i times. Let $\mathscr{E}$ equal the set of all $(m_1 + \cdots + m_N)$-tensors obtained this way, and let $\mathscr{E}^+$ be the vector subspace of $V = \{(m_1 + \cdots + m_N)\text{-tensors}\}$ generated by $\mathscr{E}$. We can always find a linear map $l\colon \mathscr{E}^+ \to \mathbf{C}$ such that $P = l \circ \phi$.

Note that l is an $\mathrm{SO}(n)$-invariant map and $\mathscr{E}^+$ is an $\mathrm{SO}(n)$-invariant subspace of V. Recall that whenever a compact group G acts on a vector space V, we can define an invariant inner product by integrating any inner product over G. We have such an inner product because $\mathrm{SO}(n)$ is compact; therefore, we can define the complementary subspace $\mathscr{E}^-$ to $\mathscr{E}^+$ with respect to this invariant inner product and split V, $V = \mathscr{E}^+ \oplus \mathscr{E}^-$. If we define l to be 0 on $\mathscr{E}^-$, then l becomes an $\mathrm{SO}(n)$-invariant map on V. By Theorem 6, $P = l \circ \phi$ is a linear combination of

(i)   terms that arise from

$$\underbrace{v^1 \otimes v^1 \otimes \cdots \otimes v^2}_{} \otimes \cdots \otimes v^N$$

by pairing as indicated and then contracting indices; this gives terms such as

$$\langle v^1, v^1 \rangle \cdot \langle v^2, v^N \rangle \cdot \cdots\,;$$

(ii)   terms that arise by tensoring with an antisymmetric n-tensor Ω

$$v^1 \otimes v^1 \otimes \cdots \otimes v^N \otimes \Omega$$

and then pairing and contracting indices; these are products
of terms like those from (i) with

$$\mathrm{Trace}\big[\,v^1 \otimes \cdots \otimes v^N \otimes \Omega\,\big],$$

which is the bracket factor.   □

PROOF OF THEOREM 5. Part (i) follows from the same argument used to prove Theorem 4. Let $P(A, B, \ldots, C)$ be an $\mathrm{SO}(n)$-invariant polynomial acting on tensors of a fixed rank, i.e. an invariant polynomial in the components of these tensors. We may assume that P is homogeneous of degree $m_1, m_2, \ldots, m_N$ in the tensors $A, B, \ldots, C$. The map ϕ is defined by

$$\phi\colon (A, B, \ldots, C) \to A \otimes \cdots \otimes A \otimes \cdots \otimes C \otimes \cdots \otimes C,$$

where the tensors are repeated the number of times given by their degrees. Again $\mathscr{E} = \mathrm{Ran}\,\phi$ generates an $\mathrm{SO}(n)$-invariant subspace $\mathscr{E}^+$ of a finite-dimensional tensor space V and V splits with respect to an $\mathrm{SO}(n)$-invariant inner product, $V = \mathscr{E}^+ \oplus \mathscr{E}^-$. Finally, we can define an $\mathrm{SO}(n)$-invariant linear map l such that

$$P \circ \phi = l|_{\mathscr{E}}.$$

By Theorem 6 we know that all invariant polynomials in $A, B, \ldots, C$ are linear combinations of

$$\mathrm{Trace}[\,A \otimes \cdots \otimes A \otimes \cdots \otimes C\,], \qquad \mathrm{Trace}[\,A \otimes \cdots \otimes A \otimes \cdots \otimes C \otimes \Omega\,].$$

This finishes part (i); part (ii) is an immediate corollary of the next theorem.   □

THEOREM 7 (HILBERT). *If G is a compact group acting linearly on a vector space V, then the ring of G-invariant polynomials P on V is finitely generated.*

PROOF. Let $\mathcal{I}$ be the ideal of polynomials generated by the G-invariant polynomials homogeneous of degree > 0. Then $\mathcal{I}$ is finitely generated; let $P_1, \ldots, P_k$ be its generators. We can assume that they are G-invariant and homogeneous of degree d_l. Let $D = \max\{d_1, \ldots, d_k\}$ and expand this list of generators to a larger finite list which generates the vector space of all G-invariant, homogeneous polynomials of degree d with $0 \leq d \leq D$. We claim that this finite list generates the ring of invariant polynomials. To prove the claim, let Q be an invariant homogeneous polynomial of degree greater than D. Then

$$Q = S_1 P_1 + \cdots + S_k P_k,$$

where $P_1, \ldots, P_k$ are G-invariant. Under the action of G the equation becomes

$$Q = S_1^g P_1 + \cdots + S_k^g P_k.$$

Since G is compact, we can average this equation over the group to obtain

$$Q = \hat{S}_1 P_1 + \cdots + \hat{S}_k P_k,$$

where the $\hat{S}_l$ are invariant, since they arise by averaging. Now we can assume that

$$\deg \hat{S}_l = \deg Q - \deg P_k,$$

because we can just delete the other terms. This allows us to complete the proof by induction on $\deg Q$; we just note that $\hat{S}_l$ are G-invariant homogeneous polynomials with lower degree than Q. $\square$

PROOF OF THEOREM 6. Recall that the theorem concerns the classification of $SO(n)$-invariant linear maps $l: \{A_{ij\cdots p}\} \to \mathbf{C}$. Since the kernel of such a map is an $SO(n)$-invariant subspace, one could consider the more general question of what are all the $SO(n)$-invariant subspaces of the space of tensors of fixed length N.

This problem amounts to writing down explicitly a decomposition of the space V of tensors into irreducible representations. A typical irreducible subspace of V consists of all tensors of the form

$$\delta_{i_1 i_2} \delta_{i_3 i_4} \cdots \delta_{i_{2s-1} i_{2s}} A_{j_1 \cdots j_m}, \quad \text{where } \left(A_{j_1 \cdots j_m} \right) \text{ has all its traces} = 0$$

and is (roughly) symmetric in certain indices and antisymmetric in others. The pieces are described precisely in terms of combinatorial objects called Young's tableaux. Irreducibility can be proved using Weyl's character formula. A beautiful explanation of these ideas is given in Weyl [**67**].

But the proof of Theorem 6 is much less complicated. We shall use induction on the dimension, and make the transition from $(n-1)$ to n dimensions by exploiting the simple case in which the tensors $(A_{j_1 \cdots j_N})$ are all symmetric. The first step is to prove

LEMMA 3. *Assume $n > 1$. If l is an $SO(n)$-invariant linear functional on the space of symmetric N-tensors $(A_{j_1 \cdots j_N})$, then l is a constant multiple of*

$$\left(A_{j_1 \cdots j_N} \right) \mapsto \operatorname{Trace}\left(A_{j_1 \cdots j_N} \right) = \sum_{i_1 \cdots i_{N/2}} A_{i_1 i_1 i_2 i_2 \cdots i_{N/2} i_{N/2}}.$$

In particular $l = 0$ if N is odd.

PROOF. We may identify $(A_{j_1\cdots j_N})$ with the polynomial

$$P(x) = \sum_{j_1\cdots j_N} A_{j_1\cdots j_N} x_{j_1} \cdots x_{j_N}.$$

Thus l is an $SO(n)$-invariant functional on the space of polynomials on $\mathbf{R}^n$ homogeneous of degree N. Since l is rotation-invariant, we may average over $SO(n)$ to write $l(P) = l(\int_{SO(n)} P^g\, dg)$. Now $\int_{SO(n)} P^g\, dg$ is rotation-invariant and homogeneous of degree N; hence it is a constant multiple of $|x|^N$ and so must vanish for N odd. For N even we see that l is determined uniquely by its action on the single polynomial $|x|^N$, so the space of all possible l is one-dimensional. $\square$

We will now prove Theorem 6 using induction and the result above concerning symmetric tensors. If $n = 1$, then $(A_{i_1\cdots i_N}) = A_{1\ldots 1}$, so the tensor space is one dimensional and l is a constant multiple of

$$A \to A_{1\cdots 1} = \begin{cases} \mathrm{Trace}\, A, & N \text{ even}, \\ \mathrm{Trace}(A \otimes \Omega), & N \text{ odd}. \end{cases}$$

So assume Theorem 6 in $(n - 1)$ dimensions, and let l be an $SO(n)$-invariant functional on the space T_N^n of N-tensors on $\mathbf{R}^n$. Consider the action of $SO(n - 1) \subseteq SO(n)$, which fixes the nth unit vector e_n. Now, given a tensor $(A_{j_1\cdots j_N}) \in T_N^n$ and a subset $E \subseteq \{1,\ldots,N\}$, the components $(A_{j_1\cdots j_N})$, $j_s = n$ for $s \in E$, $j_s \neq n$ for $s \notin E$ form an $(N - |E|)$ tensor on $\mathbf{R}^{n-1}$. In this way we can write $T_N^n \cong \Sigma_{E \subseteq \{1\cdots N\}} \oplus T_{N-|E|}^{n-1}$, and the isomorphism commutes with the action of $SO(n - 1)$.

Note that the pieces of a given tensor $(A_{j_1\cdots j_N})$ all have the form $\mathrm{Trace}(A_{j_1\cdots j_N} \otimes e_n \otimes e_n \otimes \cdots \otimes e_n)$, where the trace contracts to an $(N - |E|)$-tensor. Since l is $SO(n - 1)$-invariant on a direct sum of $T_{N-|E|}^{n-1}$, Theorem 6 in $(n - 1)$ dimensions express l as a linear combination of terms of the form

$$(i) \qquad \mathrm{Trace}'\,\mathrm{Trace}[A \otimes e_n \otimes \cdots \otimes e_n]$$

and

$$(ii) \qquad \mathrm{Trace}'\,\mathrm{Trace}[A \otimes \Omega' \otimes e_n \otimes \cdots \otimes e_n].$$

Here Trace acts on tensors on $\mathbf{R}^n$, and contracts to an $(N - |E|)$-tensor, while Trace' acts on tensors on $\mathbf{R}^{n-1}$ and contracts to a scalar. Also, Ω' is an alternating tensor on $\mathbf{R}^{n-1}$.

Now, $\Omega' = \mathrm{Trace}(\Omega \otimes e_n)$, and for a 2-tensor (A_{ij}) we have

$$\mathrm{Trace}'\, A = \sum_{i=1}^{n-1} A_{ii} = \mathrm{Trace}\, A - A_{nn} = \mathrm{Trace}\, A - \mathrm{Trace}(A \otimes e_n \otimes e_n).$$

Analogous formulas hold for higher-rank tensors, and it follows that (i), (ii) are linear combinations of terms

$$(i') \qquad \mathrm{Trace}[A \otimes e_n \otimes \cdots \otimes e_n],$$

$$(ii') \qquad \mathrm{Trace}[A \otimes \Omega \otimes e_n \otimes \cdots \otimes e_n].$$

This time the traces contract to scalars.

Set $B = \text{Trace}[A \otimes \Omega]$, so that B is a tensor and the term (ii′) may be written as $\text{Trace}[B \otimes e_n \otimes \cdots \otimes e_n]$.

So far, we have used only the $\text{SO}(n - 1)$-invariance of l; now it is time to use $\text{SO}(n)$-invariance. In view of the full $\text{SO}(n)$-invariance, e_n in (i′), (ii′) may be replaced by any other unit vector v. Thus, l is a linear combination of terms

$$\text{Trace}[A \otimes v \otimes \cdots \otimes v], \qquad \text{Trace}[B \otimes v \otimes \cdots \otimes v],$$

where the coefficients of these terms are independent of $v \in S^{n-1}$. Averaging over all $v \in S^{n-1}$, we have l expressed as a linear combination of terms

$$l'(A) = \int_{S^{n-1}} \text{Trace}[A \otimes v \otimes \cdots \otimes v] \, dv,$$

and

$$l'(B) = \int_{S^{n-1}} \text{Trace}[B \otimes v \otimes \cdots \otimes v] \, dv.$$

A glance at the definition shows that l' is an $\text{SO}(n)$-invariant functional, and that $l'(A) = l'(\hat{A})$, where

$$\hat{A}_{j_1 \cdots j_N} = \frac{1}{N!} \sum_{\pi \in S_N} A_{j_{\pi(1)} \cdots j_{\pi(N)}}$$

is the symmetrization of A. So the lemma on symmetric tensors shows that l' is a constant multiple of $\text{Trace}\,\hat{A}$, which is a combination of traces of A. Applying the same reasoning to $l'(B)$, we have shown that l is a linear combination of terms of the form $\text{Trace}\,A$, $\text{Trace}\,B$. Finally, since $B = \text{Trace}[A \otimes \Omega]$, we see that l is a linear combination of terms $\text{Trace}\,A$, $\text{Trace}[A \otimes \Omega]$. The inductive step is complete, and Theorem 6 is proved. $\square$

We need some notation in order to state the next theorem. Let

$$T^k = \{\text{tensors of rank } k \text{ on } \mathbf{R}^n\}, \qquad \mathbf{T} = T^{k_1} \otimes \cdots \otimes T^{k_p},$$

where $k_1, \ldots, k_p$ are a collection of indices. The theorem concerns $\text{SO}(n)$-invariant sets $\mathcal{E} \subseteq \mathbf{T}$. For example, a point of $\mathcal{E}$ could consist of a curvature tensor and all its covariant derivatives up to order 6, so that $\mathcal{E} \subset T^4 \oplus T^5 \oplus \cdots \oplus T^{10}$.

THEOREM 8 (WEYL). *Let P be a polynomial on $\mathbf{T}$ whose restriction to $\mathcal{E}$ is $\text{SO}(n)$-invariant. Then $P = P_1 + P_2$, where $P_1 \equiv 0$ on $\mathcal{E}$, P_2 is $\text{SO}(n)$-invariant on $\mathbf{T}$, and P_2 is generated from elements on a certain finite list.*

PROOF. The last statement follows from Theorem 5. We assume that P is homogeneous in $A_{i_1 \cdots i_{k_l}}$ of degree m_{k_l}. Let

$$\mathcal{T} = T^{k_1} \otimes \cdots \otimes T^{k_1} \otimes \cdots \otimes T^{k_p} \otimes \cdots \otimes T^{k_p},$$

where the T^{k_l} factor is included m_{k_l} times and define ϕ and l as in Theorem 4

$$\mathcal{E} \subseteq \mathbf{T} \xrightarrow{\phi} \mathcal{T} \xrightarrow{l} \mathbf{C}.$$

Let $\mathcal{E}^+ = \text{span}(\phi(\mathcal{E}))$, so that

$$P|_{\mathcal{E}} = l \circ \phi,$$

where l is an $SO(n)$-invariant linear map on $\mathcal{E}^+$. If $\mathcal{E}^-$ is the $SO(n)$-invariant complementary subspace to $\mathcal{E}^+$, then $\mathfrak{I} = \mathcal{E}^+ \oplus \mathcal{E}^-$ and we can define $l^\#$ by

$$l^\#\big|_{\mathcal{E}^+} = l, \qquad l^\#\big|_{\mathcal{E}^-} = 0,$$

so that $l^\#$ is an $SO(n)$-invariant linear map on $\mathfrak{I}$. Putting

$$P_2 = l^\# \circ \phi, \qquad P_1 = P - P_2,$$

finishes the proof, since P_2 is $SO(n)$-invariant and $P_2|_{\mathcal{E}} = P|_{\mathcal{E}}$. $\quad\square$

We turn now to Weyl's invariant theory for $SO(n, m)$, the special orthogonal group preserving the quadratic form $x_1^2 + \cdots + x_n^2 - x_{n+1}^2 - \cdots - x_{n+m}^2$. Since $SO(n, m)$ is not compact, all the proofs above involving the existence of $SO(n, m)$-invariant complementary subspaces have to be modified. This is achieved through Weyl's "*unitarian trick*", which reduces questions concerning $SO(n, m)$ to those concerning $SO(n + m)$.

PROPOSITION 7. *Let $\mathcal{E} \subset T^s$ be an $SO(n, m)$-invariant subspace. Then there exists an $SO(n, m)$-invariant complementary subspace.*

PROOF. Introduce the norms and map indicated:

$$SO(n + m) \qquad\qquad\qquad\qquad SO(n, m)$$
$$\|v\|^2 = v_1^2 + \cdots + v_{n+m}^2 \qquad \|v\|^2 = v_1^2 + \cdots + v_n^2 - v_{n+1}^2 - \cdots - v_{n+m}^2$$
$$\alpha_0 \colon (v_1 \cdots v_{n+m}) \longrightarrow (v_1, \ldots, v_n, iv_{n+1}, \ldots, iv_{n+m}),$$

and define the map $\alpha \colon T^s \to T^s$ by

$$A_{j_1 \cdots j_s} \to \sigma(j_1) \cdots \sigma(j_s) A_{j_1 \cdots j_s},$$

where

$$\sigma(j) = \begin{cases} 1, & 1 \leqslant j \leqslant n, \\ i, & n + 1 \leqslant j \leqslant n + m. \end{cases}$$

Denote the Lie algebras of $SO(n, m)$ and $SO(n + m)$ by $so(n, m)$ and $so(n + m)$, and let α_0' be the map induced on the Lie algebras by α_0. Using the diagram

$$\begin{array}{ccc} so(n + m) & \text{acts on} & T^s \\ \downarrow \alpha_0' & & \downarrow \alpha \\ so(n, m) & \text{acts on} & T^s \end{array}$$

and noting that the complexifications of $so(n, m)$ and $so(n + m)$ are isomorphic as algebras over $\mathbf{R}$, we can now prove results for $SO(n, m)$ using the analogous results for $SO(n + m)$. In particular a subspace of T^s is $SO(n, m)$-invariant if and only if it is annihilated by the complexification of $so(n, m)$; so there is an $SO(n, m)$-invariant-complement to $\mathcal{E}$ as required, finishing the proof. $\quad\square$

This finishes our tour of Weyl's invariant theory. We now see what we can say about the heat kernel (86),

$$K_t(x, x) = \sum_{i \geqslant 0} (4\pi t)^{-n/2} \phi_i(x, x) t^i.$$

By Weyl's theorem (Theorem 8), each ϕ_l is a sum of terms of the form

$$\text{Trace}\big[R_{ijkl/\sigma} \otimes \cdots \otimes R_{abcd/\tau} \big].$$

Because of the $O(n)$-invariance of the ϕ_i, terms of the form

$$\text{Trace}\big[R_{ijkl/\sigma} \otimes \cdots \otimes R_{abcd/\tau} \otimes \Omega \big]$$

never appear. By the homogeneity formula (90) of the last section, each term must satisfy

$$(2 + |\sigma|) + \cdots + (2 + |\tau|) = 2l.$$

Recall that by Theorem 2, the heat kernel for the operator $\partial/\partial t - \square_p$ takes the form

$$K_p^t(x, y) = (4\pi t)^{-n/2} e^{-\text{dist}^2(x, y)/4t} \Big\{ \sum_{i \geqslant 0} \phi_i^p(x, y) t^i \Big\}.$$

The expressions

$$S_\alpha = \partial_y^\alpha \phi_i^p(x, y)\big|_{y=x}$$

are tensor-valued polynomials in the $R_{ijkl/\sigma}$. It is not hard to check that all of our invariant theory goes through for tensor-valued polynomials (except that the statement of Theorem 7 has to be changed a little); therefore S_α is a linear combination of terms of the form

$$\text{Trace}\big[R_{ijkl/\sigma} \otimes \cdots \otimes R_{abcd/\tau} \big],$$

where now the trace is taken by contracting down until we are left with s indices, where s is the number of indices in the multi-index $\alpha = (\alpha_1, \ldots, \alpha_s)$.

If we work hard enough, we can obtain this result without using invariant theory. This is because at each stage of the computation of the $\phi_l(x, y)$, we remain within the class of functions of this form. More precisely, each function $\psi(x, y)$ appearing in the computation of $\phi_l(x, y)$ has the property that in normal coordinates $(\partial_y^\alpha \psi(x, y)|_{y=x})_{|\alpha|=p}$ is a linear combination of p-tensors of the form $\text{Trace}[R_{ijkl/\sigma} \otimes \cdots \otimes R_{abcd/\tau}]$. Later, when we consider analogous proofs of the asymptotic expansion of the Bergman kernel, the computations will take us out of the appropriate class of functions, and invariant theory will be needed to complete the proof.

9. Gilkey's invariant theory. With Gilkey's invariant theory we can explicitly calculate the polynomial involved in computing the Euler characteristic and thus prove

THEOREM 9 (CHERN-GAUSS-BONNET).

$$\chi(M) = C_n \int_M \text{Pff}(R_{ijkl}) \, d\text{vol},$$

where

$$\mathrm{Pff}(R_{ijkl})\, dx_1 \wedge \cdots \wedge dx_n = c \sum \mathrm{sgn}(i)\,\mathrm{sgn}(j) R_{i_1 i_2 j_1 j_2}\, dx_{j_1} \wedge dx_{j_2}$$
$$\wedge \cdots \wedge R_{i_{n-1} i_n j_{n-1} j_n}\, dx_{j_{n-1}} \wedge dx_{j_n}$$

for n even, and $\mathrm{sgn}(i) = \mathrm{sgn}(i_1, \ldots, i_n)$.

PROOF. We claim

$$\chi(M) = \int_M P(\partial^\alpha g_{ij})\, d\,\mathrm{vol},$$

where

(96)

    (i) P is a polynomial in $(\partial^\alpha g_{ij})$, as long as $g_{ij}(x) = \delta_{ij}$

          and $\partial_{x_k} g_{ij}(x) = 0$ (see the proof of Theorem 2, part (ii)),

    (ii) P is $O(n)$-invariant,

    (iii) P satisfies the weight condition (90) with $l = n$,

    (iv) for manifolds $M^n = M^{n-1} \times S^1$ with the product metric on
          M^n, $P \equiv 0$.

Indeed, we already have proved the first three properties. To prove (iv), recall how P was obtained. On $\Lambda^p(M)$ forms may, or may not, contain $d\theta$:

$$f\, dx_{i_1} \wedge \cdots \wedge dx_{i_p}, \qquad g\, dx_{i_1} \wedge \cdots \wedge dx_{i_{p-1}} \wedge d\theta,$$

and so $(\partial/\partial t + \square) = 0$ splits into two problems involving

$$\square(f\, dx_{i_1} \wedge \cdots \wedge dx_{i_p}), \qquad \left(\square(g\, dx_{i_1} \wedge \cdots \wedge dx_{i_{p-1}})\right) \wedge d\theta.$$

The heat kernels of both are the same, but the degrees are shifted by 1, causing the alternating sum to add to 0; thus, $P \equiv 0$. The proof is completed once the following theorem is proved. $\square$

THEOREM 10 (GILKEY). *Properties* (96) *imply*

$$P = C_n \mathrm{Pff}(R_{ijkl}).$$

PROOF. First it is easy to verify that Pff satisfies (96.i)–(96.iv); all that remains is to show that the space of such P satisfying (96.i)–(96.iv) is 1 dimensional (0 dimensional, if n is odd). We begin by proving through four observations that (96.ii), (96.iii) and (96.iv) imply $P = P(R_{ijkl})$, with no covariant derivatives involved.

Observation 1. $\partial^\alpha g_{ij}$ are independent variables for $i \leqslant j, |\alpha| \geqslant 2$.

Observation 2. Suppose $M = \tilde{M} \times S^1$ with coordinates $(x_1, \ldots, x_n)$, where $(x_2, \ldots, x_n)$ are coordinates for $\tilde{M}$, while (x_1) coordinates S^1. We claim that M has the product metric iff the equation

$$\partial^\alpha g_{ij} = 0 \quad \text{for } |\alpha| \geqslant 2$$

holds, whenever $i = 1$, or if any index in α contains a 1. To see this, note that for the product metric, $g_{1k} = g_{k1} \equiv 1$ or 0; and so $\partial^\alpha g_{1k} = \partial^\alpha g_{k1} = 0$, for any $|\alpha| \geqslant 1$, and, in particular, for $|\alpha| \geq 2$. By rotation invariance we conclude that $\partial_{x_1} g_{ij} = 0$. Conversely, assume that $\partial^\alpha g_{ij} = 0$, if $i, j = 1$ or if 1 occurs in α, for $|\alpha| \geq 2$. Then $g_{ij}(x_1, \ldots, x_n) = g_{ij}(x_2, \ldots, x_n)$ and so the metric satisfies

$$ds^2 = dx_1^2 + \sum_{i,j>1} g_{ij}\, dx_i\, dx_j,$$

as required.

Observation 3. By Observations 1 and 2, property (96.iv) forces each monomial in $P(\partial^\alpha g_{ij})$ to involve the index 1, either in α or in i, j.

Observation 4. Each monomial in $P(\partial^\alpha g_{ij})$ contains at least two occurrences of the index 1, although maybe not in the same factor $\partial^\alpha g_{ij}$; this is forced by $O(n)$-invariance and is checked by looking at reflections. Since there is nothing special about the index 1, each monomial in $\partial^\alpha g_{ij}$ is of the form

$$\partial^\alpha g_{ab} \cdots \partial^\gamma g_{kl}$$

where each of $1, 2, \ldots, n$ appears at least two times. If s is the number of factors in this monomial, then the weight property (90) implies

(i) $$(|\alpha| + |\beta| + \cdots + |\gamma|) = n$$

while also

(ii) $$(|\alpha| + |\beta| + \cdots + |\gamma|) + 2s \geq 2n,$$

since each index $1, \ldots, n$ appears at least twice. Since $|\alpha|, |\beta|, \ldots, |\gamma| \geq 2$, (i) yields $2s \leq n$, so (ii) becomes $(|\alpha| + |\beta| + \cdots + |\gamma|) \geq 2n - 2s \geq n$. If either inequality is strict, we get a contradiction with (i). Therefore $s = n/2$ and $|\alpha| = |\beta| = \cdots = |\gamma| = 2$. So our monomial has degree $n/2$ and involves only second derivatives of the metric, as claimed. In other words, $P = P(\partial^\alpha g_{ij})$ ($|\alpha| = 2$) with P homogeneous of degree $n/2$.

To prove Gilkey's theorem, we shall produce a linear functional on the space of P satisfying (96) which vanishes only at zero. This implies that the space (96) is one dimensional. Our functional will be of the simple form

$$P \to \text{coefficient of a monomial,}$$

so it is sufficient to pick out a monomial whose coefficient cannot vanish if $P \neq 0$.

We choose the monomial $g_{11/22} g_{33/44} \cdots g_{n-1,n-1/nn}$ and calculate its coefficient. Here $g_{ij/kl} = \partial^2 g_{ij}/\partial x_k \partial x_l$. Let P satisfy (96). We conclude that

1. P contains a monomial of the form $g_{1j/kl} \cdot$ (other terms).

Since P is $O(n)$-invariant, it is invariant under permutations of the indices $\{1, \ldots, n\}$, so this is obvious.

2. P contains a monomial of the form $g_{11/kl} \cdot$ (other terms). If this were false then $j \neq 1$ in step 1. By permuting indices we may take $j = 2$, so P contains a monomial $g_{12/kl} \cdot$ (other terms). One checks easily that a generic rotation in the $x_1 - x_2$ plane produces in P a monomial $g_{11/pq} \cdot$ (other terms).

3. P contains a monomial of the form $g_{11/22} \cdot$ (other terms). If $k = l$ in step 2, this is obvious by permuting indices. Otherwise, we may assume $k = 2, l = 3$, so

that P contains a monomial $g_{11/23} \cdot$ (other terms). Now performing a generic rotation in the $x_2 - x_3$ plane, we see that P contains a monomial $g_{11/22} \cdot$ (other terms), as claimed.

4. Note that the other terms in step 3 cannot contain any more 1's or 2's. Now repeat steps 2 and 3 over and over, to obtain in P a monomial $g_{11/22} \cdots g_{2k-1,2k-1/2k,2k} \cdot$ (other terms) for ever larger k. Finally, we obtain

5. P contains the monomial $g_{11/22} g_{33/44} \cdots g_{n-1,n-1/nn}$. The proof of Gilkey's theorem is complete. $\square$

Instead of using invariant theory, one can prove the Gauss-Bonnet theorem by very careful study of the construction of the heat kernels. This was the original method used by Patodi [52] to relate the heat equation to the Gauss-Bonnet theorem. Gilkey's theorem appears in [27]; this also contains a list of earlier references and a discussion of other index theorems.

CHAPTER 6.
AN OVERVIEW OF TOPICS IN SEVERAL COMPLEX VARIABLES

1. Introduction. Now that the preliminaries are finished, we will briefly discuss the topics in several complex variables that will be covered. We begin with a table comparing the topics for Riemannian manifolds that have been discussed with the analogous questions in several complex variables.

Riemannian manifold M:	*Strongly pseudoconvex domain D :*
linear analysis: Laplacian, wave equation	$\bar{\partial}, \bar{\partial}_b,$ $\square, \square_b$
geometry: normal coordinates, $\nabla_X Y$, parallel transport, geodesics, arclength, connection in principal bundle	Moser normal form, geometry of Poincaré metric, Monge-Ampère equation, Cheng-Yau theorem, formal analysis at ∂D, Cartan-Tanaka-Chern invariants, chains, parallel transport of frames along chains, parametrization
refined linear analysis: heat equation, invariance theory, Chern-Gauss-Bonnet theorem	Bergman-Szegö kernels, invariance theory for nonsemisimple groups, ?

We began our study of Riemannian geometry by first considering the simplest case, $\mathbf{R}^n$, with the Lie group of Euclidean motions. Similarly, we will begin the study of strongly pseudoconvex domains by considering the unit ball in $\mathbf{C}^n$ and the Siegel Domain and their Lie groups of linear fractional transformations, $SU(n, 1)$ and the Heisenberg group.

The study of strongly pseudoconvex domains leads us to the topic of reflections, a subject which has no analogy in the Riemannian case. Using reflections, we can give a practical classification (in other words, computable in a reasonable fashion) for certain domains with analytic boundary.

For weakly pseudoconvex domains matters are much less settled. We have theorems establishing the interior regularity of the Poincaré metric, and, if we assume that the boundary is analytic, we have theorems establishing the C^∞-regularity and subellipticity of $\bar\partial$; but the analogies for the other entries in the table are still unknown.

Now we will fill in a few of the details of this big picture.

2. Linear analysis: $\bar\partial$, $\Box$, $\bar\partial_b$, $\Box_b$. We will consider functions, $(0, 1)$-forms, and $(0, 2)$-forms on a bounded, strongly ψ-convex domain D. Recall, for example, that a $(0, 2)$-form is locally of the form

$$f_{jk}\, d\bar z_j \wedge d\bar z_k.$$

We want to solve the equation

$$(1.a) \qquad\qquad\qquad \bar\partial u = \alpha$$

on D, where α is a $(0, 1)$-form satisfying $\bar\partial\alpha = 0$. There are many solutions, since given any solution, we can produce others by adding holomorphic functions to the original solution. If we introduce the Hilbert space $L^2(D)$, we can define a good solution u by requiring that

$$(1.b) \qquad\qquad\qquad u \perp \{\text{holomorphic functions}\}.$$

We will need a general fact about a first order system of differential operators $\mathcal{L}$ on $\bar D$; for $u, v \in C^\infty(\bar D)$ and $\mathcal{L}^*$ the formal adjoint of $\mathcal{L}$, we have

$$\int_D (\mathcal{L}u)\bar v\, d\operatorname{vol} = \int_D u\,\overline{(\mathcal{L}^*v)}\, d\operatorname{vol} + \int_{\partial D} u\,\overline{(A^\#v)}\, d\operatorname{vol},$$

where $A^\#$ is a 0th order system of operators on the boundary. The domain of the adjoint $\mathcal{L}^*$ consists of those $v \in C^\infty(\bar D)$ such that the boundary term equals 0.

In our case $\mathcal{L} = \bar\partial$ and the domain of the adjoint is

$$\operatorname{Dom}(\bar\partial^*) = \{v \in C^\infty(\bar D): A^\#v = 0 \text{ on } \partial D\}.$$

Note that with $u = \bar\partial^*\omega$, then for F holomorphic,

$$\langle \bar\partial^*\omega, F\rangle = \langle \omega, \bar\partial F\rangle = 0,$$

whenever the $(0, 1)$-form $\omega \in \operatorname{Dom}(\bar\partial^*)$; in other words, $u \perp \{\text{holomorphic functions}\}$. This means that if we solve the equation

$$\bar\partial\bar\partial^*\omega = \alpha, \qquad \omega \in \operatorname{Dom}(\bar\partial^*),$$

then we can solve our original problem (1) by putting $u = \bar\partial^*\omega$.

In fact problem (1) is equivalent to the case $\bar\partial\alpha = 0$ of the system

$$(2) \qquad \begin{aligned} \Box\omega \equiv (\bar\partial\bar\partial^* + \bar\partial^*\bar\partial)\omega = \alpha, \\ \omega \in \operatorname{Dom}(\bar\partial^*), \qquad \bar\partial\omega \in \operatorname{Dom}(\bar\partial^*). \end{aligned}$$

To see this, note that $0 = \bar{\partial}\alpha = \bar{\partial}(\bar{\partial}\partial^* + \bar{\partial}^*\partial)\omega = \bar{\partial}\bar{\partial}\partial^*\bar{\partial}\omega$, and so

$$0 = \langle\bar{\partial}\omega, \bar{\partial}\bar{\partial}^*\bar{\partial}\omega\rangle = \langle\bar{\partial}^*\bar{\partial}\omega, \bar{\partial}^*\bar{\partial}\omega\rangle,$$

showing that $\bar{\partial}^*\bar{\partial}\omega = 0$. Comparing this with equation (2) gives $\bar{\partial}\bar{\partial}^*\omega = \alpha$, which by our earlier calculation is equivalent to (1). Problem (2) for general α is called the "$\bar{\partial}$-*Neumann problem*."

We turn now to the boundary manifold $\partial D = M$ and its tangent bundle TM. Define $\mathcal{T} \subseteq TM$ as the maximal subspace such that $i\mathcal{T} = \mathcal{T}$. If we fix a map

$$J: \{\text{real vectors in } \mathcal{T}\} \to \{\text{real vectors in } \mathcal{T}\}$$

such that $J^2 = -1$ (J corresponds to multiplications by $\sqrt{-1}$ in the ambient $\mathbf{C}^n$), then we can split $\mathbf{C} \otimes \mathcal{T} = \mathcal{T}^{(0,1)} \oplus \mathcal{T}^{(1,0)}$ by defining

$$\mathcal{T}^{(1,0)} = \{Z: Z = (X + iJX), \text{ some } X\},$$

$$\mathcal{T}^{(0,1)} = \{\bar{Z}: \bar{Z} = (X - iJX), \text{ some } X\}.$$

As a brief review, recall that for boundary forms

type	*description*
$(0,0)$	function
$(0,1)$	linear form mapping $\bar{Z} \to \mathbf{C}$
$(0,2)$	alternating 2-form mapping $(\bar{Z}, \bar{W}) \to \mathbf{C}$.

Now define $\bar{\partial}_b$ on $(0,0)$- and $(0,1)$-forms by

$$(\bar{\partial}_b f)(\bar{Z}) = \bar{Z}f, \qquad (\bar{\partial}_b\lambda)(\bar{Z}, \bar{W}) = c\{\bar{Z}\lambda(\bar{W}) - \bar{W}\lambda(\bar{Z}) - \lambda[\bar{Z}, \bar{W}]\}$$

so that $(\bar{\partial}_b)^2 = 0$. The analogy for problem (1) is

$$\bar{\partial}_b u = \alpha \quad \text{on } M$$

where we are given a $(0,1)$-form α satisfying $\bar{\partial}_b\alpha = 0$. As above, this is equivalent to the problem

$$\Box_b \omega \equiv (\bar{\partial}_b\bar{\partial}_b^* + \bar{\partial}_b^*\bar{\partial}_b)\omega = \alpha \quad \text{on } M.$$

This time there are no boundary conditions, for M has no boundary.

The codimension one subspace $\mathcal{T} \subset T(\partial D)_p$ and the map $J: \mathcal{T} \to \mathcal{T}$ contain all the information for complex analysis and geometry on ∂D. More generally, we define a *C-R* (*Cauchy-Riemann*) *manifold* to be a $(2n - 1)$-manifold M together with a subbundle $\mathcal{T} \subseteq TM$ with codimension one fibres, and a smoothly varying automorphism $J: \mathcal{T} \to \mathcal{T}$ with $J^2 = -\text{Id}$. As in the familiar case $M = \partial D$, we can split $\mathbf{C} \otimes \mathcal{T}$ into $\mathcal{T}^{(1,0)} \oplus \mathcal{T}^{(0,1)}$. For $M = \partial D$, the commutator of two vector fields in $\mathcal{T}^{(1,0)}$ agains lies in $\mathcal{T}^{(1,0)}$. A general *C-R* manifold with this property is called *integrable*. There is also a natural definition of strict pseudoconvexity for a *C-R* manifold. Thus, biholomorphic geometry may be thought of more generally as the study of strictly pseudoconvex integrable *C-R* manifolds.

In fact, a recent remarkable theorem of Kuranishi [**42**] asserts that every such manifold of real dimension $\geqslant 9$ arises locally as the boundary of a domain in $\mathbf{C}^n$.

This breaks down in dimension 3 (Nirenberg [**50**]); the intermediate dimensions are not yet settled. Kuranishi's proof is based on careful study of a highly degenerate boundary problem for the $\Box_b$-operator—see also Boutet de Monvel [**4**] for an earlier result.

3. The unit ball and the Siegel domain. In this section we consider two simple models for strictly pseudoconvex domains: the unit ball and the Siegel domain. The *unit ball* is defined by

$$B = \left\{ z \in \mathbf{C}^n \colon u(z) = 1 - \sum_1^n |z_k|^2 > 0 \right\}$$

and its linear fractional transformations are given by $\mathrm{SU}(n,1)$. To realize the action of $\mathrm{SU}(n,1)$, let $z_k = \zeta_k/\zeta_0$, for $(\zeta_0,\dots,\zeta_n) \in \mathbf{C}^{n+1}$, so that the inverse image of the ball is the subset of $\mathbf{C}^{n+1}$ given by

$$\left\{ \zeta \colon |\zeta_0|^2 - \sum_1^n |\zeta_k|^2 > 0 \right\}.$$

Since $\mathrm{SU}(n,1)$ is the group of linear transformations that preserve this quadratic form, an action of this group on the ball is induced by its action on $\mathbf{C}^{n+1}$.

The *Siegel domain D* is the unbounded version of the unit ball; it is defined by

$$D = \left\{ z \in \mathbf{C}^n \colon \operatorname{Re} z_1 > |z'|^2 \right\},$$

where $z' = (z_2,\dots,z_n)$. Using the same coordinates as above, we can identify D with the subset of $\mathbf{C}^{n+1}$

$$\left\{ \zeta \colon \frac{1}{2}\zeta_1\bar\zeta_0 + \frac{1}{2}\zeta_0\bar\zeta_1 - \sum_2^n |\zeta_k|^2 > 0 \right\}$$

so that once again the group $\mathrm{SU}(n,1)$ of linear transformations preserving this quadratic form acts by linear fractional transformations on D. Note that there is a linear fractional transformation of $\mathbf{C}^n$ bringing the unit ball to D.

The boundary of D has the structure of a nilpotent Lie group, with the group multiplication law

$$(z_1, z') \cdot (w_1, w') = (z_1 + w_1 + 2z' \cdot w', z' + w').$$

This is the *Heisenberg group N*. N acts on ∂D by moving the origin around; moreover, $N \subset \mathrm{SU}(n,1)$. We also have a subgroup $H^+ \subset \mathrm{SU}(n,1)$, leaving the origin in ∂D fixed:

$$H^+ = \left\{ T \in \mathrm{SU}(n,1) \colon Te = \lambda e \right\},$$

where $e = (1,0,\dots,0)$. Note that H^+ contains the subgroup of Heisenberg dilations

$$(z_1, z') \to (\delta^2 z_1, \delta z')$$

for $\delta > 0$. This later group is just a copy of $\mathbf{R}^\times$, the multiplicative group of positive reals. H^+ also contains the subgroup H defined by

$$H = \left\{ T \in \mathrm{SU}(n,1) \colon Te = e \right\}.$$

To clarify the structure of H, introduce the inversion operator i on the Siegel domain D; i is a linear fractional transformation of D which interchanges 0 and ∞. Explicitly, i is the linear fractional transformation induced by the matrix

$$\left(\begin{array}{cc|c} 0 & 1 & 0 \\ 1 & 0 & \\ \hline 0 & & \mathbf{I} \end{array}\right) \in U(n,1);$$

note that $i^2 = $ Identity. Since the Heisenberg translations $(= N)$ preserve ∞ and move 0, $i(N)i$ fixes 0; also, its derivative at $0 = \mathbf{I}$. This means that $H \supset U(n-1)$ and $H \supset iNi$; in other words $H \supset N$. In terms of Lie algebras we have

$$\mathrm{su}(n,1) = n \oplus h^+, \qquad h^+ = \mathbf{R}^\times \oplus h, \qquad h = u(n-1) \oplus n.$$

A natural family of curves in the unit sphere is preserved under linear fractional transformations; this is the family of circles called chains. A *chain* is the intersection of the sphere with a complex line, not necessarily through the origin. An example of a chain in the Siegel domain is given by the line

$$\{z \colon \mathrm{Re}\, z_1 = 0,\, z' = 0\}.$$

The unit ball carries an $SU(n,1)$-invariant metric

$$ds^2 = \sum_{j,k} \frac{\partial^2 \log(1 - |z|^2)}{\partial z_j \partial \bar{z}_k} dz_j\, d\bar{z}_k$$

and the associated volume element

$$d\,\mathrm{vol} = C_n \frac{s\,dz_1 \wedge \cdots \wedge dz_n \wedge d\bar{z}_1 \wedge \cdots \wedge d\bar{z}_n}{\left(1 - |z|^2\right)^{n+1}}.$$

With this metric the Ricci curvature is constant and negative. Finally, we claim that the Bergman kernel $K(z,w)$ and the Szegö kernel $S(z,w)$ for the unit ball B are

$$K(z,w) = \frac{n!}{\pi^n} \frac{1}{(1 - z \cdot \bar{w})^{n+1}}, \qquad S(z,w) = \frac{(n-1)!}{2\pi^n} \frac{1}{(1 - z \cdot \bar{w})^n}.$$

Indeed, recall the Bergman kernel arises from the orthogonal projection

$$\Pi \colon L^2(B) \to H(B), \qquad H(B) = \text{holomorphic functions}$$

via

$$\Pi f(z) = \int_B K(z,w) f(w)\, dw.$$

For $f \in H$ we want to show that for K given as above, $\Pi f = f$. Now

$$\Pi f(0) = \int_B K(0,w) f(w)\, dw = \int_B \frac{n!}{\pi^n} \frac{1}{(1 - 0 \cdot \bar{w})^{n+1}} f(w)\, dw$$

$$= \mathrm{Average}_B(f) = f(0).$$

Using this computation and the transitive action of SU(n, 1) on B, we can show that $(\Pi f)(z) = f(z)$, for $z \in B$, $f \in H$. This shows that $\Pi|_H$ = identity. Since $K(z, w)$ is holomorphic in z, it is easy to see that image $\Pi \subset H$. Finally because of the symmetric form of the kernel $K(z, w)$, a simple calculation shows that Π is selfadjoint. Therefore Π is a projection as required. The Szegö kernel is handled similarly.

From these simple models we can see that the function $u(z) = 1 - |z|^2$ appears many times. To get analogous formulas for general strictly pseudoconvex domains D, we will use the geometry of the Poincaré metric and the Monge-Ampère equation.

4. Geometry of strictly pseudoconvex domains: Moser normal form. Let D be a strictly pseudoconvex domain with real analytic boundary. This last assumption will be written as ∂D is C^ω. The boundary ∂D is said to be in *Moser normal form* near $0 \in \partial D$, if locally it can be represented by a convergent power series

$$\mathrm{Im}(z_n) = |z'|^2 + \sum_{\substack{|\alpha|,|\beta| > 2 \\ l > 0}} A^l_{\alpha\bar\beta} \, (\mathrm{Re}\, z_n)^l z'^\alpha \bar z'^\beta,$$

where $z' = (z_1, \ldots, z_{n-1})$, and where certain sums of the $A^l_{\alpha\bar\beta}$'s vanish. Note that when ∂D is in Moser normal form, the straight line

$$\{z: z' = 0 \text{ and } \mathrm{Im}\, z_n = 0\}$$

lies on ∂D.

THEOREM. *If D is a strictly pseudoconvex domain with C^ω boundary and $p \in D$, then there exists a biholomorphic map on a neighborhood of p such that*

$$p \to 0, \qquad \partial D \to \partial \tilde D,$$

where $\partial \tilde D$ is locally in Moser normal form.

PROOF (SKETCH). In this proof we will assign weights to terms using the table

$\mathrm{Re}\, z_n$	weight 2
$\mathrm{Im}\, z_n$	2
$z_1, \ldots, z_{n-1}$	1
$\bar z_1, \ldots, \bar z_{n-1}$	1

Since we can assume that ∂D is tangent to $\mathrm{Im}\, z_n = 0$ at $p = 0$, we can represent ∂D locally by

$$\mathrm{Im}\, z_n = \sum_{1 \leqslant j,k \leqslant n-1} \left(\lambda_{jk} z_j z_k + \bar\lambda_{jk} \bar z_j \bar z_k \right)$$

$$+ \sum_{1 \leqslant j,k \leqslant n-1} g_{j\bar k} z_j \bar z_k + (\text{terms of higher weight}),$$

which under the change of coordinates

$$z \to \hat{z}, \text{ where}$$
$$\hat{z}_1 = z_1, \ldots, \hat{z}_{n-1} = z_{n-1}, \qquad \hat{z}_n = z_1 - 2$$
$$\sum_{1 \leqslant j,k \leqslant n} \lambda_{jk} z_j z_k,$$

becomes

$$\operatorname{Im} \hat{z}_n = \sum_{1 \leqslant j,k \leqslant n} g_{j\bar{k}} z_j \bar{z}_k + (\text{terms of higher weight}).$$

Since the strict pseudoconvexity implies that the quadratic form is positive definite, we can rotate and dilate coordinates 1 through $n-1$ so that ∂D is given by

$$\operatorname{Im} z_n = |z'|^2 + (\text{terms of higher weight}).$$

To complete the proof we will use the family of maps

$$\mathcal{Q} = \left\{ \begin{array}{l} \text{biholomorphic maps } \phi\colon z \to w \text{ such that } \phi(0) = 0 \\ \text{and locally } \phi \text{ preserves surfaces of the form} \\ \operatorname{Im} z_n = |z'|^2 + (\text{terms of weight} \geqslant 3) \end{array} \right\},$$

and the subfamily

$$\mathcal{Q}_0 = \left\{ \begin{array}{l} \phi \in \mathcal{Q}\colon \phi'(0) = \mathrm{I}, \; \dfrac{\partial^2 w'}{\partial z_j \partial z_k}\bigg|_0 = 0, \text{ and} \\[2ex] \operatorname{Re}\left(\dfrac{\partial^2 w_n}{\partial z_n^2} \right) = 0, \text{ where } \phi\colon (z', z_n) \to (w', w_n) \text{ and} \\[2ex] 1 < j, k < n-1 \end{array} \right\}.$$

We can simplify an element $\phi \in \mathcal{Q}$ by composing it successively with linear fractional transformations from $\mathbf{R}^\times$, from $U(n-1)$, and from iNi.

Here we have in mind the lattice of subgroups

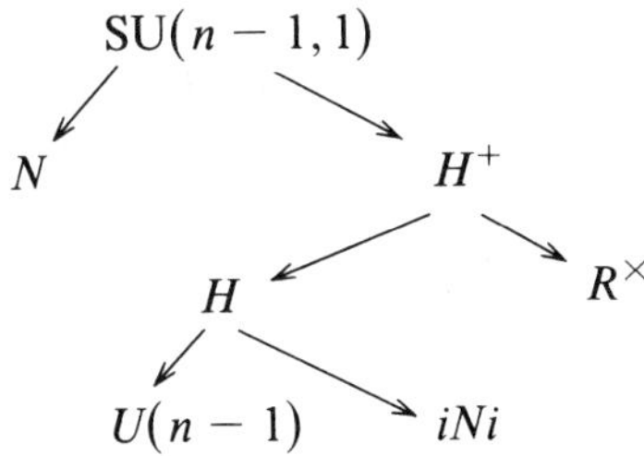

The result is

(3) $$\mathcal{Q} = H^+ \mathcal{Q}^0 = \mathcal{Q}^0 H^+.$$

In fact we have the

THEOREM (MOSER). *Locally, any $\partial D \in C^\omega$ given by*

$$\operatorname{Im} z_n = |z'|^2 + (\textit{terms of weight} \geqslant 3)$$

may be carried to

$$\operatorname{Im} z_n = |z'|^2 + \sum_{\substack{|\alpha|,|\beta|>2 \\ l\geq 0}} A^l_{\alpha\beta}(\operatorname{Re} z_n)^l z'^\alpha \bar z'^\beta$$

by exactly one member of $\mathcal{C}^0$.

Using formal power series arguments, we can prove uniqueness; convergence is established using another argument.

Note that the decomposition (3) induces an action of H^+ on normal forms: take a normal form, apply an element of H^+, and then apply the unique element of $\mathcal{C}^0$ which returns the map to normal form.

Let $\phi\colon \partial D \to \partial\tilde D$ be a map that sends $D \in C^\omega$ to Moser normal form. We know that such maps exist for strictly pseudoconvex domains by the theorem above. Recall that the straight line

$$\{\tilde z\colon \tilde z' = 0 \text{ and } \operatorname{Im} \tilde z_n = 0\}$$

lies on $\partial\tilde D$. A *chain* is defined to be the inverse image of this line under ϕ. There is a whole family of chains through a point on ∂D; these chains turn out to be the solution curves to certain second order ODE's. In other words, chains are partly the analogues of geodesics. On the sphere all chains are circles. More precisely, they are the intersections of the sphere with complex lines.

Moser normal forms are one way of getting local invariants for strictly pseudoconvex domains. Later we will see how local invariants can reduce the question of whether a domain is locally equivalent to the unit ball to the question of whether the defining function for the boundary satisfies a finite number of nonlinear PDE's.

The theorem above produces Moser normal forms for embedded submanifolds. We will now sketch an intrinsic formulation for abstract Cauchy-Riemann manifolds; this is a second means of producing local invariants. Given an integrable C-R manifold M, we will define an $\mathbf{R}^\times$-bundle E over M and a frame bundle $Y \to E$. The bundle Y carries a natural Cartan connection ω satisfying

$$d\omega + [\omega,\omega] = 0 \quad \text{in the flat model (Siegel domain),}$$
$$= \Pi \quad \text{a matrix of 2-forms, in general.}$$

Chains are defined by projecting parallel transport down to M. For embedded surfaces the coefficients of Π are the $A^l_{\alpha\bar\beta}$ for $l = 0$ and $|\alpha|,|\beta| \leq 3$. Here, we will define these bundles only for the flat model; the general case is studied in Chapter 10.

As a first step, we define the bundles $\tilde Y \to \tilde E$ and $\tilde E \to M$. Define

$$M = \{\operatorname{Re} z_1 = |z'|^2\},$$

$$\tilde E = \left\{\xi\colon \|\xi\|^2 \equiv \tfrac{1}{2}\xi_0\bar\xi_1 + \tfrac{1}{2}\xi_1\bar\xi_0 - \sum_2^n \xi_k\bar\xi_k = 0\right\},$$

where $z' = (z_2,\ldots,z_n)$, $\xi = (\xi_0,\ldots,\xi_n)$ and $z_j = \xi_j/\xi_0$ for $j \geq 1$, so that $\tilde{E} \to M$ is a $\mathbf{C}^\times$-bundle over M. Next consider frames $F = \{(e_0, e_1,\ldots,e_n)\}$ satisfying

$$e_i \in \mathbf{C}^{n+1},$$

$$\|e_0\|^2 = \|e_1\|^2 = 0, \qquad \langle e_\alpha, \bar{e}_\beta \rangle = \delta_{\alpha,\beta} \quad \text{for } \alpha, \beta > 2,$$

$$\langle e_0, \bar{e}_\alpha \rangle = \langle e_1, \bar{e}_\alpha \rangle = 0, \qquad \langle e_0, \bar{e}_1 \rangle = 1.$$

The frame F is called a Q-frame. Putting $\tilde{Y} = \{Q\text{-frames}\}$ and defining a projection $(e_0, e_1,\ldots,e_n) \to e_0$ turns $\tilde{Y} \to \tilde{E}$ into a frame bundle. To define the connection note that $\tilde{Y}$ and $U(n, 1)$ can be identified. Then the tangent space of $\tilde{Y}$ is $u(n, 1)$ and the Lie algebra $h \subset u(n, 1)$ of H can be used to define the "horizontal subspace" of the connection; in other words the connection is defined by splitting the tangent space

$$u(n, 1) = n \oplus h,$$

where n is the Lie algebra of the Heisenberg translations N. Observe next that there is a natural action of the circle on $\tilde{E}$ and $\tilde{Y}$,

$$(e_0, e_1,\ldots,e_n) \to (|z|e_0, |z|e_1,\ldots,|z|e_n),$$

for z with $|z| = 1$. The bundles E and Y are defined by modding out with respect to this action. The splitting becomes

$$su(n, 1) = n \oplus h^+,$$

and we see that, on the Lie algebra level, the Heisenberg translations N provide a natural Cartan connection satisfying the structural equations.

A third way of defining local invariants for the boundary of a strictly pseudoconvex domain D involves the Monge-Ampère equation; this is the subject of the next section.

5. The Poincaré metric and the complex Monge-Ampère equation. A *Poincaré metric* for a strictly pseudoconvex domain D is a Hermitian metric

$$ds^2 = \sum_{j,k} \frac{\partial^2 \log \psi}{\partial z_j \partial \bar{z}_k} dz_j \, d\bar{z}_k$$

with constant negative curvature

$$\text{Ric}_{j\bar{k}} = -Cg_{j\bar{k}}.$$

The existence of a Poincaré metric can be reduced to finding a solution u of the *complex Monge-Ampère equation*

$$(4) \qquad \text{Det}(\partial\bar{\partial} \log u) = cu^{-(n+1)} \, dV$$

$$= \text{volume form for the Poincaré metric,}$$

where dV is the form

$$dV = dz_1 \wedge \cdots \wedge dz_n \wedge d\bar{z}_1 \wedge \cdots \wedge d\bar{z}_n.$$

For example the solution of the equation for the unit ball is $u(z) = 1 - |z|^2$. In general solving the equation consists of two parts.

The first part is the formal study of the equation

$$u \sim \phi + O\big([\delta(z)]^n\big),$$

where ϕ is smooth and calculated from the Taylor expansion of the boundary, and $\delta(z) = \text{dist}(z, \partial D)$. We find

$$u \sim \phi + \sum_{\substack{k,l \\ k \geqslant (n+1)l}} \phi_{kl}\delta^k(z) \log \delta^l(z).$$

The second part is an existence theorem.

THEOREM (CHENG-YAU). *There exists a unique solution u of (4) with $u \in C^\infty(D)$ and $u \in C^{n+1/2-\varepsilon}(\overline{D})$.*

Returning to the example of the unit ball for a moment, we see that along with the Poincaré metric, we have the related metric on $\mathbf{C}^{n+1}$ given by

$$ds^2 = |d\zeta_0|^2 - \sum_{1 \leqslant k \leqslant n} |d\zeta_k|^2,$$

where $z = \zeta^1/\zeta_0$ and $\zeta' = (\zeta_1,\ldots,\zeta_n)$. If u is a solution of the Monge-Ampère equation (4), then the analogous metric on $D \times \mathbf{C}^\times$ is given by

$$ds^2 = \sum_{j,k>0} \frac{\partial^2}{\partial z_j \partial \bar{z}_k} \big\{|z_0|^2 u(z)\big\} \, dz_j \, d\bar{z}_k.$$

Although the metric ds^2 is nondegenerate, it is indefinite; one finds that the restriction $\widetilde{ds}^2$ of ds^2 to $\partial D \times \mathbf{C}^\times$ is degenerate. We can define a nondegenerate metric $\widetilde{\widetilde{ds}}^2$

$$\widetilde{ds}^2 = \zeta^2\big(\widetilde{\widetilde{ds}}^2\big),$$

where $(z, \theta, \zeta) \in \partial D \times S^1 \times \mathbf{R}^\times$ and $z_0 = \zeta e^{i\theta}$. In fact $\widetilde{ds}^2$ is a Lorentz metric on $\partial D \times S^1$ and the conformal class of this metric is a local invariant of ∂D.

Rays of light are invariant curves on $\partial D \times S^1$ and their projections onto ∂D define the *chains* of the boundary. It follows that we can also characterize chains by generalizing the geodesic equation of a manifold. Recall that on a Riemannian manifold, geodesics are defined by the Hamiltonian

$$H = \frac{1}{2} \sum g^{jk}(x)\xi_j\xi_k, \qquad (g^{jk}) > 0.$$

Similarly we can define chains by the Hamiltonian

$$H = \sum a_{jk}(x)\xi_j\xi_k + \sum b_l(x)\xi_l + V(x), \qquad (a_{jk}) \geqslant 0.$$

Note that (a_{jk}) will now merely be positive semidefinite; this means that chains need not have good behavior globally, or even locally. For example chains can be constructed that spiral inward; on the other hand, Burns-Schneider proved the global theorem that there are pairs of points that cannot be connected by chains.

6. Refined linear analysis. In §3 we saw that the Bergman kernel for the unit ball is

$$K(z, w) = \frac{c_n}{\left(1 - z \cdot \overline{w}\right)^{n+1}}.$$

Note that

$$K(z, z) = \frac{c_n}{\left(1 - |z|^2\right)^{n+1}}.$$

In order to write an expansion of the Bergman kernel in the general case, we will need the

LEMMA. *If $\psi(z) \in C^\infty$, then there exists an extension $\psi(z, w)$ satisfying*
   (i) $\overline{\partial}_z \psi|_{z=w} = 0$ *to infinite order,*
   (ii) $\partial_w \psi|_{z=w} = 0$ *to infinite order,*
(iii) $\psi(z, z) = \psi(z)$.

Now let $D = \{\psi(z) > 0\} \subset \mathbf{C}^n$ be a strictly pseudoconvex domain.

THEOREM. *The Bergman kernel has the expansion*
   (a) $K_D(z, z) = \phi(z)/\psi^{n+1}(z) + \tilde{\phi}(z)\log \psi(z),$
where ϕ, $\tilde{\phi}$ are smooth and ϕ is nonvanishing;
   (b) $K_D(z, w)$ *is obtained from $K_D(z, z)$ by extending ϕ, $\tilde{\phi}$, ψ as in the lemma above.*

Note the analogy between this expansion and the asymptotic expansion of the heat kernel on a Riemannian manifold,

$$K_t(z, z) \sim \frac{C_n}{t^{n/2}}\left\{1 + \sum_1^\infty \phi_k(z)t^k\right\}.$$

We can give a more precise description of the Bergman kernel using the Monge-Ampère equation. Again, a clue is provided by the unit ball, where the kernel takes the form

$$K(z, z) = \frac{C_n}{u^{n+1}(z)},$$

for $u(z)$ a solution of the Monge-Ampère equation (4). This suggests that we compare $K_D(z, z)$ and $C_n/u^{n+1}(z)$ for a general D. If D is locally in Moser normal form

$$\operatorname{Im} z_n > |z'|^2 + \sum_{p,q,r,s < n} A_{pq\overline{rs}} z_p z_q \overline{z}_r \overline{z}_s + \text{(other terms)},$$

then by a calculation of Christoffers

$$K(z_t, z_t) = \frac{C_n}{t^{n+1}}\left\{1 + \gamma_n \sum_{p,q,r,s < n} |A_{pq\overline{rs}}|^2 t^2 + O(t^3)\right\},$$

$$\frac{C_n}{u^{n+1}(z_t)} = \frac{C_n}{t^{n+1}}\left\{1 + \gamma_n' \sum_{p,q,r,s < n} |A_{pq\overline{rs}}|^2 t^2 + O(t^3)\right\},$$

where $z_t = (0, 0, \ldots, it)$ and $\gamma'_n \neq \gamma_n$. We can write this invariantly by introducing the metric

$$ds^2 = \partial\bar{\partial}\big(|z_0|^2 u(z)\big) = \partial\bar{\partial}(U).$$

7. Reflections. Let D be a strictly pseudoconvex domain with real-analytic boundary, say $D = \{r(z) > 0\}$, $r \in C^\omega$. Let $r(z, w)$ be the power series obtained from that of $r(z)$ by replacing each $\bar{z}$ by $\bar{w}$. For instance, the defining function for an ellipsoid is

$$r(z) = \sum_k \big(\lambda_k z_k^2 + \lambda_k \bar{z}_k^2 + z_k \bar{z}_k\big),$$

so

$$r(z, w) = \sum_k \big(\lambda_k z_k^2 + \lambda_k \bar{w}_k^2 + z_k \bar{w}_k\big).$$

In general, $r(z)$ is defined up to a nonvanishing real-analytic factor; hence, so is $r(z, w)$. It follows that the set

$$\{(z, w) \mid r(z, w) = 0\}$$

is invariantly associated to ∂D, i.e. it is preserved under biholomorphic mappings of domains with real-analytic boundaries.

For domains in $\mathbf{C}^1$, this amount to the Schwarz reflection principle: we reflect $z \in D$ across the boundary to w which solves $r(z, w) = 0$. In more than one dimension, we obtain a tool much more powerful than Schwarz reflection. For, the family of codimension one varieties $V_w = \{z \in \mathbf{C}^n \mid r(z, w) = 0\}$, $w \in \mathbf{C}^n$, must be preserved by biholomorphic maps. Thus, for instance, an analytic automorphism of an ellipsoid necessarily carries a variety $\{z \mid \sum_k \lambda_k z_k^2 + \lambda_k \bar{\alpha}_k^2 + z_k \bar{\alpha}_k = 0\}$ to a variety $\{z \mid \sum_k \lambda_k z_k^2 + \lambda_k \bar{\beta}_k^2 + z_k \bar{\beta}_k = 0\}$, which strongly restricts how the automorphism can look.

As a simple application of the method of reflections, we shall prove in Chapter 8 that if $\partial D = \{r(z) = 0\}$, $\partial \tilde{D} = \{\tilde{r}(z) = 0\}$ and r, $\tilde{r}$ are polynomials on $\mathbf{C}^n$ ($n > 1$), then any biholomorphic map Φ from D to $\tilde{D}$ is algebraic. To get an idea of the proof, we can show already that Φ is algebraic on each V_w.

In fact, for $z \in V_w$ we have also $w \in V_z$ (because $r(z)$ is real, hence $r(z, w) = \overline{r(w, z)}$). For typical $z \in V_w$, V_z will be nonsingular at w, so that we can associate to each $z \in V_w$ the hyperplane $T(z) =$ tangent space to V_z at w. All of this may be written down explicitly in terms of the polynomial $r(z, w)$, so that $z \to T(z)$ and its inverse are easily seen to be algebraic. (Here $T(z)$ is regarded as a point in the projective space of hyperplanes through w.) If $\Phi: D \to \tilde{D}$ and $\Phi(w) = \tilde{w}$ then we have the natural maps

$$
\begin{array}{ccc}
V_w \ni z & \overset{\text{Algebraic}}{\underset{\text{map}}{\to}} & T(z) \\[1em]
\downarrow \Phi & & \downarrow D_\Phi \\[1em]
\tilde{V}_{\tilde{w}} \ni \tilde{z} & \overset{\text{Algebraic}}{\underset{\text{map}}{\to}} & \tilde{T}(z)
\end{array}
$$

where D_Φ carries a hyperplane H of tangent vectors at w to the image of H under the differential $\Phi'(w)$. One checks that D_Φ is a projective transformation, and, in particular, algebraic. The diagram commutes, so $\Phi|_{V_w}$ must be algebraic since the other maps are all algebraic. If we are in $\mathbf{C}^1$, then we have shown that Φ is algebraic when restricted to a single point, so we have gained nothing. But in $\mathbf{C}^n$ $(n > 1)$ we know that Φ is algebraic on a codimension one variety, and a single additional trick completes the proof that Φ is algebraic. A harder application of the method of reflection is Webster's classification of ellipsoids: in $\mathbf{C}^n$ $(n > 1)$ two ellipsoids are biholomorphically equivalent only if they are already equivalent by a (complex) linear transformation of $\mathbf{C}^n$. An ellipsoid other than the ball has no biholomorphic self-maps other than linear transformations. See Chapter 8.

All of this depends on the implicit assumption that a biholomorphic map of two C^ω strictly pseudoconvex domains continues analytically past the boundary. This was originally proved by Chern-Moser theory, but Hans Lewy gave a simple proof based on the method of reflection. The ideas were extended by Nirenberg-Webster-Yang to prove C^∞ regularity of biholomorphic maps of strictly pseudoconvex domains. More recently, a very simple proof of C^∞ regularity was discovered by Bell and Ligocka. Their ideas opened up the possibility of dropping the assumption of strict pseudoconvexity.

Finally, we refer to the book by Krantz [**41**] for background material in several complex variables as well as further discussion of some of the topics mentioned here.

CHAPTER 7. ANALYSIS ON THE SIEGEL DOMAIN AND ITS BOUNDARY, THE HEISENBERG GROUP

1. Solution to $\bar{\partial}_b$, $\Box_b$ on the Heisenberg group (Folland-Stein [25]). Recall that the Heisenberg group H^n is the boundary of the Siegel domain

$$\left\{ \operatorname{Im} z_{n+1} > |z_1|^2 + \cdots + |z_n|^2 \right\} \subset \mathbf{C}^{n+1},$$

with the group operation

$$(\zeta, t) \cdot (\zeta', t') = \left(\zeta + \zeta', t + t' + 2 \operatorname{Im} \zeta \cdot \bar{\zeta}' \right),$$

where the coordinates (ζ, t) are defined by $\zeta = (z_1, \ldots, z_n) = x + iy \in \mathbf{C}^n$ and $t = \operatorname{Re} z_{n+1}$. The Lie algebra is spanned by the left-invariant vector fields

$$X_j = \frac{\partial}{\partial x_j} + 2 y_j \frac{\partial}{\partial t}, \qquad Y_j = \frac{\partial}{\partial y_j} - 2 x_j \frac{\partial}{\partial t}, \qquad T = \frac{\partial}{\partial t}, \qquad j = 1, \ldots, n,$$

which satisfy the commutation relations

$$\left[Y_j, X_k \right] = 4 \delta_{jk} T, \qquad \left[X_j, X_k \right] = \left[Y_j, Y_k \right] = \left[X_j, T \right] = \left[Y_j, T \right] = 0.$$

The Haar measure on the group is given by Lebesgue measure $d\operatorname{vol} = dx_1\, dy_1 \cdots dx_n\, dy_n\, dt$.

Define the $(0, 1)$-forms $\bar{\omega}^j$ as the duals of the vector fields $\bar{Z}_j$, where

$$Z_j = \tfrac{1}{2}(X_j - iY_j), \qquad \bar{Z}_j = \tfrac{1}{2}(X_j + iY_j),$$

and choose a metric so that

$$\langle \overline{\omega}^j, \overline{\omega}^k \rangle = \delta_{jk}.$$

We can now define $\overline{\partial}_b$ acting on functions f and $(0, 1)$-forms $\Sigma \, \phi_j \overline{\omega}^j$:

$$\overline{\partial}_b f = \sum_{j=1}^{n} \left(\overline{Z}_j f \right) \overline{\omega}^j, \qquad \overline{\partial}_b \left(\sum_{j=1}^{n} \phi_j \overline{\omega}^j \right) = \sum_{j=1}^{n} \left(\overline{Z}_j \phi_k - \overline{Z}_k \phi_j \right) \overline{\omega}^j \wedge \overline{\omega}^k.$$

For a general Cauchy-Riemann manifold, there would be other terms involving $[\overline{Z}_j, \overline{Z}_k]$. A calculation shows that the operator $\Box_b = \overline{\partial}_b \overline{\partial}_b^* + \overline{\partial}_b^* \overline{\partial}_b$ acts on $(0, 1)$-forms by

$$\Box_b \left(\sum_{j=1}^{n} f_j \overline{\omega}^j \right) = \sum_{j=1}^{n} \left(\mathcal{L}_\alpha f_j \right) \overline{\omega}^j,$$

where

$$\mathcal{L}_\alpha = -\frac{1}{2} \sum_{j=1}^{n} \left(Z_j \overline{Z}_j + \overline{Z}_j Z_j \right) + i\alpha T \quad \text{and} \quad \alpha = n - 2.$$

In other words $\Box_b$ restricted to $(0, 1)$-forms is "diagonal" and is determined by an operator $\mathcal{L}_\alpha$ that depends only on the dimension n.

In order to solve $\overline{\partial}_b$ and $\Box_b$, we will construct a fundamental solution to $\mathcal{L}_\alpha$ for $\alpha \in \mathbf{C}$ except for a dicrete set of bad α; that is, an operator with convolution kernel K_α on H^n satisfying $\mathcal{L}_\alpha K_\alpha = \delta_0$. In the coordinates (ζ, t), the operator $\mathcal{L}_\alpha$ is invariant under rotations in ζ and dilations of the form $(r\zeta, r^2 t)$. These properties imply that the kernel K_α satisfies

$$K_\alpha = \frac{\Omega(u)}{\left(|\zeta|^4 + |t|^2 \right)^{\text{power}}},$$

where $u = |\zeta|^2 + it$ and $\Omega(u)$ is homogeneous of degree 0 with respect to the standard dilations. In other words, with $u = re^{i\theta}$, Ω can be regarded as a function of θ alone. Because $\mathcal{L}_\alpha K_\alpha = 0$ away from the origin, the function $\Omega(\theta)$ satisfies a differential equation; we readily find that

$$\Omega(e^{i\theta}) = e^{ik\theta},$$

where k depends upon α.

THEOREM 1 (FOLLAND-STEIN). *The kernel*

$$K_\alpha(\zeta, t) = \frac{c_\alpha}{\left(|\zeta|^2 - it \right)^{(n+\alpha)/2} \left(|\zeta|^2 + it \right)^{(n-\alpha)/2}}$$

satisfies $\mathcal{L}_\alpha K_\alpha = \delta_0$ *and therefore can be used to invert*

$$\Box_b \left(\sum_{j=1}^{n} f_j \overline{\omega}^j \right) = \sum_{j=1}^{n} \left(\mathcal{L}_\alpha f_j \right) \overline{\omega}^j.$$

PROOF. We will not compute the constant c_α. Observe that $\mathcal{L}_\alpha K_\alpha$ is supported at the origin and is a distribution with the same homogeneity as that of the δ

function. This means that either $\mathcal{L}_\alpha K_\alpha$ is a multiple of the δ function or is zero. That is, $\mathcal{L}_\alpha K_\alpha = \gamma(\alpha)\delta_0$, and the only question is whether $\gamma(\alpha) = 0$. To evaluate $\gamma(\alpha)$, we test the distribution $\mathcal{L}_\alpha K_\alpha$ against a smooth function

$$\phi(\zeta, t) = \begin{cases} 1 & \text{for } |\zeta|^2 + |t|^2 \leq 1 \\ 0 & \text{for } |\zeta|^2 + |t|^2 \geq 2 \end{cases}.$$

Evidently, $\gamma(\alpha) = \langle \mathcal{L}_\alpha K_\alpha, \phi \rangle = \lim_{\varepsilon \to 0+} \langle \mathcal{L}_\alpha K_\alpha^\varepsilon, \phi \rangle$, where

$$K_\alpha^\varepsilon(\zeta, t) = \frac{c_\alpha}{\left(|\zeta|^2 - it + \varepsilon\right)^{(n+\alpha)/2}\left(|\zeta|^2 + it + \varepsilon\right)^{(n-\alpha)/2}}.$$

(Here we use $K_\alpha^\varepsilon \to K_\alpha$ as distributions.) Now $\int_{|\zeta|^2 + t^2 \geq 1} |\mathcal{L}_\alpha K_\alpha^\varepsilon(\zeta, t)|\, d\zeta\, dt \to 0$ as $\varepsilon \to 0+$, so

$$\gamma(\alpha) = \lim_{\varepsilon \to 0+} \langle \mathcal{L}_\alpha K_\alpha^\varepsilon, \phi \rangle = \lim_{\varepsilon \to 0+} \int_{H^n} \mathcal{L}_\alpha K_\alpha^\varepsilon(\zeta, t)\, d\zeta\, dt.$$

However, for different $\varepsilon > 0$, the kernels $\mathcal{L}_\alpha K_\alpha^\varepsilon$ are dilates of one another, and it follows that the integral on the right is independent of ε. Taking $\varepsilon = 1$, we obtain

$$\gamma(\alpha) = \int_{H^n} \mathcal{L}_\alpha K_\alpha^1(\zeta, t)\, d\zeta\, dt,$$

which can be evaluated in terms of gamma functions. The result is $\gamma(\alpha) \neq 0$ as long as $\pm\alpha \neq n, n + 2, n + 4, \dots$. Except for these discrete values of α, we obtain $1/(\gamma(\alpha))K_\alpha$ as the fundamental solution for $\mathcal{L}_\alpha$. $\square$

The idea of viewing the equations of complex analysis as the analogues of the Laplacian on the Heisenberg group is due to Stein. Theorem 1 is the simplest confirmation of this point of view.

2. $\square$ on the Siegel domain with $\bar{\partial}$-Neumann boundary conditions (Phong [53]). In this section we will consider $\square$ acting on $(0, 1)$-forms defined on the Siegel domain

$$D = \left\{ (z, z_{n+1}) \in \mathbf{C}^{n+1} \colon \operatorname{Im} z_{n+1} > |z|^2 \right\}.$$

The analysis of the operator $\square$ is difficult because of the presence of two types of homogeneity: the operator $\square$ is elliptic, while the nontrivial components of the boundary conditions have Heisenberg homogeneity. Because of this, the solution of $\square$ involves composing two kernels: one is Heisenberg group invariant on the boundary and the other is a kernel of isotropic homogeneity. To understand this complication we begin by splitting $\square$ into two pieces. To define this splitting we introduce the $(1, 0)$-forms

$$\omega^j = dz_j, \qquad j = 1, 2, \dots, n,$$

$$\omega^{n+1} = -\sqrt{2} \sum_{j=1}^n \bar{z}_j\, dz_j - \frac{i}{\sqrt{2}}\, dz_{n+1}$$

and choose a metric so that these forms are orthonormal. We can now define the dual frame

$$Z_j = \frac{\partial}{\partial z_j} + 2i\bar{z}_j \frac{\partial}{\partial z_{n+1}}, \qquad j = 1, 2, \ldots, n,$$

$$Z_{n+1} = i\sqrt{2}\, \frac{\partial}{\partial z_{n+1}}$$

and the vector field

$$T = \frac{1}{\sqrt{2}} \left(\bar{Z}_{n+1} - Z_{n+1} \right).$$

Recall that the domain of $\square$ restricted to $(0, 1)$-forms $\eta = \sum_{j=1}^{n+1} \phi_j \bar{\omega}^j$ is

$$D(\square) = \left\{ \eta : \phi_{n+1} = 0 \text{ on } \partial D;\ \bar{Z}_{n+1}\phi_j = 0 \text{ on } \partial D, j = 1, \ldots, n \right\},$$

and the action of $\square$ on $(0, 1)$-forms can be written

$$\square(\eta) = \sum_{j=1}^{n} \left(\square^+ \phi_j \right) \bar{\omega}^j + \left(\square^\# \phi_{n+1} \right) \bar{\omega}^{n+1},$$

where

$$\text{(1)} \qquad \square^+ = \square_b - Z_{n+1}\bar{Z}_{n+1}, \qquad \square_b = -\frac{1}{2} \sum_{k=1}^{n} \left(Z_k \bar{Z}_k + \bar{Z}_k Z_k \right) - i(n-2)T,$$

$$\square^\# = \square_b^* - Z_{n+1}\bar{Z}_{n+1}, \qquad \square_b^\# = \square_b - 2T.$$

The calculations below are simpler if we use the coordinates (z, t, ζ), where

$$\zeta = \operatorname{Im} z_{n+1} - |z|^2, \qquad t = \operatorname{Re} z_{n+1},$$

for $(z, z_{n+1}) \in D$, and solve the problem on surfaces $\{\zeta = \text{constant}\}$. For these coordinates,

$$Z_j = \frac{\partial}{\partial z_j} + i\bar{z}_j \frac{\partial}{\partial t}, \qquad j = 1, 2, \ldots, n,$$

$$Z_{n+1} = i\frac{1}{\sqrt{2}} \left(\frac{\partial}{\partial t} - i\frac{\partial}{\partial \zeta} \right), \qquad T = -i\frac{\partial}{\partial t}.$$

Since the operators $Z_1, \ldots, Z_n$, T do not contain $\partial/\partial \zeta$, they are tangential to the surfaces $\{\zeta = \text{constant}\}$.

Consider the equation

$$\text{(2.i)} \qquad \square \eta = \square \left(\sum_{j=1}^{n+1} \phi_j \bar{\omega}^j \right) = \left(\square^+ \oplus \square^\# \right) \left(\sum_{j=1}^{n+1} \phi_j \bar{\omega}^j \right) = \sum_{j=1}^{n+1} \psi_j \bar{\omega}^j.$$

This equation breaks into two parts

$$\text{(2.ii)} \qquad \begin{cases} \square^+ \phi_j = \psi_j & \text{in } D, \\ \bar{Z}_{n+1}\phi_j = 0 & \text{on } \partial D, j = 1, \ldots, n, \end{cases}$$

$$\text{(2.iii)} \qquad \begin{cases} \square^{\#}\phi_{n+1} = \psi_{n+1} & \text{in } D, \\ \phi_{n+1} = 0 & \text{on } \partial D. \end{cases}$$

The $\bar{\partial}$-*Neumann problem* is to find a solution of the system (2.i) or equivalently of the system (2.ii) and (2.iii). We can now state more precisely the goal of this chapter: it is to find an explicit approximate formula for the kernel associated with equation (2.i). We will see shortly that the kernel $G^{\#}$ associated with (2.iii) is easy to find. The bulk of the remainder of this chapter is devoted to finding an approximate formula for the kernel K associated with (2.ii)—for the precise statement see Theorem 2 at the beginning of §3.

Recall from §7 of Chapter 3 that when L is elliptic, the problem

$$\begin{aligned} Lu &= f \quad \text{in } D, \\ u &= g \quad \text{on } \partial D \end{aligned}$$

can be solved by transforming it into an equation for a pseudodifferential operator on ∂D; a Green's function then gives the solution u. Applying this to the boundary value problem (2.iii) gives

$$\phi_{n+1} = G^{\#}\psi_{n+1},$$

where $G^{\#}$ is the appropriate Green's function.

It is not as easy to find the kernel associated with (2.ii); i.e., the solution K of

$$\text{(3)} \qquad \begin{aligned} \square^{+} K &= \delta_p \quad \text{in } D, \\ \bar{Z}_{n+1} K &= 0 \quad \text{on } \partial D. \end{aligned}$$

We do this in several steps.

Step 1. We derive two 2nd-order equations (A, B) satisfied by K.

Step 2. We note by the symmetries of the problem that K is a function ϕ of four auxiliary variables.

Step 3. We derive a first order equation for ϕ.

Step 4. We analyze the 1st-order equation for ϕ.

Step 4 is the subject of §3.

Step 1. Since $\bar{Z}_{n+1}$ commutes with $T, Z_j, \bar{Z}_j, j = 1, \ldots, n$, equation (3) yields

$$\begin{aligned} \bar{Z}_{n+1}(\square^{+} K) &= \square^{+}\left(\bar{Z}_{n+1}K\right) = \bar{Z}_{n+1}\delta_p \quad \text{in } D, \\ \left(\bar{Z}_{n+1}K\right) &= 0 \qquad\qquad\qquad\qquad\quad \text{on } \partial D. \end{aligned}$$

This is an elliptic boundary value problem for $\bar{Z}_{n+1}K$ with Dirichlet boundary conditions. Applying the elliptic theory of §7 of Chapter 3 again, we find

$$\bar{Z}_{n+1}K = G^{+}\left(\bar{Z}_{n+1}\delta_p\right),$$

for an appropriate Green's function G^{+}. Using the definition of $\square^{+}$ from (1) gives equation (A),

$$\text{(A)} \qquad \square_b K = \delta_p + Z_{n+1}G^{+}\bar{Z}_{n+1}\delta_p.$$

We now derive the second equation (equation B) satisfied by K. Let H be the Bergman projection operator. In Chapter 6 we saw that if $\alpha = \bar{\partial}f$, then the solution of $\bar{\partial}u = \alpha$ that is perpendicular to the space of holomorphic functions is given by either of the expressions

$$u = (I - H)f = (\bar{\partial}^* \Box^{-1} \bar{\partial})f.$$

This gives us the identity

$$I - H = \bar{\partial}^* \Box^{-1} \bar{\partial}.$$

We apply this identity, component by component, to δ_p. Now

$$\bar{\partial}\delta_p = \sum_{j=1}^{n} \left(\bar{Z}_j \delta_p \right)\bar{\omega}^j + \left(\bar{Z}_{n+1}\delta_p \right)\bar{\omega}^{n+1},$$

$$\Box^{-1}\bar{\partial}\delta_p = \sum_{j=1}^{n} \left(K\bar{Z}_j \delta_p \right)\bar{\omega}^j + \left(G^\# \bar{Z}_{n+1}\delta_p \right)\bar{\omega}^{n+1}$$

and

$$\bar{\partial}^*\left(\sum_{j=1}^{n+1} \phi_j \bar{\omega}^j \right) = - \sum_{j=1}^{n+1} Z_j \phi_j$$

so that

$$(I - H)\delta_p = - \left(\sum_{j=1}^{n} Z_j K\bar{Z}_j \delta_p + Z_{n+1} G^\# \bar{Z}_{n+1}\delta_p \right),$$

or

$$\text{(B)} \qquad \sum_{j=1}^{n} Z_j K\bar{Z}_j \delta_p = - \delta_p - Z_{n+1} G^\# \bar{Z}_{n+1}\delta_p + H\delta_p,$$

which is equation (B), the second equation for the kernel K. In principle equation (B) is contained in equation (A), but in practice it provides additional information.

Step 2. Because of the symmetries of the problem it is reasonable to assume that the kernel K is invariant under Heisenberg translations and rotations in the first n coordinates and vanishes at infinity. These conditions mean that K must be of the form

$$K((z, t, \zeta), (w, s, \mu)) = \phi\big(|z - w|^2, t - s + 2\operatorname{Im}(z \cdot \bar{w}), \zeta, \mu\big),$$

for some function ϕ. We will treat ζ and μ as parameters and consider $\phi = \phi(v, \tau)$ as a function of

$$v = |z - w|^2, \qquad \tau = t - s - iz \cdot \bar{w} + iw \cdot \bar{z}.$$

Step 3. We begin by calculating how equations (A) and (B) restrict ϕ. Recall that if $Tf(z) = \int K(z, w)f(w)\, dw$ and X is a vector field with iX selfadjoint on the Heisenberg group, then

$$TXf(z) = \int K(z, w) X_w f(w)\, dw = - \int X_w K(z, w)f(w)\, dw.$$

We will use this identity below with $X = Z_j$, $\overline{Z}_j$, T. In equation (B) the term $K\overline{Z}_j$ occurs; applying the identity above with $X = \overline{Z}_j$, we find

$$K\overline{Z}_j = -\left(\frac{\partial}{\partial \overline{w}_j} - iw_j \frac{\partial}{\partial s}\right)\phi\left(|z - w|^2, t - s - iz \cdot \overline{w} + iw \cdot \overline{z}\right)$$

$$= (z_j - w_j)\phi_v + iz_j\phi_\tau - iw_j\phi_\tau = (z_j - w_j)\left(\frac{\partial}{\partial v} + i\frac{\partial}{\partial \tau}\right)\phi.$$

Similarly, we calculate

$$Z_j\psi = \left(\frac{\partial}{\partial z_j} + i\overline{z}_j \frac{\partial}{\partial t}\right)\psi\left(|z - w|^2, t - s + 2\operatorname{Im} z \cdot \overline{w}\right)$$

$$= \left(\overline{z_j - w_j}\right)\psi_v - i\overline{w}_j\psi_\tau + i\overline{z}_j\psi_\tau = \left(\overline{z_j - w_j}\right)\left\{\frac{\partial}{\partial v} + i\frac{\partial}{\partial \tau}\right\}\psi.$$

Combining these expressions produces

$$Z_j K\overline{Z}_j = \left\{\frac{\partial}{\partial v} + i\frac{\partial}{\partial \tau}\right\}\phi + |z_j - w_j|^2\left\{\frac{\partial}{\partial v} + i\frac{\partial}{\partial \tau}\right\}^2\phi$$

and so

$$\sum_{j=1}^n Z_j K\overline{Z}_j = n\left\{\frac{\partial}{\partial v} + i\frac{\partial}{\partial \tau}\right\}\phi + v\left\{\frac{\partial}{\partial v} + i\frac{\partial}{\partial \tau}\right\}^2\phi.$$

We can now write equation (B) as

$$(\text{B}') \quad v\left(\frac{\partial}{\partial v} + i\frac{\partial}{\partial \tau}\right)^2\phi + n\left(\frac{\partial}{\partial v} + i\frac{\partial}{\partial \tau}\right)\phi = -I - Z_{n+1}G^\#\overline{Z}_{n+1} + H.$$

Turning next to equation (A), we calculate

$$Z_j K = \left(\overline{z_j - w_j}\right)\left\{\frac{\partial}{\partial v} + i\frac{\partial}{\partial \tau}\right\}\phi, \qquad \overline{Z}_j K = (z_j - w_j)\left\{\frac{\partial}{\partial v} - i\frac{\partial}{\partial \tau}\right\}\phi,$$

so that

$$Z_j\overline{Z}_j K = Z_j(z_j - w_j)\left\{\frac{\partial}{\partial v} - i\frac{\partial}{\partial \tau}\right\}\phi$$

$$= \left\{\frac{\partial}{\partial v} - i\frac{\partial}{\partial \tau}\right\}\phi + |z_j - w_j|^2\left\{\frac{\partial}{\partial v} + i\frac{\partial}{\partial \tau}\right\}\left\{\frac{\partial}{\partial v} - i\frac{\partial}{\partial \tau}\right\}\phi$$

and after summing,

$$\sum_{j=1}^n Z_j\overline{Z}_j K = n\phi_v - in\phi_\tau + v\left\{\frac{\partial^2}{\partial v^2} + \frac{\partial^2}{\partial \tau^2}\right\}\phi.$$

A similar calculation shows that

$$\sum_{j=1}^n \overline{Z}_j Z_j K = n\phi_v + in\phi_\tau + v\left\{\frac{\partial^2}{\partial v^2} + \frac{\partial^2}{\partial \tau^2}\right\}\phi;$$

and therefore

$$-\frac{1}{2} \sum_{j=1}^{n} \left(Z_k \bar{Z}_k + \bar{Z}_k Z_k \right) K = -n\phi_v - v(\Delta_{v,\tau}\phi).$$

Since $-(n-2)TK = (n-2)i\phi_\tau$, we know

$$\Box_b K = -n\phi_v - v(\Delta_{v,\tau}\phi) + (n-2)i\phi_\tau;$$

and this allows us to write equation (A) as

$$\text{(A')} \qquad -v\left(\frac{\partial^2}{\partial v^2} + \frac{\partial^2}{\partial \tau^2} \right)\phi - n\frac{\partial\phi}{\partial v} + i(n-2)\frac{\partial\phi}{\partial \tau} = I + Z_{n+1}G^+ \bar{Z}_{n+1}.$$

Adding equations (A') and (B') yields

$$2iv\frac{\partial^2\phi}{\partial v\partial \tau} - 2v\frac{\partial^2}{\partial \tau^2}\phi + i(2n-2)\frac{\partial\phi}{\partial \tau} = H + Z_{n+1}(G^+ - G^{\#})\bar{Z}_{n+1},$$

and after integrating with respect to τ and dividing by $2i$, we get

$$\text{(4.i)} \quad v\left(\frac{\partial}{\partial v} + i\frac{\partial}{\partial \tau} \right)\phi + (n-1)\phi = \frac{1}{2i} \int \left(H + Z_{n+1}(G^+ - G^{\#})\bar{Z}_{n+1} \right) d\tau$$

$$= \alpha_n + \beta_n,$$

where

$$\text{(4.ii)} \qquad\qquad \alpha_n = \frac{1}{2i} \int Z_{n+1}(G^+ - G^{\#})\bar{Z}_{n+1}\, d\tau,$$

$$\text{(4.iii)} \qquad\qquad \beta_n = \frac{1}{2i} \int H\, d\tau.$$

This is our desired first-order equation for ϕ. Note that α_n has isotropic homogeneity and β_n has Heisenberg homogeneity.

3. Phong's theorem. We are trying to solve the boundary value problem (3):

$$\Box^+ K\big((z, z_{n+1}), (w, w_{n+1})\big) = \delta_{(w, w_{n+1})},$$

$$\bar{Z}_{n+1} K\big((z, z_{n+1}), (w, w_{n+1})\big) = 0 \quad \text{on } \partial D.$$

Using the coordinates

$$(z, z_{n+1}) \to (z, t, \zeta), \qquad t = \operatorname{Re} z_{n+1}, \qquad \zeta = \operatorname{Im} z_{n+1} - |z|^2,$$

$$(w, w_{n+1}) \to (w, s, \mu), \qquad s = \operatorname{Re} w_{n+1}, \qquad \mu = \operatorname{Im} w_{n+1} - |w|^2,$$

we found in the last section that

$$K = \phi\big(|z - w|^2, t - s + 2\operatorname{Im}(z \cdot \bar{w}), \zeta, \mu\big) \equiv \phi(v, \tau, \zeta, \mu),$$

where ϕ satisfies the first-order equation (4). In this section we will write down the first term of an expansion for K.

We will measure homogeneities by assigning the following weights:

	Weight 1	Weight 2
isotropic homogeneities:	τ, ζ, μ	v
Heisenberg homogeneities:	v, ζ, μ, τ	—

Now define

$$\Phi_- = 2v + |\zeta - \mu|^2 + \tau^2 \approx \text{dist}^2((z, z_{n+1}), (w, w_{n+1})),$$

$$\Phi_+ = 2v + |\zeta + \mu|^2 + \tau^2, \qquad \Psi = v + \zeta + \mu - i\tau.$$

By the table, Φ_+, Φ_- are of isotropic weight 2 and Ψ is of Heisenberg weight 1.

THEOREM 2 (PHONG). *The kernel K has the expansion*

$$K = K^0 + (\textit{terms with weaker singularities}),$$

where

$$K^0 = c_n \Phi_-^{-n} + \frac{\theta_n + \lambda_n}{v^{n-1}}$$

and the terms θ_n, λ_n satisfy (with P and Q polynomials)
  (i) $\theta_n = P(v, \tau, \zeta, \mu)/\Phi_+^{a_n}$ *has isotropic homogeneity* ≥ 2,
  (ii) $\lambda_n = Q(v, \tau, \zeta, \mu)/\Psi^{b_n}$ *has Heisenberg homogeneity* ≥ -2,
  (iii) $\theta_n + \lambda_n = 0$ *to order* $(n-1)$ *when* $v = 0$.

Moreover there is an algorithm for finding θ_n and λ_n for each dimension.

REMARK. As we have seen, the Neumann kernel $K(z, w)$ solves an elliptic PDE (well behaved under Euclidean dilations) under a boundary condition with a natural homogeneity under Heisenberg dilations. Phong's theorem explains the Heisenberg and Euclidean aspects of the kernel $K(z, w)$. In fact, K can be written as a sum $K = K_{\text{Euclidean}}(z, w) + K_{\text{Heisenberg}}(z, w)$ with each piece homogeneous under the appropriate dilations. The catch is that the two pieces $K_{\text{Euclidean}}(z, w)$ and $K_{\text{Heisenberg}}(z, w)$ have a spurious singularity which cancels when the two pieces are combined into the full kernel K. In order to write K without spurious singularities, one can state Phong's theorem in the form $K \sim \Sigma_l K_{\text{Euclidean}}^l \cdot K_{\text{Heisenberg}}^l$.

PROOF. We use the same notation and coordinates as in the last section. The equation for ϕ is

$$(5) \qquad \left[v\left(\frac{\partial}{\partial v} + i\frac{\partial}{\partial \tau} \right) + (n - 1) \right]\phi = \alpha_n + \beta_n,$$

where

$$\alpha_n = \frac{1}{2i} \int Z_{n+1}(G^+ - G^\#)\overline{Z}_{n+1}\, d\tau, \qquad \beta_n = \frac{1}{2i} \int H\, d\tau.$$

Recall that for the Siegel domain D the Bergman kernel is

$$H = c_n \left(\frac{z_{n+1} - \overline{w}_{n+1}}{2i} - z \cdot \overline{w} \right)^{-(n+2)};$$

therefore

$$\beta_n = c(\Psi)^{-(n+1)}.$$

To handle the other term, recall that

$$Z_{n+1} = i\frac{1}{\sqrt{2}} \left(\frac{\partial}{\partial t} - i\frac{\partial}{\partial \zeta} \right);$$

after computing G^+, $G^\#$ by the methods of Chapter 3, §7, we obtain

$$\alpha_n = \alpha_{n,1} + \alpha_{n,2} + (\text{terms of weaker homogeneity}),$$

where

$$\alpha_{n,1} = c(\Phi_-)^{-(n+1)}\big[(n-1)\Phi_- - 2n\nu\big],$$

$$\alpha_{n,2} = c(\overline{\Psi}_0)^2(\Phi_+)^{-(n+1)}, \qquad \Psi_0 = \zeta + \mu - i\tau.$$

Term	Isotropic Homogeneity	Heisenberg Homogeneity
α_n	yes	—
$\alpha_{n,1}; \alpha_{n,2}$	$-2n$	—
β_n	—	yes
Ψ_0	1	1

It is convenient to rewrite equation (5) in terms of $\tilde{\phi}$, where $\phi = \nu^{1-n}\tilde{\phi}$. We assume that $\tilde{\phi}$ vanishes to order $n-1$ at $\nu = 0$. Because

$$\nu\left(\frac{\partial}{\partial\nu} + i\frac{\partial}{\partial\tau}\right)(\nu^{1-n}\tilde{\phi}) = \nu^{2-n}\left(\frac{\partial}{\partial\nu} + i\frac{\partial}{\partial\tau}\right)\tilde{\phi} + (1-n)\nu^{1-n}\tilde{\phi},$$

we can write equation (5) as

$$\left[\nu\left(\frac{\partial}{\partial\nu} + i\frac{\partial}{\partial\tau}\right) + (n-1)\right]\phi = \left[\nu\left(\frac{\partial}{\partial\nu} + i\frac{\partial}{\partial\tau}\right) + (n-1)\right](\nu^{1-n}\tilde{\phi})$$

$$= \nu^{2-n}\left(\frac{\partial}{\partial\nu} + i\frac{\partial}{\partial\tau}\right)\tilde{\phi} = \alpha_n + \beta_n,$$

or

$$\left(\frac{\partial}{\partial\nu} + i\frac{\partial}{\partial\tau}\right)\tilde{\phi} = \nu^{n-2}(\alpha_n + \beta_n).$$

Neglecting terms of weaker homogeneity, this is

$$(6.\text{i}) \qquad \left(\frac{\partial}{\partial\nu} + i\frac{\partial}{\partial\tau}\right)\tilde{\phi} = \nu^{n-2}(\alpha_{n,1} + \alpha_{n,2} + \beta_n).$$

We now solve (6) by constructing solutions $\eta_n, \theta_n, \lambda_n$ for the three equations

$$(6.\text{ii}) \qquad \left(\frac{\partial}{\partial\nu} + i\frac{\partial}{\partial\tau}\right)\eta_n = \nu^{n-2}\alpha_{n,1} + (\text{weaker singularities}),$$

$$(6.\text{iii}) \qquad \left(\frac{\partial}{\partial\nu} + i\frac{\partial}{\partial\tau}\right)\theta_n = \nu^{n-2}\alpha_{n,2} + (\text{weaker singularities}),$$

$$(6.\text{iv}) \qquad \left(\frac{\partial}{\partial\nu} + i\frac{\partial}{\partial\tau}\right)\lambda_n = \nu^{n-2}\beta_n.$$

We begin with (6.ii). We claim that $c_n v^{n-1}\Phi_-^{-n}$ is a solution. This is verified by calculating

$$\left(\frac{\partial}{\partial v} + i\frac{\partial}{\partial \tau}\right)\left(c_n v^{n-1}\Phi_-^{-n}\right)$$

$$= (n-1)c_n v^{n-2}\Phi_-^{-n} - nc_n v^{n-1}\Phi_-^{-(n-1)} \cdot \left\{\left(\frac{\partial}{\partial v} + i\frac{\partial}{\partial \tau}\right)\Phi_-\right\}$$

and using the fact that

$$\left(\frac{\partial}{\partial v} + i\frac{\partial}{\partial \tau}\right)\Phi_- = \frac{\partial}{\partial v}\Phi_- + \text{(weaker singularities)}$$

$$= 2 + \text{(weaker singularities)}.$$

We turn next to the equation (6.iv):

$$\left(\frac{\partial}{\partial v} + i\frac{\partial}{\partial \tau}\right)\lambda_n = v^{n-2}\beta_n.$$

We will construct a solution λ_n by induction. If $n = 2$, the right-hand side is Ψ^{-3} and we can take $\lambda_2 = c\Psi^{-2}$. Since

$$\left(\frac{\partial}{\partial v} + i\frac{\partial}{\partial \tau}\right)\beta_n = c\beta_{n+1},$$

we can set

$$\lambda_{n+1} = c\left[(n-1)\lambda_n - v^{n-1}\beta_n\right],$$

so that

$$\left(\frac{\partial}{\partial v} + i\frac{\partial}{\partial \tau}\right)\lambda_{n+1} = c\left[(n-1)v^{n-2}\beta_n - (n-1)v^{n-2}\beta_n - v^{n-1}c\beta_{n+1}\right]$$

$$= cv^{n-1}\beta_{n+1},$$

as required.

This leaves equation (6.iii):

$$\left(\frac{\partial}{\partial v} + i\frac{\partial}{\partial \tau}\right)\theta_n = v^{n-2}\alpha_{n,2} + \text{(weaker singularities)}$$

or

$$(7) \qquad \frac{\partial}{\partial v}\theta_n = v^{n-2}\alpha_{n,2} + \text{(weaker singularities)}.$$

Again we will construct a solution θ_n using induction on the dimension n. Note that the theorem will be proved if

(i) $(\theta_n + \lambda_n)$ vanishes to order $(n-1)$ at $v = 0$.

(8) (ii) θ_n is a rational function $P(v, \tau, \zeta, \mu)/\Phi_+^{\text{power}}$, with all terms having isotropic homogeneity $\geqslant -2$.

Since $(\partial/\partial v)\theta_2 = \alpha_{2,2}$, we set

$$\theta_2 = c\left(\overline{\Psi}_0\right)^2(\Phi_+)^{-2}$$

and note that at $v = 0$, $\Phi_+ = \Psi_0\overline{\Psi}_0$ and $\Psi = \overline{\Psi}_0$, so that

$$\theta_2 + \lambda_2 = 0,$$

as required. Define

$$\theta_n' = c\left[(n-1)\theta_n - v^{n-1}\alpha_{n,2}\right].$$

It is easy to check that equation (7) and property (8.ii) are satisfied; we will modify it so that property (8.i) holds. Note, by the induction step,

$$\theta_{n+1}' + \lambda_{n+1} = c\left[(n-1)\{\theta_n + \lambda_n\} - v^{n-1}\{\alpha_{n,2} + \beta_n\}\right]$$

vanishes to order $n-1$; if we correct θ_{n+1}' with something of higher homogeneity then equation (7) and property (8.ii) will not be affected. Put

$$b_{n+1} = \left[\left(\frac{\partial}{\partial v} + i\frac{\partial}{\partial \tau}\right)^{n-1}(\theta_{n+1}' + \lambda_{n+1})\right]\Big|_{v=0},$$

which by (7) is

$$b_{n+1} = \left(\frac{\partial}{\partial v} + i\frac{\partial}{\partial \tau}\right)^{n-2}\left(v^{n-1}\alpha_{n+1,2} + (\text{weaker singularities}) + v^{n-1}\beta_{n+1}\right)\Big|_{v=0}$$

$$= \left(\frac{\partial}{\partial v} + i\frac{\partial}{\partial \tau}\right)^{n-2}$$

$$(\text{weaker singularities})\,\big|_{v=0}.$$

Since α_{n+1} has isotropic weight $-2(n+1)$, b_{n+1} has isotropic weight

$$\text{weight} > 2(n-1) - 2(n+1) - 2(n+2);$$

in other words, b_{n+1} has weight $\geqslant -2n+1$. We now define the solution θ_{n+1} by

$$\theta_{n+1} = \theta_{n+1}' + cv^{n-1}b_{n+1}^*,$$

where b_{n+1}^* is obtained from b_{n+1} by appropriately adjusting the denominator to be a power of Φ_+ (this only gives weaker singularities). Equation (7) and property (8.ii) still hold for θ_{n+1}, since $v^{n-1}b_{n+1}^*$ has isotropic weight

$$\text{weight} \geqslant 2(n-1) - 2n + 1 = -1.$$

Finally,

$$\left(\frac{\partial}{\partial v} + i\frac{\partial}{\partial \tau}\right)^{n-1}\left(\tilde{c}v^{n-1}b_{n+1}^* + \theta_{n+1}' + \lambda_{n+1}\right)\Big|_{v=0} = 0,$$

by the definition of b_{n+2}. This verifies property (8.i) and completes the proof of the theorem. $\square$

4. Approximate solution to $\square_b$ on strictly pseudoconvex domains. In this section we explain how to pass from the exact solution of $\square_b$ on the Heisenberg group to a good approximate solution for a general strictly pseudoconvex domain D with boundary $M = \{r(z) = 0\} \subseteq \mathbf{C}^{n+1}$. Our remarks summarize crudely the constructions in Folland-Stein [25]. The process is analogous to solving variable-coefficient elliptic equations by freezing coefficients. There are three main steps.

(i)   Finding how closely the boundary M may be approximated by the Heisenberg group.

(ii)   Transplanting the fundamental solution of $\square_b$ from the Heisenberg group to M.

(iii)   Extracting useful information from the transplanted solution.

Now, the sharpest answer to (i) is to place M in Moser's normal form about a point $p \in M$. Actually, what we need for $\square_b$ is much simpler: After an analytic change of coordinates in a neighborhood of p, we may suppose $p = 0$ and M osculates the Heisenberg group $H^n = \partial(\text{Siegel domain}) \subseteq \mathbf{C}^{n+1}$ to third order at 0.

Since Chapter 9 discusses Moser's normal form, we omit here the construction of the change of coordinates. (What is involved is a calculation of Taylor series to order 3.)

Thus, there is a map $\Theta_p \colon M \to H^n$ which carries p to the origin and satisfies

(a) Θ_p is a local diffeomorphism.

(b) Θ_p is analytic (i.e., annhiliated by the $\bar{\partial}_b$-operator on M) to third order at p, when H^n is viewed as sitting inside $\mathbf{C}^{n+1}$.

Moreover, we may make Θ_p depend smoothly on p. We write $\Theta(x, y)$ for $\Theta_y(x)$. Thus $\Theta \colon M \times M \to H^n$. An important property of Θ is

(c) for x close to y in M, $\Theta(x, y)$ is very close to $[\Theta(y, x)]^{-1}$ in H^n.

EXAMPLE. If already $M = H^n$, then we can take $\Theta(x, y) = xy^{-1}$.

Next we carry out step (ii) by using Θ to transplant the fundamental solution of §1 from H^n to M. We start by picking a good metric on M. The natural compatibility condition is that the metric acting on tangent vectors of type $(1, 0)$ be proportional to the Levi form, i.e.,

$$\| Z \|^2 = (\text{scalar}) \cdot \sum_{jk} \frac{\partial^2 r}{\partial z_j \partial \bar{z}_k} \xi_j \bar{\xi}_k \quad \text{when } Z = \sum_{k=1}^{n+1} \xi_k \frac{\partial}{\partial z_k} \text{ is tangent to } M = \{r = 0\}.$$

Such a metric is called a *Levi metric*. Since the purpose of solving $\square_b$ is to solve $\bar{\partial}_b$ and thus construct analytic functions, we can restrict attention to Levi metrics without serious loss of generality.

Next we pick $(0, 1)$-forms $\bar{\omega}^1, \ldots, \bar{\omega}^n$ locally on M so that the $\bar{\omega}^k$ form an orthonormal basis for the $(0, 1)$-forms at each point of M. Here "orthonormal" refers to the Levi metric. The general $(0, 1)$-form is $u = \sum_k u_k \bar{\omega}^k$, and the $\square_b$-equation for $(0, 1)$-forms may be written as

$$(9) \qquad \square_b \left(\sum_k u_k \bar{\omega}^k \right) = \sum_k f_k \bar{\omega}^k.$$

Recall that the solution for $\square_b$ on the Heisenberg group is $u_k(x) = \int_{H^n} K_\alpha(xy^{-1}) f_k(y)\, dy$ for $\alpha = n - 2$. Now we can guess an approximate solution to (9), namely

$$(10) \qquad u_k^0(x) = \int_M K_\alpha(\Theta(x, y)) f_k(y)\, d\mu(y)$$

for a suitable smooth measure μ on M. Step (ii) is complete.

Finally, we want to make use of our approximate solution. The obvious starting point is to apply $\Box_b$ to (10); one finds that $\Box_b u^0 = f + \varepsilon f$, where $\varepsilon f = \Sigma_k (\varepsilon_k f)\overline{\omega}^k$ and $\varepsilon_k f(x)$ is a sum of expressions of the form

$$(11) \qquad \int_M G(x, y) K'(\Theta(x, y)) f_l(y)\, d\mu(y)$$

with $G \in C^\infty(M \times M)$ and K' homogeneous on the Heisenberg group with a singularity weaker than δ_0, the delta mass at the origin. Now, εf may be regarded as a small perturbation of the identity, because $f = \int_M \delta_0(\Theta(x, y)) f(y)\, dy$ and δ_0 is homogeneous on H^n. Thus, the integral kernel in (11) has a weaker singularity than the identity.

There are two natural strategies to exploit this. We can either use successive approximations starting with (10) to build ever more accurate approximate solutions of $\Box_b u = f$, or we can define the analogues of Sobolev spaces for $\Box_b$ and prove that the integral operator (11) carries H^s to $H^{s'}$ ($s' > s$). Either approach yields C^∞ regularity of solutions of $\Box_b$ and much more. The result of calculations with successive approximations is as follows.

THEOREM 3. $\Box_b(\Sigma_k u_k \overline{\omega}^k) = (\Sigma_k f_k \overline{\omega}^k)$ *may be solved modulo* C^∞-*smoothing errors by* $u_j(x) = \Sigma_k \int_M K_{jk}(x, y) f_k(y)\, d\mu(y)$ *with* $K_{jk}(x, y) = \delta_{jk} K_\alpha(\Theta(x, y)) +$ *an asymptotic series of weaker singularities.*

The natural Sobolev spaces for $\Box_b$ on the Heisenberg group are

$$H^s_{\text{Heisenberg}} = \left\{ f \in L^2(H^n) \;\middle|\; \left(\prod_k X_k^{\mu_k}\right)\left(\prod_k Y_k^{\nu_k}\right) T^\alpha f \in L^2 \right.$$

$$\left. \text{whenever } \sum_k \mu_k + \sum_k \nu_k + 2\alpha \leqslant s \right\}$$

for s a positive integer. For s not a positive integer the definition is harder and will not be given here. Note that these Sobolev spaces are natural for the non-Euclidean dilations on the Heisenberg group.

On a strictly pseudoconvex boundary M, we define H^s_M for s a positive integer to consist of all $f \in L^2(M)$ such that $Z_1 Z_2 \cdots Z_m \overline{W}_1 \overline{W}_2 \cdots \overline{W}_l T_1 \cdots T_q f \in L^2$ whenever the Z's and W's are of type $(1, 0)$, the T's are arbitrarily smooth vector fields, and $m + l + 2q \leqslant s$. Again there is a natural extension to s not a positive integer.

THEOREM 4. *The mapping* $f \to \int_M K_\alpha(\Theta(x, y)) f(y)\, d\mu(y)$ *carries* H^s_M *to* H^{s+2}_M, *while* $f \to \varepsilon f$ *carries* H^s_M *to* H^{s+1}_M.

COROLLARY. $\Box_b u = f \in H^s_M$ *implies* $u \in H^{s+2}_M$.

See Folland-Stein [**25**] for the proofs.

We should note that Phong [**53**] gave an approximate solution of the Neumann equation $\Box u = \omega$ for $\overline{\partial}$ on general strictly pseudoconvex domains D. The transplantation from H^n to D follows the same philosophy as for $\Box_b$, but this time is harder technically because of the presence of the boundary and the two interacting homogeneities on D. As in the Heisenberg case, the integral kernel for Phong's approximate solution of $\Box u = \omega$ is a sum of products of purely "Euclidean" and purely "Heisenberg" kernels. The approximate solution u^0 satisfies the boundary conditions exactly, and $\Box u^0 = \omega + \varepsilon\omega$ where ε has a weak singularity.

So far we have described how to use the Heisenberg group for strictly pseudoconvex domains. However, we left out a very important ingredient in the application of nilpotent Lie groups to a broader class of problem: the idea of "lifting" (Rothschild-Stein [54]), which we now briefly describe.

The key step in solving $\Box_b u = f$ on a strictly pseudoconvex boundary was to approximate $\Box_b$ by a translation-invariant operator on the Heisenberg group. For more general differential equations, this idea apparently cannot be applied. Consider the Grusin operator $L = (\partial/\partial x)^2 + (x\partial/\partial y)^2$ on $\mathbf{R}^2$; this is elliptic at $x \neq 0$ and degenerate at $x = 0$. On the other hand, a translation-invariant operator on a Lie group looks the same at all points and so cannot provide a good approximation to L. The way out of this dilemma is to introduce a new variable z and study the "lifted" operator $\tilde{L} = (\partial/\partial x)^2 + (x\partial/\partial y + \partial/\partial z)^2$ on functions on $\mathbf{R}^3$. The new operator $\tilde{L}$ behaves the same at all points; in fact, after a change of coordinates, one can realize $\tilde{L}$ as a translation-invariant operator on the Heisenberg group. Now, $\tilde{L}$ and L are equal when they act on functions independent of z. So any regularity theorem for $\tilde{L}$ automatically yields a regularity theorem for L. Also, one can build a fundamental solution for L by starting with the (essentially known) fundamental solution for $\tilde{L}$ and integrating out the extra variables z.

A similar construction can be carried out for any Hörmander operator [32] $L = \Sigma_j X_j^2 + X_0$, where the X_j's are noncommuting vector fields. At any point p we can introduce extra variables $z_1, z_2, \ldots, z_N$ and lift L to an operator $\tilde{L}$ which is well approximated about p by a translation-invariant operator $\tilde{L}_0$ on a nilpotent Lie group. From this follow sharp regularity theorems for L, and a parametrix for L may be written as the integral of a fundamental solution for $\tilde{L}_0$ over the z_k. A complete discussion of these matters appears in [54].

Before the work of Folland-Stein and Phong, the C^∞-regularity for $\bar{\partial}$, $\bar{\partial}_b$, $\Box$, $\Box_b$ on strictly pseudoconvex domains was established by Kohn [39] using energy estimates. More recently, by a powerful extension of his earlier techniques, Kohn [40] and his students Catlin [9] and D'Angelo [13] (see also Diederich-Pflug [14]) achieved a deep understanding of C^∞-regularity of $\bar{\partial}$-Neumann on weakly pseudoconvex domains. Roughly speaking, there is C^∞-regularity for the $\bar{\partial}$-Neumann problem for $(0,1)$-forms on D if and only if no complex-analytic curve sits inside the boundary of D. It would be a major achievement to write down the basic fundamental solutions for the Bergman or Szegö kernel for any nontrivial class of weakly pseudoconvex domains. Only a few simple examples are known.

We should mention also the important work of Trèves [62] and Tartakoff [60] on real-analyticity of solutions of $\bar{\partial}$. To understand this requires the real-analytic analogue of Fourier integral operators, which we have not treated here.

Finally, we refer the reader to Chapter 12, where the idea of transplantation recurs in connection with the Bergman and Szegö kernels.

CHAPTER 8. ELLIPSOIDS AND THE METHOD OF REFLECTIONS

1. Introduction. Let D, D' be two bounded strictly pseudoconvex domains in $\mathbf{C}^n$ with real-analytic boundaries, say

$$(1) \qquad D = \left\{ z \in \mathbf{C}^n \,\big|\, r(z) > 0 \right\}, \qquad D' = \left\{ z' \in \mathbf{C}^n \,\big|\, r'(z') > 0 \right\}$$

where r, r' are real-valued convergent power series in z, $\bar{z}$. We shall study here whether D and D' are biholomorphic. Recall this means $\Phi: D \to D'$ for some complex analytic one-to-one, onto map Φ. The known methods of classifying domains under biholomorphic maps start with the following result.

THEOREM 1. *If D, D' are strictly pseudoconvex with smooth boundaries, then any biholomorphic map $\Phi: D \to D'$ is smooth up to the boundary. If the boundaries ∂D, $\partial D'$ are real-analytic, then Φ continues analytically to a biholomorphic map of a neighborhood of $\overline{D}$ onto a neighborhood of $\overline{D'}$.*

A simple proof of Theorem 1 will be given in §4. Now in the real-analytic case we can regard Φ as defined in a neighborhood of D, and therefore

$$(2) \qquad \left\{ \begin{array}{c} r'(z') = \mu(z)r(z) \\ z' = \Phi(z) \end{array} \right\},$$

where $\mu(z)$ is a convergent power series in z, $\bar{z}$, and $\mu(z) \neq 0$ for z near D.

The idea of the method of reflections is to pass from $r(z)$ to the analytic function $r(z, \bar{w})$ of $z, \bar{w} \in \mathbf{C}^n$ such that $r(z, \bar{z}) = r(z)$. The power series for $r(z, \bar{w})$ is obtained from that of $r(z)$ just by substituting $\bar{w}$ in place of $\bar{z}$. In particular, $r(z, \bar{w})$ is uniquely determined, by $r(z)$. From (2) we obtain

$$(3) \qquad \left\{ \begin{array}{c} r'(z', \bar{w}') = \mu(z, \bar{w})r(z, \bar{w}) \\ z' = \Phi(z),\ w' = \Phi(w) \end{array} \right\},$$

where $\mu(z, \bar{w}) \neq 0$ for z, w near the same boundary point $p \in \partial D$. In particular $r'(z', \bar{w}') = 0$ if and only if $r(z, \bar{w}) = 0$ for $z' = \Phi(z)$, $w' = \Phi(w)$, z and w near $p \in \partial D$. To $w \in \mathbf{C}^n$ near p we associate the analytic variety $Q_w = \{ z \,|\, r(z, \bar{w}) = 0\}$. The family $\{Q_w \,|\, w \in \mathbf{C}^n\}$ of codimension-one varieties in $\mathbf{C}^n$ is thus invariantly associated to the domain D. That is, if $\Phi: D \to D'$ is biholomorphic and $\{Q_w\}$, $\{Q'_{w'}\}$ are the families of varieties associated to D, D', then $\Phi: Q_w \to Q'_{\Phi(w)}$. In $\mathbf{C}^n$ $(n > r)$, these subvarieties are a powerful tool to classify domains. It was S. Webster who realized how much could be done by this technique, which is called the method of reflections.

In §2 we prove

THEOREM 2 (WEBSTER). *Assume the functions $r(z)$, $r'(z')$ defining the domains D, $D' \subseteq \mathbf{C}^n$ in (1) are polynomials. If $n \geqslant 2$ and $\Phi \colon D \to D'$ is biholomorphic, then Φ is an algebraic map, i.e. its components are algebraic functions.*

Given specific D, D' it is hard to decide whether a biholomorphic map exists. In §3 we consider the classification of ellipsoids, i.e. domains in $\mathbf{C}^n$ ($n \geqslant 2$) of the form

$$(4) \qquad \sum_{j,k=1}^{n} \left(a_{jk} x_j x_k + b_{jk} y_j y_k + c_{jk} x_j y_k \right) < d,$$

where $z_j = x_j + \sqrt{-1}\, y_j$ and the quadratic form is positive. We prove

THEOREM 3 (WEBSTER). *If two ellipsoids D, $D' \subseteq \mathbf{C}^n$ ($n \geqslant 2$) are biholomorphic, then already they are equivalent by a linear transformation of $\mathbf{C}^n$. If D, D' are not biholomorphic to the unit ball, then any biholomorphic map $\Phi \colon D \to D'$ is linear.*

Theorems 2 and 3 will be proved by the method of reflections.

EXAMPLE. Consider the ellipsoid

$$\left\{ r(z) = |z|^2 + \sum_{j=1}^{n} A_j \left(z_j^2 + \bar{z}_j^2 \right) - 1 < 0 \right\}.$$

Then

$$r(z, \bar{w}) = z \cdot \bar{w} + \sum_{j=1}^{n} A_j \left(z_j^2 + \bar{w}_j^2 \right) - 1$$

and

$$w \mapsto Q_w = \left\{ z \in \mathbf{C}^n \,\Big|\, z \cdot \bar{w} + \sum_{j=1}^{n} A_j z_j^2 = 1 - \sum_{j=1}^{n} A_j \bar{w}_j^2 \right\}.$$

An analytic automorphism of this ellipsoid must carry each Q_w to some $Q_{w'}$. This places strong restrictions on the automorphism.

We conclude this section with a quick review of algebraic and rational maps. Recall that a (multiple-valued) analytic function $w = F(z_1 \cdots z_n)$ defined in the complement of a variety $V \subseteq \mathbf{C}^n$ is called *algebraic* if the polynomial equation $P(w, z_1, \ldots, z_n) = 0$ holds, whenever $w = F(z_1, \ldots, z_n)$, $z_1, \ldots, z_n \in \mathbf{C}^n \setminus V$. Thus w can be obtained from $(z_1, \ldots, z_n)$ by solving a polynomial equation. A map $\Phi \colon (z_1, \ldots, z_n) \to (w_1, \ldots, w_n)$ is called algebraic if each w_j is an algebraic function of $(z_1, \ldots, z_n)$. The composite of two algebraic maps is again algebraic; if Φ is algebraic and invertible, then Φ^{-1} is algebraic. We omit the proofs. A map $\Phi \colon (z_1, \ldots, z_n) \to (w_1, \ldots, w_n)$, single valued and well defined outside a subvariety $V \subseteq \mathbf{C}^n$ is called *rational* if each w_j is a rational function of $(z_1, \ldots, z_n)$. Φ is called *birational* if Φ and Φ^{-1} are both rational. The next proposition is needed in §3; we prove it here since it clarifies the relation between algebraic and rational maps.

PROPOSITION 1. *Let $F: \mathbf{C}^n \setminus V \to \mathbf{C}^1$ be a single-valued algebraic function. Then F is rational.*

PROOF. When $w = F(z)$, $z \in \mathbf{C}^n \setminus V$; we have

$$P(w, z) = \sum_{k=0}^{m} P_k(z)w^k \equiv 0,$$

by definition. (Note that $m \geq 1$, $P_m \not\equiv 0$.) This gives us a bound

$$|w| \leq C(1 + |z|)^M / P_m(z).$$

So $G(z) \equiv P_m(z)F(z)$ is a single-valued analytic function of polynomial growth on $\mathbf{C}^n \setminus V$. By the removable singularity theorem, G extends to an analytic function of polynomial growth on $\mathbf{C}^n$. So G must be a polynomial, and $F(z) = G(z)/P_m(z)$ is a rational function. $\square$

2. The proof of Webster's theorem on algebraic mappings. In this section we prove Theorem 2. We return to the notation of §1. Thus $\Phi: D \to D'$ is biholomorphic and extends to a neighborhood of D; we have the families $\{Q_w\}$, $\{Q'_{w'}\}$ associated to D, D', and we know that $\Phi: Q_w \to Q'_{w'}$ for $w' = \Phi(w)$, w near ∂D.

The first step in proving Theorem 2 is to show that the restriction of Φ to Q_w is algebraic. To see this, we begin with an observation: Since $r(z, \bar{z})$ is real, it follows by analytic continuation that $r(z, \bar{w}) = r(w, \bar{z})$. In particular $y \in Q_w$ means $r(y, \bar{w}) = 0$, hence $r(w, \bar{y}) = 0$ so that also $w \in Q_y$. Now to $y \in Q_w$ we associate $T_w(Q_y)$, the tangent space of Q_y at w. $T_w(Q_y)$ is a hyperplane in $\mathbf{C}^n$, given in coordinates $(p_1, \ldots, p_n)$ as $\{(\zeta_1, \ldots, \zeta_n) \in \mathbf{C}^n \mid \Sigma_1^n p_k \bar{\zeta}_k = 0\}$. Since $(p_1, \ldots, p_n)$ and $(\lambda p_1, \ldots, \lambda p_n)$ define the same hyperplane ($\lambda \neq 0$), we may regard $T_w(Q_y)$ as a point in complex projective space $\mathbf{CP}^{n-1}$. So we have a natural map $\sigma_w: Q_w \to \mathbf{CP}^{n-1}$.

PROPOSITION 2. *The map σ_w is a local diffeomorphism of Q_w and an algebraic map.*

PROOF. Immediately from the definition we get

$$\sigma_w(y) = (p_1, \ldots, p_n) \quad \text{with } p_k = \partial r(y, \bar{w})/\partial \bar{w}_k,$$

so that $\sigma_w: y \to (p_1, \ldots, p_n)$ is a polynomial map. To see that σ_w is a diffeomorphism locally near ∂D, we check the differential $\sigma'_w: T_y(Q_w) \to T(\mathbf{CP}^{n-1})$ at $y = w = z^0 \in \partial D$. We can pick coordinates in $\mathbf{C}^n$ so that

$$\frac{\partial r}{\partial z_k}(z^0, \overline{z^0}) = \begin{cases} 1 & \text{if } k = n \\ 0 & \text{if } k \neq n \end{cases}.$$

Then

$$T_{z^0}(Q_{z^0}) = \{(\zeta_1, \ldots, \zeta_n) \mid \zeta_n = 0\},$$

and

$$\sigma_{z^0}(z^0) = (p_1, \ldots, p_n) = (0, 0, \ldots, 0, 1) \in \mathbf{CP}^{n-1},$$

so that p_k/p_n $(k = 1,\ldots,n-1)$ serve as local coordinates on $\mathbf{C}P^{n-1}$. One computes in these coordinates that σ_w' is the matrix $(\partial^2 r(z^0, \overline{z^0})/\partial z_j \partial \overline{w}_k)_{i \leq j, k \leq n-1}$, which is nonsingular by the strict pseudoconvexity of D. Thus σ_w is a local diffeomorphism. $\quad\square$

Now we can show that $\Phi\big|_{Q_w}$ is algebraic. In fact we have a commutative diagram:

$$
\begin{array}{ccc}
Q_w & \xrightarrow{\ \ \Phi\ \ } & Q_{w'} \\
\downarrow \sigma_w & & \downarrow \sigma_{w'} \\
\mathbf{C}P^{n-1} & \xrightarrow[\Phi'(w)]{} & \mathbf{C}P^{n-1}
\end{array}
$$

The map on the bottom takes hyperplanes in $T_w(\mathbf{C}^n)$ to hyperplanes in $T_{w'}(\mathbf{C}^n)$ by composing with the differential $\Phi'(w)$. In homogeneous coordinates $(p_1,\ldots,p_n)$, this map is linear; in local coordinates p_k/p_n, it is a linear fractional transformation. So the map $\Phi'(w)$ is certainly algebraic. Proposition 2 shows that σ_w, $\sigma_{w'}$ are also algebraic. Consequently $\Phi\big|_{Q_w} = \sigma_{w'}^{-1}\Phi'(w)\sigma_w$ is algebraic, as claimed.

Now we shall prove that Φ is algebraic. We begin with $D \subseteq \mathbf{C}^2$. Fix $z^0 \in \partial D$ and two points p^1, p^2 on Q_{z^0} in general position (to be defined soon). For z near z^0, we have

$$Q_z \cap Q_{p^1} = \{w^1\}, \qquad Q_z \cap Q_{p^2} = \{w^2\}.$$

Consider

$$(5) \qquad\qquad D \to Q_{p^1} \times Q_{p^2}, \qquad z \mapsto (w^1, w^2),$$

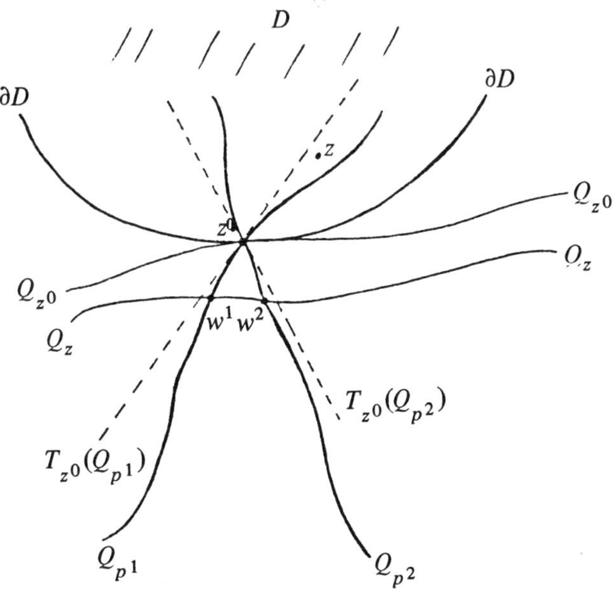

FIGURE 1

which is defined for z near z^0. The points p^1, p^2 are said to be in *general position* if (5) is a local diffeomorphism. Indeed for z sufficiently close to z^0, the map (5) is well approximated by $D \to T_{z^0}(Q_{p^1}) \times T_{z^0}(Q_{p^2})$, $z \mapsto (\overline{w}_1, \overline{w}_2)$, where $\overline{w}_i$ is defined by $Q_z \cap T_{z^0}(Q_{p^i}) = \{\overline{w}_i\}$. Proposition 2 shows that we may pick p^1, p^2 to make the tangent spaces $T_{z^0}(Q_{p^i})$ transverse, and a computation of the derivative of (5) confirms that p^1, p^2 are in general position. See Figure 1. Note that (5) is an algebraic map, since w^i is found by solving the simulataneous polynomial equations

$$r(w^i, \bar{z}) = 0, \qquad r(w^i, \bar{p}^i) = 0.$$

Putting $p^{i\prime} = \Phi(p^i)$ and repeating the construction yields an algebraic map $D' \to Q_{p^{1\prime}} \times Q_{p^{2\prime}}$, which is a local diffeomorphism. Since $\Phi|_{Q_{p^i}}$ is known to be algebraic, we have the commutative diagram

$$
\begin{array}{ccc}
D & \xrightarrow{\;\;\Phi\;\;} & D' \\
\downarrow & & \downarrow \\
Q_{p^1} \times Q_{p^2} & \xrightarrow[\Phi|_{Q_{p^1}} \times \Phi|_{Q_{p^2}}]{} & Q_{p^{1\prime}} \times Q_{p^{2\prime}}
\end{array}
$$

with all maps except Φ known to be algebraic. Hence Φ is algebraic as required. This completes the proof when $n = 2$.

For $D \subseteq \mathbf{C}^n$, fix $z^0 \in \partial D$ and choose $p^1, \ldots, p^n \in Q_{z^0}$ in general position (to be defined shortly). For each $i = 1, \ldots, n$, the variety $\mathcal{Q}_i = \bigcap_{j \neq i} Q_{p^j}$ is one dimensional, so we can define w^i by $Q_z \cap \mathcal{Q}_i = \{w^i\}$ for $z \in D$ near z^0. As before, the map $D \to \mathcal{Q}_1 \times, \ldots, \times \mathcal{Q}_n$, $z \mapsto (w^1, \ldots, w^n)$ is defined for z near z^0, and the points p^i are said to be in general position if it is a local diffeomorphism. This map is algebraic, and the rest of the argument is the same as in the two-dimensional case. This completes the proof of Theorem 2. $\qquad \square$

3. Ellipsoids and biholomorphic maps between them. Recall the boundary of an ellipsoid in $\mathbf{C}^n$ has the form

$$(6) \qquad \sum_{j,k=1}^{n} \left(a_{jk} x_j x_k + b_{jk} y_j y_k + c_{jk} x_j y_k \right) = 1,$$

where $z_j = x_j + \sqrt{-1}\, y_j \in \mathbf{C}$. We can rewrite this using the conjugate variables z_j, $\bar{z}_j$ as $\langle z, \bar{z} \rangle + Q(z, z) + \overline{Q}(\bar{z}, \bar{z}) = 1$, where $\langle \cdot \rangle$ is a Hermitian form and $Q(\cdot, \cdot)$ is a quadratic form. The positivity of the quadratic form (6) implies that $\langle, \rangle$ is positive-definite. (Apply (6) to $e^{i\theta}z$ and average over θ.) Therefore, we can diagonalize it by a complex linear transformation, and our ellipsoid becomes

$$\sum_{j=1}^{n} |z_j|^2 + Q(z, z) + \overline{Q}(\bar{z}, \bar{z}) = 1.$$

Now consider the restriction of the real quantity $Q(z, z) + \overline{Q}(\bar{z}, \bar{z})$ to the unit sphere in $\mathbf{C}^n$. After a unitary transformation, we may assume this quantity is maximized at $z = (1, 0, \ldots, 0) \in$ Unit Sphere. This unitary transformation puts $Q(z, z) + \overline{Q}(\bar{z}, \bar{z})$ in the special form $A_1(z_1^2 + \bar{z}_1^2) + Q'(z', z') + \overline{Q}'(\bar{z}', \bar{z}')$, where $(z_1, \ldots, z_n) = (z_1, z')$. We can apply the same maximization argument to $Q' + \overline{Q}'$ in fewer variables, and proceed inductively. This proves

PROPOSITION 3. *The boundary of an ellipsoid $D \subseteq \mathbf{C}^n$ can be written*

$$\sum_{j=1}^n |z_j|^2 + \sum_{j=1}^n A_j\left(z_j^2 + \bar{z}_j^2\right) - 1 = 0$$

after a complex linear transformation.

PROOF OF THEOREM 3. Suppose that $\Phi\colon D \to D'$ is a biholomorphic map between ellipsoids in $\mathbf{C}^n$. We may assume D, D' are given as in Proposition 3 with coefficients $\{A_i\}$, $\{A_i'\}$ respectively. The proof of Webster's theorem proceeds in several steps.

Step 1. Φ is algebraic.

Step 2. Φ is birational.

Step 3. The coefficients $\{A_i\}$ and $\{A_i'\}$ agree.

Step 4. Φ is linear unless all the A_j vanish.

We already know that Φ is algebraic: this follows from Theorem 2. We begin with

Step 2. Since Φ is algebraic, we can think of it as a globally-defined, possibly multiple-valued map $\Phi\colon \mathbf{C}^n \setminus V \to \mathbf{C}^n$, where V is a subvariety of $\mathbf{C}^n$. By a careful analytic continuation, we shall show that Φ has a single-valued branch on $\mathbf{C}^n \setminus V$. Proposition 1 will then imply that Φ is rational. Applying the same argument to Φ^{-1} shows that Φ is birational, as desired. So it is enough to define Φ as a single-valued function globally on $\mathbf{C}^n \setminus V$.

Let $\Phi(z) = (w_1, \ldots, w_n)$ satisfy polynomial equations

$$(7) \qquad P^j(z, w_j) = \sum_{k=0}^{m_j} P_k^j(z)(w_j)^k = 0, \qquad z \in \mathbf{C}^n \setminus V.$$

After possibly enlarging V (throw in $z \in \mathbf{C}^n$ where $w_j \mapsto P^j(z, w_j)$ has discriminant zero) we may assume that locally there is only one solution w of the equations $P^j(z, w_j) = 0$.

To show that Φ is globally single valued, we consider the analytic continuation of Φ along a path $\gamma \subseteq \mathbf{C}^n \setminus V$ connecting z_{initial} and z_{final}. That is, we continue each coordinate function w_j along the path according to the equation $P^j(z, w_j) = 0$; denote this by $z'(t) = \Phi(\gamma(t))$. If we can prove that $z'_{\text{final}} = \Phi(z_{\text{final}})$ is independent of the path γ, then we have shown that $\Phi(z)$ is single valued as desired.

It is sufficient to prove this for γ a generator of $\pi_1(\mathbf{C}^n \setminus V)$. To describe a set of generators, recall that a complex line l in $\mathbf{C}^n$, in general position, intersects the variety V in a finite number of points $B_i\colon l \cap V = \{B_1, \ldots, B_n\}$. It is not difficult to show that $\pi_1(\mathbf{C}^n \setminus V)$ is generated by those loops $t \to \gamma(t) \in \mathbf{C}^n \setminus V$ that originate at a fixed basepoint z_{initial}, travel to a neighborhood of B_i, encircle B_i once inside l, and then return to z_{initial} along the original path to $z_{\text{final}} = z_{\text{initial}}$.

(See Figure 2.) We can assume that $z_{\text{initial}} \in \partial D$, where D is our ellipsoid. Let $\tilde{V} = \{y \mid Q_y \subseteq V\}$; this is a proper subvariety of $\mathbf{C}^n$. Note that we can arrange γ so that

$$(8) \qquad\qquad \gamma(t) \notin \tilde{V}.$$

We will need (8) later.

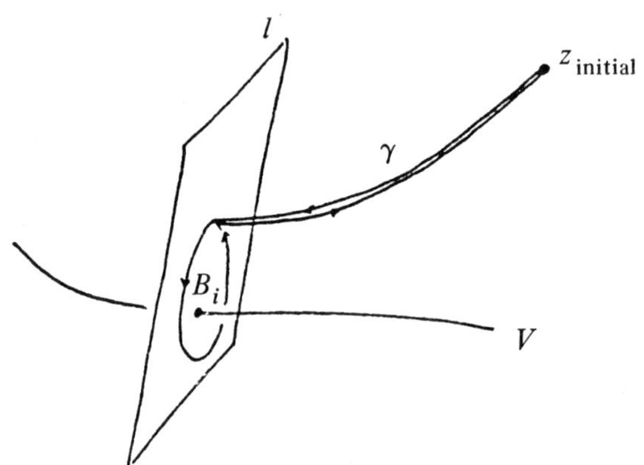

FIGURE 2

LEMMA 1 *Let w be a point of $Q_{z_{\text{initial}}}$ near z_{initial}. After possibly reparametrizing γ, we can find a closed loop $\gamma^{\#} = \{t \to \gamma^{\#}(t)\} \subseteq \mathbf{C}^n \setminus V$ starting at w and satisfying*

$$(9) \qquad \begin{array}{ll} \text{(i)} & r\left(\gamma(t), \overline{\gamma^{\#}(t)}\right) \equiv 0, \\[2mm] \text{(ii)} & \gamma^{\#} \text{ is homotopically trivial in } \mathbf{C}^n \setminus V. \end{array}$$

We shall prove this shortly. For now, we show that Lemma 1 implies $z'_{\text{initial}} = z'_{\text{final}}$, where $z'_{\text{initial}} = \Phi(z_{\text{initial}})$. Indeed, let $w'(t) = \Phi(\gamma^{\#}(t))$ be the analytic continuation of Φ along $\gamma^{\#}$; this is a closed loop by virtue of (9ii). Now (9i) yields $w' = w'_{\text{final}} \in Q'_{z'_{\text{final}}}$. Thus, if w' belongs to a neighborhood of z'_{initial} in $Q'_{z'_{\text{initial}}}$, then w' belongs to $Q'_{z'_{\text{final}}}$. Since any small piece of $Q'_{z'}$ already determines z', we can conclude that $z'_{\text{initial}} = z'_{\text{final}}$. Thus Φ is single valued, hence birational. All that remains is to prove Lemma 1.

PROOF OF LEMMA 1. We can assume that γ makes a very small loop around the point B_1. Thus $\gamma(t)$ travels from z_{initial} to l during the time interval $0 \leqslant t \leqslant t_0$, then $\gamma(t)$ loops around B_1 during $t_0 \leqslant t \leqslant t_1$, and finally $\gamma(t)$ returns from l to z_{initial} during $t_1 \leqslant t \leqslant 1$. We first define $\gamma^{\#}(t)$ for $0 \leqslant t \leqslant t_0$. To do so, it is enough to assume $\gamma^{\#}(t)$ has been constructed for $0 \leqslant t \leqslant \tau$, and then shows how to extend $\gamma^{\#}(t)$ into $0 \leqslant t \leqslant \min\{\tau + h, t_0\}$ with h independent of τ. There are two cases.

Easy case. If $\gamma^{\#}(\tau)$ is far from V, then we just continue $\gamma^{\#}$ in any reasonable fashion to satisfy (9i), and $\gamma^{\#}$ will continue to avoid V.

Delicate case. If $\gamma^{\#}(\tau)$ is near V, then by reparametrizing γ we may assume that $\gamma(t)$ remains constant for a short time, starting at $t = \tau$. During that short time, we let $\gamma^{\#}(t)$ move inside $Q_{\gamma(\tau)}$, starting at $\gamma^{\#}(\tau)$ and ending at some point far from V. This is possible by virtue of (8). Now we can proceed as in the easy case. So $\gamma^{\#}(t)$ can be defined for $0 \leqslant t \leqslant t_0$. Moreover, by repeating the argument of the Delicate case above, we can suppose that $\gamma^{\#}(t_0)$ is far from V.

Next we define $\gamma^{\#}(t)$ for $t_0 \leqslant t \leqslant t_1$. Since the loop in $\gamma(t)$ is small, we have $t_1 - t_0 \ll 1$, so we can continue $\gamma^{\#}(t)$ into $0 \leqslant t \leqslant t_1$ so that (9i) holds, and $\gamma^{\#}(t)$ still misses V. Although $\gamma(t_0) = \gamma(t_1)$, it may happen that $\gamma^{\#}(t_1) \neq \gamma^{\#}(t_0)$. To correct this, we reparametrize γ so that $\gamma(t)$ is constant for a short time starting at $t = t_1$; during this short time we can let $\gamma^{\#}(t)$ move back to $\gamma^{\#}(t_0)$, staying within $Q_{\gamma(t_0)}$. Thus, we have defined $\gamma^{\#}(t)$, $0 \leqslant t \leqslant t_1$. Note that $\gamma^{\#}(t_0) = \gamma^{\#}(t_1)$ now, and the loop $\gamma^{\#}(t)$, $t_0 \leqslant t \leqslant t_1$ is contained in a small ball in $\mathbf{C}^n \setminus V$. Finally, we have to define $\gamma^{\#}(t)$ for $t_1 \leqslant t \leqslant 1$. This is easy—since $\gamma(t)$ retraces its initial path during $t_1 \leqslant t \leqslant 1$, we can simply let $\gamma^{\#}(t)$ do the same. Now we have defined $t \to \gamma^{\#}(t) \in \mathbf{C}^n \setminus V$ for $0 \leqslant t \leqslant 1$. We know that (9i) holds, and (9ii) is obvious, since the part of $\gamma^{\#}$ from $t = t_0$ to $t = t_1$ can be shrunk to a point. $\quad\square$

Step 3. In this step we prove $\{A_i\} = \{A_i'\}$. Recall that

$$D = \left\{z \in \mathbf{C}^n \mid r(z, \bar{z}) = 0\right\}, \qquad D' = \left\{z' \in \mathbf{C}^n \mid r'(z', \bar{z}') = 0\right\},$$

and if $z' = \Phi(z)$, $w' = \Phi(w)$, then

$$(10) \qquad\qquad r'(z', \bar{w}') = \mu(z, \bar{w})r(z, \bar{w}).$$

So far, we have been using Q_w, defined by the vanishing of $r(z, \bar{w})$. Now we are going to make use of the varieties defined by the vanishing of a linear combination $\Sigma_j \lambda_j r(z, \bar{w}_j)$. The key observation that makes this possible is the following.

LEMMA 2. *If* Φ *is birational, then* $\mu(z, \bar{w}) = \beta(z)\overline{\beta(w)}$ *for a rational function* $\beta(\cdot)$.

PROOF. Since $\mu(z, \bar{w}) = r'(\Phi(z), \overline{\Phi(w)})/r(z, w)$ with Φ rational, we know that $\mu(z, \bar{w})$ is rational. Similarly if we define $\mu'(z', \bar{w}')$ by $r(z, \bar{w}) = \mu'(z', \bar{w}')r'(z', \bar{w}')$, then $\mu'(z', \bar{w}')$ is rational and satisfies $\mu(z, \bar{w})\mu'(z', \bar{w}') = 1$. Since Φ is rational, $z' = F(z)/g(z)$, where $F(z) = (f_1(z), \ldots, f_n(z))$ and $f_i(z)$, $g(z)$ are polynomials. Therefore

$$(11) \qquad r(z, \bar{w})\mu(z, \bar{w}) = r'\left(\frac{F(z)}{g(z)}, \frac{\overline{F(w)}}{\overline{g(w)}}\right) = \frac{P_0(z, \bar{w})}{Q(z)\,\overline{Q(w)}}$$

for some polynomials P_0, Q.

We claim $r(z, \bar{w})$ divides $P_0(z, \bar{w})$. For, we know that $\mu(z, \bar{w})$ has no singularities for z, w near $p_0 \in \partial D$. Therefore if $r(z, \bar{w}) = 0$ and z, w are near p_0, it follows that $P_0(z, \bar{w}) = 0$. In other words, P_0 vanishes in an open subset of $\{r = 0\}$. However, $\{r = 0\}$ is connected, and so by analytic continuation $r(z, \bar{w}) = 0$

implies $P_0(z, \overline{w}) = 0$. The Hilbert Nullstellensatz [43, p. 256] shows that r divides P_0^k for some k; since r is irreducible, r divides P_0 as claimed. (One can get away without using the Nullstellensatz—we leave this to the reader.)

Now since r divides P_0, we can rewrite (11) in the form

$$(12) \qquad \mu(z, \overline{w}) = \frac{P(z, \overline{w})}{Q(z)\,\overline{Q(w)}}.$$

Similarly,

$$(13) \qquad \mu'(z', \overline{w}') = \frac{P'(z', \overline{w}')}{Q'(z')\,\overline{Q'(w')}}.$$

Since $\mu(z, w)\mu'(z', w') = 1$, we get from (12), (13) that

$$1 = \frac{P(z, \overline{w})P'(z', \overline{w}')}{Q(z)\,\overline{Q(w)}\,Q'(z')\,\overline{Q'(w')}} = \frac{P(z, \overline{w})\tilde{P}(z, \overline{w})}{Q(z)\,\overline{Q(w)}\,Q'(z')\,\overline{Q'(w')}\,S(z)\,\overline{S(w)}},$$

for some polynomials $\tilde{P}, S$. Therefore $P(z, \overline{w})\tilde{P}(z, \overline{w}) = R(z)\overline{R(w)}$ for some rational function R. Clearing denominators gives

$$P(z, \overline{w})\tilde{\tilde{P}}(z, \overline{w}) = \tilde{S}(z)\,\overline{\tilde{S}(w)}, \quad \text{for polynomials } \tilde{\tilde{P}}, \tilde{S}.$$

Since the polynomials form a unqiue factorization domain, it follows that $P(z, \overline{w}) = A(z)\overline{B(w)}$. From (12) we obtain $\overline{P(z, \overline{w})} = P(w, \overline{z})$, which implies $A = B$. Substituting back into (12), we now have

$$\mu(z, \overline{w}) = \left(\frac{A(z)}{Q(z)}\right) \cdot \left(\overline{\frac{A(w)}{Q(w)}}\right) = \beta(z)\overline{\beta(w)}. \qquad \square$$

Now we can pass to linear combinations of the $r(\,\cdot\,, w_j)$.

LEMMA 3. Φ *carries the hypersurface* $\{z \mid \Sigma_j \lambda_j r(z, \overline{w}_j) = 0\}$ *to a hypersurface* $\{z' \mid \Sigma_j \lambda'_j r(z', \overline{w}'_j) = 0\}$.

PROOF. Consider the spaces

$$H = \text{linear span } \{r(\,\cdot\,, \overline{w}) \text{ as } w \text{ varies}\},$$

$$H' = \text{linear span } \{r'(\,\cdot\,, \overline{w}') \text{ as } w' \text{ varies}\}$$

These spaces consist of second-degree polynomials and are thus finite dimensional. We can pick generators of H, $r(\,\cdot\,, \overline{w}_1),\ldots,r(\,\cdot\,, \overline{w}_m)$ so that w_j is neither a zero nor a pole of $\beta(w)$ in Lemma 2. Set $w'_j = \Phi(w_j)$. By enlarging our list, we may assume that the $r'(\,\cdot\,, \overline{w}'_j)$ span H'. Now by Lemma 2,

$$\sum_{j=1}^{m} \lambda'_j r'(z', \overline{w}'_j) = 0 \quad \text{if and only if} \quad \sum_{j=1}^{m} \lambda'_j \beta(z)\,\overline{\beta(w_j)}\,r(z, \overline{w}_j) = 0,$$

i.e.

$$(14) \qquad \beta(z) \cdot \sum_{j=1}^{m} \lambda_j r(z, \overline{w}_j) = 0, \quad \text{where } \lambda_j = \lambda'_j \overline{\beta(w_j)}.$$

Assume for the moment that $\beta(z) = 0$. This gives $\mu(z, \overline{w}_j) = \beta(z)\beta(w_j) = 0$ for $j = 1,\ldots,m$, since $\beta(w_j)$ never has a pole; therefore

$$r'(z', \overline{w}_j') = \mu(z, \overline{w}_j)r(z, \overline{w}_j) = 0, \qquad j = 1,\ldots,m.$$

But that is a contradiction, since the $r'(\,\cdot\,, \overline{w}_j')$ span H'. This means $\beta(z)$ is never zero when $z' = \Phi(z)$ is well defined, i.e. for $z \in \mathbf{C}^n \setminus V$. Therefore (14) implies

$$\sum_{j=1}^m \lambda_j' r'(z', \overline{w}_j') = 0 \quad \text{if and only if} \quad \sum_{j=1}^m \lambda_j r(z, \overline{w}_j) = 0.$$

which is the claim of the Lemma. $\square$

Denote by T the isomorphism between H and H' used in the proof of Lemma 3, i.e.

$$(15) \qquad T: H' \to H, \qquad \sum_{j=1}^m \lambda_j' r'(z', \overline{w}_j') \to \sum_{j=1}^m \left(\lambda_j' \overline{\beta(w_j)}\right) r(z, \overline{w}_j).$$

Sitting inside H' and H are the linear polynomials. We shall prove that T takes linear polynomials to linear polynomials. Lemma 3 then shows that Φ carries hyperplanes to hyperplanes. This is the next main step in the proof of Theorem 3. In order to give a useful characterization of the linear polynomials, we will introduce inner products so that T becomes unitary. This requires the map i: $\mathbf{C}^n \to H$ which sends w to $r(\,\cdot\,, \overline{w})$.

LEMMA 4. *There is a unique inner product $\langle\,\cdot\,,\,\cdot\,\rangle$ on H satisfying*

$$(\mathrm{i})\ \langle\,,\,\rangle \text{ is nondegenerate,}$$

$$(\mathrm{ii})\ \mathrm{r}(z, \overline{w}) = \langle\, \mathrm{i}z, \mathrm{i}w\,\rangle.$$

PROOF. Let $\{\psi_\alpha\}_{\alpha \leq m}$ be a basis of H. Then we can find $\overline{\chi_\alpha(w)}$ so that

$$r(z, \overline{w}) = \sum_{\alpha=1}^m \overline{\chi_\alpha(w)}\, \psi_\alpha(z).$$

Since $r(z, \overline{w}) = \overline{r(w, \overline{z})}$, this gives

$$r(w, \overline{z}) = \sum_{\alpha=1}^m \chi_\alpha(w) \overline{\psi_\alpha(z)},$$

or after changing notation

$$(17) \qquad r(z, \overline{w}) = \sum_{\alpha=1}^m \chi_\alpha(z) \overline{\psi_\alpha(w)}.$$

This means $H \subseteq$ linear span$\{\chi_\alpha(z), \alpha = 1,\ldots,m\}$; since $\dim H = m$, $\{\chi_\alpha(z)\}_{\alpha \leq m}$ is a basis of H. The change of basis equations $\chi_\alpha = \Sigma_\beta g_{\alpha\beta}\psi_\beta$, where $(g_{\alpha\beta})$ is an invertible matrix, give us the formula

$$r(z, \overline{w}) = \sum_{\alpha,\beta=1}^m g_{\alpha\beta}\psi_\beta(z) \overline{\psi_\alpha(w)}.$$

Thus (16) holds if we define $\langle \, , \, \rangle$ by

$$(18) \qquad \left\langle \sum_{\alpha} \lambda_\alpha \chi_\alpha, \sum_{\beta} \tilde{\lambda}_\beta \chi_\beta \right\rangle = \sum_{\alpha, \beta} g_{\alpha\beta} \bar{\lambda}_\alpha \tilde{\lambda}_\beta.$$

The uniqueness of the inner product is immediate, since the image of i spans H.
$\square$

Lemma 4 also defines a natural inner product on H'. Let

$$h' = \sum_j \lambda'_j r'(\cdot, \bar{w}'_j) = \sum_j \lambda'_j i(w'_j), \qquad \tilde{h}' = \sum_k \tilde{\lambda}'_k r'(\cdot, \bar{w}'_k) = \sum_k \tilde{\lambda}'_k i(w'_k),$$

so that

$$\langle h', \tilde{h}' \rangle = \sum_{jk} \bar{\lambda}'_j \tilde{\lambda}'_k r'(w'_j, \bar{w}'_k).$$

Since

$$Th' = \sum_j \left(\lambda'_j \overline{\beta(w_j)} \right) r(\cdot, w_j) = \sum_j \left(\lambda'_j \overline{\beta(w_j)} \right) i(w_j),$$

$$T\tilde{h}' = \sum_k \left(\tilde{\lambda}'_k \overline{\beta(w_k)} \right) r(\cdot, \bar{w}_k) = \sum_k \left(\tilde{\lambda}'_k \overline{\beta(w_k)} \right) i(w_k),$$

we have

$$\langle Th', T\tilde{h}' \rangle = \sum_{jk} \bar{\lambda}'_j \beta(w_j) \tilde{\lambda}'_k \overline{\beta(w_k)} \, r(w_j, \bar{w}_k)$$

$$= \sum_{jk} \bar{\lambda}'_j \tilde{\lambda}'_k r'(w'_j, \bar{w}'_k) = \langle h', \tilde{h}' \rangle,$$

showing that T is unitary with respect to $\langle \, , \, \rangle$.

Next note that T carries $r(\cdot, \bar{w}) \in H$ to $\overline{\beta(w)} r'(\cdot, w') \in H'$, whenever $w' = \Phi(w)$ is well defined. So if we set

$$\mathcal{Q} = \text{closure of } \left\{ \lambda r(\cdot, \bar{w}) \,\big|\, \lambda \in \mathbf{C}, w \in \mathbf{C}^n \right\} \subseteq H,$$

$$\mathcal{Q}' = \text{closure of } \left\{ \lambda' r(\cdot, w') \,\big|\, \lambda' \in \mathbf{C}, w' \in \mathbf{C}^n \right\} \subseteq H',$$

it follows that $T(\mathcal{Q}) \subseteq \mathcal{Q}'$. Thus, T preserves both an inner product and a distinguished subset $\mathcal{Q}$.

Let us now specialize to the case of the ellipsoid of Proposition 3. We find that the general element $h \in H$ is given by $h = \chi_0 \psi_0 + \chi_* \psi_* + \Sigma_1^n \chi_j \psi_j$, where the χ's are arbitrary numbers, and $\psi_0 = 1$, $\psi_* = \Sigma_1^n A_j z_j^2$, $\psi_j = z_j$. The ψ's form a basis for H except for the ball (i.e. all $A_j = 0$) in which case ψ_* may evidently be deleted. Since we know that dim H is a biholomorphic invariant for ellipsoids, it follows that $\{ |z|^2 + \Sigma_j 2 A_j \text{Re}(z_j^2) < 1 \}$ cannot be biholomorphic to the ball unless all $A_j = 0$. From now on assume some $A_j \neq 0$.

Next we write down the inner product $\langle \, , \, \rangle$ and the subset $\mathcal{Q} \subseteq H$. Routine computations yield

$$\|h\|^2 = \sum_i^n |\chi_j|^2 - |\chi_0|^2 + 2\,\text{Re}\left(\chi_0 \bar{\chi}_* \right),$$

$$h \in \mathscr{Q} \quad \text{if and only if} \quad \sum_i^n A_j \chi_j^2 - \chi_*^2 - \chi_* \chi_0 = 0.$$

Now $T\colon H \to H'$ preserves $\|h\|^2$ and $\mathscr{Q} = \{h \in H \mid (h, h) = 0\}$, where $(h, h) = \Sigma_1^n A_j \chi_j^2 - \chi_*^2 - \chi_* \chi_0$. A linear transformation preserving the zero set $\mathscr{Q}$ of (h, h) must fix (h, h) up to a nonvanishing scalar factor β. In other words, T preserves $\langle \, , \, \rangle$ exactly and $(\, , \,)$ up to a factor β.

Since T preserves two quadratic forms, it is natural to look at the eigenvectors and eigenvalues of $(\, , \,)$ with respect to $\langle \, , \, \rangle$. That is, we seek $\omega_j \in H$, $\lambda_j \in \mathbf{C}$ such that

$$(19) \qquad (\xi, \omega_j) = \lambda_j \langle \xi, \omega_j \rangle.$$

This is slightly different from an ordinary eigenvalue problem, since $(\, , \,)$ is bilinear and $\langle \, , \, \rangle$ sesquilinear. Multiplying ω_j by $\rho e^{i\theta}$ has the effect of multiplying λ_j by $e^{2i\theta}$, so it is the $|\lambda_j|$ that have invariant meaning. Since T preserves the two inner product (up to factor β), it follows that the eigenvalues $\{\lambda_j\}$, $\{\lambda_j'\}$ corresponding to the two ellipsoids coincide up to a factor of $|\beta|$.

It is simple enough to compute the λ's explicitly. Substituting our formulas for $(\, , \,)$ and $\langle \, , \, \rangle$ into (19), we obtain the equations

$$(20) \qquad \lambda \bar{\chi}_j = A_j \chi_j, \qquad (1 \leqslant j \leqslant n),$$

$$(21) \qquad \lambda\left(-\bar{\chi}_0 + \bar{\chi}_*\right) = -\tfrac{1}{2}\chi_*,$$

$$(22) \qquad \lambda \bar{\chi}_0 = -\chi_* - \tfrac{1}{2}\chi_0.$$

If any $\chi_j \neq 0$, then (20) yields $|\lambda_j| = A_j$; and for each $|\lambda_j| = A_j$ there is an obvious solution w_j of (20)–(23) with a single nonzero component in w_j.

If all $\chi_j = 0$ $(1 \leqslant j \leqslant n)$, then we are left with the task of solving (21), (22). After elementary manipulation one finds two additional λ's, namely the roots λ_+ of $4\lambda^2 - 8\lambda + 1 = 0$. Thus our list of all $|\lambda_j|$'s is $\{A_1, A_2, \ldots, A_n, \lambda_+, \lambda_-\}$. This list therefore differs from $\{A_1', A_2', \ldots, A_n', \lambda_+, \lambda_-\}$ only by a factor of $|\beta|$. Since the quadratic form defining an ellipsoid is positive, we have $0 \leqslant A_j < 1$, while $\lambda_- < 1 < \lambda_+$. Hence $\lambda_+ = \max\{A_1, A_2, \ldots, A_n, \lambda_+, \lambda_-\} = \max\{A_1', \ldots, A_n', \lambda_+, \lambda_-\}$, so $|\beta| = 1$ and $\{A_1, \ldots, A_n\} = \{A_1', \ldots, A_n'\}$. The two ellipsoids D, D' are therefore the same, and Step 3 is complete. $\square$

We omit Step 4 (linearity of Φ). It is clear that the information from Step 3 places powerful constraints on the map T, therefore on Φ. For instance if $A_1, A_2, \ldots, A_n, \lambda_+, \lambda_-$ are distinct, then T must be diagonal.

Our discussion of Steps 1 and 2 follows the paper of Webster [64]. Step 3 follows an unpublished proof by Webster. The discussion of Steps 3 and 4 in [64] is actually shorter than here, but the ideas here obviously apply to much more general domains than the ellipsoids, so we thought them worth presenting.

4. Smoothness of biholomorphic maps. In this section we prove that a biholomorphic map of strictly pseudoconvex domains is smooth up to the boundary. The original proof in [21] was long and hard, but E. Ligocka and S. Bell [2] found a very simple proof using only the smoothness of solutions of the $\bar{\partial}$-Neumann problem. (See also Nirenberg-Webster-Yang [51] for a proof based on reflections.) Nowadays one can deal also with classes of weakly pseudoconvex domains. We are grateful to Bell for providing the following particularly simple argument in the strictly pseudoconvex case.

Proof of Theorem 1. Suppose $f: D_1 \to D_2$ is a biholomorphic mapping between smooth bounded strictly pseudoconvex domains contained in $\mathbf{C}^n$. Let P_1 and P_2 denote the Bergman orthogonal projections associated to D_1 and D_2, respectively, and let $u = \det[f']$. A simple Hilbert space argument using the fact that $|u|^2$ is equal to the *real* Jacobian determinant of f viewed as a mapping on $\mathbf{R}^{2n}$ reveals that the Bergman projections transform under f according to the formula $u((P_2\phi) \circ f) = P_1(u(\phi \circ f))$.

A key element in the proof of the theorem is the fact that if h is a holomorphic function on D_2 that is in $C^\infty(\overline{D}_2)$, then it is possible to construct a function ϕ in $C^\infty(\overline{D}_2)$ that vanishes to infinite order on ∂D_2 such that $P_2\phi = h$. Indeed, the C^∞ version of the Cauchy-Kowalewski theorem asserts that the Cauchy problem

$$\Delta\psi = h \qquad \text{on } D_2,$$
$$\psi = \nabla\psi = 0 \quad \text{on } \partial D_2$$

can be solved modulo functions which vanish to infinite order on ∂D_2, i.e., there is a function ψ in $C^\infty(\overline{D}_2)$ such that $\psi = \nabla\psi = 0$ on ∂D_2 and such that $h - \Delta\psi$ vanishes to infinite order on ∂D_2. Now $\Delta\psi$ is orthogonal to holomorphic functions on D_2 via integration by parts. Hence, $\phi = h - \Delta\psi$ is a function in $C^\infty(\overline{D}_2)$ vanishing to infinite order on ∂D_2 such that $P_2\phi = P_2 h - P_2\Delta\psi = h$.

One more auxiliary fact is required before the theorem can be proved; namely, that if $\phi \in C^\infty(\overline{D}_2)$ vanishes to infinite order on ∂D_2, then $u(\phi \circ f)$ is in $C^\infty(\overline{D}_1)$. Since the components of f are bounded holomorphic functions on D_1, the classical Cauchy estimates yield that there are constants C_α independent of z such that $|\partial^\alpha f_k/\partial z^\alpha(z)| \leq C_\alpha \operatorname{dist}(z, \partial D_1)^{-|\alpha|}$ for all z in D_1. Hence, to see that all the derivatives of $u(\phi \circ f)$ are bounded functions on D_1, it suffices to prove that

$$\operatorname{dist}(f(z), \partial D_2) \leq C\operatorname{dist}(z, \partial D_1).$$

The proof of this classical estimate is due to Henkin and is quite simple. Let r be a defining function for D_1 that is strictly plurisubharmonic in a neighborhood of $\overline{D}_1$. Then $r \circ f^{-1}$ is a negative subharmonic function on D_2. Since $r \circ f^{-1}$ attains its maximum value, zero, on ∂D_2, the classical Hopf lemma implies that $(r \circ f^{-1})(z) \leq -C\operatorname{dist}(z, \partial D_2)$. The desired inequality now follows because $-r(w) \leq C'\operatorname{dist}(w, \partial D_1)$.

The proof of the theorem can now be completed. Let h be a holomorphic function on D_2 that is in $C^\infty(\overline{D}_2)$. Let ϕ be a function in $C^\infty(\overline{D}_2)$ vanishing to infinite order on ∂D_2 such that $P_2\phi = h$. The transformation formula for the Bergman projections yields that $u(h \circ f) = u((P_2\phi) \circ f) = P_1(u(\phi \circ f))$. Kohn's formula, $P_1 = I - \bar{\partial}^* N\bar{\partial}$, which relates the Bergman projection P_1 to the $\bar{\partial}$-Neumann operator N, together with Kohn's subelliptic estimates for N, reveal that P_1 preserves the space $C^\infty(\overline{D}_1)$. Hence, since $u(\phi \circ f)$ is in $C^\infty(\overline{D}_1)$, the relation $u(h \circ f) = P_1(u(\phi \circ f))$ implies that $u(h \circ f)$ is in $C^\infty(\overline{D}_1)$. Let $h = 1$ to obtain that u is in $C^\infty(\overline{D}_1)$. Let $h = z_k$, the kth coordinate function of $\mathbf{C}^n$, to obtain that uf_k is in $C^\infty(\overline{D}_1)$. Hence f extends smoothly to ∂D_1 where u does not vanish. But u cannot vanish on ∂D_1. To see this, observe that the same arguments can be applied to the inverse mapping $F = f^{-1}$. Hence, $U = \det[F']$ is in $C^\infty(\overline{D}_2)$. Now, the relation

$$U(z) = 1/u(F(z))$$

implies that u is bounded away from zero. This completes the proof of the theorem. $\quad\square$

CHAPTER 9. MOSER'S NORMAL FORM

Suppose we are given a strictly ψ-convex domain in $\mathbf{C}^{n+1} = \{(z_1,\ldots,z_n, w)\}$, where $w = u + iv$. We know that the boundary has a defining function satisfying the Levi condition. In the case we consider first, this defining function is given by a formal power series, which we can assume is centered at the origin. The goal is to find coordinates so that the boundary near the origin is in Moser's normal form

$$(*) \qquad v = |z|^2 + \sum_{|\alpha|,|\beta| \geqslant 2} \sum_{l=0}^{\infty} A^l_{\alpha\bar\beta} u^l z^\alpha \bar z^\beta,$$

where the coefficients $A^l_{\alpha\bar\beta} = A^l_{j_1,\ldots,j_r\bar k_1,\ldots,\bar k_s}$ satisfy

(a) $\sum_{p=1}^n A^l_{pj_2\bar p\bar k_2} = 0$, for all $l, j_2, \bar k_2$.

(b) $\sum_{p,q=1}^n A^l_{pqj_3\bar p\bar q} = 0$, for all l, j_3.

(c) $\sum_{p,q,r=1}^n A^l_{pqr\bar p\bar q\bar r} = 0$, for all l,

and the indices are tensor indices varying from 1 to n. Requirement (a) says that Trace $A_{22} = 0$; (b) that Trace$^3 A_{23} =$ Trace$^2 A_{32} = 0$ and (c) that Trace$^3 A_{33} = 0$. Of course the coordinate change will also be given by formal power series. In the last section of the chapter we will consider the question of when these series actually converge.

1. Moser's theorem for formal power series. We are interested not only in bringing a boundary to normal form but also in the maps that preserve this form. Define a *biholomorphism of normal forms* to be a biholomorphic map that takes the surface $(*)$ in normal form to any other hypersurface in normal form and preserves the origin.

THEOREM 1 (MOSER). (1) *Let ∂D be a strictly pseudoconvex real analytic boundary and let p be a point in ∂D. There is a biholomorphic map defined in a neighborhood of p which places ∂D in Moser's normal form and takes p to the origin.*

(2) *Suppose ∂D is already in normal form, and $p = 0$. Then the class of all biholomorphic maps which fix 0 and carry ∂D to a boundary in normal form is*

parametrized by $\mathcal{H}$, the group of linear fractional transformations of the hyperquadric which fix the origin.

PROOF. After translating and rotating, we can assume that the power series is centered at 0 and that the tangent space at 0 is $v = 0$. This means that for a real-valued F our surface is given by

$$v = F(z, \bar{z}, u),$$

(1)
$$F(0,0,0) = 0, \qquad \frac{\partial F}{\partial z_k} = \frac{\partial F}{\partial \bar{z}_k} = \frac{\partial F}{\partial u} = 0 \quad \text{at } (0,0,0).$$

The *weights* of the terms in the formal power series are determined by the table

term	weight
$z, \bar{z}$	1
u, v	2

Equation (1) means that F has no terms of weight 1 and 3 terms of weight 2,

$$\langle z, \bar{z} \rangle, \qquad Q(z), \qquad \overline{Q}(\bar{z}),$$

where $\langle , \rangle$ is a nondegenerate quadratic form and Q is a quadratic form. The form $\langle , \rangle$ is nondegenerate because the boundary was assumed to be strictly ψ-convex. Equation (1) becomes

$$v = \langle z, \bar{z} \rangle + Q(z) + \overline{Q}(\bar{z}) + (\text{terms of weight} \geqslant 3).$$

After the change of variables

$$(z, w) \to (z^*, w^*), \qquad z^* = z, \qquad w^* = w - 2iQ(z),$$

the variable v^* becomes

$$v^* = \operatorname{Im} w^* = \operatorname{Im} w - 2\operatorname{Re} Q(z) = v - Q(z) - \overline{Q}(\bar{z})$$

so that in the new coordinates

(2)
$$v = \langle z, \bar{z} \rangle + (\text{terms of weight} \geqslant 3).$$

In order to study the surface (2), we introduce the group $\mathcal{G}_1$ consisting of its biholomorphic transformations; in other words $\mathcal{G}_1$ consists of all transformations given by formal power series that take surfaces of the form (2) to surfaces of the form (2).

LEMMA 1. *Any* $\Phi \in \mathcal{G}_1$ *can be uniquely factored as* $\Phi = \Psi \circ \phi$, *where* ϕ *is a fractional linear transformation in* $\mathcal{H}$ *and* Ψ *agrees with the identity in the initial part of its Taylor expansion; in other words, at the origin*

$$\frac{\partial w^*}{\partial z_k} = 0, \qquad \frac{\partial w^*}{\partial w} = 1, \qquad \frac{\partial^2 w^*}{\partial z_j \partial z_k} = 0, \qquad \frac{\partial z_k^*}{\partial z_j} = \delta_{jk}, \qquad \operatorname{Re}\left(\frac{\partial^2 w^*}{\partial w^2}\right) = 0$$

where $\Psi(z, w) = (z^*, w^*)$.

PROOF OF LEMMA 1. Recall that the group

$$\mathcal{H} = \text{fractional linear transformations of } H \text{ which fix the origin}$$

consists of compositions of

(i) Heisenberg dilations,

(ii) unitary rotations in z,

(iii) translations at ∞ parametrized by $(v_1, \ldots, v_n, \zeta) \in \mathbf{C}^n \times \mathbf{R}$.

The translations (iii) are given by

$$(z_k, w) \mapsto (z_k^*, w^*) = \frac{(z_k + \bar{v}_k w/2i, \, w)}{1 + \Sigma_k v_k z_k + w\big(\zeta - (i/4)\Sigma_k |v_k|^2\big)},$$

where (z_k, w) is an abbreviation for $(z_1, \ldots, z_k, \ldots, z_n, w)$.

We are given $\Phi \in \mathcal{G}_1$ mapping

$$\Phi \colon (z_k, w) \to (z_k^*, w^*)$$

which satisfies

$$z_k^* = 0, \qquad w^* = 0, \qquad \partial w^*/\partial w = \lambda > 0, \qquad \partial w^*/\partial z_k = 0$$

at the origin. Since λ is positive and real, we can find a dilation such that

$$\Phi = (\text{better } \Phi) \circ (\text{Heisenberg dilation}),$$

where (better Φ): $(z_k, w) \mapsto (z_k^*, w^*)$ satisfies $\partial w^*/\partial w = 1$. This means

$$w^* = w + \sum a_{jk} z_j z_k + (\text{terms of weight} \geq 3),$$

$$(3.) \quad z_k^* = \sum_j c_{kj} z_j + \beta_k w + (\text{terms quadratic in } z) + (\text{terms of weight} \geq 3).$$

From the definition of v^* and equation (2), we get

$$v^* = \operatorname{Im} w^*, \qquad v^* = |z^*|^2 + (\text{terms of weight} \geq 3);$$

comparing the terms of weight 2 in these equations to those in equation (3) shows $(\partial^2 w^*/\partial z_j \partial z_k)\big|_0 = 0$, so that $a_{jk} = 0$. Similarly, (c_{kj}) must be a unitary matrix; therefore, for a suitable rotation

$$(\text{better } \Phi) = (\text{still better } \Phi) \circ (\text{unitary rotation}),$$

where (still better Φ): $(z_k, w) \mapsto (z_k^*, w^*)$ and satisfies $\partial z_j^*/\partial z_k = \delta_{jk}$. Equation (3) now becomes

$$w^* = w + (\text{terms of weight} \geq 3),$$

$$z_k^* = z_k + \beta_k w + (\text{terms quadratic in } z) + (\text{terms of weight} \geq 3).$$

The first term of the Taylor series of the translations (iii), with $\zeta = 0$, is

$$(z_k, w) \mapsto (z_k + \bar{v}_k w/2i, \, w).$$

Choosing the parameters $(v_1, \ldots, v_n, 0)$ of the translation so that $\bar{v}_k/2i = \beta_k$ allows us to factor (still better ϕ) as

$$(\text{still better } \Phi) = (\text{much better } \Phi) \circ (\text{translation } (v_k, 0) \text{ at } \infty),$$

where (much better Φ): $(z_k, w) \mapsto (z_k^*, w^*)$ satisfies

$$w^* = w + (\text{terms of weight} \geq 3),$$

$$v^* = z_k + (\text{terms quadratic in } z) + (\text{terms of weight} \geq 3).$$

Finally we need to consider translations parametrized by $(0, \zeta)$; these take the form

$$(z_k, w) \mapsto \left(\frac{z_k}{1 + \zeta w}, \frac{w}{1 + \zeta w} \right).$$

For appropriate ζ, the second component can be expanded as $w/(1 + \zeta w) = w - \zeta w^2 + \cdots$, which shows that we can find a factorization

$$(\text{much better } \Phi) = \Psi \circ (\text{translation } (0, \zeta) \text{ at } \infty)$$

so that the map $\Psi \colon (z_k, w) \mapsto (z_k^*, w^*)$ satisfies $\operatorname{Re}(\partial^2 w^*/\partial w^2) = 0$. Reviewing the choices made shows that we have constructed a unique Ψ satisfying the conditions of the lemma. $\square$

Let $\mathcal{G}_0$ be the subgroup of $\mathcal{G}_1$ consisting of all holomorphic changes of coordinates about 0

$$\Psi \colon (z, w) \mapsto (z^*, w^*)$$

satisfying

(i) through first order, $\Psi = I$,
(ii) $(\partial^2 w^*/\partial z_j \partial z_k)\big|_0 = 0$,
(iii) $\operatorname{Re}(\partial^2 w^*/\partial w^2)\big|_0 = 0$.

The lemma shows that $\mathcal{G}_1 = \mathcal{G}_0 \circ \mathcal{H}$, uniquely.

In the next section we will prove

THEOREM 2 (MOSER). *If ∂D is given by*

$$v = |z|^2 + (\textit{terms of weight} \geqslant 3),$$

then there is precisely one transformation in $\mathcal{G}_0$ which carries ∂D to a hypersurface in normal form.

COROLLARY. *The only biholomorphic transformations (given by formal power series) taking the hyperquadric H to itself are the fractional linear transformations.*

Clearly now Theorem 1 is reduced to the more precise Theorem 2.

2. Proof of Moser's theorem (Theorem 2). Recall that the boundary ∂D is given by

$$(4) \qquad\qquad v = |z|^2 + F(z, \bar{z}, u),$$

where F is a formal power series of weight $\geqslant 3$, $(z, w) \in \mathbf{C}^{n+1}$ and $w = u + iv$. We will find a tranformation

$$z_k^* = z_k + f_k(z, w), \qquad w^* = w + g(z, w)$$

and a formal power series F^* of weight $\geqslant 3$ satisfying
(i) at the origin,

$$\frac{\partial f_k}{\partial z_l} = \frac{\partial f_k}{\partial w} = \frac{\partial g}{\partial z_l} = \frac{\partial g}{\partial w} = \frac{\partial^2 g}{\partial z_k \partial z_l} = \operatorname{Re}\left(\frac{\partial^2 g}{\partial w^2} \right) = 0;$$

(ii)

$$(5) \quad |z|^2 + 2\operatorname{Re}\langle f, z\rangle + \langle f, f\rangle + F^*(z + f, \bar{z} + \bar{f}, u + \tfrac{1}{2}(g + \bar{g}))$$
$$= |z|^2 + F(z, \bar{z}, u) + \frac{1}{2i}(g - \bar{g}),$$

where f, g are evaluated at (z, w) and $w = u + i|z|^2 + iF(z, \bar{z}, u)$;

(iii) the formal power series F^* satisfies the trace conditions (a), (b), (c). Condition (i) implies that the transformation $(z_k, w) \mapsto (z_k^*, w^*)$ is an element of $\mathcal{G}_0$; conditions (ii) and (iii) imply that (4) is equivalent to

$$v^* = |z|^2 + 2\operatorname{Re}\langle f, z\rangle + \langle f, f\rangle + F^*(z + f, \bar{z} + \bar{f}, u + \tfrac{1}{2}(g + \bar{g}))$$

or

$$v^* = |z^*|^2 + F^*(z^*, \bar{z}^*, u^*),$$

where F^* satisfies the Moser trace conditions (a), (b), (c). We will prove Theorem 2 by showing that given any F, there exists precisely one F^*, and functions f_k, g, such that conditions (i), (ii) and (iii) hold. Let

$$I = \left\{ \begin{array}{lcll} \text{terms in (5) of weight:} & \nu & \text{in } F, F^* \\ & \nu & \text{in } g \\ & \nu - 1 & \text{in } f \end{array} \right\},$$

$$II = \left\{ \begin{array}{lcll} \text{terms in (5) of weight:} & < \nu & \text{in } F, F^* \\ & < \nu & \text{in } g \\ & < \nu - 1 & \text{in } f \end{array} \right\}.$$

LEMMA 2. *The terms of weight ν in* (5) *depend linearly on elements of* I, *nonlinearly on elements of* II, *and not at all on any other terms in f_k and g.*

PROOF OF LEMMA 2. We begin with three observations about the Taylor series expansions of g and f_k. Consider the Taylor series

$$g(z, w)\Big|_{w = u + i|z|^2 + iF(z, \bar{z}, u)} = \sum_\alpha \frac{1}{\alpha!}\left(\partial_w^\alpha g\big|_{u + i|z|^2}\right)\left(iF(z, \bar{z}, u)\right)^\alpha$$

$$= \sum_\alpha \frac{1}{\alpha!}\left(\text{wt. } a_\alpha + [\text{wt.}(a_\alpha) + 1] + \cdots\right)(\text{wt. } 3 + \text{wt. } 4 + \cdots)^\alpha$$

and note

(i) a term of weight ν comes from the product of a term of weight ν_0 in $(\partial_w^\alpha g\big|_{u + i|z|^2})$ with terms of weight $\nu_1, \ldots, \nu_{|\alpha|}$ in $(iF(z, \bar{z}, u))^\alpha$ satisfying $\nu_0 + \nu_1 + \cdots + \nu_{|\alpha|} = \nu$.

(ii) The first term of the series arises from a term of weight $(\nu_0 + 2|\alpha|)$ in g; furthermore, $\nu_1 \geqslant 3, \ldots, \nu_{|\alpha|} \geqslant 3$.

(iii) If $|\alpha| > 0$, then $(\nu_0 + 2|\alpha|)$, $\nu_1, \ldots, \nu_{|\alpha|}$ are all $< \nu$. Observations (i) and (ii) are obvious; to verify (iii), consider

$$\nu = \nu_0 + \nu_1 + \cdots + \nu_{|\alpha|} \geqslant \nu_0 + 3|\alpha| > \nu_0 + 2|\alpha|.$$

Exactly the same observations apply to the Taylor series expansions of f_k. Applying these observations to the Taylor series

$$F^*\big(z + f, \bar{z} + \bar{f}, u + \tfrac{1}{2}(g + \bar{g})\big)$$

$$= \sum_{\alpha,\beta,k} \frac{1}{\alpha!\beta!k!} \partial_z^\alpha \partial_{\bar{z}}^\beta \partial_u^k F^* \big|_{(z,\bar{z},u)} \left(f^\alpha \bar{f}^\beta \left(\frac{g + \bar{g}}{2} \right)^k \right),$$

shows that

$$F^*\big(z + f, \bar{z} + \bar{f}, u + \tfrac{1}{2}(g + \bar{g})\big) = \{\text{terms of wt. } \nu \text{ in } F^*(z, \bar{z}, u)\}$$

$$+ \left\{ \text{expression in terms of wt.} \begin{array}{ll} < \nu & \text{in } F^* \\ < \nu - 1 & \text{in } f \\ < \nu & \text{in } g \end{array} \right\}.$$

Similarly, if we let F_ν, F_ν^*, g_ν, f_ν denote terms of weight ν in the functions F, F^*, g, f, respectively, then (5) gives

$$F_\nu + \operatorname{Im} g_\nu \big|_{w=u+i|z|^2} + \{\text{expression in } F_\mu, g_\mu \text{ for } \mu < \nu\}$$

$$= 2\operatorname{Re}\langle f_{\nu-1}, z \rangle \big|_{w=u+i|z|^2} + \left\{ \text{expression in} \begin{array}{l} f_\mu \text{ for } \mu < \nu - 1 \\ \\ F_\mu \text{ for } \mu < \nu \end{array} \right\}$$

$$+ F_\nu^*(z, \bar{z}, u) + \{\text{expression in } F_\mu^*, g_\mu, F_\mu \text{ for } \mu < \nu\}.$$

and this proves the lemma. $\square$

Given all the F_ν, we can prove the theorem by using induction on the weight ν to solve for F_ν^*, $f_{\nu-1}$, g_ν. Assume that we have already determined the functions for $0, 1, \ldots, \nu - 1$; then by (5)

$$F_\nu + \operatorname{Im} g_\nu \big|_{w=u+i|z|^2} = 2\operatorname{Re}\langle f_{\nu-1}, z \rangle \big|_{w=u+i|z|^2} + F_\nu^*$$

$$+ \{\text{already determined function}\}.$$

We will solve this equation by using the linear operator

$$L^\nu \colon (f_{\nu-1}, g_\nu) \mapsto \big(\operatorname{Im} g_\nu - 2\operatorname{Re}\langle f_{\nu-1}, z \rangle\big) \big|_{w=u+i|z|^2}.$$

Let $\mathfrak{N}_\nu$ denote a complement to the range of L^ν. Since $F_\nu - \{\text{already determined function}\}$ is given and since one can show that L^ν is injective, the equation

$$F_\nu - \{\text{already determined function}\} = F_\nu^* - L^\nu(f_{\nu-1}, g_\nu)$$

uniquely determines $F_\nu^* \in \mathfrak{N}_\nu$ and $f_{\nu-1}$, g_ν. Taking the direct sum over ν gives a linear operator acting on formal power series

$$L \colon (f, g) \to \big(\operatorname{Im} g - 2\operatorname{Re}\langle f, z \rangle\big) \big|_{w=u+i|z|^2},$$

with the property that, if $\mathfrak{N} = \bigoplus_\nu \mathfrak{N}_\nu$, then $\mathfrak{N}$ is a complement to the range of L. If we follow the prescription above, we get unique formal power series f, g and F^* with $F^* \in \mathfrak{N}$ and satisfying equation (5). This leaves conditions (i) and (iii): condition (i) was incorporated in the weight analysis at the beginning of the proof, and an examination of the linear algebra shows that $\mathfrak{N}$ may be taken to be precisely the set of F^* which satisfy trace condition (iii). This finishes our sketch of the proof of Moser's theorem. The reader should note that we did not check the two crucial points of linear algebra: L^ν is injective, and a complement for its range is the space of F^* satisfying the trace conditions. These verifications are the hard parts of Moser's discussion of formal power series.

3. The convergence of the Moser normal form. We will prove in this section that for a surface given by a convergent power series the formal argument of the last section actually gives a convergent change of coordinates into Moser normal form. Note that a surface in normal form always contains the line

$$L = \{z = 0, w \text{ real}\}.$$

The idea of the proof is to expand the power series not around a point but around a curve; and then to find a transformation that maps this curve into the line L.

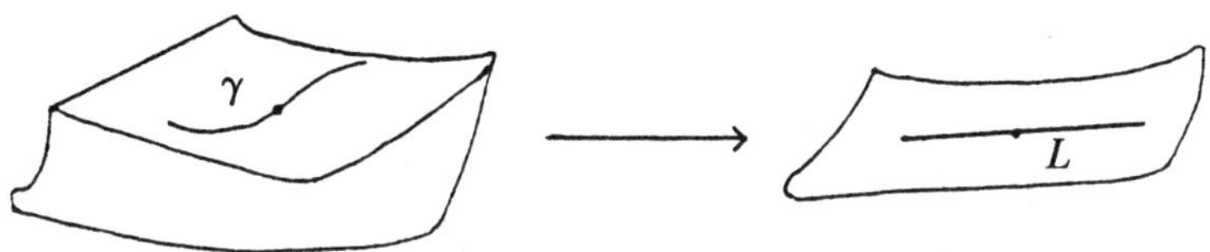

FIGURE 3

Let $t \mapsto \gamma(t)$ denote a real analytic curve in the given surface and assume that $\gamma'(0)$ is transverse to the "holomorphic" part of the tangent space $T(\partial D)$. (That is, $\gamma'(0)$ is tangent to ∂D but does not remain so after multiplication by $\sqrt{-1}$.) Let $\{e_\alpha\}$ be vectors of type $(1, 0)$ varying analytically along γ and spanning the holomorphic part of the tangent space $T(\partial D)$ at each point of γ.

Step 1. As a first step, we place the boundary ∂D into a partial normal form

$$v = \langle z, z \rangle_u + F(z, \bar{z}, u),$$

where $\langle , \rangle_u$ is a Hermitian form depending upon u and F is a convergent power series, all of whose terms contain at least two z_j's and at least two $\bar{z}_j$'s. Recall that $(z, w) \in \mathbf{C}^n \times \mathbf{C}$ and $w = u + iv$. We use the term "partial normal form" to indicate the fact that although the boundary satisfies many of the conditions defining a normal form, no requirements are made yet about the trace conditions. We also demand that the change of coordinates be biholomorphic, that the curve $\gamma = \{t \mapsto \gamma(t)\}$ be sent to the line $L = \{t \mapsto (z = 0, w = t)\}$, and that the frame $\{e_\alpha\}$ be sent to the frame $\{\partial/\partial z_\alpha\}$. Given γ, $\{e_\alpha\}$, we claim that there is precisely one coordinate change meeting these demands.

Step 2. The purpose of this step is to replace the Hermitian form $\langle , \rangle_u$ with one that is independent of u. In fact, we claim that given a curve $\gamma(t)$ and a frame $\{e_\alpha\}$ satisfying

$$\langle e_\alpha, e_\beta \rangle_{\text{Levi form}} = \mu(t)\delta_{\alpha\beta},$$

we can find a unique transformation described by a convergent process taking the boundary into partial normal form

$$v = z \cdot \bar{z} + F(z, \bar{z}, u),$$

the curve γ into the line L, and the frame $\{e_\alpha\}$ into the frame $\{\zeta \partial/\partial z_\alpha\}$, for a suitable scalar factor ζ.

Step 3. In this step we show that by suitably choosing the parametrized curve γ and the frame $\{e_\alpha\}$ we can satisfy the trace conditions. In fact the vanishing of $\mathrm{Trace}^2 F_{32}$ is a second order system of ODE's for the unparametrized curve γ. If we pick a γ that is a solution of this system, then the requirement that $\mathrm{Trace}\, F_{22}$ vanish becomes the condition that a framing of this curve satisfy a certain system of ODE's. Finally, suppose that we chose such a frame. The remaining requirement is that $\mathrm{Trace}^2 F_{33}$ vanish: this becomes a single 3rd-order ODE involving the parametrizations, which can also be satisfied. This completes the sketch of the proof of the convergence of the Moser normal form; we will now fill in a few details.

Proof of Step 1. Say $\gamma(0)$ is the origin and the holomorphic part of $T_{\partial D}\big|_0$ is $\{(z, w) \in \mathbf{C}^n \times \mathbf{C} \mid w = 0\}$. Assume that the curve γ is given by $t \mapsto (p(t), q(t)) \in \mathbf{C}^n \times \mathbf{C}$, where p and q are analytic functions and satisfy $(p(0), q(0)) = 0$. The transversality condition becomes $dq/dt \neq 0$. Suppose that we extend p, q, to complex values of t and use a local change of variables φ such that

$$(z^*, w^*) \mapsto (z, w) = (z^* + p(w^*), q(w^*)).$$

Then φ is biholomorphic and maps the line L to the curve γ via

$$\{t \mapsto (z^* = 0, w^* = t)\} \underset{\varphi}{\to} \{t \mapsto (z = p(t), q(t) = w)\};$$

in other words, the transformation φ^{-1} straightens out the curve γ, allowing us to assume from the start that γ is given by

$$t \mapsto (0, \mathrm{Re}\, w = t, \mathrm{Im}\, w = 0).$$

So far, since the straight line γ lies on ∂D, our surface is given by a power series G satisfying

$$v = G(z, \bar{z}, u), \qquad G(0, 0, u) \equiv 0.$$

We claim that we can find a transformation

$$w^* = w + g(z, w), \qquad g(0, w) \equiv 0, \qquad z^* = z,$$

given by a convergent power series g, that transforms the boundary into

$$v^* = G^*(z^*, \bar{z}^*, u^*),$$

where G^* contains no pure powers $z^{*\alpha}$ or $\bar{z}^{*\alpha}$; in other words, $G^*(z^*, 0, u^*) \equiv 0$ and $G^*(0, \bar{z}^*, u^*) \equiv 0$. Note that the condition $g(0, w) \equiv 0$ means that the line γ is preserved. To verify this claim note that

$$v^* = v + \frac{1}{2i}(g(z, w) - \bar{g}(z, w)) = G^*\!\left(z, \bar{z}, u + \frac{1}{2}(g(z, w) + \bar{g}(z, w))\right),$$

whenever $w = u + iv = u + iG(z, \bar{z}, u)$, and this implies

$$G(z, \bar{z}, u) + \frac{1}{2i}(g(z, w) - \bar{g}(z, w)) = G^*\left(z, \bar{z}, u + \frac{1}{2}(g(z, w) + \bar{g}(z, w))\right).$$

Treating z and $\bar{z}$ as independent and putting $\bar{z} = 0$ gives

$$G(z, 0, u) + \frac{1}{2i}g(z, u + iG(z, 0, u)) = G^*\left(z, 0, u + \frac{1}{2}g(z, u + iG(z, 0, u))\right),$$

since $g(0, w) \equiv 0 \Rightarrow g = \Sigma_{|\alpha| \geqslant 1} c_{\alpha j} z^\alpha w^j \Rightarrow g(\bar{z}, \bar{w})\big|_{\bar{z}=0} = 0$. Consider the equation

$$(6) \qquad\qquad G(z, 0, u) + \frac{1}{2i}g(z, u + iG(z, 0, u)) \equiv 0.$$

This implies $G^*(z, 0, u + \frac{1}{2} g(z, u + iG(z, 0, u))) \equiv 0$, which after the real analytic transformation

$$(z, u) \mapsto \left(z, u + \tfrac{1}{2}g(z, u + iG(z, 0, u))\right),$$

becomes $G^*(z^*, 0, u^*) \equiv 0$, which is what we want to prove; in other words, if we can satisfy (6), we are done. Note that

$$(z, u) \mapsto (z, u + iG(z, 0, u))$$

is also a real analytic transformation; this reduces (6) to

$$\tilde{G}(\tilde{z}, \tilde{u}) + \frac{1}{2i}g(\tilde{z}, \tilde{u}) \equiv 0,$$

which is solved by putting $g = -2i\tilde{G}$. Retracing our steps shows that we have determined a power series $G^*(z^*, \bar{z}^*, u^*)$ which contains no pure powers z^{*^α} or $\bar{z}^{*^\alpha}$.

    This means that we can assume that

$$(7) \qquad\qquad v = \langle z, z \rangle_u + F(z, \bar{z}, u), \qquad F(z, 0, u) \equiv 0,$$

where F contains no terms in $z, \bar{z}$ of type $(1, 1)$. Our second claim is that we can find a biholomorphic change of coordinates

$$z^* = z + f(z, w), \qquad w^* = w,$$

with $f(0, w) \equiv 0$ (to preserve the line γ) and $(\partial f/\partial z_\alpha)(0, w) = 0$, so that (7) is transformed into a surface of the same form, but this time with an F^* involving only $z^\alpha \bar{z}^\beta$, where $|\alpha|, |\beta| \geqslant 2$. To verify this claim consider

$$\begin{aligned}
v &= \langle z^*, z^* \rangle_u + F^*(z^*, \bar{z}^*, u) \\
&= \langle z, z \rangle_u + 2\operatorname{Re}\langle f, z \rangle_u + \langle f, f \rangle_u + F^*(z^*, \bar{z}^*, u).
\end{aligned}$$

Comparing this to (7) gives

$$F^*(z^*, \bar{z}^*, u) = F - 2\operatorname{Re}\langle f, z \rangle_u - \langle f, f \rangle_u.$$

Note that the last term on the right contains the appropriate powers as long as $f(0, w) = (\partial f/\partial z_\alpha)(0, w) = 0$. Given F, we will choose f so as to kill the remaining terms of type $(m, 1)$, for $m > 1$; that is so that $(\partial F^*/\partial \bar{z}_\beta)(z, 0, u) \equiv 0$. This will happen if $(\partial/\partial \bar{z}_\beta)F^*(z^*, 0, u) \equiv 0$, i.e. $\Sigma_\alpha g_{\alpha\beta}(u)f_\alpha(z, w) + \partial F/\partial \bar{z}_\beta = 0$, and if $(g_{\alpha\beta}(u)) > 0$ (which holds since ∂D is strictly pseudoconvex). So we can find the required f by inverting the matrix $(g_{\alpha\beta})$, and this proves the claim.

We now consider the condition on the frame. To review—the boundary is given by (7), with all terms in F containing at least two z's and two $\bar{z}$'s, and γ is the curve $\{z = 0, w = t\}$. Introduce the coordinate transformation

$$z^* = c(w)z, \qquad c(w) \text{ a matrix,}$$
$$w^* = w;$$

this acts on the frame by

$$e_\alpha \mapsto c(t)e_\alpha\big|_{\gamma(t)}.$$

It is easy to find matrices $c(w)$ such that

$$c(t)e_\alpha\big|_{\gamma(t)} = \frac{\partial}{\partial z_\alpha},$$

and it is not much harder to show that these matrices are unique. This concludes Step 1.

Proof of Step 2. Writing (7) as

$$v = \theta(u)|z|^2 + F,$$

suggests the coordinate transformation

$$z^* = c(w)z, \qquad c(w) \text{ a number,}$$
$$w^* = w,$$

giving, for appropriate $c(w)$, $v = |z|^2 + F$. The desired result now follows immediately.

Proof of Step 3. Given a boundary ∂D, a parametrized curve $\gamma: t \to (p(t), q(t)) \in \partial D$ and a frame $(e_\alpha(t))$ defined along γ, we have constructed a biholomorphic map $\Phi: \partial D \to \partial\tilde{D}$, where $\partial\tilde{D} = \{v = |z|^2 + \tilde{F}(z, \bar{z}, u)\}$ and $\tilde{F} = \Sigma_{|\alpha|,|\beta|\geqslant 2} \tilde{F}_{\alpha\bar{\beta}}(u)z^\alpha\bar{z}^\beta$. By following the construction carefully, we can in principle read off the Taylor series of $\tilde{F}$ from those of ∂D, γ, (e_α). In particular, the trace conditions Trace $\tilde{F}_{22}(u) = 0$, Trace^2 $\tilde{F}_{32}(u) = 0$, Trace^3 $\tilde{F}_{33}(u) = 0$ are really conditions on the Taylor series of ∂D, $\gamma(t)$, $e_\alpha(t)$ about the points on the curve γ. Regarding ∂D as fixed, and $\gamma(t)$, $e_\alpha(t)$ as arbitrary, we find after much work that

(a) Trace^2 $\tilde{F}_{32}(u) = 0$ means that the unparametrized curve γ satisfies a second-order ordinary differential equation. The standard existence theorem for ODE's therefore shows that we can pick the unparametrized curve to satisfy the trace condition on $\tilde{F}_{32}$. In particular, the initial direction of γ (transverse to the complex part of $T(\partial D)$) may be prescribed arbitrarily, and the frame $e_\alpha(t)$ and parametrization of γ have no effect on the trace condition for $\tilde{F}_{32}$. γ is called a "chain".

(b) Assume the unparametrized curve has been picked to satisfy Trace^2 $\tilde{F}_{32}(u) = 0$. Then Trace $\tilde{F}_{22}(u) = 0$ if and only if the frame $e_\alpha(t)$ satisfies a first-order ordinary differential equation along γ. Again applying the standard existence theorem for ODE's we conclude that the frame $e_\alpha(t)$ may be picked to satisfy the trace condition on $\tilde{F}_{22}$. We are free to specify the frame at $t = 0$; the propagation of the frame along the curve is then uniquely determined.

(c) Finally, suppose the unparametrized curve γ and the frame (e_α) have been picked as above. Then the last trace condition $\mathrm{Trace}^3\,\tilde{F}_{33}(u) = 0$ becomes a third-order ordinary differential equation for the parametrization of γ. In fact, the equation for the parameter τ involves a Schwartzian derivative, so that τ is uniquely determined up to linear fractional transformations.

Thus, we may transform ∂D to Moser's normal form by convergent power series, by picking first the unparametrized curve γ, then the frame e_α, and finally the parametrization of γ. The choices can all be made properly by solving ordinary differential equations. We may pick arbitrarily an initial direction for γ, an initial frame e_α, and a linear fractional transformation of the line. The number of free parameters agrees exactly with the dimension of the isotropy group $\mathcal{H}$, so all the formal power series transformations to normal form arise from suitable γ, e_α, τ and thus converge. This completes Moser's convergence proof. It is remarkable that one can get away without using anything more than existence of solutions of ODE's.

Note that the convergence proof made repeated use of the assumption that the boundary ∂D was real-analytic. In fact, various series, e.g. $(p(t), q(t))$, are defined initially for real t but are then used for complex t. One can easily give a reasonable definition of Moser's normal form for C^∞ boundaries, but it is also easy to write down smooth strictly pseudoconvex domains which cannot be placed in normal form. Chains may still be defined on C^∞ boundaries as solution curves for a family of second-order differential equations; the equations are obtained by osculating the C^∞ boundary to high order with a real-analytic boundary.

The theorems of this chapter are due to Moser [48] and Chern and Moser [11].

CHAPTER 10. CHERN'S THEOREM

The first half of this chapter contains a description of the relevant bundles, groups and forms, beginning with the flat case and then considering the general case. The second half of the chapter consists of a statement of the theorem and a brief outline of its proof.

1. The flat case. In this chapter we will change notation slightly and denote the boundary of the Siegel domain as

$$Q = \left\{ (\omega, z) \in \mathbf{C} \times \mathbf{C}^n \colon \mathrm{Im}\,\omega = \sum_1^n |z_\alpha|^2 \right\}.$$

Using the quadratic form

$$\langle v, v' \rangle = \frac{-i}{2}\left(v_{n+1}\bar{v}_0' - v_0\bar{v}_{n+1}' \right) - \sum_1^n v_\alpha \bar{v}_\alpha'$$

in $\mathbf{C}^{n+2}$ to define

$$E^* = \left\{ v \in \mathbf{C}^{n+2} \colon \langle v, v \rangle = 0 \right\},$$

allows us to construct a line bundle

$$E^* \to Q, \qquad v \mapsto \left(\frac{v_{n+1}}{v_0}, \frac{v}{v_0} \right) = (\omega, z).$$

There are several groups naturally associated with this bundle:

(i) The indefinite unitary group $U(n + 1, 1)$ consisting of matrices over $\mathbf{C}^{n+2}$ preserving the quadratic form $\langle \, , \, \rangle$.

(ii) The group of Heisenberg translations $N \subset U(n + 1, 1)$. For $(\omega', z') \in Q$, the translation $T_{(\omega', z')} \in N$ is defined by

$$T_{(\omega', z')} \colon (\omega, z) \mapsto (\omega + \omega' + 2iz \cdot z', z + z')$$

and is identified with the matrix

$$\left(
\begin{array}{c|ccc|c}
1 & 0 & \cdots & 0 & 0 \\
\hline
z_1' & & & & 0 \\
\vdots & & I & & \vdots \\
z_n' & & & & 0 \\
\hline
\omega' & 2i\bar{z}_1' & \cdots & 2i\bar{z}_n' & 1
\end{array}
\right) \in U(n + 1, 1).$$

(iii) The Heisenberg dilations $\mathbf{R}^\times \subseteq U(n + 1, 1)$. For $t \in \mathbf{R}^\times$, the Heisenberg dilation D_t acts on Q by

$$D_t \colon (\omega, z) \mapsto (t^2 \omega, tz)$$

and is identified with the matrix

$$\left(
\begin{array}{c|c|c}
t^{-1} & 0 & 0 \\
\hline
0 & I & 0 \\
\hline
0 & 0 & t
\end{array}
\right) \in U(n + 1, 1).$$

(iv) The Heisenberg rotations $U(n) \subseteq U(n + 1, 1)$. The matrix $(U_{\alpha\beta}) \in U(n)$ is associated with the rotation $R_{U_{\alpha\beta}}$ which acts on Q by

$$R_{U_{\alpha\beta}} \colon (\omega, z) \mapsto \left(\omega, \sum_\beta U_{\alpha\beta} z_\beta \right)$$

and is identified with the matrix

$$\left(
\begin{array}{c|c|c}
1 & 0 & 0 \\
\hline
0 & U_{\alpha\beta} & 0 \\
\hline
0 & 0 & 1
\end{array}
\right).$$

(v) The isotropy groups H^+ and H. These are defined by

$$H^+ = \{\text{group of LFT of } Q \text{ fixing the origin}\}$$
$$= \{T \in U(n + 1, 1) \colon Te^0 = \lambda e^0, \text{ for some } \lambda \in \mathbf{C}, \text{ modulo } S^1\},$$

$$H = \{\text{group of LFT } \phi \text{ of } Q \text{ fixing } 0 \text{ with } |\det \phi'(0)| = 1\}$$
$$= \{T \in U(n+1,1): Te^0 = e^0\},$$

where LFT = linear fractional transformation, S^1 is thought of as the unit circle in $\mathbf{C}$, and $(e^0,\ldots,e^{n+1})$ is the standard basis of $\mathbf{C}^{n+2}$. If we think of $H \subseteq U(n+1,1)$ as matrices, then H consists of matrices of the form

$$\begin{pmatrix} 1 & v_1 \cdots v_n & s + i\sum |v_\alpha|^2 \\ 0 & & \dfrac{i}{2}\sum_\alpha u_{1\alpha}\bar{v}_\alpha \\ \vdots & u_{\alpha\beta} & \vdots \\ 0 & & \dfrac{i}{2}\sum_\alpha u_{n\alpha}\bar{v}_\alpha \\ 0 & 0 \cdots 0 & 1 \end{pmatrix},$$

where $(u_{\alpha\beta})$ is unitary, $v_\alpha \in \mathbf{C}$ and $s \in \mathbf{R}$. The group H^+ is generated by H and the Heisenberg dilations and consists of matrices of the obvious form. Recall that the Lie algebras of $U(n+1,1)$, N, H^+, H are denoted by $u(n+1,2)$, n, h^+, h and were described in a previous chapter, where it was shown that

$$(1) \qquad\qquad u(n+1,1) \cong n + h^+ .$$

The next step is to construct a frame bundle

$$\begin{array}{c} Y^* \\ \pi \downarrow \\ E^* \end{array},$$

where Y^* consists of all frames $(z_0,\ldots,z_{n+1})$ such that

$$\langle z_j, z_k \rangle = \langle e^j, e^k \rangle$$

and the projection is

$$(z_0,\ldots,z_{n+1}) \mapsto z_0.$$

Here $z_j \in \mathbf{C}^{n+2}$ and z_0 satisfies $\|z_0\|^2 = \|e^0\|^2 = 0$. If we associate the frame $(z_0,\ldots,z_{n+1})$ with the matrix which maps

$$(e^0,\ldots,e^{n+1}) \mapsto (z_0,\ldots,z_{n+1}),$$

then the space Y^* can be identified with $U(n+1,1)$.

The splitting (1) defines a natural connection in the frame bundle. At a point $p \in Y^*$, the tangent space splits

$$T_p U(n+1,1) \cong u(n+1,1) \cong n + h^+ ,$$

and we can use

$$h^+ \cong T_p\big(\pi^{-1}(\text{point of } E^*)\big) = \text{``verticle part of } T_p Y^* \text{''}$$

to define a connection; in other words, the connection is defined by putting n as the horizontal part of $T_p Y^*$.

The final step is to mod out by the S^1 action consisting of multiplication by a complex number with modulus 1. This gives bundles

$$
\begin{array}{c}
Y \\
\downarrow \\
E \\
\downarrow \\
Q
\end{array}
$$

where Y is a principal H-bundle over E; that is, the fibres are isomorphic to H. Roughly speaking, in the general case Chern tries to find bundles over C-R manifolds and connections that resemble those in the flat case.

2. The general case. Let M be a $(2n + 1)$-dimensional real manifold which is strictly ψ-convex and has an integrable C-R structure $J^2 = -I$. This defines forms of type $(1, 0)$ and $(0, 1)$ on M:

$$
\left.
\begin{aligned}
\omega(JX) = i\omega(X) \quad &\text{type} \quad (1, 0) \\
= -i\omega(X) \quad &\text{type} \quad (0, 1)
\end{aligned}
\right\},
$$

where $X \in HM \equiv \{\text{complex part of } TM\}$.

Using the C-R structure, we can construct an $\mathbf{R}^+$ bundle over M

$$
\begin{array}{c}
E \\
\pi^1 \downarrow \\
M
\end{array}
$$

by specifying the fibre E_p, for $p \in M$:

$$
E_p = \{\text{real covectors } \omega \in T_p^*M : \omega \perp H_p M\}.
$$

A $(1, 0)$- or $(0, 1)$-form on E is defined to be the pullback of a $(1, 0)$- or $(0, 1)$-form on M.

In order to define the analogue of the bundle of frames, first we need to define tautological forms. The tautological 1-form $\hat{\omega}$ on E is defined by

$$
\hat{\omega}(X) = \left\langle \omega, \pi_*^1 X \right\rangle,
$$

for $X \in T_{(q, \omega)} E$ and $q \in M$. This makes sense since $\omega \in T_q^*M$ and $\pi_*^1 X \in T_q M$.

LEMMA 1. *The tautological 1-form $\hat{\omega}$ on E satisfies*

$$
\text{(S1)} \qquad d\hat{\omega} = i\sum_\alpha \omega^\alpha \wedge \overline{\omega}^\alpha + \hat{\omega} \wedge \phi,
$$

where ω^α are $(1, 0)$-forms on E and ϕ is a real 1-form on E.

The lemma is an immediate consequence of the strict ψ-convexity of M. After these preliminaries, we can define a bundle

$$
\text{(2)} \qquad
\begin{array}{c}
Y \\
\pi^2 \downarrow \\
E
\end{array}
$$

by defining the fibre to be

$$Y_q = \{\, (1,0)\text{-covectors } \omega^\alpha \text{ and } 1\text{-covectors } \phi \text{ at } q \in E\colon (S1) \text{ holds}\,\}.$$

With the obvious projection this becomes a bundle.

LEMMA 2. *The bundle $Y \to E$ is a principal H-bundle.*

PROOF. We want to define an action of H on the fibres. Given a point $(q, \omega^{\alpha*}, \phi^*) \in Y_q$ and

$$\begin{pmatrix} 1 & v_1 \cdots v_n & s + i \sum |v^\alpha|^2 \\ 0 & & \frac{i}{2} \sum u_{1\alpha} \bar{v}_\alpha \\ \vdots & u_{\alpha\beta} & \vdots \\ 0 & & \frac{i}{2} \sum u_{n\alpha} \bar{v}_\alpha \\ 0 & 0 \cdots 0 & 1 \end{pmatrix} \in H,$$

define

(3)
$$T(\omega^{\alpha*}, \phi^*) = (\omega^\alpha, \phi),$$

where

$$\omega^\alpha = v_\alpha \hat{\omega} + \sum_\beta u_{\alpha\beta} \omega^{\beta*},$$

$$\phi = s\hat{\omega} + i \sum_\alpha \bar{v}_\alpha \omega^{\alpha*} - i \sum_\alpha v_\alpha \overline{\omega^{\alpha*}} + \phi^*.$$

It is easy to verify that $(q, \omega^\alpha, \phi) \in Y_q$ and that the bundle is principal. $\square$

Our next task is to define the connection; this too involves tautological forms. Let $\xi = (q, \omega^\alpha, \phi) \in Y$ and $X \in T_\xi Y$. The tautological 1-forms $\overset{+}{\omega}$, $\overset{+}{\omega}{}^\alpha$ and $\overset{+}{\phi}$ on Y are defined by

$$\overset{+}{\omega} = (\pi^2)^* \hat{\omega}, \qquad \overset{+}{\omega}{}^\alpha(X) = \langle \omega^\alpha, \pi^2_* X \rangle, \qquad \overset{+}{\phi}(X) = \langle \phi, \pi^2_* X \rangle,$$

where $\hat{\omega}$ is the tautological 1-form on E. This is well defined since $\omega^\alpha \in T_q^* E$ and $\pi^2_* X \in T_q E$. Note that (S1) holds on Y:

(S1)
$$d\overset{+}{\omega} = i \sum \overset{+}{\omega}{}^\alpha \wedge \overline{\overset{+}{\omega}{}^\alpha} + \overset{+}{\omega} \wedge \overset{+}{\phi}.$$

Recall that a connection for the principal H-bundle (2) is an h-valued 1-form on Y that is invariant under the H-action (3) in the fibre; in other words, a connection is a matrix (π_{pq}) of 1-forms such that $(\pi_{pq}(X)) \in h$, for $X \in TY$. The "horizontal subspace" of the connection is the kernel of this 1-form.

In the flat case $M = Q$ described in the first section, we used the splitting (1) of the tangent space to define a connection in the principal H-bundle $Y \to E$. This determines the connection 1-form (π_{pq}). Not only does the first structural equation (S1) hold in the flat case, but also the following equations:

$$(S2)_{\text{flat}} \qquad d\overset{+}{\omega}{}^{\alpha} = \sum_{\beta} \overset{+}{\omega}{}^{\beta} \wedge \phi_{\beta\bar{\alpha}} + \overset{+}{\omega} \wedge \bar{\phi}_{\alpha},$$

$$(S3)_{\text{flat}} \qquad d\overset{+}{\phi} = i\sum \overset{+}{\omega}{}^{\beta} \wedge \phi_{\beta} + i\sum \bar{\phi}_{\beta} \wedge \overline{\overset{+}{\omega}{}^{\beta}} + \overset{+}{\omega} \wedge \psi,$$

where the forms $\overset{+}{\omega}$, $\overset{+}{\omega}{}^{\alpha}$, ϕ_{α}, $\phi_{\alpha\beta}$ and ψ are defined by the connection according to

$$(4) \qquad \overset{+}{\omega} = \frac{1}{2}\pi_{n+1,0}, \qquad \overset{+}{\omega}{}^{\alpha} = \pi_{\alpha 0}, \qquad \overset{+}{\phi} = -\pi_{00} + \pi_{n+1,n+1},$$

$$\phi\alpha = 2\bar{\pi}_{\alpha,n+1}, \qquad \psi = -4\pi_{0,n+1}, \qquad \phi_{\alpha\bar{\beta}} = \pi_{\beta\alpha} - \delta_{\alpha\beta}\pi_{00}.$$

That is, knowning the connection is equivalent to knowing all the forms in (S1), (S2) and (S3). Finally, in the flat case we can check that the following equations are also valid:

$$(S4)_{\text{flat}} \qquad d\phi_{\beta\bar{\alpha}} = \sum_{\gamma} \phi_{\beta\bar{\gamma}} \wedge \phi_{\gamma\bar{\alpha}} + i\,\overline{\overset{+}{\omega}{}^{\beta}} \wedge \bar{\phi}_{\alpha}$$

$$- i\phi_{\beta} \wedge \overset{+}{\omega}{}^{\alpha} - i\delta_{\beta\alpha}\left(\sum_{\sigma} \phi_{\sigma} \wedge \overset{+}{\omega}{}^{\sigma} \right) - \frac{1}{2}\delta_{\beta\alpha}\psi \wedge \overset{+}{\omega},$$

$$(S5)_{\text{flat}} \qquad d\bar{\phi}_{\alpha} = \overset{+}{\phi} \wedge \bar{\phi}_{\alpha} + \sum_{\beta} \bar{\phi}_{\beta} \wedge \phi_{\beta\bar{\alpha}} - \frac{1}{2}\psi \wedge \overset{+}{\omega},$$

$$(S6)_{\text{flat}} \qquad d\psi = \overset{+}{\phi} \wedge \psi + 2i\sum_{\beta} \bar{\phi}_{\beta} \wedge \phi_{\beta},$$

$$(T1) \qquad \phi_{\alpha\bar{\beta}} + \bar{\phi}_{\beta\bar{\alpha}} - \delta_{\alpha\bar{\beta}}\,\overset{+}{\phi} = 0 \quad \text{and} \quad \psi \text{ is real.}$$

In the general case in which M is no longer assumed to be Q, rather than define the connection immediately, we will first write down the appropriate equations that the tautological forms satisfy. Equations (S1), (S2) and (S3) are the same as (S1), (S2)$_{\text{flat}}$ and (S3)$_{\text{flat}}$. Equation (S4) is obtained by adding

$$+ \sum_{\rho\sigma} S_{\beta\rho\bar{\alpha}\bar{\sigma}} \overset{+}{\omega}{}^{\rho} \wedge \overline{\overset{+}{\omega}{}^{\sigma}} + \sum_{\rho} V_{\beta\bar{\alpha}\rho} \overset{+}{\omega}{}^{\rho} \wedge \overset{+}{\omega} - \sum_{\rho} \bar{V}_{\beta\bar{\alpha}\rho} \overset{+}{\omega}{}^{\rho} \wedge \overset{+}{\omega}$$

to the right-hand side of (S4)$_{\text{flat}}$; similarly, (S5) arises by adding

$$- \sum_{\beta\gamma} V_{\beta\bar{\alpha}\gamma} \overset{+}{\omega}{}^{\beta} \wedge \overset{+}{\omega}{}^{\gamma} + \sum_{\beta\sigma} \bar{V}_{\alpha\bar{\beta}\sigma} \overset{+}{\omega}{}^{\beta} \wedge \overline{\overset{+}{\omega}{}^{\alpha}} + \sum_{\beta} P_{\beta\bar{\alpha}} \overset{+}{\omega}{}^{\beta} \wedge \overset{+}{\omega} + \sum_{\beta} Q_{\beta\bar{\alpha}} \overline{\overset{+}{\omega}{}^{\beta}} \wedge \overset{+}{\omega}$$

to the right-hand side of (S5)$_{\text{flat}}$. Finally the correction term that yields (S6) from (S6)$_{\text{flat}}$ is another complicated expression that involves not only the tensors S, V,

P and Q but also R. These tensors are required to satisfy the symmetry conditions

$$S_{\beta\gamma\bar{\alpha}\bar{\sigma}} = S_{\gamma\beta\bar{\sigma}\bar{\alpha}} = S_{\gamma\beta\bar{\alpha}\bar{\sigma}} = \bar{S}_{\alpha\sigma\bar{\gamma}\bar{\beta}},$$

(T2) $$\sum_\alpha S_{\alpha\gamma\bar{\alpha}\bar{\sigma}} = 0, \qquad \sum_\rho V_{\beta\bar{\rho}\rho} = 0, \qquad \mathrm{Re}\left(\sum_\alpha P_{\alpha\bar{\alpha}}\right) = 0.$$

THEOREM 1 (CHERN). (i) *In the situation above we can find unique 1-forms* $\phi_{\beta\bar{\alpha}}$, ϕ_α *and* ψ *on Y and functions*

$$S_{\beta\gamma\bar{\alpha}\bar{\sigma}}, \qquad V_{\alpha\bar{\beta}\sigma}, \qquad P_{\beta\bar{\alpha}}, \qquad Q_{\bar{\beta}\bar{\alpha}}, \qquad R_\sigma$$

which satisfy (S1)-(S6) *and* (T1), (T2).

(ii) *There exists a unique connection* (π_{pq}) *on Y so that* (4) *holds.*

(iii) *All the forms and functions above can be read off at a given point of Y just by knowning finitely many terms in the Taylor expansion of the C-R-structure at the given point of M.*

3. The ideas in the proof of Chern's theorem. We begin this section with some general comments about Cartan's method of equivalence and end with a brief description of how this method is used to prove Chern's theorem. As background, we show how the method of equivalence can be used in Riemannian geometry.

Given a map

$$(M, \text{geometric structure}) \to (M', \text{same type of geometric structure}),$$

Cartan's method provides a means of determining whether locally there exists a diffeomorphism $f\colon M \to M'$ that preserves this structure.

EXAMPLE. A G-structure on a manifold M consists of a coframe θ_j defined up to the equivalence

(5) $$\theta_j^* = \sum_k \mu_{jk}\theta_k,$$

where

$$(\mu_{jk}) \in G \subseteq \mathrm{GL}(n, \mathbf{R}).$$

Assume now that $G = \mathrm{I}$. We say that f preserves the G-structure if

$$f^*(\theta_j') = \theta_j, \qquad j = 1,\ldots,n.$$

If we define c_{jkl} and c'_{jkl} by

$$d\theta_j = \sum_{k,l} c_{jkl}\theta_k \wedge \theta_l, \qquad d\theta_j' = \sum_{k,l} c'_{jkl}\theta_k' \wedge \theta_l',$$

then $c_{jkl} = c'_{jkl}$, whenever f preserves the G-structure. This is because $f^*(d\theta_j') = d\theta_j$. There are two extreme cases of interest.

(i) c_{jkl} contain a coordinate system,

(ii) c_{jkl} are all constant.

Given a general G, we would like to reduce to the case $G = \mathrm{I}$.

Step 1. The group G can be reduced by using the (μ_{jk}) in (5) to determine certain of the c_{jkl}. If this procedure does not reduce G to I, then we go on to Step 2.

Step 2. In this step we "prolong" the G-structure on M to a G'-structure on $M \times G$. The G'-structure is defined by a coframe $\{\omega_j, \omega_{jk}\}$, where

$$\omega_j = \sum_k \mu_{jk}\theta_k, \qquad \omega_{jk} = d\mu_{jk} + \text{(other terms)}.$$

We now apply Step 1 with the new G'-structure. Note that this procedure does not necessarily stop, since G' may be larger than G.

As an illustration of this method, consider the Riemannian case in which the coframes determine a metric

$$ds^2 = \theta_1^2 + \cdots + \theta_n^2,$$

which is defined up to the equivalence

$$\theta_j^* = \mu_{jk}\theta_k, \qquad \text{where } \mu_{jk}\mu_{ik} = \delta_{ij}.$$

Here, and throughout the remainder of the section, we will use the summation convention. The coframe determines the structure constants c_{jkl} by

$$d\theta_j = c_{jkl}\theta_k \wedge \theta_l,$$

and we calculate that

$$d\theta_j^* = d\mu_{jk} \wedge \mu_{ik}\theta_i^* + \mu_{jk}c_{klm}\theta_l \wedge \theta_m.$$

Applying Step 1 to this equation gets us into trouble, because of the appearance of the $d\mu_{jk}$. Instead we prolong according to Step 2 by introducing the new coframe $\{\omega_j, \omega_{jk}\}$ defined by

$$\omega_j = \mu_{jk}\theta_k, \qquad d\omega_j = \omega_{ji} \wedge \omega_i + T_{jkl}\omega_k \wedge \omega_l,$$

where

$$\omega_{ij} + \omega_{ji} = 0, \qquad T_{jkl} + T_{jlk} = 0.$$

T_{jkl} is the torsion tensor. Using the equivalence (5) we get

$$(6) \qquad d\omega_j = \omega_{ji} \wedge \omega_i + T_{jkl}\omega_k \wedge \omega_l = \omega_{ji}^* \wedge \omega_i + T_{jkl}^*\omega_k \wedge \omega_l$$

or

$$\left(\omega_{ji} - \omega_{ji}^*\right) \wedge \omega_i + \left(T_{jkl} - T_{jkl}^*\right)\omega_k \wedge \omega_l = 0,$$

$$\left[\omega_{ji} - \omega_{ji}^* + \left(T_{jki} - T_{jki}^*\right)\omega_k\right] \wedge \omega_i = 0,$$

which by linear algebra yields

$$\omega_{ji} - \omega_{ji}^* = a_{jik}\omega_k, \qquad \text{where } a_{jik} + a_{ijk} = 0,$$

by the skew symmetry of the ω_{ij}. (The linear algebra used here is called the *Cartan Lemma*.) We can now apply Step 1 using the two equations

$$\omega_{ji} = \omega_{ji}^* + a_{jik}\omega_k,$$

$$\left(\tfrac{1}{2}(a_{jik} - a_{jki}) + T_{jki} - T_{jki}^*\right)\omega_k \wedge \omega_l = 0$$

for G'; this requires that we determine the a_{ijk} such that $T_{ijk}^* = 0$ in the $nn(n + 1)/2 + nn(n - 1)/2 = n^3$ equations

$$a_{jik} + a_{ijk} = 0, \qquad a_{jik} - a_{jki} = -2T_{jki}.$$

This system can be solved using the trick of "rotating indices", in other words, by suitably permuting the indices. We find that

$$(7) \qquad d\omega_{ij} = \omega_{ik} \wedge \omega_{kj} + R_{ijkl}\omega_k \wedge \omega_l,$$

where R_{ijkl} is the curvature tensor. Equations (6) and (7) are the Cartan structural equations, which we have derived through his method of equivalence.

We now turn to the proof of Chern's theorem. Let M^{2n+1} be a C-R manifold and let $\mathcal{I}$ be a closed complex differential ideal generated by 1-forms. In the case of the hypersurface Q considered in the first section, $\mathcal{I}$ is the ideal generated by the restrictions of $(1,0)$-forms to Q. Let $\mathcal{I}_1 = \{1\text{-forms in } \mathcal{I}\}$, so that

$$\mathcal{I}_1 + \bar{\mathcal{I}}_1 = \Lambda^1, \qquad \mathcal{I}_1 \cap \bar{\mathcal{I}}_1 = (\theta),$$

where Λ^1 denotes the 1-forms on M and θ is the annihilator of the maximal complex tangent space HM. Choose forms θ^α so that $\{\theta, \theta^\alpha\}$ is a basis for $\mathcal{I}$ and consider the equivalence

$$\theta^* = u\theta, \qquad \theta^{\alpha*} = v^\alpha\theta + u^\alpha_\beta\theta^\beta, \qquad \theta^{\bar{\alpha}*} = v^{\bar{\alpha}}\theta + u^{\bar{\alpha}}_{\bar{\beta}}\theta^{\bar{\beta}},$$

where the notation of the first section has been changed slightly so that $u_{\alpha\beta}$ is now denoted u^α_β. Because of the integrability condition $d\mathcal{I} \subset \mathcal{I}$, we can write

$$d\theta = ih_{\alpha\bar{\beta}}\theta^\alpha \wedge \theta^{\bar{\beta}} \quad \text{mod } \theta.$$

Once again we must prolong; but this time instead of prolonging by looking at $M \times G$, we will prolong "along θ". We can define a basis $\{\omega, \omega^\alpha, \omega^{\bar{\alpha}}, \phi\}$ for a coframe on $E = M \times \mathbf{R}$ by using the equations

$$(8) \qquad \begin{aligned} \omega &= u\theta, \\ d\omega &= iuh_{\alpha\bar{\beta}}\theta^\alpha \wedge \theta^{\bar{\beta}} + \omega \wedge (-du/u + \text{something}) \\ &= ig_{\alpha\bar{\beta}}\omega^\alpha \wedge \omega^{\bar{\beta}} + \omega \wedge \phi, \\ \omega^\alpha &= \text{multiples of } \theta^\alpha. \end{aligned}$$

The basis is defined up to the equivalence

$$\omega^* = \omega, \qquad \omega^{\alpha*} = v^\alpha\omega + u^\alpha_\beta\omega^\beta, \qquad \omega^{\bar{\alpha}*} = v^{\bar{\alpha}}\omega + u^{\bar{\alpha}}\omega + u^{\bar{\alpha}}_{\bar{\beta}}\omega^{\bar{\beta}},$$

$$\phi^* = s\omega + ig_{\rho\bar{\sigma}}\left(u^\rho_\beta v^{\bar{\sigma}}\omega^\beta - u^{\bar{\sigma}}_{\bar{\beta}}v^\rho\omega^{\bar{\beta}}\right) + \phi.$$

In order to apply Step 1 to this prolongation we need an expression for $d\phi$ and $d\phi^\alpha_\beta$. To begin this calculation, differentiate (8) to get

$$0 = i\left(dg_{\alpha\bar{\beta}} - g_{\gamma\bar{\beta}}\phi^\gamma_\alpha + g_{\alpha\bar{\gamma}}\phi^{\bar{\gamma}}_{\bar{\beta}} - g_{\alpha\bar{\beta}}\phi\right)\omega^\alpha \wedge \omega^{\bar{\beta}} + \omega \wedge (d\phi + \text{something}).$$

A bit of linear algebra shows us that the ϕ^γ_α can be chosen so that

$$dg_{\alpha\bar{\beta}} - g_{\gamma\bar{\beta}}\phi^\gamma_\alpha + g_{\alpha\bar{\gamma}}\phi^{\bar{\gamma}}_{\bar{\beta}} - g_{\alpha\bar{\beta}}\phi = 0;$$

which means $\omega \wedge (d\phi + \text{something}) = 0$ or $d\phi = -(\text{something})$. In fact, we find

$$d\phi = i\omega_{\bar{\beta}} \wedge \phi^{\bar{\beta}} + i\phi_{\bar{\beta}} \wedge \omega^{\bar{\beta}} + \omega \wedge \psi,$$

where $\omega_{\bar{\beta}} = g_{\alpha\bar{\beta}}\omega^{\alpha}$, etc. In order to find a similar expression for $d\phi_{\beta}^{\alpha}$, note that (8) and the integrability condition give

$$d\omega^{\alpha} = \omega^{\beta} \wedge \phi_{\beta}^{\alpha} + \omega \wedge \phi^{\alpha}.$$

Differentiating this and working very hard yields the desired expression for $d\phi_{\beta}^{\alpha}$. Using Step 1 now, we find, after many uses of Cartan's Lemma, that G has been reduced to the identity. This concludes the sketch of the proof of Chern's theorem.

The theorems in this chapter are due independently to Chern-Moser [11] and Tanaka [58, 59] who built upon the early work of Cartan in two variables [8].

4. Applications. We explain briefly how to use Theorem 1 to define chains, and also to write down finitely many differential equations on ψ which hold if and only if $D = \{\psi(z) > 0\}$ is locally biholomorphic to the ball. To find the differential equations, we use a result of Chern in [11] that the C-R manifold M is flat (i.e. locally biholomorphic to a sphere) if and only if the "curvatures" $(S_{\beta\gamma\bar{\alpha}\bar{\sigma}})$, $(V_{\alpha\bar{\beta}\sigma})$, $(P_{\beta\bar{\alpha}})$, $(Q_{\bar{\beta}\bar{\alpha}})$, (R_{σ}) in Theorem 1 all vanish. This is analogous to the fact that a Riemannian metric is locally isometric to R^n if and only if the curvature R_{ijkl} vanishes. Since $\psi \to [(S_{\beta\gamma\bar{\alpha}\bar{\sigma}})$ etc. associated to the boundary $M = \{\psi(z) = 0\}]$ is a nonlinear differential operator by Theorem 1, part (iii), it follows that the boundaries $\{\psi = 0\}$ locally biholomorphic to the sphere have been characterized by finitely many differential equations on ψ.

Regarding the chains, we start by recalling that the connection in Theorem 1 canonically identifies the tangent space $T_p Y$ at each point with the Lie algebra $u(n + 1, 1)$. Therefore, any element $\xi \in u(n + 1, 1)$ gives rise to a vector field on Y. Integrating the vector field, we get family of parametrized curves in Y. Finally, projecting from Y down to M, we associate to each $\xi \in u(n + 1, 1)$ a distinguished family of parametrized curves on M. For the particular element

$$\xi_0 = \begin{pmatrix} 0 & & & \\ \vdots & & 0 & \\ \vdots & & & \\ 0 & & & \\ 1 & 0 & \cdots & 0 \end{pmatrix} \in u(n + 1, 1)$$

(infinitesimal Heisenberg translation by $(0, 0, \ldots, 0, \omega)$) the family of curves obtained by this procedure agrees with Moser's chains. Since the main step here is to integrate a vector field, it is clear that chains are given by solutions of ordinary differential equations. Curiously, the parametrization of chains given above in terms of Y does not agree with Moser's parametrization. Theorem 1 also allows us to define a notion of parallel transport of frames, which agrees with Moser's definition in Chapter 9. In two dimensions, these ideas go back to E. Cartan [8], who also studied the distinguished curves corresponding to other ξ in the Lie algebra.

The results on chains stated here are proved by computing the connection in the special case of a boundary in normal form. Details are formidable.

CHAPTER 11. THE COMPLEX MONGE-AMPÈRE EQUATION

The first section discusses the formal properties of a complex Monge-Ampère equation, while the second section sketches a proof of the Cheng-Yau theorem. This theorem establishes existence, uniqueness, C^∞ interior regularity, and some regularity up to the boundary for solutions of the basic equation. The final section describes chains and local invariants of pseudoconvex boundaries as an application of the formal study.

1. The Monge-Ampère equation. The problem is to find on a given strictly pseudoconvex domain $D \subseteq \mathbf{C}^N$ a complete Kähler-Einstein metric

$$ds^2 = \sum_{jk} g_{j\bar{k}}(z)\, dz_j\, \overline{dz_k}\,.$$

A Kähler metric ds^2 is called Kähler-Einstein if $\mathrm{Ric}_{j\bar{k}} = -kg_{j\bar{k}}$, where k is a constant which we take here to be positive. Recall that the Ricci tensor for a Kähler metric is given by

$$\mathrm{Ric}_{j\bar{k}} = c_n \frac{\partial^2}{\partial z_j \partial \bar{z}_k} \log \det(g_{\mu\bar{\nu}}).$$

So we are looking for solutions $g_{j\bar{k}}$ of

$$(1) \qquad g_{j\bar{k}} = (\text{const}) \frac{\partial^2}{\partial z_j \partial \bar{z}_k} \log \det(g_{\mu\bar{\nu}}).$$

Setting $\zeta = \det(g_{\mu\bar{\nu}})$, we see that solving (1) is equivalent to solving the complex Monge-Ampère equation

$$(\text{M-A1}) \qquad \det\left(\frac{\partial^2}{\partial z_j \partial \bar{z}_k} \log \zeta \right) = (\text{const})\zeta \quad \text{in } D,$$
$$\zeta \to \infty \qquad \text{at } \partial D,$$

since if ζ solves (M-A1) then $g_{j\bar{k}} = (\text{const})(\partial^2/\partial z_j \partial \bar{z}_k) \log \zeta$ is a solution of the Kähler-Einstein equation (1).

EXAMPLE. For the unit ball $D = \{ |z| < 1 \} \subseteq \mathbf{C}^n$,

$$(2) \qquad \zeta = c_n / \left(1 - |z|^2 \right)^{n+1}$$

is a solution of (M-A1).

This suggests that for a more general domain D, we try a solution of the form

$$(3) \qquad \zeta = c_n / \left(u(z) \right)^{n+1},$$

where $u(z) = 0$ at ∂D to exactly first order. Rewriting the Monge-Ampère equation (M-A1) in terms of $u(z)$ gives

$$(\text{M-A2}) \qquad \begin{array}{ll} J(u) = 1 & \text{in } D, \\ u = 0 & \text{at } \partial D, \end{array}$$

where

$$
J(u) = (-1)^n \det
\begin{vmatrix}
u & \dfrac{\partial u}{\partial \bar{z}_1} & \cdots & \dfrac{\partial u}{\partial \bar{z}_n} \\
\dfrac{\partial u}{\partial z_1} & & & \\
\vdots & & \dfrac{\partial^2 u}{\partial z_j \partial \bar{z}_k} & \\
\dfrac{\partial u}{\partial z_n} & & &
\end{vmatrix} .
$$

In constructing the formal solution to this equation, we shall see that "much" of the Taylor series of $u(z)$ at ∂D is specified infinitesimally. This is in contrast to the Dirichlet problem for the Laplace equation

$$
\begin{aligned}
\Delta u = 0 & \quad \text{on } \mathbf{R}^{n+1}_+, \\
u = f & \quad \text{on } \mathbf{R}^n,
\end{aligned}
\tag{4}
$$

where $\mathbf{R}^{n+1}_+ = \{(x_1, \ldots, x_n, t) \in \mathbf{R}^{n+1} \mid t \geq 0\}$.

In (4), $\partial u / \partial t$ evaluated at the boundary is not determined locally by f, although it is determined by f globally. Once it is determined, all the higher derivatives $(\partial / \partial t)^k u \big|_{\mathbf{R}^n}$ on the boundary are fixed, since for example, $(\partial^2 / \partial t^2) u \big|_{\mathbf{R}^n} = -\Delta_x f$. In this formal sense the Monge-Ampère equation (M-A2) is better than the Laplace equation.

Now we shall look for a "formal" solution of (M-A2), that is a function $u(z)$ vanishing at ∂D, which satisfies

$$
J(u) = 1 + O\big(\big[\operatorname{dist}(z, \partial D)\big]^s\big), \quad \text{in } D,
\tag{5$_s$}
$$

with s as large as possible. To do so, we exploit two simple identities for the J-operator, which are proved by applying row and column operations to the definition. The identities are

$$
J(\phi\psi) = \phi^{n+1} J(\psi) + \psi(\text{Remainder}),
\tag{6.i}
$$

$$
J(\psi + \phi\psi^s) = J(\psi) \cdot \big[1 + c_{n,s}\phi\psi^{s-1}\big] + \psi^s \cdot (\text{Remainder}) \qquad (s \geq 2)
\tag{6.ii}
$$

where the remainders are smooth functions on $\bar{D}$. Now let ψ be a smooth defining function for D; this means $D = \{\psi > 0\}$ and $\psi' \neq 0$ on ∂D. The strict pseudo-convexity of D shows that $J(\psi) > 0$ on ∂D. Hence we may take $\phi = [J(\psi)]^{-(1/n+1)}$ in (6.i) to obtain a smooth function $u_1 = [J(\psi)]^{-(1/n+1)}\psi$ vanishing on ∂D, so that $J(u_1) = 1 + O(\psi)$. So we have solved (5) with $s = 1$.

Next suppose we have found a smooth function u_{s-1} which vanishes at ∂D and satisfies $J(u_{s-1}) = 1 + O(\psi^{s-1})$ $(s \geq 2)$. Thus, $J(u_{s-1}) = 1 + \eta(u_{s-1})^{s-1}$ for an $\eta \in C^\infty(\bar{D})$. Setting $u_s = u_{s-1} + \phi(u_{s-1})^s$ and using (6.ii) gives

$$
J(u_s) = \big[1 + \eta(u_{s-1})^{s-1}\big] \cdot \big[1 + c_{n,s}\phi(u_{s-1})^{s-1}\big] + O(\psi^s).
$$

So if we put

$$\phi = \frac{-\eta}{c_{n,s}} = \frac{1 - J(u_{s-1})}{c_{n,s}(u_{s-1})^{s-1}},$$

then we get

$$(7) \qquad J(u_s) = 1 + O(\psi^s), \qquad u_s = u_{s-1} \cdot \left[1 + \frac{1 - J(u_{s-1})}{c_{n,s}}\right].$$

As long as $c_{n,s} \neq 0$, this provides an inductive procedure to pass from solutions of $(5)_{s-1}$ to solutions of $(5)_s$. It happens that $c_{n,s} = s(n + 2 - s)$, which vanishes exactly for $s = (n + 2)$, so our inductive procedure stops there, yielding a smooth function u which vanishes at ∂D and satisfies $J(u) = 1 + O(\psi^{n+1})$. Note that u arises by applying a nonlinear differential operator to any defining function ψ; in fact $u = P_{n+1}P_n P_{n-1} \cdots P_2 P_1 \psi$ where $P_1(\psi) = [J(\psi)]^{-(1/n+1)}\psi$, and $P_s: u_{s-1} \to u_s$ by (7). An easy induction on s using (6.i), (6.ii) shows that the condition $J(u) = 1 + O(\psi^s)$ specifies u up to an error $O(\psi^{s+1})$. So our formal solution u is determined modulo $O(\psi^{n+2})$. For more details see [22].

The fact that $c_{n,s} = 0$ for $s = n + 2$ shows that the actual solution to (M-A2) is not smooth up to the boundary for general strictly pseudoconvex D, since we will already get a contradiction by Taylor expanding to order $(n + 2)$ about a boundary point. Rather, we expect logarithmic terms, so that the full solution to (M-A 2) should have the following asymptotic expansion near the boundary:

$$(8) \qquad u \sim \phi\psi + \sum_{p,q \geq 1} \phi_{pq}\psi^p(\psi^{n+1}\log\psi)^q, \quad \text{with } \phi, \phi_{pq} \in C^\infty(\overline{D}).$$

Part of the information contained in ϕ, ϕ_{pq} is determined locally by ∂D, the rest is determined globally. It would be very interesting to understand this precisely; a beautiful result in this direction has recently been obtained by R. Graham [29].

From (8) we obtain formally $u \in C^{n+2-\varepsilon}(\overline{D})$. The Cheng-Yau theorem gives us almost this much regularity for the solution of (M-A2).

2. The Cheng-Yau theorem. The goal of this section is to sketch the proof of the following

THEOREM 1 (CHENG-YAU) [10]. *The solution u of (M-A2) exists and is unique. Moreover u is C^∞ in the interior of D and belongs to $C^{n+(3/2)-\varepsilon}(\overline{D})$.*

Recall that $u \in C^{n+(3/2)-\varepsilon}(\overline{D})$ if $(\partial/\partial x)^\alpha u \in \text{Lip}(\frac{1}{2} - \varepsilon)$ for $|\alpha| \leq n + 1$.

To prove the theorem, we start with the approximate solution ψ of (M-A2) constructed in §1. Thus ψ is smooth up to the boundary and $J(\psi) = e^{-F}$ where F vanishes to order $(n + 1)$ at ∂D. We shall look for solutions of (M-A2) of the form $u = \psi e^v$; we hope that v will vanish to high order at ∂D so that $u \in C^{n+(3/2)-\varepsilon}(\overline{D})$. We shall study v in terms of the approximate Poincaré metric $ds^2 = \sum_{j,k} g_{j\bar{k}} dz_j d\bar{z}_k$, $g_{j\bar{k}} = (\partial^2/\partial z_j \partial\bar{z}_k)\log 1/\psi$. In terms of the unknown v and the metric $g_{j\bar{k}}$, equation (M-A2) may be rewritten in the form

$$\det(g_{j\bar{k}} + v_{j\bar{k}}) = e^{F}e^{(n+1)v}\det(g_{j\bar{k}}) \quad \text{in } D,$$

(M-A3)
$$(g_{j\bar{k}} + v_{j\bar{k}}) > 0 \quad \text{in } D,$$

$$v \to 0 \quad \text{at } \partial D,$$

where subscripts on v denote differentation. Again we note that F is smooth and tends rapidly to zero at ∂D, and we hope that v will do the same.

To solve (M-A3) we use the "continuity method"; that is, we introduce a parameter $0 \leqslant s \leqslant 1$ and look for solutions v^{s} of

$$(9) \qquad \det(g_{j\bar{k}} + v_{j\bar{k}}^{s}) = e^{sF}e^{(n+1)v^{s}}\det(g_{j\bar{k}}).$$

For $s = 0$ the equation has a trivial solution $v^{s} = 0$, while for $s = 1$ we recover (M-A3). We attempt to vary v^{s} continuously as s moves from 0 to 1 and still satisfy (9). Differentiating (9) in the parameter s leads to a simple linearized equation:

$$(10) \quad \left\{-(n+1)I + \Delta'\right\}\frac{\partial V}{\partial s} = (\text{given functions defined in terms of } v),$$

where Δ' is the Laplacian in the metric $g_{j\bar{k}}' = g_{j\bar{k}} + v_{j\bar{k}}^{s}$. Since $c > 0$, the maximum principle lets us invert $\{-(n+1)I + \Delta'\}$, so that (10) is a kind of ODE for $s \to v^{s}$. Hence we can continue to solve (9) for small time s.

What we need in order to continue the solution all the way to $s = 1$ are good a priori bounds on v^{s}. For, if v^{s} is defined for $s \leqslant s_{0}$ and $v^{s_{0}}$ satisfies good bounds independent of ε_{0}, then (10) may be used to extend v^{s} to $s \leqslant s_{0} + \varepsilon$ with ε independent of s_{0}. Thus we can get to $s = 1$.

Now, proving good bounds on solutions of (9) is equivalent to proving good bounds for solutions of (M-A3).

To summarize: We assume v is a bounded, smooth solution of (M-A3), and we try to prove good a priori estimates for v and its derivatives. If we can do this, then the existence and regularity will follow by the continuity method.

We now explain how to estimate solutions of (M-A3) in the interior. The estimates are based on tricks rooted in a long literature on the real and complex Monge-Ampère equations. An important part is played by the

MAXIMUM PRINCIPLE. Let u be a smooth, bounded, real-valued function in D. Then there exists a sequence $x_{1}, x_{2}, \dots$ in D such that

(i) $u(x_{i}) \to \sup_{D} u$,

(ii) $\|u'(x_{i})\| \to 0$,

(iii) $\limsup_{i \to \infty}$ (highest eigenvalue of $u''(x_{i})) \leqslant 0$.

Here the derivatives u', u'' are covariant derivatives with respect to $g_{j\bar{k}}$, and norms and eigenvalues are also taken with respect to $g_{j\bar{k}}$. The maximum principle can be proved easily by subtracting a barrier function from u.

We begin now with a 0th order estimate for solutions of (M-A3). Apply the maximum principle to v. Part (iii) above yields

$$0 \leqslant (g_{j\bar{k}} + u_{j\bar{k}})(x_{i}) \leqslant (1 + \varepsilon)(g_{j\bar{k}})(x_{i}) \quad \text{for large } i,$$

so (M-A3) and (i) imply

$$(1 + \varepsilon)^n \geqslant \limsup_{i \to \infty} e^{F(x_i)} e^{(n+1)v(x_i)} \geqslant e^{-\inf_D F} e^{(n+1)\sup_D v}.$$

Thus $\sup_D v \leqslant C_1$, where C_1 depends only on F. An analogous application of the maximum principle to $-v$ shows that $\sup_D(-v) \leqslant C_2$, where C_2 depends only on F. Hence we get an apriori bound

$$|v(z)| \leqslant C \quad \text{in } D.$$

Next we estimate the second derivatives $v_{j\bar{k}}$ by applying the maximum principle to $e^{-cv}(n + \Delta v)$, where Δ, Δ' denote the Laplacians in the $g_{j\bar{k}}$, $g'_{j\bar{k}} = g_{j\bar{k}} + v_{j\bar{k}}$ metrics respectively. To do this, we shall study $\Delta'[e^{-cv}(n + \Delta v)]$. In principle, this involves derivatives of v up to fourth order. However, the expression

$$\Delta\big[\det(g_{j\bar{k}} + v_{j\bar{k}})\big]$$

contains the same fourth-order terms yet equals $\Delta[e^{(n+1)v}e^F]$ by (M-A3). Thus, we may eliminate the fourth-order terms from $\Delta'[e^{-cv}(n + \Delta v)]$. After picking coordinates to diagonalize $g_{j\bar{k}}$ and $g'_{j\bar{k}}$, we can compute that

$$(11) \qquad \Delta'\big[e^{-cv}(n + \Delta v)\big] \geqslant e^{-cv}\Big\{(n + 1)\Delta v + \Delta F - n^2 \inf_{i \neq l} R_{i\bar{i}l\bar{l}}\Big\}$$
$$- c'e^{-cv}(n + \Delta v)$$
$$+ \Big(c + \inf_{i \neq l} R_{i\bar{i}l\bar{l}}\Big)e^{-cv}e^{c'F}(n + \Delta v)^{n/(n-1)},$$

where $R_{i\bar{j}k\bar{l}}$ is the curvature associated to $g_{j\bar{k}}$. Although the derivation of (11) is messy, it is elementary. In fact, once $\Delta[e^{-cv}(n + \Delta v)]$ is expressed in terms of derivatives of v up to third order, as explained above, we may simply regard the derivatives v, v_j, $v_{j\bar{k}}$, $v_{j\bar{k}}$ as independent variables, and then (11) becomes a messy elementary inequality for numbers. This involves algebra, but no calculus.

Now (11) and the maximum principle show that $e^{-cv}(n + \Delta v) \leqslant C_3$. For otherwise, at a suitable sequence x_i, the left side of (11) will have a negative lim inf, while the right side will remain large and positive. Here we pick $c \geqslant 1 + \|R_{i\bar{j}k\bar{l}}\|$; recall that we already have an a priori bound on $|v|$, and note that the $(n + \Delta v)^{n/(n-1)}$ term will dominate all the other terms on the right. So we know that $n + \Delta v \leqslant C$ a priori. Recalling that $0 \leqslant (g_{j\bar{k}} + v_{j\bar{k}})$, we obtain at once $\|v_{j\bar{k}}\| \leqslant C$, the norm taken in $g_{j\bar{k}}$. In view of (M-A3) we obtain also a lower bound on the eigenvalues of $(g_{j\bar{k}} + v_{j\bar{k}})$. So we know a priori the following:

$$(12) \qquad |v| \leqslant C, \qquad c(g_{j\bar{k}}) \leqslant (g_{j\bar{k}} + v_{j\bar{k}}) \leqslant C(g_{j\bar{k}}).$$

These estimates in a ball of radius 1 in $(g_{j\bar{k}})$ yield also $\|v'\| \leqslant C$ in a ball of radius $1/2$, the norm being taken in $(g_{j\bar{k}})$. (This follows from elliptic theory.) Consequently, we have also $\|v_j\| \leqslant C$. Observe that we have not yet estimated derivatives of the type v_{jk} or $v_{\bar{j}\bar{k}}$.

Next we estimate the mixed third derivatives of v, using the quantity

$$(13) \qquad S = \sum g'^{i\bar{r}}g'^{\bar{j}s}g'^{k\bar{t}}v_{,ij\bar{k}}\,v_{,r\bar{s}t}.$$

Here a comma denotes covariant differentiation in the g' metric. The analogue of S for real Monge-Ampère equations was introduced by Calabi. One calculates $\Delta'S$ and then applies the maximum principle to S. In principle, $\Delta'S$ involves derivatives of v up to order 5. However, by differentiating (M-A3) three times, we can re-express $\Delta'S$ in terms of derivatives of v up to order 4. Remarkably, the fourth-order derivatives now enter $\Delta'S$ as a sum of squares, and it follows that

$$(14) \qquad \Delta'S \geqslant c_1 S - c_2, \qquad c_1 \text{ and } c_2 \text{ positive constants.}$$

To make the estimate, one uses (12) and (13) to bound junk terms. The algebra is complicated, but again one needs no further calculus once the fifth derivatives have been eliminated from $\Delta'S$. Now (14) and the maximum principle show at once that S is bounded a priori, $S \leqslant C_4$. In view of (12), (13) we now know that $\|v_{,jk l}\| \leqslant C_5$, where the norm and covariant derivatives are now taken in terms of $(g_{j\bar{k}})$.

So we have estimates for v, v', and the terms of mixed type in v'', v'''. Finally, we can use the elliptic theory of Chapter 3 in each ball of radius one in $(g_{j\bar{k}})$, to conclude that $v \in \mathrm{Lip}(3 - \varepsilon)$ with a priori bounds. So we can use Schauder theory (note that the equation is elliptic because of the lower bound for $(g_{j\bar{k}} + v_{j\bar{k}})$ in (12)) to conclude that $v \in C^\infty(D)$ with a priori bounds $\|v_{,j\bar{k} \cdots p}\| \leqslant C$ on each of its covariant derivatives. As usual the norm and the covariant differentiation are in terms of $(g_{j\bar{k}})$. This concludes the discussion of boundary regularity.

We now know enough estimates to solve (M-A3) by the continuity method, and to prove interior regularity. But so far we know little about the boundary behavior of u because our estimates are defined in terms of a metric which degenerates at the boundary.

3. Proof of boundary regularity. Recall that v solves (M-A3) with $F = O(\psi^{n+1})$, ψ a defining function for D. In this section, we shall show that

$$(15) \qquad \|v_{,jk \cdots p}\| \leqslant C_\varepsilon \psi^{n+(3/2)-\varepsilon}.$$

Because of relations such as $|v_j| \leqslant C \|v_{,j}\| / \psi$ ($|\cdot| = $ Euclidean length, $\|\cdot\| = $ length in $g_{j\bar{k}}$), the estimate (15) implies $v \in C^{n+(3/2)-\varepsilon}(\bar{D})$, which is the boundary regularity asserted by the theorem. We may assume that $-\psi$ is strictly plurisubharmonic.

Consider now the linearized form of (M-A3):

$$(n + 1)v + F = \int_0^1 \frac{d}{ds}\Big[\log \det(g_{j\bar{k}} + sv_{j\bar{k}})\Big]\, ds = \sum_{pq}\left(\int_0^1 A^{p\bar{q}}\, ds\right) v_{p\bar{q}},$$

where $(A^{p\bar{q}})$ is the inverse of $(g_{j\bar{k}} + sv_{j\bar{k}})$. From our earlier estimates we know that as matrices $(A^{p\bar{q}}) \sim (g^{jk})$, so the linearized equation takes the form

$$\Delta_A v - (n + 1)v = F.$$

This is a well-behaved elliptic equation on each ball of unit radius with respect to $g_{j\bar{k}}$. Therefore all the estimates (15) for derivatives of v follow by standard elliptic theory if we can prove

$$(16) \qquad |v| \leqslant C_\varepsilon \psi^{n+(3/2)-\varepsilon}.$$

(If a harmonic function is bounded on a ball, then all derivatives are bounded on the inner half of the ball.) So the Cheng-Yau theorem is reduced to proving (16).

LEMMA 1. $|v| \leqslant C\psi$.

PROOF. Since $-\psi$ is plurisubharmonic, $-\Delta_A\psi > 0$, and so

$$\Delta_A(v - C\psi) - (n+1)(v - C\psi) \geqslant 0.$$

By the maximum principle we can pick $\{z_i\}$ so that

$$(v - C\psi)(z_i) \to \sup_D (v - C\psi), \qquad \limsup_{i\to\infty} \Delta_A(v - C\psi)(z_i) \leqslant 0.$$

Comparing the last two inequalities gives $v - C\psi \leqslant 0$, i.e., $v \leqslant C\psi$. Repeating the argument with $v - C\psi$ replaced by $-v - C\psi$ yields $-v \leqslant C\psi$, so the lemma is proved. $\square$

LEMMA 2. $n + \Delta v = n + O(\psi)$, where Δ is defined using the $g_{j\bar{k}}$ metric.

PROOF. Let $\lambda_1,\ldots,\lambda_n$ be the eigenvalues of $(v_{j\bar{k}})$ with respect to $(g_{j\bar{k}})$. Using (M-A3) and Lemma 1 we have

$$\prod_1^n (1 + \lambda_k) = e^{(n+1)v}e^F = 1 + O(\psi).$$

Since $(1/n)\Sigma_1^n(1 + \lambda_k) \geqslant \Pi_1^n(1 + \lambda_k)^{1/n}$, we have

$$(17) \qquad (n + \Delta v) \geqslant n + O(\psi),$$

establishing half of the desired estimate. Consider now (11) with $c = 1$,

$$\Delta'\left[e^{-v}(n + \Delta v)\right] \geqslant e^{-v}(n + \Delta v) - ne^{-v} - e^{-v}\Delta F - n^2\left(1 + \inf_{i\neq l} R_{i\bar{i}l\bar{l}}\right)e^{-v}$$

$$+ \left(1 + \inf_{i\neq l} R_{i\bar{i}l\bar{l}}\right)e^{-v}\exp\left(\frac{-(n+1)v + F}{n-1}\right)(n + \Delta u)^{n/(n-1)}.$$

Near the boundary, we expect $(g_{j\bar{k}})$ to look like the Poincaré metric for the ball, for which $\inf_{i\neq l} R_{i\bar{i}l\bar{l}} = -1$. In fact, a computation gives $(1 + \inf_{i\neq l} R_{i\bar{i}l\bar{l}}) = O(\psi)$. Therefore the last two terms in the previous inequality are $O(\psi)$; here we make use of (12) from §2. So we have

$$(18) \qquad \Delta'\left[e^{-v}(n + \Delta v)\right] \geqslant e^{-v}(n + \Delta v) - ne^{-v} - C\psi,$$

and now (17) yields

$$\Delta'\left[e^{-v}(n + \Delta v)\right] \geqslant -C'\psi.$$

Since $-\psi$ is strictly plurisubharmonic, we have $-\Delta'\psi \approx -\Delta\psi \geqslant c\psi$ in view of the way $(g_{j\bar{k}})$ degenerates. Therefore $\Delta'[e^{-v}(n + \Delta v) - C''\psi] \geqslant 0$ for large C''. We conclude that $e^{-v}(n + \Delta v) - C''\psi$ cannot take its maximum in the interior of D. Now apply the maximum principle to $e^{-v}(n + \Delta v) - C''\psi$. For a sequence $\{z_i\}$ which we now know approaches the boundary, $e^{-v}(n + \Delta v) - C''\psi$ approaches its sup, while

$$\limsup_{i \to \infty} \Delta'\{e^{-v}(n + \Delta v) - C''\psi\} \leq 0.$$

Since

$$\Delta'\{e^{-v}(n + \Delta v) - C''\psi\} \geq \{e^{-v}(n + \Delta v) - C''\psi\} - ne^{-v} + O(\psi),$$

by (18), it follows that $0 \geq \sup_D\{e^{-v}(n + \Delta v) - C''\psi\} - n$. (Here we used Lemma 1 and the fact that $\{z_i\} \to \partial D$.) Another application of Lemma 1 yields $n \geq (n + \Delta v) - C'''\psi$ in D, which together with (17) proves Lemma 2. $\square$

Lemma 2 immediately gives an estimate of the mixed second derivatives v_{ij}. In fact, retaining the notation of the proof of Lemma 2, we saw that $\prod_1^n(1 + \lambda_k) = 1 + O(\psi)$, while Lemma 2 says that $(1/n)\Sigma_1^n(1 + \lambda_k) = 1 + O(\psi)$. Thus the numbers $(1 + \lambda_k)$ have practically the same arithmetic and geometric mean, so they must be almost equal. That is, $1 + \lambda_k = 1 + O(\psi)$, so

$$(19) \qquad \qquad \|v_{,j\bar{k}}\| = O(\psi),$$

where the covariant derivatives and norm are taken in terms of $(g_{j\bar{k}})$. Together with Lemma 1, this yields

$$(20) \qquad \qquad \|v'\| = O(\psi),$$

where the norm is taken in terms of $(g_{j\bar{k}})$. We get this using elliptic theory on a ball of radius one in $(g_{j\bar{k}})$.

LEMMA 3. *If* $\|v'\|^2 = \Sigma_{jk} g^{j\bar{k}} v_j \bar{v}_k$ *is the square of the norm of the gradient of v in the $(g_{j\bar{k}})$-metric, then*

$$\|v'\|^2 \leq C_\varepsilon \psi^{2n+1-\varepsilon}.$$

PROOF. A long calculation using (19) shows that

$$\left\{\Delta' \frac{\|v'\|^2}{\psi^\beta} - C'\psi^\alpha\right\} \geq C_{n,\beta}\left\{\frac{\|v'\|^2}{\psi^\beta} - C'\psi^\alpha\right\}$$

$$+ \tilde{C}_{n,\alpha}\psi^\alpha - C\|F'\|^2\psi^{-1-\beta} - C\psi^{4n},$$

where, for certain ranges of α and β, the constants $C_{n,\beta}$ and $\tilde{C}_{n,\alpha}$ are positive and $\tilde{C}_{n,\alpha}\psi^\alpha$ swamps the last two terms near ∂D. We claim that for suitable α and β,

$$(21) \qquad \qquad \|v'\|^2 \leq C\psi^\beta \text{ implies } \|v'\|^2 \leq C'\psi^{\alpha+\beta}.$$

To see this, consider the "annulus" $\{0 \leq \psi(z) \leq \varepsilon\}$ and apply the maximum principle to the function $\{(\|v'\|^2)/\psi^\beta - C'\psi^\alpha\}$, where C' is large, depending on ε. If the maximum occurs on the inner boundary $\{\psi(z) = \varepsilon\}$, then that minimum is negative, so (21) is obvious. Otherwise, the maximum principle gives us a sequence on which $\{(\|v'\|^2)/\psi^\beta - C'\psi^\alpha\}$ approaches its sup over the annulus, while

$$0 \geq C_{n\beta}\left\{(\|v'\|^2)/\psi^\beta - C'\psi^\alpha\right\} + \left\{\tilde{C}_{n\alpha}\psi^\alpha - C\|F'\|^2\psi^{-1-\beta} - C\psi^{4n}\right\},$$

with the second term in braces positive. Consequently, $(\|v'\|^2)/\psi^\beta - C'\psi^\alpha \leq 0$ is the annulus $\{0 \leq \psi \leq \varepsilon\}$, so again (21) holds. Thus for suitable α, β estimate (21) is now proved in all cases.

Now starting with (20) and using (21) repeatedly, we can prove $\|v'\|^2 \leq C_s\psi^s$ for ever larger s, until at last we arrive at the conclusion of Lemma 3. $\square$

We now are ready to finish the proof of the Cheng-Yau theorem by verifying (16). By Lemmas 1 and 3,

$$\|v'\| \leq C_\varepsilon\psi^{n+(1/2)-\varepsilon}, \qquad |v| \leq C\psi;$$

therefore we may recover $v(z)$ by integrating v' over a path from z to ∂D, and the resulting estimate is (16). The Cheng-Yau theorem is proved. $\square$

Further work on the complex Monge-Ampère equations (M-A1)–(M-A3) has been carried out recently by Lee and Melrose [**44**].

THEOREM (LEE-MELROSE [**44**]). *The solution u of* (M-A2) *has the asymptotic expansion*

$$(22) \quad u \sim \psi \sum_{k \geq 0} \eta_k\left(\psi^{n+1}\log\psi\right)^k \ near \ \partial D = \{\psi = 0\} \subseteq \mathbf{C}^n \ \ with \ \eta_k \in C^\infty(\bar{D}).$$

This means that for large N, $u - \psi\Sigma_{k=0}^N \eta_k(\psi^{n+1}\log\psi)^k$ has many continuous derivatives on $\bar{D}$ and vanishes to high order at ∂D.

We include only the most superficial sketch of the proof. The idea is reminiscent of the Schauder theory in Chapter 3. Starting with partial information on u (e.g. the Cheng-Yau theorem), we write down a linearized equation and hope to get out more than we put in by making a careful study of linear PDE with less-than-perfect coefficients. Then we use the new information on u to improve the coefficients of the linearized PDE and repeat the same argument over and over. It is easy enough to write down a linearized equation. In (M-A3), we just expand the determinant by minors to obtain

$$(23) \qquad L_v = \sum_{jk} a_{jk}\frac{\partial^2 v}{\partial z_j \partial\bar{z}_k} = g \quad \text{where the } (a_{j\bar{k}}) \text{ are the minors.}$$

Of course $a_{j\bar{k}}$ and g depend on v.

The work comes in understanding (23). We have to assume u is given to high order by (22) near ∂D, then calculate the $(a_{j\bar{k}})$. Basically, L looks like the Laplacian in the Poincaré metric on the ball. In particular, L is elliptic in D but degenerates at the boundary. The heart of the proof is to understand how solutions of such degenerate elliptic linear equations behave near the boundary. See also R. Graham [**29**] for an analysis of (23) from a Heisenberg-group point of view.

4. Chains. We begin this section by showing how to use the Poincarè metric to get local invariants and distinguished curves on the boundary ∂D of a strictly pseuodconvex domain D. We know from the formal study of the first section that the volume form of the Poincarè metric on D is given by

$$d\,\mathrm{vol} = c_n \frac{dz_1 \wedge \cdots \wedge dz_n \wedge d\bar{z}_1 \wedge \cdots \wedge d\bar{z}_n}{\left(u(z)\right)^{n+1}},$$

where $u(z)$ vanishes to exactly first order at ∂D. The procedure in the first section gives us the Taylor expansion of u to order $(n+1)$ at ∂D.

EXAMPLE. For the unit ball $D = \{1 - |z|^2 > 0\}$ we have $u(z) = 1 - |z|^2$ as the solution to (M-A2). If we introduce projective coordinates $\xi_0 = z_0$, $\xi_k/\xi_0 = z_k$ $(1 \leqslant k \leqslant n)$ on $\mathbf{C}^1 \times D$, then we can define the function

$$U(z_0, z_1, \ldots, z_n) = |z_0|^2 u(z) = |\xi_0|^2 - \sum_1^n |\xi_k|^2,$$

which gives us an indefinite metric

$$ds^2 = \sum_{j,k \geqslant 0} \frac{\partial^2 U}{\partial z_j \partial \bar{z}_k} dz_j d\bar{z}_k = |d\xi_0|^2 - \sum_1^n |d\xi_k|^2.$$

The linear fractional transformations of the ball arise as isometries of this metric.

In the general case we can follow the same procedure by introducing the function

$$(24) \quad U(z_0, z_1, \ldots, z_n) = |z_0|^2 u(z), \qquad (z_0, z_1, \ldots, z_n) \in (\mathbf{C} \setminus \{0\}) \times D$$

and defining a metric

$$(25) \qquad ds^2 = \sum_{jk \geqslant 0} \frac{\partial^2 U}{\partial z_j \partial \bar{z}_k} dz_j d\bar{z}_k,$$

which turns out to be nondegenerate.

Now (24), (25) are invariantly defined. More precisely, every biholomorphic map $\Phi \colon D \to \tilde{D}$ lifts to a biholomorphic map $\Phi^{\#} \colon (\mathbf{C} \setminus \{0\}) \times D \to (\mathbf{C} \setminus \{0\}) \times \tilde{D}$ which preserves (24), (25). The rule for lifting is

$$\Phi^{\#} \colon (z_0, z) \to (\tilde{z}_0, \tilde{z}) = \left(z_0 [\det \Phi'(z)]^{1/(n+1)}, \Phi(z)\right),$$

and the verification that (24), (25) are preserved by $\Phi^{\#}$ is trivial.

Consequently, we have a list of tensors invariantly attached to points of $(\mathbf{C} \setminus \{0\}) \times D$, namely the curvature for (25) and its covariant derivatives. Although we cannot compute (25) inside the domain, we do know its Taylor series to high order at ∂D. So the tensors $R_{j\bar{k}l\bar{m},p,\ldots,\bar{q}}$ can be computed on $(\mathbf{C} \setminus \{0\}) \times \partial D$, as long as the defining formulas do not involve too many differentiations of the metric $(g_{j\bar{k}})$.

EXAMPLE. $\|R_{j\bar{k}l\bar{m}}\|^2$ is a scalar invariant attached to points of $(\mathbf{C} \setminus \{0\}) \times \partial D$; it can be computed explicitly in terms of the Taylor expansion of ∂D. If D is in Moser's normal form,

$$D = \left\{ \mathrm{Im}\, z_1 > |z'|^2 + \sum A_{p\bar{q}r\bar{s}} z_p \bar{z}_q z_r \bar{z}_s + \text{higher order terms} \right\},$$

then at $(z_0, z_1, \ldots, z_n) = (z_0, 0, \ldots, 0) \in (\mathbf{C} \setminus \{0\}) \times \partial D$ we can calculate that

$$\left\| R_{j\bar{k}l\bar{m}} \right\|^2 = c_n |z_0|^4 \sum \left| A_{p\bar{q}r\bar{s}} \right|^2.$$

More generally, we can write down scalar invariants

$$(26) \quad \Omega = \text{Trace}\left(R_{a\bar{b}c\bar{d}, p\bar{q}, \ldots, r} \otimes \cdots \otimes R_{a'\bar{b}'c'\bar{d}', p'q', \ldots, \bar{r}'} \right) = |z^0|^{2s} P\left(A_{\alpha\bar{\beta}}^l \right),$$

for $(z_0, z_1, \ldots, z_n) = (z_0, 0)$ and ∂D in normal form. Here s is a positive integer and P is a polynomial in the coefficients $A_{\alpha\bar{\beta}}^l$ in Moser's normal form. Ratios of two Ω's with the same "weight" s will be scalar invariants attached to a boundary point. More generally, if $\Omega_1, \ldots, \Omega_m$ are finitely many expressions of the form (26) with weights $s_1, \ldots, s_m$, then to each boundary point $z \in \partial D$ we pick any z^0 and set $F(z) = (\Omega_1(z^0, z), \ldots, \Omega_m(z^0, z))$. $F(z)$ will be a vector in $\mathbf{C}^m$; if we identify $(\xi_1, \ldots, \xi_m) \in \mathbf{C}^m$ with $(\lambda^{s_1}\xi_1, \ldots, \lambda^{s_m}\xi_m)$ for $\lambda > 0$, then $F(z)$ is an invariant taking values in the quotient space $\mathbf{C}^m / \mathbf{R}^\times$.

Returning to (26), the invariance of Ω means that $P(A_{\alpha\bar{\beta}}^l)$ has a transformation law under biholomorphic maps. If $\phi \colon \partial D \to \partial \tilde{D}$ is biholomorphic, where

$$\partial D = \left\{ \text{Im } z_1 = |z'|^2 + \sum A_{\alpha\bar{\beta}}^l (\text{Re } z_1)^l z'^\alpha \bar{z}'^\beta \right\}$$

and

$$\partial \tilde{D} = \left\{ \text{Im } z_1 = |z'|^2 + \sum B_{\alpha\bar{\beta}}^l (\text{Re } z_1)^l z'^\alpha \bar{z}'^\beta \right\}$$

are in normal form, then

$$(27) \qquad\qquad P\left(A_{\alpha\bar{\beta}}^l \right) = \left| \det \Phi'(0) \right|^{2s/(n+1)} P\left(B_{\alpha\bar{\beta}}^l \right).$$

CONJECTURE. Every polynomial satisfying (27) arises from some Ω as in (26).

This problem seems natural when we try to attach boundary invariants to domains. It is also what is needed for a precise asymptotic expansion for the Bergman kernel; see Chapter 12.

In fact we have to be careful in stating the conjecture, since $R_{j\bar{k}l\bar{m}, \ldots, \bar{q}} \bar{q}$ of very high rank cannot be computed locally. One finds that the Ω of (26) can be computed locally basically if $s \leqslant n$. So we really have a prescription to write down polynomials satisfying (27) only when $s < n$. For general s we cannot even write down nontrivial examples, let alone classify all possibilities. The algebra in Chapter 12 proves our conjecture when $s \leqslant n - 20$.

We have seen how to attach invariant tensors and scalars to a boundary point using (M-A2). Now we shall see how to find distinguished curves. We use polar coordinates $z_0 = \rho e^{i\theta}$ so that $(\mathbf{C} \setminus \{0\}) \times D \simeq \mathbf{R}^+ \times S^1 \times \partial D$. Let us restrict the indefinite metric (24) to $\mathbf{R}^+ \times S^1 \times \partial D$; this is natural since (24) cannot be computed in the interior. The result after some small calculation is

$$(28) \qquad\qquad \partial\bar{\partial} U \big|_{\mathbf{R}^+ \times S^1 \times \partial D} = \rho^2 Q$$

with

$$(29) \qquad Q = \frac{1}{(n+1)} \left(\frac{\partial u - \bar{\partial} u}{i} \right) d\theta + \sum_{j,k=1}^{n} \frac{\partial^2 u}{\partial z_j \partial \bar{z}_k} dz_j d\bar{z}_k.$$

One checks that Q is a nondegenerate Lorentz metric on $S^1 \times \partial D$. Now the left-hand side of (28) is a biholomorphic invariant, and ρ is a scalar factor. Therefore the conformal class of Q is a biholomorphic invariant. In other words, the following is true: Let $\Phi: D \to \tilde{D}$ be a biholomorphic map and let $Q, \tilde{Q}$ be the Lorentz metrics defined on $S^1 \times \partial D$, $S^1 \times \partial \tilde{D}$ by (29). Then Φ lifts to a map $\Phi^{\#}$: $S^1 \times \partial D \to S^1 \times \partial \tilde{D}$ which carries Q to another Lorentz metric that differs from $\tilde{Q}$ by a scalar factor.

Using the Lorentz metric (29), we can easily find distinguished curves in $S^1 \times \partial D$.

PROPOSITION 1. *Two conformally equivalent Lorentz metrics have the same light rays.*

PROOF. Recall that a light ray of a Lorentz metric $ds^2 = \sum g_{jk} dx_j dx_k$ is just a Hamiltonian path for the Hamiltonian $H = \frac{1}{2} \sum g^{jk}(x) p_j p_k$ satisfying $H = 0$, where p_j and x_j are conjugate variables. Under a conformal change of metric $\widetilde{ds}^2 = \rho(x) \sum g_{jk}(x) dx_j dx_k$, the new Hamiltonian becomes $\tilde{H} = \rho^{-1}(x) H$. We already saw in Chapter 2 that multiplying a Hamiltonian H by a nonvanishing factor changes the zero-energy Hamiltonian trajectories only by a change of clock. Therefore the unparametrized light rays of ds^2, $\widetilde{ds}^2$ are the same.   □

Because of this proposition, we can take the light rays as distinguished curves on $S^1 \times \partial D$. If we define *chains* as the image of light rays under the projection $S^1 \times \partial D \to \partial D$, then these too become biholomorphic invariants.

Note that the calculation (29) requires that we know u to second order at ∂D, i.e., we need u up to errors $O(\psi^3)$. Recall that the formal procedure of §1 defined inductively

$$u_1 = [J(\psi)]^{-1/(n+1)} \psi, \qquad u_2 = u_1 \cdot [1 + c_n(1 - J(u_1))],$$

which involves derivatives of ψ up to order 4. Since $u = u_2 + O(\psi^3)$, we can use u_2 in place of u to calculate Q (29) and find the chains.

Consider the Hamiltonian $H(x_j, \Theta; p_j, p_\Theta)$ for light rays on $\partial D \times S^1$, where (x_j, Θ) are coordinates on $\partial D \times S^1$ and (p_j, p_Θ) are the conjugate variables. Since θ does not appear explicitly in (29), H is independent of Θ, and it follows at once from Hamilton's equations that p_Θ remains constant along a light ray. Consequently, the chains in ∂D are Hamiltonian trajectories for

$$(30) \qquad H = \sum_{ij} a^{ij}(x) p_i p_j + \sum_j b^j(x) p_\Theta p_j + V(x) p_\Theta^2.$$

Here $(a^{ij}(x))$ is positive semidefinite with a one-dimensional nullspace, $p_\Theta = $ arbitrary constant, and the chains are precisely the trajectories on which $H = 0$.

In principle, H can be computed. Using H we can get global information about chains.

EXAMPLE 1. The chains are well behaved globally on a boundary of the form

$$\partial D = \{\operatorname{Im}(z_1) = \phi(z')\}.$$

The point is that ∂D is translation-invariant in $\operatorname{Re} z_1$. Therefore, using coordinates $x_1, \ldots, x_{2n-1}$ on ∂D with $x_1 = \operatorname{Re} z_1$, we obtain the chains from a Hamiltonian (30) with no x_1-dependence. The conjugate momentum p_1 is therefore conserved along a chain. From $p_1 = \text{const.}$, $H = 0$ and the form (30) of H, one can easily bound all the momenta, $|p_j| \leqslant \text{const.}$ (This amounts to checking that $\xi_1 \neq 0$, where $(\xi_1, \ldots, \xi_{2n-1})$ generates the nullspace of $a^{ij}(x)$ in (30).) Since the Hamiltonian vector field blows up only as $p \to \infty$, the trajectories are globally well behaved. See Example 4 to find what can go wrong if ∂D isn't translation-invariant.

EXAMPLE 2. If $\partial D = \{\operatorname{Im} z_1 = \phi(|z'|)\}$, then the chains on ∂D can be completely understood. In fact the ordinary differential equations for chains may be uncoupled and reduced to the calculation of integrals. To see this, use $y = \operatorname{Re} z_1$, $r = |z_1|$, $w = z'/|z'|$ as coordinates in ∂D. All the coordinates except r are cyclic for the Hamiltonian (30) so the problem is reduced to Hamilton's equations in one-space variable, which is trivial.

EXAMPLE 3 (BURNS-SCHNEIDER [7]). Take the hyperquadric $Q = \{\operatorname{Im} z_1 = |z'|^2\}$ with the origin removed, and identify (z_1, z') with $(\lambda^{2k} z_1, \lambda^k z')$ for all integers $-\infty < k < +\infty$ and a fixed $\lambda > 1$. In this way we obtain a C-R manifold M which looks locally like Q. It is easy to see that the two points (z_1, z') and $(-\bar{z}_1, -z')$ cannot be connected by a chain on M, simply because the chain which joins them in Q passes through the origin. M can be identified with the boundary of a domain by the mapping $(z_1, z') \to (z_1^{(\pi i)/(\ln \lambda)}, z_1^{-(1/2)} z')$; the powers of z_1 make sense, since $\operatorname{Im} z_1 > 0$ on M.

So we see that the analogy between geodesics and chains cannot be carried too far.

EXAMPLE 4. Chains may also have local pathologies. For instance, on $\partial D = \{\operatorname{Im} z_1 = |z'|^2 + (\operatorname{Re} z_1)|z'|^8\}$ there are chains which spiral in towards the origin. Note that $\partial D \subseteq \mathbf{C}^2$ is the simplest domain in normal form not covered by Example 1. To understand how the chains look, we introduce coordinates $y = \operatorname{Re} z_1$, $re^{i\theta} = z_2$ on ∂D, and then switch to new variables $\theta, s, v, p_\theta, p_s, p_v$ by means of the canonical transformation

$$(31) \qquad r = p_v^{-1/2} s, \qquad y = -\tfrac{1}{2} p_v^{-1} p_s s + v, \qquad \theta \text{ unchanged},$$

$$p_r = p_v^{1/2} p_s, \qquad p_y = p_v, \qquad p_\theta \text{ unchanged}.$$

This is somewhat analogous to Sundmann's regularization of the 3-body problem in Chapter 2. Calculating the Hamiltonian (30) and composing with (31), we find that

$$(32) \quad H = \left[p_s^2 + s^2 + \frac{p_\theta^2}{s^2} + (2p_\theta - 6p_\Theta) \right]$$

$$+ v p_v^{-3} \left[16s^8 + 16 p_\theta s^6 + (192 p_\theta p_\Theta - 432 p_\Theta^2) s^4 \right] + O(p_v^{-4}).$$

(To arrive at (32), we multiplied H by a nonvanishing factor p_v^{-1}; this of course has no effect on the unparametrized chains.)

What is important here is the behavior of orbits with p_v large. One checks by starting at (31) that a spiral chain on ∂D corresponds to a Hamiltonian path for (32) on which $H = 0$, s and p_s remain bounded, v tends to zero, and p_v tends to infinity. Now for large p_v, the first term in brackets gives the main contribution to H in (32); the rest is a small perturbation. So, approximately,

$$H \approx p_s^2 + s^2 + p_\theta^2/s^2 + (2p_\theta - 6p_\Theta).$$

(The formula would be exact if ∂D were the hyperquadric.) This Hamiltonian is trivial to understand: v, p_v, p_θ, p_Θ remain constant, while s and p_s undergo a simple periodic motion corresponding to a particle in a one-dimensional potential well. Adding in the small corrections in (32) to obtain the exact Hamiltonian, we can perform a first-order perturbation expansion to see more accurately how v, p_v, etc. really behave. One finds that p_v, rather than remaining constant, increases to infinity very slowly for large time, while s and p_s remain bounded. On most orbits $v \to \infty$, but by picking the right initial conditions one gets $v \to 0$. So spiral chains exist on ∂D. No one knows how prevalent spiral chains are for generic domains.

For more details of the calculations in this section see [**22**].

CHAPTER 12. THE BERGMAN KERNEL

1. The asymptotic expansion. In this section we will briefly discuss some in the ideas in the proof of

THEOREM 1. *If $D = \{\psi(z) > 0\} \subseteq \mathbf{C}^n$ is a strongly pseudoconvex domain with $\psi' \neq 0$ on ∂D, then the Bergman kernel $K(z, w)$ may be written on the diagonal in the form*

$$(1) \qquad K(z, z) = \phi(z)/\psi^{n+1}(z) + \tilde{\phi}(z) \log \psi(z)$$

where $\phi, \tilde{\phi} \in C^\infty(\overline{D})$ and $\phi|_{\partial D} \neq 0$. Off the diagonal, we may extend ψ, ϕ, $\tilde{\phi}$ to functions $\phi(z, w)$, $\tilde{\phi}(z, w)$ with the properties

1. *$\psi(z, w)$ is almost analytic in z, $\overline{w}$ in the sense that $\overline{\partial}_z \psi(z, w)$ and $\partial_w \psi(z, w)$ vanish to infinite order at $z = w$.*

2. *$\psi(z, z) = \psi(z)$.*

3. *Similarly for $\phi(z, w)$ and $\tilde{\phi}(z, w)$.*

The Bergman kernel has the form

$$K(z,w) = \phi(z,w)/\psi^{n+1}(z,w) + \tilde{\phi}(z,w)\log\psi(z,w)$$

for $(z,w) \in \Omega_\varepsilon = \{|z - w| < \varepsilon, \ \mathrm{dist}(z, \partial D) < \varepsilon\}$ and $\varepsilon > 0$ small. Outside Ω_ε, $K(z,w)$ is smooth up to the boundary as a function of (z, w). There is an analogous expansion for the Szegö kernel.

We give here a brief sketch of the original proof in [21] and of the alternate proof in [5]. Later we shall look more closely at the Bergman kernel by trying to compute bthe functions ϕ and $\tilde{\phi}$. The proof of Boutet-Sjöstrand [5] for the Szegö kernel is roughly as follows: fix $\partial D \subseteq \mathbf{C}^n$ and define the operator $T_{\partial D}: L^2(\partial D) \to L^2(\partial D)$ by $f \mapsto \bar{\partial}_b f$. The Szegö kernel gives the orthogonal projection from $L^2(\partial D)$ to the kernel of $T_{\partial D}$. Let $B \subseteq \mathbf{C}^n$ be the unit ball. Then microlocally in $T^*(\partial D)$ there exists a canonical change of coordinates $\Phi: T^*(\partial D) \to T^*(\partial B)$ so that the induced Fourier integral operator $U: L^2(\partial D) \to L^2(\partial B)$ satisfies

$$(2) \qquad T_{\partial D} = \varepsilon \cdot U^{-1} T_{\partial B} U,$$

where $T_{\partial B}$ is the $\bar{\partial}_b$-operator on the sphere, and ε is an elliptic $(n-1) \times (n-1)$ matrix of pseudodifferential operators of order zero. This reduces the study of the Szegö kernel for ∂D to the known simple case $D = $ unit ball.

To achieve (2), we first pick Φ and the principal symbol of ε so that (2) holds modulo lower-order errors. In view of Ergorov's theorem, this is purely a question about the symplectic geometry of the principal symbol of $T_{\partial D}$. Once (2) holds modulo lower-order errors, we may then add lower-order corrections to ε so that (2) holds modulo a smoothing error. The smoothing error has to be carried along in the proof, but it never causes real trouble, so we shall ignore it.

Now in principle, (2) gives the Szegö projection $K_{\partial D}$ as $U^{-1} K_{\partial B} U$. In practice one has to work to compose these operators explicitly to get a recognizable result. The calculations may be put in the context of Fourier integral operators with complex phase functions. For the ball, $\psi_B(z) = 1 - |z|^2, \psi_B(z, w) = 1 - z \cdot \bar{w}$,

$$(3) \qquad K_{\partial B}(z,w) = \frac{C_n}{(1 - z\cdot\bar{w})^n} = \tilde{c}_n \int_0^\infty \lambda^{n-1} e^{-\lambda\psi_B(z,w)}\, d\lambda.$$

Thus, $K_{\partial B}$ is a Fourier integral operator with phase function $-\lambda\psi_B(z, w)$. Since U and U^{-1} are also Fourier integral operators, the calculus of operators with complex phase expresses $K_{\partial D}$ as a Fourier integral operator. The result is

$$K_{\partial D}(z,w) = \tilde{c}_n \int_0^\infty a(z,w,\lambda) e^{-\lambda\psi(z,w)}\, d\lambda, \quad \text{where}$$

$$(4) \qquad a(z,w,\lambda) \sim \sum_{l \geq 0} \lambda^{n-1-l} a_l(z,w) \quad \text{and} \quad a_l(z,w) \in C^\infty(\bar{D} \times \bar{D}),$$

which of course agrees with (3) for the unit ball.

Now we can simply read off from (4) that

$$K_{\partial D}(z, w) \sim \sum_{l \geq 0} \left(\int_0^\infty \lambda^{n-1-l} e^{-\lambda \psi(z,w)} \, d\lambda \right) a_l(z, w)$$

$$= \sum_{\substack{l \geq 0 \\ l \neq n}} c_l (\psi(z, w))^{l-n} a_l(z, w) + (\text{const}) \log \psi(z, w) \cdot a_n(z, w)$$

$$= \frac{\phi(z, w)}{(\psi(z, w))^n} + \tilde{\phi}(z, w) \log \psi(z, w),$$

where $\phi = \sum_{l \neq n, \, l \geq 0} c_l \psi^l a_l$ and $\tilde{\phi} = (\text{const}) a_n$. A calculation of the principal symbol in (4) is enough to show that $\phi(z, z)\big|_{\partial D} \neq 0$. So the analogue of Theorem 1 for the Szegö kernel is proved.

In fact, the proof as presented in [5] is slightly different from our explanation. Rather than reducing $\bar{\partial}_b$ on D to $\bar{\partial}_b$ on the sphere, Boutet-Sjöstrand exhibit Fourier integral operators to reduce both these operators to the same "model" equation. The model is a variant of (57a) of Chapter 4.

To adapt the argument of Boutet-Sjöstrand to the Bergman kernel, one has to deal with the $\bar{\partial}$-Neumann problem instead of the $\bar{\partial}_b$-equation. However, we learned in Chapter 3 how a boundary-value problem on D reduces to a pseudodifferential equation on ∂D. The pseudodifferential equation arising from the $\bar{\partial}$-Neumann problem may be analyzed by the same methods we just explained for $\bar{\partial}_b$, and Theorem 1 for the Bergman kernel again results from a calculation with Fourier-integral operators with complex phase.

The original proof of Theorem 1 [21] starts with the following observation: if D_1, D_2 are strictly pseudoconvex domains which share a piece of a common boundary, then on a small neighborhood of a point in the common boundary, the two Bergman kernels differ by a function of (z, w) which is smooth up to the boundary.

This may be easily read off from the smoothness of solutions to the $\bar{\partial}$-Neumann problem in D_2. We form $u = \partial_z^\beta \{K_{D_1}(z, \cdot) \chi_{D_1} - K_{D_2}(z, \cdot)\}$, which is clearly orthogonal to analytic functions on D_2, while $\bar{\partial} u = \alpha$ is supported in $\partial D_1 \setminus \partial D_2$. So u solves a $\bar{\partial}$-Neumann problem, and it follows that u is smooth up to the boundary away from the singular support of α. The argument is essentially due to Kerzman [37], who was also the first to prove the smoothness of the Bergman kernel off the boundary diagonal.

Returning to the proof of Theorem 1, let H_D denote the Hilbert space of analytic functions in $L^2(D)$ and $H_D^\perp$ the space of all distributions f on D satisfying $\int_D f(z) \overline{g(z)} \, dz = 0$ for all $g \in H_D$. Suppose that we have found a decomposition of the Dirac delta function δ_w at w:

$$(5) \qquad \delta_w = F_w(z) + G_w(z) + \varepsilon_w(z), \qquad F_w \in H_D, \, G_w \in H_D^\perp,$$

where ε_w is an appropriately small error. Thinking now of a function $H(z, w)$ on $D \times D$ as an operator $Hf(z) = \int_D H(z, w) f(w) \, dw$ on $L^2(D)$, we can rewrite the last equation as

$$f(z) = (Ff)(z) + (Gf)(z) + (\varepsilon f)(z),$$

where $Ff \in H_D$ and $Gf \in H_D^\perp$. Since εf is small, we can interpret $F(z, w)$ as an approximation to the Bergman kernel $K(z, w)$. If for example $\|\varepsilon f\|_{L^2(D)} \leqslant \frac{1}{2}\|f\|_{L^2(D)}$, then we can recover the exact Bergman kernel by iteration, using a sequence of successive approximations.

We now explain how to make the decomposition (5). Take a point $w \in D$ near the boundary, and let w^* be the boundary point closest to w. Now, D is biholomorphically equivalent to the unit sphere to third order at w^*. Therefore we can find a strictly pseudoconvex comparison domain $D_{\text{small}} \subseteq D$ which is tangent to ∂D to third order at w^*, and which locally near w^* is biholomorphic to a piece of the unit ball. By cutting down D_{small}, we may assume that D_{small} is biholomorphic to a domain $B_{\text{small}} \subseteq$ Unit Ball whose boundary agrees with the unit sphere in a neighborhood of the image of w^*. It is easy to write down D_{small} and the biholomorphic map explicitly. Note that $w \in D_{\text{small}}$ if initially w was close to ∂D.

Next we look at the Bergman kernel for D_{small}. We already know the Bergman kernel for the ball, so the observation at the start of the proof shows that in effect we know the Bergman kernel for B_{small}. In view of the explicit biholomorphic mapping of D_{small} to B_{small}, we have the Bergman kernel $K_{D_{\text{small}}}(z, w)$ given as

$$K_{D_{\text{small}}}(z, w) = K_{\text{Explicit}}(z, w) + \tilde{\varepsilon}(z, w),$$

where the error $\tilde{\varepsilon}$ is smooth up to the boundary of D_{small}.

To make the decomposition (5), we now write

$$\delta_w = \left[K_{\text{Explicit}}(\cdot, w) \right] + \left[\delta_w - K_{D_{\text{small}}}(\cdot, w)\chi_{D_{\text{small}}} \right]$$
$$+ \left[\tilde{\varepsilon}(\cdot, w)\chi_{D_{\text{small}}} - K_{\text{Explicit}}(\cdot, w)\chi_{D - D_{\text{small}}} \right]$$
$$\equiv F_w + G_w + \varepsilon_w.$$

The definition of the Bergman kernel gives $K_{D_{\text{small}}}(\cdot, w)$ only as an analytic function on D_{small}, but inspection of the explicit formula shows at once that F_w continues analytically into all of D. So $F_w \in H_D$. Evidently $G_w \in H_D^\perp$ by the reproducing property of $K_{D_{\text{small}}}$. Finally (except for the smoothing error $\tilde{\varepsilon}$) the kernel ε_w is supported in the thin region $D - D_{\text{small}}$, and it follows that ε is a small operator on L^2.

Thus, we have our decomposition (5), and we can read off the Bergman projection K_D by successive approximation: $K_D = \Sigma_{l \geqslant 0} \pm F\varepsilon^l$, where F, G are the operators in (5). It remains to calculate $F\varepsilon^l$, which amounts to doing complicated multiple integrals. We omit the details.

2. Christoffers' Theorem. How accurately can we compute the functions ϕ, $\tilde{\phi}$ in Theorem 1? We may as well work on the diagonal $z = w$, since e.g. $\phi(z, w)$ is given to infinite order about $z = w$ by the expansion

$$\phi(z, w) \sim \Sigma_\alpha (1/\alpha!)(\bar{\partial}^\alpha\phi)(z)\, \overline{(w - z)^\alpha}.$$

Now, if D_1 and D_2 are strictly pseudoconvex domains whose boundaries coincide near a point p, we saw in §1 that their two Bergman kernels $K_{D_1}(z, z)$, $K_{D_2}(z, z)$ differ by a smooth function near p. So in the expansion

$$(1) \qquad K_D(z, z) = \frac{\phi(z)}{\psi^{n+1}(z)} + \tilde{\phi}(z) \log \psi(z),$$

we see that ϕ is determined up to nth order at the boundary by local information, while the rest of ϕ is determined globally.

Similarly, $\tilde{\phi}$ is determined to infinite order by local information; again, since $K_D(z, z)$ is determined locally modulo smooth functions on $\bar{D}$. By looking carefully at either of the two proofs of Theorem 1 we can get more: for $p \in \partial D$ and $K_D(z, z)$ given by (1), we have $\partial^\alpha \phi(p)$ ($|\alpha| \leqslant n$) and $\partial^\beta \tilde{\phi}(p)$ (arbitrary β) determined uniquely by the Taylor series of ∂D at p. For fixed α, β, only finitely many derivatives of ∂D enter into the formula.

Thus, the maps $S: \psi \to \phi$ and $T: \psi \to \tilde{\phi}$ may be regarded as nonlinear differential operators. It would be interesting to write them down explicitly.

Two basic properties of S and T are immediate from the definition:

(A) If $\eta \neq 0$ on $\{\psi = 0\}$, then $S(\eta\psi) = \eta^{n+1}S(\psi) + O(\psi^{n+1})$ and $T(\eta\psi) = T(\psi) + O(\psi^M)$, for any M.

Indeed, $\eta\psi$ and ψ define the same domain and so lead to the same Bergman kernel.

(B) If Φ is a biholomorphic map, then $S(\psi \circ \Phi) = |\det \Phi'|^{-2}S(\psi) \circ \Phi + O(\psi^{n+1} \circ \Phi)$, while $T(\psi \circ \Phi) = |\det \Phi'|^{-2}T(\psi) \circ \Phi + O(\psi^M \circ \Phi)$ for any M.

These invariance properties are very strong. No one knows how to write down a single nonzero operator T with the required properties. However, we shall see below how to write down all possible operators S satisfying (A) and (B). It appears that among all possible S, the choice which appears in the Bergman kernel is rather typical.

So we now are trying to write down expressions which transform like the Bergman kernel. The first example is $K_{\text{Poincaré}}(z) = C_n/(u(z))^{n+1}$, where u is the formal solution of the Monge-Ampère equation (M-A2) of the preceding chapter. We have already seen that the map which takes ψ to u on $\{\psi \geqslant 0\}$ is a nonlinear differential operator and that u changes by $O(\psi^{n+1})$ when ψ is multiplied by a nonvanishing factor η.

So we might guess that $\phi/\psi^{n+1} = K_{\text{Poincaré}}$. This is unfortunately false. In fact, suppose that the boundary ∂D is in Moser's normal form,

$$\text{Im } z_n = |z|^2 + \sum_{|\alpha|,|\beta| \geqslant 2} \sum_{l \geqslant 0} A^l_{\alpha\bar{\beta}}(\text{Re } z_n)^l z^\alpha \bar{z}^\beta,$$

where $\text{Trace}(A_{22}) = 0$, $\text{Trace}^2(A_{32}) = 0$, $\text{Trace}^3(A_{33}) = 0$. Let us compare the Bergman kernel with $K_{\text{Poincaré}} = C_n/u^{n+1}$ as we approach $0 \in \partial D$ from above. Set $z_t = (0, 0, \ldots, 0, it)$ so that $z_t \in D$ for $t > 0$.

THEOREM 2 (CHRISTOFFERS [12]). *As $t \to 0^+$ we have*

$$(6) \qquad K_D(z_t, z_t) = C_n \frac{1}{t^{n+1}} + \frac{\gamma_n}{t^{n-1}} \sum_{pqrs} \left| A^0_{p\bar{q}r\bar{s}} \right|^2 + o(t^{1-n}),$$

$$(7) \quad K_{\text{Poincaré}}(z_t, z_t) = \frac{C_n}{u^{n+1}(z)} = C_n \frac{1}{t^{n+1}} + \frac{\gamma'_n}{t^{n-1}} \sum_{pqrs} \left| A^0_{p\bar{q}r\bar{s}} \right|^2 + O(t^{1-n}),$$

where γ_n, γ'_n are constants depending on the dimension, and $\gamma'_n \neq \gamma_n$.

Thus, $\phi/\psi^{n+1} \neq C_n/u^{n+1}$. The proof of Christoffers' theorem involves hard calculations. In principle we already know how to calculate u, so (7) is just a matter of hard work. To prove (6), we use Theorem 1 to express $K_D(z, w)$ in terms of $\phi(z)$, $\tilde{\phi}(w)$. (Recall, we can express $\phi(z, w)$ in terms of $\phi(z)$ and similarly for $\tilde{\phi}$.) We then write down the reproducing formula $F_\varepsilon(z) \int_D K_D(z, w) F_\varepsilon(w) dw$ for explicit F_ε such as $F_\varepsilon(z) = (z_n + i\varepsilon)^{-(n+k)}$. Letting $\varepsilon \to 0^+$, we obtain from the reproducing formula some information about the derivatives of $\phi(z)$ at the origin. By picking enough F_ε, we get enough information to recover ϕ, ϕ', ϕ'', at the origin, from which we read off (6). The details are very complicated.

Christoffers' calculations give $K_D(z, z)$ up to an error $o(\psi^{-n+1})$ when $z = z_t$ and D is in normal form. By putting Theorem 2 in an invariant form, we can read off $K_D(z, z)$ modulo $o(\psi^{-n+1})$ for general D and z. For the invariant version of Theorem 2, we recall from the preceding chapter the function

$$U(z_0, z_1, \ldots, z_n) = |z_0|^2 u(z_1, \ldots, z_n) \quad \text{on } (\mathbf{C} \setminus \{0\}) \times D$$

and the indefinite metric

$$ds^2 = \sum_{j,k \geq 0} \frac{\partial^2 U}{\partial z_j \partial \bar{z}_k} dz_j \, \overline{dz_k} \quad \text{with curvature } R_{p\bar{q}r\bar{s}}.$$

A calculation shows that $\| R_{p\bar{q}r\bar{s}} \|^2 U^2 = (\text{const}) t^2 \Sigma_{pqrs} | A^0_{p\bar{q}r\bar{s}} |^2$, where the norm on the left is taken in ds^2. So Christoffers' theorem shows that when $z = z_t$ and D is in normal form, the Bergman kernel is given by

$$(8) \qquad K_D(z, z) = \frac{C_n}{u^{n+1}} \left\{ 1 + \gamma''_n \| R_{p\bar{q}r\bar{s}} \| U^2 + o(\psi^2) \right\}.$$

As the two sides of (8) both transform invariantly, we know that (8) holds for general strictly pseudoconvex domains, since it holds in normal form. After $K_{\text{Poincaré}} = C_n/u^{n+1}$, formula (8) is our second example of a nonlinear differential operator which transforms like the Bergman kernel.

After seeing formula (8) and recalling Weyl's invariant theory for the heat equation, we can make a conjecture on how to calculate the Bergman kernel.

CONJECTURE. Let $R_{p\bar{q}r\bar{s},\sigma\cdots\bar{\tau}}$ be the various covariant derivatives of the curvature in the metric ds^2 as above. Then the Bergman kernel is given by

$$(9) \qquad K(z,z) = \frac{C_n}{u^{n+1}}\left[1 + \sum_{s\geq 2} P_s\left(R_{\alpha\bar{\beta}\gamma\bar{\delta},\rho\cdots\bar{\sigma}}\right)U^s\right]$$

$$\text{modulo } C^\infty(\overline{D}) + (\log\psi)\cdot C^\infty(\overline{D}),$$

where P_s is a linear combination of traces of tensor products of the $R_{\alpha\bar{\beta}\gamma\bar{\delta},\rho\cdots\bar{\sigma}}$.

Let us see what this means for the U^2 coefficient, which we would need to compute $K(z,z)$ modulo $o(\psi^{1-n})$. For each expression $\Omega = \text{Trace}[R_{\alpha\bar{\beta}\gamma\bar{\delta},\rho\cdots\bar{\sigma}} \otimes \cdots \otimes R_{p\bar{q}r\bar{s},\mu\cdots\bar{\lambda}}]$ there is natural degree of homogeneity, namely the power of $|z_0|^2$ which appears in it. (Here z_0 is the extra variable in $\mathbf{C}\setminus\{0\}$.) This has to balance the homogeneity of U^2 so that $|z_0|^2$ does not appear in $K(z,z)$. The only invariant Ω with the proper degree to balance U^2 turns out to be $\|R_{p\bar{q}r\bar{s}}\|^2$. So the conjecture implies formula (8), thought it does not predict the value of the constant γ_n''. The conjecture is true—we shall sketch a proof in the next section. Before getting involved in the proof, we should note some important differences between the Bergman kernel and the heat kernel, for which of course the analogue of our conjecture is true.

For the heat kernel, one can simply follow its construction step by step and check explicitly that we never depart from the desired form

$$K_t(x,x) \sim \frac{C_n}{t^{n/2}}\left\{1 + \sum_{s\geq 1} P_s\left(R_{ijkl,p\cdots q}\right)t^s\right\},$$

with P_s linear combinations of traces of tensor products of the $R_{ijkl,p\cdots q}$. If we try the same direct aproach for the Bergman kernel, we come immediately to a serious difficulty. Both the known proofs of Theorem 1 depend strongly on making arbitrary choices; either a canonical transformation Φ in the Boutet-Sjöstrand proof, or an approximating domain D_{small} in the original argument. Therefore, if we look at an intermediate step in the construction of the Bergman kernel, we shall see many irrelevant terms which fail to transform correctly under biholomorphic maps. It is only at the end that the terms all combine in the right way to produce an invariant Bergman kernel. Since the intermediate steps bear little resemblance to the final result, it is hard to imagine how to use the "direct approach" to prove the conjecture on the Bergman kernel. Perhaps one should look for an invariant proof of Theorem 1. The phase function $\lambda\psi(z,w)$ in the Boutet-Sjöstrand proof bears a tantalizing similarity to our $U = |z_0|^2 u(z)$.

The asymptotics of the heat kernel can also be understood by using Weyl's invariant theory for the group $O(n)$. Unfortunately, if we try using invariant theory for the Bergman kernel, then we find that the relevant group is H, the group of linear fractional transformations of the ball which fix a given point on the boundary. This group is not semisimple so Weyl's invariant theory does not apply. In the next section we shall study invariant theory for H, exploiting the fact that H sits inside the semisimple group $U(n, 1)$. Although the algebra is not completely understood, we shall get far enough to prove the conjecture (9).

3. Parabolic invariant theory. In this section we prove some results of invariant theory from [23] that can be used to attack the conjecture of the previous section. Let H be a subgroup of semisimple Lie group G. Typically G will consist of matrices preserving an indefinite quadratic form, and H will be the set of matrices in G which fix a given vector e in the light cone. There are several different settings for invariant theory.

Problem 1. Given an action of G on a vector space X, find all G-invariant polynomials on X. This case was discussed in Chapter 5. For our purposes, it is well understood.

Problem 2. Given an action of H on X, find all H-invariant polynomials on X. This problem is too hard.

Problem 3. Given an action of G on X find all H-invariant polynomials on X. This may be reduced back to Problem 1 using a little algebraic geometry, which we shall explain in a moment. The results are due to Weizenbock [65], Seshadri [56], and Hochschild-Mostow [31].

Problem 4. Given an action of G on X and a vector subspace $Y \subseteq X$ which is H-invariant but not G-invariant, find all H-invariant polynomials on Y.
This is what we need for the Bergman kernel.

Let us start with an example of Problem 3. Take $G = O(n, m)$ acting on R^{n+m}, fix a vector e in the light cone, and set $H = \{T \in G \mid Te = e\}$. Recall from Chapter 5 that the G-invariant polynomials in the components of vectors $v_1, \ldots, v_N$ are already generated by the obvious examples $\langle v_j, v_k \rangle$. The analogous result for H is as follows.

PROPOSITION 1. *All the H-invariant polynomials in the components of vectors $v_1, \ldots, v_N$ are generated by the obvious examples $\langle v_j, v_k \rangle$ and $\langle v_j, e \rangle$.*

Here, we are in the setting of Problem 3, where X is a direct sum of N copies of R^{n+m} and G acts on X by $T(v_1, \ldots, v_N) = (Tv_1, \ldots, Tv_N)$. To prove the proposition, we introduce the concept of a *normal variety* $V \subseteq \mathbf{C}^n$. The variety V is called normal if whenever a rational function $F = P/Q$ is holomorphic on $V \setminus$ (singular points of V) we can find a polynomial S whose restriction to V agrees with F.

EXAMPLE 1. $V = \{(z, w) \mid z^2 = w^3\}$ is not normal. The points of V are given by $z = t^3$, $w = t^2$; $t = z/w = w^2/z$ is evidently a rational function of (z, w) holomorphic on V away from the singular point $(0, 0)$. However, any polynomial $S(z, w)$ takes the form $S(t^3, t^2)$ on V, and we certainly cannot write t in this form.

EXAMPLE 2. The variety defined by setting a quadratic form $\langle r, r \rangle$ equal to zero is normal. This is a standard result in elementary algebraic geometry.

PROOF OF PROPOSITION 1. Let $P(v^1, \ldots, v^N)$ be an H-invariant polynomial. Using P we shall define a function $F(v^0, v^1, \ldots, v^N)$ for v^0 in the light cone and $v^1, \ldots, v^N$ arbitrary. We first pick a matrix $T_{v^0} \in G$ which carries v^0 back to e,

$(T_{v^0})v^0 = e$. This can be done since G acts transitively on the light cone, but the choice of T_{v^0} is determined only modulo left multiplication by matrices in H. Now set

$$F(v^0, v^1, \ldots, v^N) = P(T_{v^0}v^1, \ldots, T_{v^0}v^N).$$

This is independent of the choice of T_{v^0} since P is H-invariant. We can easily check that F is a G-invariant function defined on

$$V = \left\{ (v^0, v^1, \ldots, v^N) \,\middle|\, \|v^0\|^2 = 0 \right\} \subset R^{n+m} \oplus X.$$

If F were a polynomial, then by Weyl's invariant theory we would know that F agrees on V with a polynomial generated by $\langle v^j, v^k \rangle$, $\langle v^j, v^0 \rangle$, $\langle v^0, v^0 \rangle = 0$. However, when $v^0 = e$ we can take $T_{v^0} = \text{Identity}$ in the definition of F, so that $F(e, v^1, \ldots, v^N) = P(v^1, \ldots, v^N)$ and consequently P is generated by $\langle v^j, v^k \rangle$, $\langle v^j, e \rangle$, proving the proposition. We are now going to show that F is a polynomial.

It is easy to write down explicit matrices $T_{v^0} \in G$ taking v^0 to e, whose entries are of the form (Polynomial in v^0)/$(v_l^0)^{\text{power}}$, where v_l^0 is any component of v^0. From the definition of F we see therefore that

$$(10) \qquad F(v^0, v^1, \ldots, v^N) = \frac{P_l(v^0, v^1, \ldots, v^N)}{(v_l^0)^{\text{power}}} \quad \text{on } V$$

for suitable polynomials P_l. While we do not yet know that F is a polynomial, we see at once that F is a rational function with no singularities away from $\{v^0 = 0\}$. The different right-hand sides of (10) agree on V, and it follows by an easy analytic continuation that they still agree on the complex variety

$$V^c = \left\{ (v^0, v^1, \ldots, v^N) \in \mathbf{C}^{(n+m)N} \,\middle|\, \langle v^0, v^0 \rangle = 0 \right\}.$$

Thus F extends from V to a rational function on V^c, and furthermore this rational function is holomorphic on V^c away from the singular set $\{v^0 = 0\}$. Since V^c is defined by the vanishing of one quadratic form, it is normal and so F agrees on V with a polynomial. The proposition is proved. $\square$

The ideas just explained give a framework for solutions of Problem 4. We suppose that G is a group of linear transformations on a vector space $V = \mathbf{R}^n$ or $\mathbf{C}^n$ while H is the subgroup of those $T \in G$ which fix a vector $e \in V$. An obvious recipe to make H-invariant polynomials $P(y)$ on Y is to start with a G-invariant polynomial $Q(v, x)$ defined on $V \oplus X$ and then set $P(y) = Q(e, y)$. Such a $P(y)$ will be called a *Weyl invariant*. Weyl invariants can be understood by the methods of Chapter 5, since we are dealing with G-invariant polynomials. For instance if $G = O(n, m)$ or $U(n, m)$ and X consists of tensors, then Q must be a linear combination of traces of tensor products. This is just what we need to write down the Bergman kernel. We now have to face the main issue: is every H-invariant polynomial on Y necessarily a Weyl invariant? The method of proof of the proposition shows that the answer is yes, provided the following variety is normal:

$$(11) \qquad V = \{(Te, Ty) \mid T \in G,\ y \in Y\}.$$

(Strictly speaking we have to complexify V. Let us ignore this point.) If $Y = X$ (Problem 3) then again V will be defined by the vanishing of a single quadratic form and so must be normal. For Y such as we need to study the Bergman kernel, the analysis of the singularity of V is very complicated, and V is not normal.

To see what actually happens when we deal with Problem 4, we present a nontrivial example. Take $G = O(n, 1)$ the group of matrices preserving the quadratic form $\|v\|^2 = 2v_0 v_1 - \Sigma_{k \geqslant 2} v_k^2$, set $e = (1, 0, \ldots, 0) \in$ Light Cone, and set $H = \{T \in G \mid Te = e\}$. We fix a positive integer g, and define for $N > g$ the vector spaces $X_N = \{$symmetric N-tensors $R = (R_{j_1 j_2, \ldots, j_N})\}$, $Y_N = \{R \in X_N \mid (R_{j_1, j_2, \ldots, j_g, 0, \ldots, 0}) = 0\}$. G acts in the obvious way on X_N and $Y_N \subseteq X_N$ is an H-invariant subspace. So we are in the setting of Problem 4, and we ask whether every H-invariant polynomial P on Y_N is a Weyl invariant. The best answer we know depends on the degree of P.

To state the result, we have to bring in a new ingredient, namely the relation of the different Y_N. There is a natural projection $\Pi\colon X_{N+1} \to X_N$ defined by setting $(\pi R)_{j_1 \cdots j_N} = R_{j_1 \cdots j_N 0}$. Note that π commutes with the action of H and carries Y_{N+1} to Y_N. By composing, we get projections $\pi_N^{N'}\colon Y_{N'} \to Y_N$ for $N' > N$, and these commute with the action of H. So if P is an H-invariant polynomial on Y_N, then $P \circ \pi_N^{N'}$ is an H-invariant polynomial on $Y_{N'}$ $(N' > N)$. Furthermore, it is easy to check that if P is a Weyl invariant, then $P \circ \pi_N^{N'}$ is a Weyl invariant. Now we can state our result.

THEOREM 3. *Let P be a homogeneous H-invariant polynomial on Y_N.*
(A) *If $\deg P < n - 1$, then P is a Weyl invariant.*
(B) *If $\deg P > n - 1$, then $P \circ \pi_N^{N'}$ is a Weyl invariant for large $N' > N$.*

To prove the theorem, we pass to the varieties V_N given by (11). Explicitly

$$V_N = \left\{ (v, R) \in R^{n+1} \oplus X_N \ \middle| \ \begin{array}{l} \text{(i) } \|v\|^2 = 0 \text{ and} \\[2mm] \text{(ii) } \displaystyle\sum_{k_1, \ldots, k_{N-g}} v_{k_1} \cdots v_{k_{N-g}} R_{j_1 \cdots j_g k_1 \cdots k_{N-g}} = 0 \end{array} \right\}.$$

Requirement (ii) means $\mathrm{Trace}[v \otimes \cdots \otimes v \otimes R] = 0$. There are G-invariant projections $\pi\colon V_{N+1} \to V_N$ defined by sending (v, R) to $(v, \tilde{R})$ with $\tilde{R}_{j_1 \cdots j_N} = \Sigma_k v_k R_{j_1 \cdots j_N k}$, i.e. $\tilde{R} = \mathrm{Trace}[v \otimes R]$. The main step in proving Theorem 3 is to show the following:

STABLE NORMALITY OF V_N. *Let $F(v, R)$ be a rational function on V_N holomorphic off the singular set. Suppose $F(v, R)$ is homogeneous of degree d in R.*
(A) *If $d < n - 1$ then F is the restriction of a polynomial to V_N.*
(B) *If $d > n - 1$ then F need not be the restriction of a polynomial. However, for large $N' > N$, $F \circ \pi_N^{N'}$ is the restriction of a polynomial to $V_{N'}$.*

The projections $\pi_N^{N'}$ collapse $V_{N'}$ near the singular set, so there is a chance of making a nice transcendental proof of stable normality. So far, the only known proof is rather nasty and technical.

At last we can discuss the example needed for the Bergman kernel. Here $G = U(n, 1)$, the group of complex matrices preserving $\|v\|^2 = v_0 \bar{v}_1 + v_1 \bar{v}_0 - \Sigma_{k \geqslant 2} v_k \bar{v}_k$, while H consists of all $T \in G$ which fix the vector $e = (1, 0, 0, \ldots, 0)$. We set $X_N = \Sigma_{0 < p, q \leqslant N} \oplus X_{pq}$, where X_{pq} consists of all tensors $(R_{i\bar{j}k\bar{l},\mu_1 \cdots \mu_p \bar{\nu}_1 \cdots \bar{\nu}_q})$ of rank $(p + 2, q + 2)$, and

$$\varepsilon_N = \{ R \in X_N \mid R_{i\bar{j}k\bar{l},\text{etc.}} \text{ arise as the covariant derivatives at}$$
$(1, 0, \ldots, 0)$ of the metric

$$ds^2 = \sum_{jk \geqslant 0} \frac{\partial^2 U}{\partial z_j \partial \bar{z}_k} dz_j \, \overline{dz_k}$$

associated to some domain in Moser's normal form$\}$.

The main problem for the Bergman kernel is to understand the H-invariant polynomials on ε_N. It turns out that ε_N sits in X_N as a submanifold, and that the study of H-invariant polynomials on ε_N can be easily reduced to $Y_N = $ Tangent space to ε_N at the origin. Roughly speaking,

$$Y_N = \left\{ \left(R_{j_1 \cdots j_N \bar{k}_1 \cdots \bar{k}_N} \right) \mid R_{j_1 0 \cdots 0 \bar{k}_1 \cdots \bar{k}_N} = R_{j_1 \cdots j_N \bar{k}_1 \bar{0} \cdots \bar{0}} = 0 \right\},$$

so that our problem is rather analogous to the setting of Theorem 3. Although we expect an analogous result here, the study of H-invariants on Y_N turns out to be much harder than Theorem 3. After much work, one can derive the following partial analogoue of Theorem 3.

THEOREM 4. *Every H-invariant polynomial of degree $n \leqslant 20$ on Y_N is a Weyl invariant.*

To prove this requires the analogue of part (A) of stable normality. The details are extraordinarily hard. We next show how Theorem 4 yields the Bergman kernel. With the notation of §§1, 2, we have

THEOREM 5. $K_D(z, z) = (C_k/u^{n+1})\{1 + \Sigma_s P_s U^s\} + O(\psi^{-20})$, *where P_s is a linear combination of traces of tensor products of the $(R_{\alpha\bar{\beta}\gamma\bar{\delta},p \cdots \bar{\sigma}})$.*

PROOF. By induction on $l \leqslant n - 20$, we shall prove that

$$(12) \qquad K_D(z, z) = \frac{C_n}{u^{n+1}} \left\{ 1 + \sum_{s \leqslant l} P_s U^s + O(\psi^{l+1}) \right\}.$$

For $l = 0$ this is Theorem 1, while for $l = 2$ it is Theorem 2. So assume (12) for l and try to prove it for $l + 1$. From the known case of (12) and since $K_D(z, z) = \phi/u^{n+1} + O(\log \psi)$ with $\phi \in C^\infty(\bar{D})$ by Theorem 1, we obtain trivially

$$(13) \qquad K_D(z, z) = \frac{C_n}{u^{n+1}} \left\{ 1 + \sum_{s \leqslant l} P_s U^s + \tilde{P}_{l+1} U^{l+1} + O(\psi^{l+2}) \right\}$$

for some smooth function $\tilde{P}_{l+1}$. The problem is to show that $\tilde{P}_{l+1}$ can be taken as a linear combination of traces of tensor products of $(R_{\alpha\bar{\beta}\gamma\bar{\delta},\text{etc.}})$.

Now, equation (13) shows that $\tilde{P}_{l+1}$ is uniquely defined on the boundary ∂D. Moreover, if ∂D is in Moser's normal form with coefficients $(A^l_{\alpha\bar{\beta}})$, then by going carefully through either of the two proofs of Theorem 1, we can check that the value of $\tilde{P}_{l+1}$ at the origin is a polynomial in the $A^l_{\alpha\bar{\beta}}$. Let us write $\tilde{P}_{l+1}(0) = Q(A^l_{\alpha\bar{\beta}})$. Furthermore, (13) and the transformation laws for the Bergman kernel and the Monge-Ampère equation show that the polynomial Q has its own transformation law under the action of the isotropy group H on normal forms. More precisely, suppose D and $\tilde{D}$ are in normal form with coefficients $(A^l_{\alpha\bar{\beta}})$ and $(\tilde{A}^l_{\alpha\bar{\beta}})$ respectively, and suppose D and $\tilde{D}$ are equivalent by a biholomorphic map Φ fixing the origin. Then $Q(A^l_{\alpha\bar{\beta}}) = |\det \Phi'(0)|^s Q(\tilde{A}^l_{\alpha\bar{\beta}})$ for a suitable s depending on l. We shall say that Q is a *boundary invariant of weight* s.

Using Theorem 4, we can understand arbitrary boundary invariants of weight s for a range of s. The point is simply to switch over from the $A^l_{\alpha\bar{\beta}}$ to the covariant derivatives $(R_{\alpha\bar{\beta}\gamma\bar{\delta},p\,\cdots\,\bar{\sigma}})$ evaluated at $(1,0)$ (which in projective space lies over the origin). One checks that $A^l_{\alpha\bar{\beta}}$ may be expressed as a polynomial in the $R_{\alpha\bar{\beta}\gamma\bar{\delta},\text{etc.}}$, so that our boundary invariant Q may now be written as $Q = P(R_{\alpha\bar{\beta}\gamma\bar{\delta},\text{etc.}})$. The transformation law for boundary invariants shows that P has a certain homogeneity and is H-invariant on

$$\varepsilon_N = \left\{ \text{tensors}(R_{\alpha\bar{\beta}\gamma\bar{\delta},\text{etc.}}) \text{ arising from a domain in normal form} \right\}.$$

The homogeneity of P imposes a bound on its degree, and Theorem 4 can be used to show that P is a linear combination of traces of tensor products of the $(R_{\alpha\bar{\beta}\gamma\bar{\delta},\text{etc.}})$. Therefore Q is a boundary invariant of weight s if and only if Q is a linear combination of traces of $(R_{\alpha\bar{\beta}\gamma\bar{\delta},\text{etc.}})$ arising from the domain with coefficients $(A^l_{\alpha\bar{\beta}})$.

Now we can apply our discussion of boundary invariants to the polynomial Q determined by P_{l+1} in (13). For a suitable $P_{l+1} = $ linear combination of traces of tensor products of $(R_{\alpha\bar{\beta}\gamma\bar{\delta},\text{etc.}})$, we have $P_{l+1}(0) = P_{l+1}$ for domains in normal form. Equation (13) shows that

$$K_D(z,z) = \frac{C_n}{u^{n+1}} \left\{ 1 + \sum_{s \leq l+1} P_s U^s + O(\psi^{l+2}) \right\} \quad \text{as } z \to 0$$

for D in normal form. Since both sides transform invariantly under biholomorphic maps, the same is true when z approaches an arbitrary boundary point of an arbitrary strictly pseudoconvex domain. This is exactly equation (12) with l replaced by $l+1$, so the induction is complete. $\square$

Essentially, we have found all the nonlinear differential operators of the form S in §2. The discussion fails completely when applied to T (the log term), both because our Monge-Ampère function carries an ambiguity $O(\psi^{n+2})$ and because the invariant theory will now involve an analogue of the hard case $\deg P = n-1$ in Theorem 3. R. Graham has recently made discoveries related to the problem of writing down T.

We should mention that an analogue of Theorem 5 holds also for the Szegö kernel.

Throughout this chapter, the analogy between the heat kernel and the Bergman kernel has shown itself ever more clearly. Is there an analogue of the Gauss-Bonnet theorem for Chern-Moser invariants?

REFERENCES

1. V. I. Arnold, *Mathematical methods of classical mechanics*, Springer-Verlag, Berlin and New York, 1978.

2. S. Bell and E. Ligocka, *A simplification and extension of Fefferman's theorem on biholomorphic mappings*, Invent. Math. **57** (1980), 283–289.

3. M. Berger, P. Gauduchon and E. Mazet, *Le spectre d'une variété Riemannienne*, Springer-Verlag, Berlin and New York, 1971.

4. L. Boutet de Monvel, *Integration des équations de Cauchy-Riemann induites formelles*, Séminaire Goulaouic-Lions-Schwartz (1974–1975).

5. L. Boutet de Monvel and J. Sjöstrand, *Sur la singularité des noyaux de Bergman et de Szegö*, Soc. Math. de France Astérisque **34–35** (1976), 123–164.

6. L. Boutet de Monvel, to appear.

7. D. Burns and S. Schneider, personal communication.

8. E. Cartan, *Sur la géométrie pseudo-conforme des hypersurfaces de deux variables complexes*. I, II, Oeuvres II, **2**, 1231–1304; Oeuvres III, **2**, 1217–1238.

9. D. Catlin, *Necessary conditions for subellipticity and hypoellipticity for the $\bar\partial$-Neumann problem on pseudoconvex domains*, in [**26**].

10. S. Y. Cheng and S-T. Yau, *On the regularity of the Monge-Ampère equation* $\det(\partial^2 u/\partial x^i \partial x^j) = F(x, u)$, Comm. Pure Appl. Math. **30** (1977), 41–68.

11. S. S. Chern and J. Moser, *Real hypersurfaces in complex manifolds*, Acta Math. **133** (1974), 219–271.

12. H. Christoffers, Thesis, Univ. of Chicago, 1980.

13. J. P. D'Angelo, *Orders of contact of real and complex subvarieties*, Illinois J. Math. **26** (1982), 41–51.

14. K. Diederich and P. Pflug, *Necessary conditions for hypoellipticity of the $\bar\partial$-problem*, in [**26**].

15. J. J. Duistermaat and V. Guillemin, *The spectrum of positive elliptic operators and periodic geodesics*, Invent. Math. **29** (1975), 39–79.

16. J. J. Duistermaat and J. Sjöstrand, to appear.

17. Yu. V. Egorov, *Subelliptic pseudodifferential operators*, Dokl. Akad. Nauk SSSR **188** (1969), 20–22.

18. ______, *On canonical transformations of pseudodifferential operators*, Uspehi Mat. Nauk **25** (1969), 235–236.

19. ______, *Subelliptic operators*, Uspehi Mat. Nauk **30:2** (1975), 71–114 = Russian Math. Surveys **30:2** (1975), 59–118.

20. ______, *Subelliptic operators*, Uspehi. Mat. Nauk **30:3** (1975), 57–104 = Russian Math. Surveys **30:3** (1975), 55–105.

21. C. Fefferman, *The Bergman kernel and biholomorphic mappings of pseudoconvex domains*, Invent. Math. **26** (1974), 1–65.

22. ______, *Monge-Ampère equations, the Bergman kernel, and geometry of pseudoconvex domains*, Ann. of Math. (2) **103** (1976), 395–416; erratum **104** (1976), 393–394.

23. ______, *Parabolic invariant theory in complex analysis*, Adv. in Math. **31** (1979), 131–262.

24. G. B. Folland and J. J. Kohn, *The Neumann problem for the Cauchy-Riemann complex*, Princeton Univ. Press, Princeton, N. J., 1972.

25. G. B. Folland and E. M. Stein, *Estimates for the $\bar\partial_b$ complex and analysis on the Heisenberg group*, Comm. Pure Appl. Math. **27** (1974), 429–522.

26. J. E. Fornaess (editor), *Recent developments in several complex variables*, Princeton Univ. Press, Princeton, N. J., 1981.

27. P. Gilkey, *The index theorem and the heat equation*, Publish or Perish, Berkeley, Calif., 1974.

28. H. Goldstein, *Classical mechanics*, Addison-Wesley, Reading, Mass., 1950.

29. R. Graham, to appear.

30. V. Guillemin, *Some classical theorems in spectral theory revisited*, in [**35**].

31. G. Hochschild and G. D. Mostow, *Unipotent groups in invariant theory*, Proc. Nat. Acad. Sci. USA **70** (1973), 646–648.

32. L. Hörmander, *Hypoelliptic second order differential equations*, Acta Math. **119** (1967), 141–171.

33. ______, *The spectral function of an elliptic operator*, Acta Math. **121** (1968), 193–218.

34. ______, *Fourier integral operators*. I, Acta Math. **127** (1971), 79–183.

35. L. Hörmander (editor), *Seminar on singularities of solutions of linear partial differential equations*, Princeton Univ. Press, Princeton, N. J., 1979.

36. L. Hörmander, *Subelliptic operators*, in [**35**].

37. N. Kerzman, *The Bergman kernel function: differentiability at the boundary*, Math. Ann. **195** (1972), 149–158.

38. S. Kobayashi and K. Nomizu, *Foundations of differential geometry*, Wiley, New York, 1969.

39. J. J. Kohn, *Harmonic integrals on strongly pseudoconvex manifolds*. I, II, Ann. of Math. (2) **78** (1963), 112–148; ibid **79** (1964), 450–472.

40. ______, *Subellipticity of the $\bar{\partial}$-Neumann problem on pseudoconvex domains: sufficient conditions*, Acta Math. **142** (1979), 79–122.

41. S. G. Krantz, *Function theory of several complex variables*, Wiley, New York, 1982.

42. M. Kuranishi, *Strongly pseudoconvex CR structures over small balls*. I, II, Ann. of Math. (2) **115** (1982), 451–500; ibid **116** (1982), 1–64.

43. S. Lang, *Algebra*, Addison-Wesley, Reading, Mass., 1971.

44. S. Lee and R. Melrose, personal communication.

45. V. P. Maslov, *Theory of perturbations and asymptotic methods*, Moskov. Gos. Univ., Moscow, 1965. (Russian)

46. A. Melin and J. Sjöstrand, *Fourier integral operators with complex valued phase functions*, Lecture Notes in Math., vol. 459, Springer-Verlag, Berlin and New York, pp. 120–223.

47. S. Minakshisundaram and A. Pleijel, *Some properties of the eigenfunctions of the Laplace-operator on Riemannian manifolds*, Canad. J. Math. **1** (1949), 242–256.

48. J. Moser, *Holomorphic equivalence and normal forms of hypersurfaces*, Proc. Sympos. Pure Math., vol. 27, part 2, Amer. Math. Soc., Providence, R. I., 1975, pp. 109–112.

49. A. Nagel and E. M. Stein, *Lectures on psudo-differential operators*, Princeton Univ. Press, Princeton, N. J., 1979.

50. L. Nirenberg, *On a problem of Hans Lewy*, Uspehi. Mat. Nauk **292** (176) (1974), 241–251.

51. L. Nirenberg, S. Webster and P. Yang, *Local boundary regularity of holomorphic mappings*, Comm. Pure Appl. Math. **33** (1980), 305–338.

52. V. K. Patodi, *Curvature and eigenforms of the Laplace operator*, J. Differential Geom. **5** (1971), 233–249.

53. D. H. Phong, *On integral representations of the Neumann operator*, Proc. Nat. Acad. Sci. USA **76** (1979), 1554–1558.

54. L. P. Rothschild and E. M. Stein, *Hypoelliptic differential operators and nilpotent groups*, Acta Math. **137** (1976), 247–320.

55. M. Sato, T. Kawai and M. Kashiwera, *Microfunctions and pseudodifferential equations*, Lecture Notes in Math., vol. 287, Springer-Verlag, Berlin and New York, 1973.

56. C. Seshadri, *On a theorem of Weizenböck in invariant theory*, J. Math. Kyoto Univ. **1** (1962), 403–409.

57. E. M. Stein, *Singular integrals and differentiability properties of functions*, Princeton Univ. Press, Princeton, N. J., 1970.

58. N. Tanaka, *On the pseudo-conformal geometry of hypersurfaces of the space of n complex variables*, J. Math. Soc. Japan **14** (1962), 397–429.

59. ______, *Graded Lie algebras and geometric structures*, Proc. U. S.-Japan Seminar in Differential Geometry, Nippon Hyoronsha, 1966, pp. 147–150.

60. D. S. Tartakoff, *A survey of some recent results in C^∞ and real analytic hypoellipticity for partial differential equations with applications to several complex variables*, in [**26**].

61. M. E. Taylor, *Pseudodifferential operators*, Princeton Univ. Press, Princeton, N. J., 1981.

62. F. Trèves, *Analytic hypoellipticity of a class of pseudodifferential operators with double characteristics and applications to the $\bar{\partial}$-Neumann problem*, Comm. Partial Differential Equations **3** (1978), 475–642.

63 ______, *Introduction to pseudodifferential and Fourier integral operators*, 2 vols., Plenum Press, New York, 1980.

64. S. M. Webster, *On the mapping problem for algebraic real hypersurfaces*, Invent. Math. **43** (1977), 53–68.

65. R. Weizenböck, *Über die invarianten von linearen gruppen*, Acta Math. **58** (1932), 230–250.

66. R. O. Wells, Jr., *The Cauchy-Reimann equations and differential geometry*, Bull. Amer. Math. Soc. (N.S.) **6** (1982), 187–199.

67. H. Weyl, *Classical groups*, Princeton Univ. Press, Princeton, New Jersey 1946.

DEPARTMENT OF MATHEMATICS, RUTGERS UNIVERSITY, NEW BRUNSWICK, NEW JERSEY 08903

DEPARTMENT OF MATHEMATICS, PRINCETON UNIVERSITY, PRINCETON, NEW JERSEY 08544

PROGRAM IN APPLIED MATHEMATICS, PRINCETON UNIVERSITY, PRINCETON, NEW JERSEY 08544

Proceedings of Symposia in Pure Mathematics
Volume 39 (1983), Part 1

Poincaré and Algebraic Geometry

PHILLIP A. GRIFFITHS[1]

ABSTRACT. A few of the contributions of Poincaré to algebraic geometry are described, with emphasis on his late work on normal functions. A very brief description of some recent works is also given.

Although the subject of algebraic geometry was not one of Poincaré's major preoccupations, his work in the field showed characteristic insight and brilliance and certainly has had a lasting effect. I shall briefly describe his major contributions, with special emphasis on the two papers on normal functions which constituted Poincaré's last published mathematical work.

In general I have tried to follow standard current notations. Unless mentioned to the contrary, homology and cohomology are with $\mathbf{Q}$ coefficients.

At the end there is a list of some of Poincaré's major papers in algebraic geometry together with the other bibliographical items referred to in the text.

As will be clear from the discussion below, the decision to emphasize Poincaré's work on normal functions reflects my own particular interest.

(a) Let me begin by recalling the period during which Poincaré worked. This was a particularly active time in algebraic geometry, especially for the study of algebraic surfaces. In the preceding thirty years, beginning with Riemann's thesis the theory of algebraic functions of one variable, or as we now know it, algebraic curves, had been developed by Riemann, Max Noether, and many others to the point where—aside from moduli questions —the theory had assumed much the form that we find it today. Beginning in the 1890's Poincaré's colleague, E. Picard, had begun his monumental work on transcendental algebraic geometry in higher dimensions, especially on surfaces, that among other things led to the Picard-Fuchs equation and Picard-Lefschetz transformation, the algebraic de Rham theorem for smoothly compactifiable affine varieties and subsequent theory of single and double integrals of the second kind on surfaces, and the Picard variety and Picard number. Meanwhile, in Italy the birational theory of algebraic surfaces was well underway in the work of Castelnuovo, Enriques, Severi (at a slightly later time), and the other members of the Italian school.

It is clear that Poincaré was well abreast of these developments, and it seems to me that his own work in algebraic geometry was frequently in

---

Reprinted from Bulletin Amer. Math. Soc. (N.S.) 6 (1982), 147–159.

1980 *Mathematics Subject Classification.* Primary

[1] Research partially supported by NSF Grant MCS7707782.

response to this activity. For example, as will be explained below, his idea of normal functions was devised to give the first complete proof of one of the main results of Picard and Castelnuovo-Enriques, the theorem that the Picard variety of a surface has dimension equal to the irregularity.

(b) Poincaré wrote a number of papers, several in collaboration with Picard, concerning abelian varieties and Jacobians of curves. As was characteristic of much of his work, several of these papers involved interaction between various branches of mathematics. For example, his studies on abelian functions led to the Cousin problems in several complex variables, and his work on translation-type surfaces in abelian varieties was instrumental in the theory of webs about which Chern will speak.

In algebraic geometry proper, at least two results from the area of abelian varieties and Jacobians are noteworthy. One is

(1) *Poincaré's complete reducibility theorem* [5]: *Given an abelian subvariety A' of an abelian variety A, there exists a finite covering*

$$\tilde{A} \xrightarrow{\pi} A$$

and abelian subvariety $A'' \subset \tilde{A}$ such that

$$\tilde{A} \cong \pi^{-1}(A') \times A''.$$

This is a clear forerunner of the modern version that states: *the category of polarized Hodge structures $/\mathbf{Q}$ is semisimple.*

A second result concerns the Jacobian variety J of a smooth algebraic curve C of genus g. Recall that there is a holomorphic mapping $u: C \to J$ defined by

$$(2) \qquad u(p) = \left(\int_{p_0}^{p} \omega_1, \ldots, \int_{p_0}^{p} \omega_g \right)$$

where $p_0 \in C$ is a base point and $\omega_1, \ldots, \omega_g$ is a basis for the holomorphic differentials on C. If C_d denotes the set of all unordered d-tuples of points

$$D = p_1 + \cdots + p_d,$$

it may be shown that C_d has a natural complex manifold structure to which the mapping (2) may be extended by the formation of abelian sums. Thus

$$(3) \qquad u: C_d \to J$$

is defined by

$$u(p_1 + \cdots + p_d) = \left(\sum_i \int_{p_0}^{p_i} \omega_1, \ldots, \sum_i \int_{p_0}^{p_i} \omega_g \right).$$

DEFINITION. *For $1 \leqslant d \leqslant g$ we define the subvariety $W_d \subset J$ to be the image variety of the proper holomorphic mapping* (3).

Equivalently, W_d is the translation-type subvariety of J given by sums of d points on the curve $u(C)$; i.e.

$$W_d = \{ u(p_1) + \cdots + (p_d) : p_i \in C \}.$$

The *Jacobi inversion theorem* states that

$$(4) \qquad W_g = J.$$

More precisely, for $d = g$ the mapping (3) is birational.

Next, if $\Theta \subset J$ is the divisor of zeroes of Riemann's theta function, then *Riemann's theorem* is

$$(5) \qquad W_{g-1} = \Theta + \kappa,$$

where the right-hand side of (5) is translation of Θ by the vector κ of Riemann constants.

Poincaré gave the following extension of (4) and (5) to all $d \leq g$:

(6) POINCARÉ'S FORMULA [6]. *If $\theta \in H^2(J)$ and $w_d \in H^{2(g-d)}(J)$ are the fundamental cohomology classes of Θ and W_d respectively, then*

$$(7) \qquad w_d = \theta^{g-d} / (g-d)!$$

When $d = g$ this implies that

$$u: C_g \to J$$

has degree one, and when $d = g - 1$ (7) gives that

$$w_{g-1} = \theta.$$

Riemann's theorem (5) is then a consequence of the fact that the Jacobian of a curve is a principally polarized abelian variety.

It is perhaps worth pointing out that (6) could not have even been properly formulated before Poincaré's introduction of homology. This is also true of the next topic dealing with residues of double integrals.

In a similar vein we also note that Picard's great work [3] on algebraic functions of two variables would not have been possible had not Poincaré concurrently developed the necessary topological concepts.

(c) The next of Poincaré's investigations in algebraic geometry that I would like to discuss centers around what is now called the Poincaré residue operator. It is an elementary fact that, on the complex projective line $\mathbf{P}^1$ with linear coordinate z, the cohomology of the punctured sphere

$$\mathbf{P}^1 - \{a_1, \ldots, a_N\},$$

a_ν distinct, is described by rational differential forms having poles at the a_ν. More precisely, if we set

$$p(z) = \prod_{\nu=1}^{N} (z - a_\nu)$$

then every cohomology class is represented by a unique form

$$w = q(z)dz / p(z), \qquad \deg q(z) \leq N - 2,$$

that has simple poles at the a_ν.

Motivated in part by his work on topology and in part by the aforementioned project of Picard concerning periods of rational integrals on algebraic surfaces, Poincaré [7] considered rational forms

$$(8) \qquad \omega = \frac{q(x, y)\, dx \wedge dy}{p(x, y)}, \qquad \deg q(x, y) \leq N - 3,$$

on the projective plane $\mathbf{P}^2$. Here, (x, y) are affine coordinates, $p(x, y)$ is an irreducible polynomial of degree N defining an algebraic curve D, and the degree restriction in (8) is equivalent to saying that ω does not have a pole on the line at infinity. Poincaré showed that the periods

$$(9) \qquad \iint_\Gamma \omega, \qquad \Gamma \in H_2(\mathbf{P}^2 - D),$$

are all of the form

$$\int_\gamma \operatorname{Res} \omega, \qquad \gamma \in H_1(D),$$

where

$$(10) \qquad \operatorname{Res} \omega = q\,dx/(\partial p/\partial y) = -q\,dy/(\partial p/\partial x) \quad \text{on } p(x,y) = 0$$

defines a differential called the *Poincaré residue*. Here γ is a 1-cycle in $D^* = D \setminus \{\text{singular locus}\}$, and the cycles Γ and γ are related by $\Gamma = \tau(\gamma)$ where

$$(11) \qquad \tau: H_1(D^*) \to H_2(\mathbf{P}^2 - D)$$

is the mapping that assigns to γ the boundary of that part of the solid ε-tube around D^* that lies over γ.

Nowadays one considers the *Poincaré residue sheaf sequence*

$$(12) \qquad 0 \to \Omega_X^n \to \Omega_X^n(D) \overset{\operatorname{Res}}{\to} \Omega_D^{n-1} \to 0$$

where X is a smooth n-dimensional algebraic variety and $D \subset X$ is a divisor that, for simplicity of explanation, we assume to be smooth. Poincaré's case corresponds to $X = \mathbf{P}^2$, and the mapping (10)

$$\omega \to \operatorname{Res} \omega$$

is

$$\operatorname{Res}: H^0(\Omega_X^n(D)) \to H^0(\Omega_D^{n-1}).$$

His results concerning the periods of double integrals (9) are encoded in the long exact cohomology sequence of (12), together with Poincaré-Lefschetz duality and Hodge theory.

It is interesting to note that in his paper [7] on residues Poincaré raised the issue concerning periods

$$(13) \qquad \iint_\Gamma \frac{q(x,y)\,dx \wedge dy}{p(x,y)^{k+1}}, \qquad \deg q(x,y) \leqslant (k+1)N - 3,$$

whose integrand has a higher order pole along C. Contrary to the one variable case, by subtracting exact forms it is not always possible to reduce the order of pole of ω to one, and Poincaré mused on but did not entirely come to grips with this fact. It is now easy to explain the reason why: For a smooth hypersurface $D \subset \mathbf{P}^n$ of degree N given in affine coordinates by

$$p(x_1, \ldots, x_n) = 0,$$

we consider the mapping on cohomology

$$\operatorname{Res}: H^n(\mathbf{P}^n - D) \to H^{n-1}(D)$$

dual to the "tube over cycle mapping"

$$\tau: H_{n-1}(D) \to H_n(\mathbf{P}^n - D)$$

encountered in the case $n = 2$ in (11). The rational n-forms

$$\Omega = \frac{q(x)\, dx_1 \wedge \cdots \wedge dx_n}{p(x)^{k+1}}, \qquad \deg q(x) \leqslant (k+1)n - (n+1),$$

have poles of order $k + 1$ along D, and they generate a vector subspace

$$F^{n-k}H^n(\mathbf{P}^n - D) \subset H^n(\mathbf{P}^n - D).$$

It can be proved (cf. [8]) that *the images*

$$(14) \qquad \operatorname{Res}(F^{n-k}H^n(\mathbf{P}^n - D)) = F^{(n-1-k)}H^{n-1}_{\mathrm{prim}}(D)$$

define the Hodge filtration on the primitive cohomology of D.

This result, which generalizes to any sufficiently ample smooth divisor D in a smooth variety X, provides a useful link between the transcendental Hodge theory of D and the algebro-geometric notion of linear systems on X.

(d) We now turn to Poincaré's magnificent works [9 and 10] on normal functions. As mentioned above, these papers appeared just before his premature death while still at the height of his mathematical powers. Upon reading the last section of the second paper it is clear, at least with the benefit of hindsight, that he was "within ε" of the Lefschetz $(1, 1)$ theorem, and indeed, as we shall see, his normal functions provided the tool for the proof of this famous result.

We consider an algebraic family $\{C_t\}_{t \in T}$ of algebraic curves of which a general member is smooth. In practice, T will usually also be a curve, and we shall denote by $t_1, \ldots, t_N$ the *critical points* for which the curves $C_{t_1}, \ldots, C_{t_N}$ are singular. For an important concrete example, we consider a smooth algebraic surface $S \subset \mathbf{P}^n$ and take $\{C_t\}_{t \in T}$ to be a general pencil of hyperplane sections. In coordinates one may think of the projection of S in a general $\mathbf{P}^3$ to be given by

$$p(x, y, z) = 0$$

in such a way that the curve C_t is the section by the plane

$$(15) \qquad z = t.$$

Associated to $\{C_t\}_{t \in T}$ is the family of Jacobian varieties

$$\mathcal{J} = \bigcup_{t \in T} J(C_t),$$

where care must be taken to insert the *generalized Jacobian* of C_t over the critical points. When this is done we obtain a complex manifold $\mathcal{J}$ that is a fibre space

$$(16) \qquad \pi: \mathcal{J} \to T$$

of commutative complex Lie groups.

DEFINITION. *A normal function* v *is a holomorphic cross-section*

$$t \to v(t) \in J(C_t)$$

of the fibre space (16).

To see how such a normal function might arise we consider an algebraic curve D on S that meets each plane section (15) in points, and we write the intersection as

$$D \cdot C_t = \sum_i p_i(t)$$

where

$$p_i(t) = (x_i(t), y_i(t), t).$$

We assume chosen a rationally varying set

$$\omega_\alpha(t) = \frac{q_\alpha(x, y, t)\, dx}{p_y(x, y, t)}, \qquad \alpha = 1, \ldots, g = \text{genus } C_t,$$

of abelian differentials that are holomorphic and linearly independent on a general C_t, and then the curve D defines a normal function ν_D by

$$(17) \qquad \nu_D(t) = \left(\sum_i \int_{p_0}^{p_i(t)} \omega_1(t), \ldots, \sum_i \int_{p_0}^{p_i(t)} \omega_g(t) \right)$$

where p_0 is a base point of the pencil. Intuitively speaking, normal functions represent an extension to curves varying algebraically on parameters of the theory of abelian integrals on a fixed curve.

In this vein, a fundamental result is the following extension of the Jacobi inversion theorem (4):

(18) POINCARÉ'S EXISTENCE THEOREM. *In case the fibre space* (16) *arises from a general pencil of curves on a surface, any normal function ν arises from a global algebraic curve D on the surface; i.e., $\nu = \nu_D$.*

As we shall see below (cf. (19)), normal functions are essentially linear data (analogous to sections of vector bundles), and the conversion to a geometric object provided by (18) is indeed quite remarkable.

For an application, we recall that the *irregularity q* of an algebraic surface is equal to number of linearly independent closed holomorphic 1-forms on the surface. Keep in mind also that the development being described occurred some 25 years before Hodge theory and the equality $h^0(\Omega_S^1) = h^1(O_S)$. Using (18) Poincaré gave the first complete proof of the existence of ∞^q algebraically equivalent but linearly inequivalent curves on the surface, as follows: Using the complete reducibility theorem (1) we choose the $\omega_\alpha(t)$ to be divided into two sets

$$\underbrace{\omega_1(t), \ldots, \omega_q(t)}_{\text{fixed part}}; \qquad \underbrace{\omega_{q+1}(t), \ldots, \omega_g(t)}_{\text{variable part}}$$

where the first set consists of the restrictions to C_t of the holomorphic 1-forms on S (Picard proved that this restriction is injective), and where the remainder is a complementary set as furnished by (1). The periods of the fixed part are constant, and from this it follows that for any algebraic curve D on S the first q abelian sums

$$\sum_i \int_{p_0}^{p_i(t)} \omega_1(t) = c_1, \ldots, \sum_i \int_{p_0}^{p_i(t)} \omega_q(t) = c_q$$

give a constant vector $c = (c_1, \ldots, c_q)$. Moreover, we may vary the normal function by varying the constant vector c while leaving fixed the entries of $\nu_D(t)$ corresponding to the variable part. Appealing now to (18), Poincaré established the existence of ∞^q curves $D(c)$ that are easily seen to be linearly inequivalent.

(e) To illustrate the fertility of normal functions, we shall very briefly recount some of the developments that have occurred since Poincaré's papers and shall mention a couple of rather general and seemingly difficult outstanding problems.

The first and foremost development is the Lefschetz (1, 1) theorem [12]. As above, on a smooth surface S we consider a general pencil $\{C_t\}$ of hyperplane sections. The following was proved by Lefschetz (cf. [11 and 13] for expository accounts):

(19) *To each normal function ν these is associated a topological invariant, called its fundamental class*

$$\eta(\nu) \in H^2(S, \mathbf{Z});$$

that has the following properties:

(i) *In case $\nu = \nu_D$ comes from an algebraic curve D on the surface,*

$$\eta(\nu_D) = \eta(D)$$

is the fundamental class of the curve; and

(ii) *a given class $\eta \in H^2(S, \mathbf{Z})$ is the fundamental class $\eta(\nu)$ for some normal function ν if, and only if, η has Hodge type* (1, 1).

Denoting by $\mathrm{Pic}^0(S)$ the q-dimensional Picard variety of S, the situation may be summarized by the following exact sequence:

$$
(20) \qquad
\begin{aligned}
0 &\to \mathrm{Pic}^0(S) \to \{\text{group of normal functions}\} \\
&\xrightarrow{\eta} H^2_{\mathrm{prim}}(S, \mathbf{Z}) \to H^{0,2}(S)
\end{aligned}
$$

where $H^2_{\mathrm{prim}}(S, \mathbf{Z})$ is the primitive part of $H^2(S, \mathbf{Z})$.

The Lefschetz (1,1) theorem, which characterizes the fundamental classes of algebraic 1-cycles on S by their Hodge types, is an immediate corollary of (20) together with Poincaré's existence theorem.

The remaining developments that we shall describe deal with higher-dimensional varieties. To a smooth complex algebraic variety V we associate the following groups:

$$Z^n = \{\text{algebraic cycles of codimension } n \text{ on } V\},$$
$$\cup$$
$$Z^n_h = \{\text{cycles homologous to zero}\},$$
$$\cup$$
$$Z^n_a = \{\text{cycles algebraically equivalent to zero}\},$$
$$\cup$$
$$Z^n_r = \{\text{cycles rationally equivalent to zero}\}.$$

The nth *intermediate Jacobian* $J^n(V)$ consists of the real torus

$$H^{2n-1}(V, \mathbf{R})/H^{2n-1}(V, \mathbf{Z})$$

provided with the complex structure indicated by the following splitting of $H^{2n-1}(V, \mathbf{R}) \otimes \mathbf{C} \cong H^{2n-1}(V, \mathbf{C})$ into complex conjugate subspaces:

$$\begin{cases} H^{2n-1}(V, \mathbf{C}) = H'(V) \oplus H''(V), \\ H'(V) = H^{2n-1,0}(V) \oplus \cdots \oplus H^{n,n-1}(V), \\ H''(V) = \overline{H'}(V). \end{cases}$$

Choosing a basis $\omega_1, \ldots, \omega_g$ for $H''(V)^*$, there is a mapping

$$u\colon Z_h^n \to J^n(V)$$

given by

$$(21) \qquad u(z) = \left(\int_\Gamma \omega_1, \ldots, \int_\Gamma \omega_g \right)$$

where $z \in Z^n$ is an algebraic n-cycle that is homologous to zero and Γ is a chain with $\partial\Gamma = z$. For divisors on curves, (21) reduces to the usual abelian sums. An easy generalization of one-half of Abel's theorem implies that

$$u(Z_r^n) = 0,$$

so that there is an induced mapping

$$(22) \qquad u\colon Z_h^n / Z_r^n \to J^n(V).$$

For simplicity of notation we shall henceforth concentrate on the crucial case when

$$\dim V = 2m - 1, \qquad n = m,$$

and we shall drop the superscript n (for example, one may think of algebraic curves on a threefold). As was the case for curves, one may speak of algebraic families $\{V_t\}_{t \in T}$ whose general member is smooth, and also (with some care) of the associated family $\{J(V_t)\}_{t \in T}$ of intermediate Jacobians. Provided that the singularities of the V_{t_α} are not too bad it can be shown that, again as was the case for curves, we arrive at a holomorphic fibre space

$$\pi\colon \mathcal{J} \to T$$

of commutative complex Lie groups with

$$\pi^{-1}(t) = J(V_t)$$

for $t \neq t_1, \ldots, t_N$ a noncritical parameter point.

Suppose now that X is a smooth $2m$-dimensional variety and $\{V_t\}$ is a general pencil of hyperplane sections. An algebraic m-cycle D on X is said to be *primitive* in case

$$D \cdot V_t = Z_t$$

is homologous to zero on V_t. There is then a corresponding normal function ν_D defined by

$$\nu_D(t) = u_t(z_t) \in J(V_t).$$

As is perhaps suggested by the formal nature of the exact sequence (20), the Lefschetz theorem (19) goes over pretty much verbatim, and we shall give in (23) below that part of the generalization of (20) that is relevant to our purposes, referring to [**14** and **15**] for precise statements and details of the proof

(23)

$$\left\{\begin{array}{l} \text{group of} \\ \text{normal functions} \end{array}\right\} \to H^{2m}_{\text{prim}}(X, \mathbf{Z}) \to \underbrace{H^{m-1,m+1}(X) \oplus \cdots \oplus H^{0,2m}(X)}.$$

A corollary of (23) is

(24) *A class* $\eta \in H^{2m}_{\text{prim}}(X, \mathbf{Z})$ *is the fundamental class of a normal function if, and only if, η has Hodge type* (m, m).

In particular, what is missing for the construction of algebraic cycles is a generalization of Poincaré's existence theorem (18).

Now Poincaré's existence theorem was given as a version of the Jacobi inversion theorem with dependence on parameters, but the latter entirely changes character in higher dimensions. For example, in [**8**] it is proved that

(25) *If V is a generic member of a sufficiently ample pencil* $\{V_t\}$ *on X, and if* $m \geqslant 2$, *then the mapping*

$$(26) \qquad\qquad u\colon Z_a/Z_r \to J(V)$$

is zero.

A consequence is that the *subgroup of invertible points* in $J(V)$, defined to be the image of the mapping (22), will be countable. That it is not always zero may be seen using (24) (cf. [**8**]).

In general we shall use the notations

$$J_0(V) = \text{image of the mapping (22)}$$
$$= \left\{\begin{array}{l} \text{``continuous part'' of the} \\ \text{invertible subgroup of } J(V); \end{array}\right.$$
$$\Theta(V) = Z_h/Z_a.$$

(27) PROBLEM. (i) *Is the group $\Theta(V)$ finitely generated*?

(ii) *Is the induced mapping*

$$u\colon \Theta(V) \to J(V)/J_0(V)$$

injective?

If (ii) were true, then (i) would be a Mordell-Weil type question and would consequently seem to be more tractable. As it stands the above problem is probably too difficult, and so it might be profitable to try some special cases, such as a smooth quintic hypersurface in $\mathbf{P}^4$ or one of Clemens' double solids [**16**]. We remark that a normal function analogue of (27) is true (cf. [**17**]).

Having commented on the apparent demise, in higher dimensions, of the Jacobi inversion theorem for sufficiently general varieties (however, cf. [**18**]), it may still be asked if the analogue of Poincaré's existence theorem could be established by some other method? According to (24), the assumption that an integral cohomology class has Hodge type (m, m) has at least led to the construction of a global analytic variety somewhere, namely the graph of the normal function ν. Moreover, in case $\nu = \nu_D$ for some primitive algebraic m-cycle D it may be proved that the knowledge of ν in a sense determines the hypersurfaces of sufficiently high degree that pass through D or a cycle homologous to it [13]. Unfortunately, as discussed at the end of [13] this construction does not seem likely to be of much help in constructing cycles. Perhaps more promising is the possibility of relating the infinitesimal properties of ν_D to the geometry of D, and we shall briefly describe this recent development.

As initial motivation, suppose that $X \subset \mathbf{P}^N$ is a smooth algebraic surface and $D_0 \subset X$ is a divisor of degree zero (e.g., the difference of lines from the two rulings of a quadric surface in $\mathbf{P}^3$). Write $D_0 = D_0^+ - D_0^-$ as the difference of effective divisors and denote by $\sigma(D_0) = D_0^+ + D_0^-$ the support of D_0. If $C \in |\mathcal{O}_X(d)|$ is a general section of X by a hypersurface of degree d, then for $d \gg 0$ the divisors $D_0^\pm \cdot C$ are effective special divisors, and the variational theory of special linear systems (cf. Chapter 7 of [20]) shows that there are a priori conditions on the tangent spaces to the graph of ν_{D_0}. Moreover, in a very few cases these infinitesimal conditions may be used to single out the adjoint hypersurfaces of large degree that pass through one or more of the curves $\sigma(D)$ where $D = D^+ - D'$ is homologous to D_0 and $\deg D^+ = \deg D_0^+$ (cf. [19]).

As another motivation, suppose that $X \subset \mathbf{P}^5$ is a smooth hypersurface of degree d that contains a 2-plane $\Lambda \cong \mathbf{P}^2$. Then

$$D = d \cdot \Lambda - (\mathbf{P}^3 \cdot X)$$

will be a primitive algebraic 2-cycle lying in X, and we may consider the normal function corresponding to the linear sections

$$V_H = H \cap X, \qquad H \in \mathbf{P}^{5*},$$

of X. In (c) above we discussed how the holomorphic 1-forms on $J(V_H)$ can be represented, via the isomorphism

$$H^{1,0}(J(V_H)) \cong H'(V_H),$$

by residues of the form

$$\mathrm{Res}\left(\frac{q(x)\, dx_1 \wedge dx_2 \wedge dx_3 \wedge dx_4}{p(x)^2} \right), \qquad \deg q(x) \leqslant 2d - 5,$$

where $p(x) = 0$ is the defining equation of $V_H \cap H$. It is certainly plausible that the derivative of the individual integrals in (21) should "change character" where the polynomial $q(x)$ vanishes on the line $\Lambda \cap V_H$.

We shall now sketch the definition of an infinitesimal invariant $\delta\nu$ of a normal function ν and shall then give one illustrative result. Let $\{V_s\}_{s \in S}$ be a family of smooth varieties of dimension $2m - 1$ and $\nu(s) \in J(V_s)\ (= J^m(V_s))$ a normal function. Even in the case $m = 1$ of curves of genus g, the construction of $\delta\nu$ is somewhat complicated due to the fact that, even though the Siegel generalized upper-half-plane $\mathcal{H}_g$ is a homogeneous complex manifold, the versal family of principally polarized abelian varieties over $\mathcal{H}_g$ is not (otherwise, all theta divisors would look alike). Because the construction is local we may assume that $S \subset \mathbf{C}^N$ is a polydisc and we may then naturally identify all the groups $H^{2m-1}(V_s, \mathbf{Z})/(\text{torsion})$ with a fixed lattice $\mathbf{Z}^{2g}$ having complexification

$$\mathbf{C}^{2g} \cong H^{2m-1}(V_s, \mathbf{C}) = H'(V_s) \oplus H''(V_s).$$

Then $H'(V_s) \subset H^{2m-1}(V_s, \mathbf{C})$ gives a holomorphically varying g-plane

$$\Lambda_s \subset \mathbf{C}^{2g},$$

and we may consider the resulting map

$$(28) \qquad\qquad \varphi: S \to G(g, 2g)$$

given by $\varphi(s) = \Lambda_s$ (when $m = 1$, $\varphi(s)$ is just the period matrix of V_s under the natural embedding $\mathcal{H}_g \subset G(g, 2g)$). Since $J(V_s)$ is the middle-dimensional intermediate Jacobian there is a natural identification

$$H^{2m-1}(V, \mathbf{C})/H'(V_s) \cong H'(V_s)^*,$$

and so the differential of (28) is given by

$$(29) \qquad\qquad \varphi_*: T_s(S) \to \operatorname{Hom}(\Lambda_s, \Lambda_s^*).$$

We define

$$\Xi_s \subset \mathbf{P}(T_s(S))$$

by the condition

$$\xi \in \Xi_s \Leftrightarrow \det \varphi_*(\xi) = 0$$

where $\xi \in \mathbf{P}(T_s(S))$ is a tangent direction. Setting

$$\Xi = \bigcup_{s \in S} \Xi_s$$

the quotient spaces

$$(30) \qquad\qquad \mathbf{C}^{2g}/\operatorname{span}\{\Lambda_s, \varphi_*(\xi)\Lambda_s\}$$

fit together to define a coherent sheaf $\mathcal{F}$ over the analytic variety Ξ. The normal function is defined by giving

$$\nu(s) \in \mathbf{Z}^{2g} \setminus \mathbf{C}^{2g}/\Lambda_s$$

for each $s \in S$, and we choose an arbitrary lifting $v(s) \in \mathbf{C}^{2g}$ of $\nu(s)$. Any other lifting is

$$\tilde{v}(s) = v(s) + \lambda(s) + \zeta(s)$$

where $\lambda(s) \in \Lambda_s$ and $\zeta(s) \in \mathbf{Z}^{2g}$. It follows that, at the point $(s, \xi) \in \mathbf{P}(T_s(S))$, the projection of $dv(s)/d\xi$ into the quotient space (30) depends only on v, up to a scalar resulting from lifting ξ to a tangent vector in $T_s(S)$. We define the infinitesimal invariant

$$(31) \qquad\qquad \delta v \in H^0(\Xi, \mathcal{F}(1))$$

to be the resulting section of $\mathcal{F} \otimes \mathcal{O}(1)$, where $\mathcal{O}(1)$ is the tautological line bundle over $\mathbf{P}(T(S))$. The details and differential geometric motivation for

398 P. A. GRIFFITHS

this construction may be found in [19], where also a proof of the following result may be found.

Let C be a smooth curve and $D = D^+ + D^-$ a fixed divisor of degree zero with support $\sigma(D) = D^+ + D^-$. For our family $\{C_s\}_{s \in S}$ we take the Kuranishi space (local moduli space) centered at $C_{s_0} = C$. Then there is a natural identification

$$T_{s_0}(S) \cong H^1(C, \Theta)$$

where Θ is the tangent sheaf of C. We note that

$$\mathbf{P}H^1(C, \Theta) \cong \mathbf{P}^{3g-4}$$

is the embedding space for the bicanonical mapping

$$\varphi_{2K}: C \to \mathbf{P}H^1(C, \Theta),$$

and it may be proved that (cf. [19])

$$\Xi_{s_0} = \bigcup_{E \in |K|} \overline{\varphi_{2K}(E)}$$

where $|K|$ is the canonical linear series on C and $\overline{\varphi_{2K}(E)}$ is the linear span of the points $\varphi_{2K}(p_i)$ where $E = p_1 + \cdots + p_{2g-2}$. We remark that if $\xi \in \mathbf{P}(T_{s_0}(S))$ is in $\varphi_{2K}(E)$, then

$$H^0(C, K(-E)) \subset \ker \varphi_*(\xi).$$

Moreover, given $\omega \in H^0(C, K(-E))$ and any vector f in the quotient space (30), the pairing (Q denoting the cup-product on $H^1(C)$)

$$(32) \qquad\qquad Q(\omega, f)$$

is well defined since $\varphi_*(\xi)$ is a symmetric transformation, and consequently

$$\mathrm{span}\{\Lambda_{s_0}, \varphi_*(\xi)\Lambda_{s_0}\} = \Lambda_{s_0}^\perp.$$

Suppose now we consider families of divisors D_s on C_s with $D_{s_0} = D$ and denote by ν_{D_s} the resulting normal function. By what we just said, to know the value of the infinitesimal invariant (31) at $s = s_0$ it will suffice to know the quantities

$$Q(\omega, \delta\nu_{D_s}) \quad \text{at } s = s_0,$$

where $\xi \in \overline{\varphi_{2K}(E)}$ and $\omega \in H^0(C, K(-E))$. These quantities are evaluated in [19], and in particular the conditions that

$$(33) \qquad\qquad Q(\omega, \delta\nu_{D_s}) = 0 \quad \text{at } s = s_0$$

are determined. Here we mention the two conclusions:

(i) the geometric condition

$$(34) \qquad\qquad (\omega) \geqslant \sigma(D)$$

implies (33), and

(ii) if D_s is chosen arbitrarily subject to $D_s = D$, then (33) and (34) are equivalent. Thus, in a certain sense $\delta\nu$ gives an extension to a family of variable curves of the Brill-Noether matrix (cf. [20]) that plays the pivotal role in the theory of special divisors on a fixed curve.

Although somewhat complicated to describe, we have mentioned this result as perhaps indicating that Poincaré's concept of normal functions has signficance that may play a role in questions of current interest in algebraic geometry.

REFERENCES

(a) *Some sample works of the period.*

1. P. Appell and E. Goursat, *Théorie des fonctions algébriques et de leurs intégrales*, Gauthier-Villars, Paris, 1929.

2. G. Castelnuovo and F. Enriques, *Grundeigeshaften der Algebraischen Flächen*, Encyklop. der Math. Wissenschaften, 3 (1903), 635–768.

3. E. Picard and G. Simart, *Théorie des fonctions algébriques de deux variables independentes*, Gauthier-Villars, Paris, 1897–1906.

(b) *References on abelian varieties.*

4. P. Griffiths and J. Harris, *Principles of algebraic geometry*, Wiley, 1978 (cf. Chapter 2; our notations concerning curves and Jacobians are the same as this reference).

5. H. Poincaré, *Sur les fonctions abéliennes*, Amer. J. Math. **8** (1886), 289–342.

6. ______, *Remarques diverses sur les fonctions abéliennes*, J. de Math., $5^{\underline{e}}$ series, **1** (1895), 219–314.

(c) *References on residues.*

7. H. Poincaré, *Sur les résidues des intégrales doubles*, Acta Math. **9** (1887), 321–380.

8. P. Griffiths, *On the periods of certain rational integrals*. I, II, Ann. of Math. (2) **90** (1969), 460–495; 496–541.

(d) *References on normal functions.*

9. H. Poincaré, *Sur les courbes tracées sur les surfaces algébriques*, Ann. Sci. de l'École Norm. Sup. **27** (1910), 55–108.

10. ______, *Sur les courbes tracées sur une surface algébrique*, Sitz. der Berliner Math. Gesellschaft. **10** (1911), 28–55.

(e) *References on recent developments.*

11. O. Zariski, *Algebraic surfaces*, Springer-Verlag, Berlin and New York, 1971 (cf. especially Chapter VII).

12. S. Lefschetz, *l'Analysis situs et la géometrié algébrique*, Gauthier-Villars, Paris, 1924.

13. P. Griffiths, *A theorem concerning the differential equations satisfied by normal functions associated to algebraic cycles*, Amer. J. Math. **101** (1979), 94–131.

14. S. Zucker, *Generalized intermediate Jacobians and the theorem on normal functions*, Invent. Math. **33** (1976), 185–222.

15. ______, *Hodge theory with degenerating coefficients: L_2 cohomology in the Poincaré metric*, Ann. of Math. (2) **109** (1979), 415–476.

16. C. H. Clemens, *Double solids*, Adv. in Math. (to appear).

17. P. Griffiths, *Periods of integrals on algebraic manifolds*. III, Publ. Math. Inst. Hautes Études Sci. **38** (1970), 125–180.

18. S. Zucker, *The Hodge conjecture for cubic fourfolds*, Compositio Math. **34** (1977), 199–209.

19. P. Griffiths, *An infinitesimal invariant of normal functions* (to appear).

20. E. Arbarello, M. Cornalba, P. Griffiths, and J. Harris, *Topics in algebraic curves*, Princeton Math. Series, Princeton Univ. Press, Princeton, N. J. (to appear)

DEPARTMENT OF MATHEMATICS, HARVARD UNIVERSITY, CAMBRIDGE, MASSACHUSETTS 02138

Proceedings of Symposia in Pure Mathematics
Volume 39 (1983), Part 1

Physical Space-Time and Nonrealizable CR-Structures

ROGER PENROSE

Abstract. Space-time views leading up to Einstein's general relativity are described in relation to some of Poincaré's early ideas on the subject. The basic geometry of twistor theory is introduced as it arises both from Minkowski space-time and the more general curved Einstein models. It is shown how this provides a CR-structure (this being, in essence, another of Poincaré's pioneering concepts) in a natural way. Nonrealizable CR-structures can arise, and an example is presented, due to C. D. Hill, G. A. J. Sparling and the author, of a complex manifold-with-boundary which cannot be extended as a complex manifold beyond its C^∞ boundary.

1. Introductory remarks. The nature of physical geometry was something that held considerable interest for Poincaré, and he often referred to it in his more popular writings. Moreover, it was Poincaré who first clearly understood the physical transformation group of special relativity—arising as a group of symmetries of Maxwell theory—and he suggested that the relativity principle concerned might hold for physics generally. This dates back to 1899, six years before Einstein's first paper on relativity (cf. Poincaré (1906), (1954) for details). It is fitting, therefore, that this symmetry group should be now very commonly referred to as the Poincaré group (otherwise known as the inhomogeneous Lorentz group, where simply "Lorentz group" now normally refers only to the related homogeneous group).

Poincaré also had interesting and, to some extent, insightful things to say about the possibility that physical space might have a non-Euclidean geometry. But here his instincts appear ultimately to have let him down. For in "Science and Hypothesis" Poincaré (1905) wrote

> "What then, are we to think of the question: Is Euclidean geometry true? It has no meaning. We might as well ask if the metric system is true, and if the old weights and measures are false; if Cartesian coordinates are true and polar coordinates false. One geometry cannot be more true than another; it can only be more convenient. Now, Euclidean geometry is, and will remain, the most convenient: 1st, because it is the simplest, and it is not so only because of our mental habits or because of the kind of direct intuition that we have of

Reprinted from Bulletin Amer. Math. Soc. (N. S.) 8 (1983), 427–448.

1980 *Mathematics Subject Classification.* Primary 83C05, 53C20.

Euclidean space; it is the simplest in itself, just as a polynomial of the first degree is simpler than a polynomial of the
second degree; 2nd, because it sufficiently agrees with the
properties of natural solids, those bodies which we can compare and measure by means of our senses."

It is, of course, no discredit to Poincaré that he should have had no inkling of
that extraordinary development, which emerged three years after his death,
namely Einstein's *general* theory of relativity. Nevertheless, I find it rather
remarkable that Poincaré, with his deep geometric and philosophical insights,
should apparently have had such a rigid feeling that geometries other than
Euclidean could have no chance of providing accurate and useful descriptions of
physical space according to some future theory. His remarks on the conventional-
ity of geometry and on the fact that the geometry we use in physics is of necessity
an idealization are, indeed, quite profound and insightful. But he seems to have
made a rather bad mistake in his reasoning, his critical faculties being perhaps
dulled by an erroneous (but natural enough) intuitive presumption that in "our
world" the geometry actually *is* Euclidean!

According to modern cosmology, it is quite on the cards that the large-scale
spatial geometry of the universe may indeed be closely in accord with
Lobachevskian geometry—that geometry which held such a fascination for
Poincaré the mathematician, yet which had been rejected as inevitably physically
inappropriate by Poincaré the philosopher! It would be interesting to know how
Poincaré would have reacted to present-day cosmology. Perhaps there is an object
lesson here for all of us.

Even if observation finally turns against its present slight preference for the
Lobacheveskian spatial geometry—and also tells against models with a positively
curved spatial geometry—leading us to believe that the large-scale spatial structure is Euclidean after all, we cannot now go back to flatness for the structure on
a more localized level. Tests of general relativity are now sufficiently good to
provide direct measurements of deviations from flatness in physical geometry, the
geometry being defined in a sense that I shall describe in the next section.
However, it is the geometry of *space-time*, rather than that of space, which has a
clear physical interpretation. And we shall see that the space-time geometry of
(Poincaré's!) *special* relativity, though flat, is yet *not* the geometry of Euclid.

In §2, I shall outline the contemporary and now well-established (Einstein)
view of curved space-time geometry. Then in §3, I shall indicate how this
geometry may be looked at in a different way, and how this new viewpoint
enables another of Poincaré's pioneering innovations, namely the study of the
instrinsic structure of boundaries of complex domains, to be applied in an
unexpected context in physics. Finally, in §4, I shall indicate how, in a sense,
these ideas also enable the physics to repay its debt to the mathematics and

provide what appears to be the first established example of a complex manifold which cannot be locally extended beyond its C^∞-smooth boundary.

2. Structure of space-time. It is to Minkowski (1908) whom we owe the idea that physical geometry should be a 4-dimensional space-time geometry, rather than a 3-dimensional spatial one. In relativity theory, there is no absolute 3-geometry, but the 4-geometry of space-time is an absolute objective physical structure. In fact, when viewed in the Minkowskian way, relativity theory becomes a theory of the absolute. What have become relative are merely the older ideas of separate space and time. One further point about Minkowski's absolute space-time geometry that will emerge is that it is most directly described in terms of measured time-intervals rather than of distances in the ordinary sense, so that the geometry is really a "chronometry". This point of view has been emphasized particularly by Synge (1960) and Bondi (1965) and it leads to a considerable clarification of the meanings of the mathematical structures involved.

In order to appreciate fully the nature of the space-time provided by Einstein's general relativity, it may be helpful first to see how the older physical theories can be described in a space-time setting. We shall consider, in turn, five alternative space-time structures, which I shall refer to (cf. Penrose (1968)) as

A: Aristotelean space-time,

G: Galilean space-time,

N: Newtonian space-time,

M: Minkowskian space-time,

E: Einsteinian space-time.

The structure of the Aristotelean space-time A is given by its expression as a product

$$(2.1) \qquad\qquad A = \mathbb{E}^1 \times \mathbb{E}^3,$$

where $\mathbb{E}^n$ denotes Euclidean n-space equipped with its flat metric (and, if desired, with an orientation), whence each $\mathbb{E}^n$ possesses a $\frac{1}{2}n(n+1)$ group of symmetries. For any two events $a, b \in A$, there is both an absolute time-difference $T(a, b) \in \mathbb{R}$ and an absolute spatial separation $D(a, b) \in \mathbb{R}$ (these being the distance functions in $\mathbb{E}^1$ and $\mathbb{E}^3$, respectively). It is therefore meaningful to say, of two events, whether they are simultaneous in A (i.e. $T(a, b) = 0$); but it is also meaningful to say whether they have the same position (i.e. $D(a, b) = 0$) irrespective of their simultaneity. For in Aristotelean physics the state of rest is distinguished among all motions, and this is reflected in the fact that the canonical copies of $\mathbb{E}^1$ in $\mathbb{E}^1 \times \mathbb{E}^3$ constitute a distinguished family of curves in A representing the world-lines of particles at rest, i.e. which have the same position at all times. The space A has a 7-parameter symmetry group.

The structure of Galilean space-time G, on the other hand is not that of a product but merely a fibre bundle with $\mathbb{E}^3$ fibres and an $\mathbb{E}^1$ base space, the projection map

$$(2.2) \qquad\qquad G \to \mathbb{E}^1$$

being the assignment of a "time" to any event in G. Thus, of any two events $a, b \in G$, it is meaningful to speak of their time-difference $T(a, b) \in \mathbb{R}$ and to say when the events are simultaneous (i.e. $T(a, b) = 0$), but their spatial separation *only* has meaning if they *are* simultaneous. This reflects the fact that in Galilean physics there is no invariantly defined state of rest, and any two nonsimultaneous events will be viewed as having the same spatial location with respect to *some* suitably moving reference system. The bundle structure of G is not, however, G's whole structure. We need also to single out a family of curves in G which represent *inertially moving particles*; and these curves may be referred to as *straight lines* in G. The simplest way to specify the particular straight line structure that G possesses is, perhaps, to say that G is an *affine space* and that the fibration (2.2) is *compatible* with this affine structure (in the sense that the fibres are all parallel, that the $\mathbb{E}^3$ fibres and $\mathbb{E}^1$ base all inherit their correct affine structures, and that parallel straight lines, transverse to the $\mathbb{E}^3$'s, provide metric-preserving maps between the $\mathbb{E}^3$'s). The total structure that all this assigns to G is somewhat weaker than the structure of A. The symmetry group of G contains that of A as a subgroup, and it is a 10-parameter group referred to as the *Galilei group*.

We pass now to Newtonian space-time N, the idea here (due to É. Cartan (1923), (1924), cf. also Friedrichs (1928), Trautman (1966)) being to treat Newtonian gravitational theory as a geometric theory in the spirit of Einstein's general relativity. We recall, for example, that a *uniform* Newtonian gravitational field permeating the whole of space would be totally undetectable. All particles would accelerate together under this field, so that relative to the particles themselves, the field would appear not to be there at all. It is only deviations from uniformity, i.e. "tidal forces", which are physically detectable, and the idea is that in the general case these supply a kind of curvature for N. As a fibre bundle, N has a structure identical with that for G,

$$(2.3) \qquad\qquad N \to \mathbb{E}^1$$

(so simultaneity and time-difference, and lack of a spatial separation concept for nonsimultaneous events, are the same as before)—but now there is a different family of curves representing inertial motions. "Inertial motion" now means "free motion under gravity", and we can refer to the corresponding world-lines in N as "geodesics". In fact, while not extremal curves, these geodesics are, indeed, the geodesics of a certain torsion-free linear connection Γ on N. The space N also possesses a *co*metric $\mathbf{g}^*$ (symmetric contravariant 2-valent tensor) which is semi-definite but degenerate (having matrix-rank 3), where $\mathbf{g}^*$ has the defining property of inducing the correct Euclidean metric on each $\mathbb{E}^3$ fibre. The connection Γ preserves $\mathbf{g}^*$, but this property does not define Γ uniquely. The different possible inequivalent Γ's correspond to the different possible physically inequivalent gravitational fields. Thus, unlike G and A, which have canonical structures, there are many *different* Newtonian space-times N. Indeed, G is a special case, given when Γ is flat. In general, N has no symmetries.

Minkowskian space-time M, the space-time of special relativity, on the other hand, possesses, like G and A, a canonical and uniform structure. Like G (and like A), M is an affine space, and inertial motions in M are described by lines which are straight with respect to this affine structure. These lines may also be thought of as the geodesics of a certain flat connection Γ, where Γ is now the unique torsion-free connection preserving a certain flat nondegenerate pseudometric $\mathbf{g}$, of Lorentzian signature $(+,-,-,-)$. (It is the nondegeneracy of $\mathbf{g}$ that now enables it to determine Γ uniquely.) The pseudometric $\mathbf{g}$ serves to define a quadratic squared "distance" function S of signature $(+,-,-,-)$, on pairs $a, b \in M$ ($S(a, b) \in \mathbb{R}$). Thus M is a pseudo-Euclidean space. The physical meaning of S is that an (ideal) inertial (i.e. unaccelerated) clock which moves from the event a to the event b will register a time-interval between a and b equal to $\{S(a, b)\}^{1/2}$. For a physical clock (or, indeed, for any massive particle) $S(a, b) > 0$, and we say that the separation between a and b is *timelike*. When $S(a, b) = 0$ we say that the separation is *null*, this being the case when a light ray connects a to b. When $S(a, b) < 0$ we say that the separation is *spacelike*, and in this case a and b will appear as simultaneous in some suitably moving inertial reference system, the distance between them (using units for which the light-speed is unity) being $\{-S(a, b)\}^{1/2}$. The symmetry group of M is the 10-parameter *Poincaré group*.

The structure of Einsteinian space-time E bears the same relation to M as N does to G. Thus there are many inequivalent spaces E, their different structures providing the various inequivalent gravitational fields. The structure, in each case, is given by a pseudometric $\mathbf{g}$, with the same $(+,-,-,-)$ signature as M, but now $\mathbf{g}$ is generally not flat. The inertial motions in E (i.e. free motions under gravity) are given, as with M, by geodesics of the unique torsion-free connection Γ preserving $\mathbf{g}$. But as with N, a *curvature* of Γ provides the physically detectable "tidal force". As with M, the pseudometric $\mathbf{g}$ provides the definition of time-interval as measured by ideal clocks, but because $\mathbf{g}$ is not flat, we normally think of the interval as defined only between infinitesimally separated points, the interval (in the timelike or null case) being given by $\{\mathbf{g}(\delta x, \delta x)\}^{1/2}$, where δx is the tangent vector which "connects" the point to its infinitestimally separated neighbour. The world-line of a massive particle has tangent vectors which are all timelike (i.e. $\mathbf{g}(\delta x, \delta x) > 0$), the tangent vectors being all null (i.e. $\mathbf{g}(\delta x, \delta x) = 0$) in the case of a massless particle. The time-interval between finitely-separated events a and b, as measured by a clock carried from a to b by such a particle is the pseudo-Riemannian "distance", given by

$$\int_a^b \{\mathbf{g}(dx, dx)\}^{1/2}.$$

Note that this time-interval depends upon the path through space-time from a to b. If a and b are not too far apart, this time-interval is a *maximum* for the geodesic (i.e. inertial path) which locally connects a to b. The metric **g** determines (via Γ) a Riemann curvature tensor **R**. The direct physical interpretation of **R** is obtained from the Jacobi equation, which describes, in terms of **R**, the geodesic derivation of (say) timelike geodesics, i.e. the tidal effect on inertially moving particles. In the general case, E has no symmetries.

In order to provide a physical theory of gravity, Einstein had to supplement the above general geometric structure by his *field equations*. These state that the trace-reversed Ricci tensor constructed from **R** is a constant multiple of the physical energy-momentum tensor, describing the density of matter. Where there is no matter, the energy-momentum tensor is zero and the space-time is Ricci-flat; and where matter is present, the Ricci tensor is locally determined by this matter density. The remainder of the curvature, namely that measured by the Weyl conformal curvature tensor **C**, is governed by differential equations and is thus constrained, in a nonlocal way, by the distribution of matter. In the limit when velocities are small and the fields are weak (in the sense of small gravitational potentials) the theory goes over into the older Newtonian theory. Various small deviations from Newtonian theory have been observationally established, and these are all consistent with the more accurate Einstein theory. In fact, Einstein's theory must now be regarded as an excellently-tested theory of gravity, having no serious rival (at least none for which there is any forseeable prospect of observing a discrepancy with Einstein's theory). The mathematics of Einsteinian manifolds —and including that of the important special case provided by Minkowski geometry—is thus something of especial interest for physics.

3. Twistor geometry and CR-structure. Let us next examine how the geometry of Minkowski space M may be reformulated in a new way (cf. Penrose (1975), (1977); Penrose and Ward (1980); also Wells (1979)). We shall see how this relates M to the theory of boundaries of complex manifolds. An analogous construction for the more general curved Einstein spaces will be considered afterwards.

Choose standard Minkowski coordinates $x = (\tau, \xi, \eta, \zeta) \in \mathbb{R}^4$, for M, where S is given by

$$S(x, \hat{x}) = (\tau - \hat{\tau})^2 - (\xi - \hat{\xi})^2 - (\eta - \hat{\eta})^2 - (\zeta - \hat{\zeta})^2$$

$$= \det\left\{ \begin{pmatrix} \tau + \zeta & \xi + i\eta \\ \xi - i\eta & \tau - \zeta \end{pmatrix} - \begin{pmatrix} \hat{\tau} + \hat{\zeta} & \hat{\xi} + i\hat{\eta} \\ \hat{\xi} - i\hat{\eta} & \hat{\tau} - \hat{\zeta} \end{pmatrix} \right\}$$

where throughout this work "i" stands for $\sqrt{-1}$ (and a bar denotes complex conjugation).

Let $(W_0, W_1, W_2, W_3) \in \mathbb{C}^4$ be coordinates for an element W of an associated complex vector space $\mathbb{T}$, called (dual) twistor space where W and x are said to be *incident* whenever

$$(3.2) \qquad (W_2, W_3) = \frac{1}{i\sqrt{2}}(W_0, W_1)\begin{pmatrix} \tau + \zeta & \xi + i\eta \\ \xi - i\eta & \tau - \zeta \end{pmatrix}.$$

This relation can only hold if $\Sigma(W) = 0$, where the $(+, +, -, -)$-signature Hermitian form Σ is defined by

$$(3.3) \qquad \Sigma(W) \equiv W_0 \overline{W_2} + W_1 \overline{W_3} + W_2 \overline{W_0} + W_3 \overline{W_1}$$

(because the Hermiticity of the (2×2) matrix in (3.2) entails that postmultiplication of (3.2) by the conjugate transpose of (W_0, W_1) yields a purely imaginary result).

The 3-complex-dimensional projective space $\mathbb{PT}.$, associated with $\mathbb{T}.$, whose points are labelled by the three complex ratios

$$W_0 : W_1 : W_2 : W_3$$

consists of two open complex manifolds $\mathbb{PT}_+$ and $\mathbb{PT}_-$, given when $\Sigma(W) > 0$ and $\Sigma(W) < 0$, respectively, together with their common boundary, the 5-real-dimensional manifold $\mathbb{PT}_0$, given when $\Sigma(W) = 0$. Thus the points of $\mathbb{PT}.$ for which (3.2) holds for some $x \in M$, all lie on $\mathbb{PT}_0$. They do not constitute quite the whole of $\mathbb{PT}_0$, however, since points of the projective line I, given by $W_0 = W_1 = 0$, admit no solution for x in (3.2).

Suppose we choose a fixed

$$\mathbf{W} \in \mathbb{PT}_0 - I$$

(where from now on I use boldface capital letters such as $\mathbf{W}$ to denote the point of the projective space $\mathbb{PT}.$, rather than $\mathbb{T}.$ or $\mathbb{C}^4$). Then we can solve (3.2) for the Minkowski point x. The solution is not unique, however, the freedom being given by

$$\begin{pmatrix} \tau + \zeta & \xi + i\eta \\ \xi - i\eta & \tau - \zeta \end{pmatrix} \mapsto \begin{pmatrix} \tau + \zeta & \xi + i\eta \\ \xi - i\eta & \tau - \zeta \end{pmatrix} + k\begin{pmatrix} W_1 \overline{W_1} & -W_1 \overline{W_0} \\ -W_0 \overline{W_1} & W_0 \overline{W_0} \end{pmatrix}$$

for arbitrary real k. Note that the final matrix on the right has rank unity and so represents, by (3.1), a *null* (i.e. zero Minkowski length) vector in M. Thus, the points x incident with the given W constitute a *null geodesic* (straight line) in M which, without confusion, we can also label by $\mathbf{W}$. Indeed, we may think of $\mathbb{PT}_0 - I$ *as* the space of null geodesics $\mathbf{W}$ in M. Our construction has shown how this space may be imbedded as a 5-dimensional real submanifold of a complex projective 3-space.

To interpret a point $x \in M$, conversely in terms of $\mathbb{PT}.$, we fix x in (3.2) and allow $\mathbf{W}$ to vary. This gives us a 2-complex-dimensional linear subspace in $\mathbb{T}.$, i.e.

a complex projective line in $\mathbb{PT}.$. Clearly this line lies in $\mathbb{PT}_0 - I$ and, moreover, all lines lying entirely in $\mathbb{PT}_0 - I$ arise in this way from points in M. Thus

> *the projective lines lying entirely in* $\mathbb{PT}_0 - I$
> *represent the points of Minkowski space M.*

Note that if we allow the coordinates τ, ξ, η, ζ to become complex i.e. x to become a point of the *complexification* $\mathbb{C}M$ of M, then the above construction yields a projective line lying entirely in $\mathbb{PT}. - I$. We may also consider projective lines in $\mathbb{PT}_0$ [resp. $\mathbb{PT}.$] which meet I. These provide "points at infinity" for M [resp. $\mathbb{C}M$], the totality of all lines in $\mathbb{PT}_0$ [resp. $\mathbb{PT}.$] describing the standard conformal compactification $M^{\#}$ [resp. $\mathbb{C}M^{\#}$] of M [resp. $\mathbb{C}M$].

The intrinsic structure of projective lines in $\mathbb{PT}_0 - I$ can be illustrated in a very graphic "physical" way. The family of light rays $\mathbf{W}$ through (i.e. incident with) a fixed point $x \in M$ represents the family of points on the corresponding complex projective line in $\mathbb{PT}_0 - I$. A complex projective line has the topology S^2 and, moreover, the holomorphic structure of a Riemann sphere. Now imagine an observer situated at x. His field of vision will be represented by the light rays through x, i.e. by the family of $\mathbf{W}$'s under consideration. Clearly the topology of the observer's entire field of vision is indeed S^2. But, more subtly, the *holomorphic* structure is also relevant. This shows up in the allowed transformations between various observers with different velocities, all of whom pass through the same space-time event x. Their fields of vision are related to one another, at x, by transformations which preserve the holomorphic structure of the Riemann sphere, i.e. their fields of vision are *conformally* related to one another. This is a well-known property of Lorentz transformations (Penrose (1959), Terrell (1959)). Indeed, the connected component of the *Lorentz group* may be regarded as the group of all *holomorphic self-transformations* of this Riemann sphere.

By virtue of its imbedding in the complex manifold $\mathbb{PT}.$, the entire (real) hypersurface $\mathbb{PT}_0$ inherits further holomorphic structure. This structure, referred to as a (maximal[1]) CR-structure falls short by just one (real) dimension of defining $\mathbb{PT}_0$ as a complex manifold. We shall see the physical interpretation of this shortly.

The structure of a (maximal) CR-manifold $\mathcal{V}$ of, say, dimension $2n + 1$ provides that in the tangent space T_p, at each point $p \in \mathcal{V}$, a $2n$-real-dimensional subspace $H_p \subset T_p$ is singled out, referred to as the *holomorphic tangent space* at p. The space H_p is to be regarded as a *complex* vector space of n dimensions, spanned by complex vectors

$$\mathbf{z}_1 = \mathbf{x}_1 + i\mathbf{y}_1, \ldots, \mathbf{z}_n = \mathbf{x}_n + i\mathbf{y}_n.$$

---

[1]Some authors use the term "CR-structure" also in the extended sense of the structure induced on any real submanifold of a complex manifold, not necessarily a hypersurface. The word "maximal" is to emphasize that I am concerned only with the hypersurface dimensionality here.

The real-linear operator J, satisfying $J^2 = -1$, is determined by its action on the related real basis $\mathbf{x}_1,\ldots,\mathbf{x}_n,\mathbf{y}_1,\ldots,\mathbf{y}_n$ for H_p by

$$(3.4) \qquad J\mathbf{x}_r = -\mathbf{y}_r, \qquad J\mathbf{y}_r = \mathbf{x}_r \qquad (r = 1,\ldots,n)$$

so that

$$(3.5) \qquad J\mathbf{z}_r = i\mathbf{z}_r \qquad (r = 1,\ldots,n).$$

It is the operator J that defines the complex structure of H_p when we regard H_p as a real vector space. One further tangent vector $\mathbf{u} \in T_p$, independent of $\mathbf{x}_1,\ldots,\mathbf{x}_n$, $\mathbf{y}_1,\ldots,\mathbf{y}_n$, is needed to complete a basis for the whole of T_p. We assume that in some neighbourhood $\mathfrak{N}_p$ of p in $\mathcal{V}$ a smooth choice of such basis vectors is made and we define the *integrability relations* for the CR-structure of $\mathcal{V}$, defined by J, to be the Lie bracket relation

$$(3.6) \qquad [\mathbf{z}_r, \mathbf{z}_{r'}] = \text{complex linear combinations of } \mathbf{z}\text{'s}$$

$(r, r' = 1,\ldots,n)$. It is clear that this integrability relation is independent of the particular choice of basis satisfying (3.5), and depends only on the choice of J and the holomorphic subspaces.

Note that what distinguishes the structure of $\mathcal{V}$ from that of a complex manifold (of dimension n) is merely the presence of the extra real dimension defined by the extra real basis vector $\mathbf{u}$.

In the case of a *complex manifold* we know from the Newlander-Nirenberg (1957) theorem that integrability relations like (3.6) are all that are required. No analyticity assumptions are needed to ensure that holomorphic coordinates can be locally introduced. the derivatives with respect to which providing complex tangent vectors compatible with the given complex structure J in the sense of (3.5). The analogous property for a (maximal) CR-manifold $\mathcal{V}$ of real dimension $2n + 1$ would be the existence of a complex $(n + 1)$-manifold $\mathcal{C}$, with complex structure J, in which $\mathcal{V}$ can be realized as a real hypersurface, the CR-structure on $\mathcal{V}$ being that induced by J. However, counterexamples given by Nirenberg [(1973), (1974)] show that the integrability relations (3.6) are insufficient to ensure that the CR-structure of $\mathcal{V}$ can be so realized in all cases.

It should be made clear that the difficulty does not lie in a possible omission of necessary additional integrability relations of the usual (differential) kind. Indeed, if the CR-structure is taken to be real-analytic, then the integrability relations (3.6) are actually sufficient to ensure that $\mathcal{V}$ can be realized in the above way. But for a C^∞ CR-structure this need not be so. In this respect, the problem of realizing CR-structures is analogous to that of the Lewy insoluble differential equation (1957). We shall see later how this can be made more explicit.

Let us return to $\mathbb{PT}_0$, which has a CR-structure explicitly realized by its imbedding in the complex manifold $\mathbb{PT}$.. Here $n = 2$. Thus, at any point $\mathbf{W}$ of $\mathbb{PT}_0$ the 5-real-dimensional tangent space $T_{\mathbf{W}}$ has a 2-complex-dimensional holomorphic subspace $H_{\mathbf{W}}$ whose complete structure is defined by a certain operator J. Let us first identify the location of $H_{\mathbf{W}}$ in physical terms. The point $\mathbf{W} \in \mathbb{PT}_0$ is interpreted as a light ray in M. We may picture this by fixing the "time" τ to take a specific value, say $\tau = 0$, so our picture becomes that of ordinary Euclidean 3-space U. Then the "photon" described by $\mathbf{W}$ is pictured as a *point* in U (its location at $\tau = 0$) together with a *direction* at that point defining its velocity (the magnitude of the velocity being unity: the "speed of light"), i.e. we represent $\mathbf{W}$ by a unit vector $\mathbf{w}$ at some point q of U. To picture $T_{\mathbf{W}}$ we envisage making a small displacement of q and $\mathbf{w}$. If this displacement is such that q is moved in a direction orthogonal to the direction of $\mathbf{w}$, then we get something in the subspace $H_{\mathbf{W}}$. We refer to these slightly displaced light rays as being *abreast* with one another. The subspace $H_{\mathbf{W}}$ divides $T_{\mathbf{W}}$ into two remaining pieces, representing light rays that are slightly ahead or lag slightly behind the original ray $\mathbf{W}$. These properties are easily seen to be independent of the "time" τ, i.e. independent of the particular Euclidean (spacelike) hyperplane U that is chosen for our description.

To picture the effect of J, imagine a 2-plane element π, at q, which is orthogonal to the direction $\mathbf{w}$. We are concerned only with displacements of q and $\mathbf{w}$ for which q is moved within π, since these correspond to real vectors in $H_{\mathbf{W}}$ (i.e. "abreast" displacements of $\mathbf{W}$). Such displacements are represented by pairs of vectors $(\mathbf{r}, \mathbf{v})$ in π, where $\mathbf{r}$ gives the displacement of q and $\mathbf{v}$ measures the change in $\mathbf{w}$ (clearly also orthogonal to $\mathbf{w}$, since $\mathbf{w}$ is a unit vector). If we keep the light ray $\mathbf{W}$ and the "neighbouring" ray to which it is displaced both fixed, but vary the time τ, we easily find that the functions $\mathbf{r}(\tau)$, $\mathbf{v}(\tau)$ are related by

$$(3.7) \qquad d\mathbf{r}/d\tau = \mathbf{v}, \qquad d\mathbf{v}/d\tau = 0.$$

Now the effect of J turns out to be simply to *rotate both* $\mathbf{r}$ *and* $\mathbf{v}$ *through a right angle in the plane* π, in a left-handed sense about the direction of $\mathbf{w}$. Clearly (3.7) is invariant under this operation, so J applies to the displacements of light rays in their entirety and not simply to light rays relative to particular points on them. (Moreover this operation is easily seen to be independent of the slope of the hyperplane U.) Thus, the CR-structure of $\mathbb{P}\mathbb{T}_0$ is seen to have direct interpretation in terms of physical space-time geometry.

Let us next consider what happens when Minkowski space M is replaced by a more general Einsteinian space-time manifold $\mathfrak{M}$. We shall suppose, for simplicity that $\mathfrak{M}$, with its pseudometric $\mathbf{g}$, is *globally hyperbolic* (Leray (1952)). According to a result of Geroch (1970) this property can be stated as the existence of a spacelike hypersurface $\mathfrak{U}$ in $\mathfrak{M}$ which intersects every null geodesic (light ray) in $\mathfrak{M}$ in precisely one point. ("Spacelike", in this context, means that its normal vectors are everywhere timelike, i.e. that its induced metric is everywhere negative definite.)

It turns out that the space $\mathbb{P}\mathfrak{J}_0$ of null geodesics in $\mathfrak{M}$ acquires a CR-structure relative to the hypersurface $\mathfrak{U}$, where the holomorphic subspaces to the tangent spaces in $\mathbb{P}\mathfrak{J}_0$ and the operator J, now denoted $J_{\mathfrak{U}}$, are defined essentially as before. Thus we may represent any $\mathbf{W} \in \mathbb{P}\mathfrak{J}_0$ by the point $q \in \mathfrak{U}$ at which $\mathfrak{U}$ intersects the null geodesic $\mathbf{W}$, together with the unit vector $\mathbf{w}$ in $\mathfrak{U}$ at q in the direction of the orthogonal projection into $\mathfrak{U}$ of the future-pointing null direction of **W**. The holomorphic tangent space to $\mathbb{P}\mathfrak{I}_0$ at **W** is provided by those displacements of **W** for which q moves in a direction orthogonal to **w**, and the (real) vectors in this space therefore correspond to pairs $(\mathbf{r}, \mathbf{v})$, as before, where **r** and **v** are (real) tangent vectors to $\mathfrak{U}$ at q lying in the 2-plane element π orthogonal to **w**. The effect of $J_{\mathfrak{U}}$ is, as before, to rotate **r** and **v** through a right angle, in a left-handed sense about **w**. It then turns out that the integrability relations (3.6) for $J_{\mathfrak{U}}$ are automatically satisfied (Penrose (1975), LeBrun (1980)) and the required CR-structure, denoted $\text{CR}_{\mathfrak{U}}$, is thereby obtained.

However, unlike the CR-structure for $\mathbb{P}\mathbb{T}_0$, $\text{CR}_{\mathfrak{U}}$ will in general depend upon the location of $\mathfrak{U}$ in $\mathfrak{M}$. The holomorphic tangent spaces, regarded as real vector subspaces of the tangent spaces to $\mathbb{P}\mathfrak{I}_0$, are in fact independent of $\mathfrak{U}$, but their complex structures, as defined by $J_{\mathfrak{U}}$ will generally vary. These properties follow from the Jacobi equation, which suitably replaces the second of equations (3.7). Only when the metric **g** of $\mathfrak{M}$ is conformally flat will $\text{CR}_{\mathfrak{U}}$ be completely independent of $\mathfrak{U}$.

By way of clarification of this point it may be remarked that even if $\mathfrak{M}$ is *stationary* (i.e. time-independent, in the sense of possessing a timelike killing vector), so that isometries of $\mathfrak{M}$ exist carrying $\mathfrak{U}$ into a succession of geometrically equivalent spacelike hypersurfaces $\mathfrak{U}'$, $\mathfrak{U}''$, ..., $\subset \mathfrak{M}$, the structures $\text{CR}_{\mathfrak{U}}$, $\text{CR}_{\mathfrak{U}'}$, $\text{CR}_{\mathfrak{U}''}$, ... on $\mathbb{P}\mathfrak{I}_0$ will generally be distinct. This is because a particular point **W** of $\mathbb{P}\mathfrak{I}_0$ represents a nonstationary object (namely a null geodesic) in $\mathfrak{M}$, so it is related differently to each of $\mathfrak{U}$, $\mathfrak{U}'$, $\mathfrak{U}''$, ... However, in this case the isometries of $\mathfrak{M}$ will carry **W** into null geodesics $\mathbf{W}'$, $\mathbf{W}''$, ..., so $\text{CR}_{\mathfrak{U}}$ at **W** will agree with $\text{CR}_{\mathfrak{U}'}$, at $\mathbf{W}'$ and with $\text{CR}_{\mathfrak{U}''}$ at $\mathbf{W}''$, etc. Thus the structures $\text{CR}_{\mathfrak{U}}, \text{CR}_{\mathfrak{U}'}$, ... will in this case be *intrinsically* identical even though they represent generally distinct CR-structures on the given space $\mathbb{P}\mathfrak{I}_0$.

The fact that CR-structures (maximal, and of given dimension $2n + 1 > 0$) *can* be locally *distinct* from one another is, in effect, an observation that dates back to an important paper of Poincaré (1907). What Poincaré actually showed was that the Riemann mapping theorem, which states that any smoothly bounded simply-connected region $\mathfrak{D}$ in the Argand plane $\mathbf{C}^1$ is holomorphically identical with the unit disc, has no direct analogue in higher complex dimension. Roughly, the argument is to show that the (smooth) boundary of a region in, say, $\mathbf{C}^2$ contains intrinsic holomorphically invariant information about its "shape". (This does not occur for a smooth curve in $\mathbf{C}^1$, as the Riemann mapping theorem shows.) The gist of Poincaré's argument can be simplified to the following "physicists proof". Consider the freedom involved in specifying the (smooth) real hypersurface boundary of a region in $\mathbf{C}^2$. This is provided by one real function of three real variables (e.g. Im ζ_1 in terms of Re ζ_1, Re ζ_2, Im ζ_2). For the *intrinsic* structure of this boundary, we must factor out by the freedom provided by the allowed local holomorphic maps of $\mathbf{C}^2$ to itself. This is provided (locally) by two holomorphic functions of two complex variables. But a holomorphic function is determined by its (analytic) values on any real environment, e.g. on its values where ζ_1 and ζ_2 are real. The real and imaginary parts there are each, in effect, independent real analytic functions, so the freedom to be factored out by is that of four real functions of two real variables. The amount of intrinsic freedom in the structure of the boundary is therefore

$$\frac{1 \text{ real function of 3 real variables}}{4 \text{ real functions of 2 real variables}}.$$

However, any finite number of functions of two variables must be regarded as "peanuts" in the context of free functions of three variables, i.e. it is completely swamped by the three-variables' worth of freedom and makes no contribution net count. Thus there must be "intrinsic invariants" of the boundary shape whose functional freedom is such as to be dependent upon three real variables. More generally, for a boundary in $\mathbb{C}^{n+1}$ (and so for a $(2n + 1)$-dimensional CR-manifold) the freedom for the invariants is functions of $2n + 1$ variables ($n > 0$).

The detailed form of these invariants was investigated thoroughly by É. Cartan (1932), Tanaka (1962) and Chern & Moser (1974). This study may be regarded as the analogue, for CR-manifolds, of the study of invariants (and covariant tensorial objects) for ordinary Riemannian geometry.

It turns out that the structures $\mathrm{CR}_{\mathfrak{U}}$ that we have just been considering do in fact differ, in general, from the original CR-structure of $\mathbb{PT}_0$ that arose in relation to Minkowski space. There is the further point that by choosing $\mathfrak{M}$ and $\mathfrak{U}$ to be suitably nonanalytic, the structure $\mathrm{CR}_{\mathfrak{U}}$ can be made to be *nonrealizable* as a real hypersurface in a complex 3-manifold, analogously to the Nirenberg counterexamples referred to earlier. This was first demonstrated by C. R. LeBrun (1980), (1982). Some of the relevant ideas and a slightly different, but related, class of counterexamples had been put forward earlier by G. A. J. Sparling (and involved a suggestion by the present author). A development of Sparling's original line of thinking will be presented in the next section.

4. Boundaries of cohomology classes and complex manifolds. The first and most important invariant of CR-structure is the signature of the Levi form. One way of defining the Levi form for a $(2n + 1)$-dimensional CR-manifold $\mathcal{V}$ is as follows. Suppose $\mathcal{V}$ is given, locally, as a real hypersurface

$$\Sigma(\zeta_0, \dots, \zeta_n) = 0$$

in a complex n-manifold $\mathcal{C}$, where Σ is a real (at least C^2) function of local holomorphic coordinates $\zeta_0, \dots, \zeta_n$ for $\mathcal{C}$. Then the Hermitian form whose matrix is

$$\frac{\partial \Sigma}{\partial \zeta_j \partial \bar{\zeta}_k},$$

but restricted to the holomorphic tangent space H_p, defines the Levi form at p. Alternatively an entirely intrinsic definition can be given which does not depend upon an imbedding in a complex $(n + 1)$-manifold and so applies to nonrealizable CR-structures. Let the complex basis vectors $\mathbf{z}_1, \dots, \mathbf{z}_n$ and real basis vector $\mathbf{u}$

be as in §3 (cf. (3.5), (3.6)). Then the Lie brackets

$$(4.1) \qquad\qquad \left[\mathbf{z}_j, \bar{\mathbf{z}}_k\right] = iL_{jk}\mathbf{u} + \text{terms in } \mathbf{z}\text{'s and } \bar{\mathbf{z}}\text{'s}$$

serve to define the matrix L_{jk} of the Levi form intrinsically.

Though the matrix of the Levi form is clearly not an invariant, its signature—by which I mean the number of plus signs, minus signs and zeros that it acquires when unitarily reduced to diagonal form—is invariant. A remark about the *sign* of the Levi form should also be made here. Suppose we are concerned with the (smooth) boundary $\mathcal{V}$ of a complex manifold; then we should take that complex manifold on the side on which Σ is negative. Thus, for example, for the unit ball in $\mathbb{C}^{n+1}$, we can take

$$\Sigma = \zeta_0\bar{\zeta}_0 + \cdots + \zeta_n\bar{\zeta}_n - 1$$

and the Levi form is positive definite $(+ + \cdots +)$, this being the situation in which the manifold is referred to as (holomorphically) strictly *pseudoconvex*. Correspondingly, if we use the intrinsic method (4.1), we choose the vector $\mathbf{u}$ so that $J\mathbf{u}$ points *outwards* away from the manifold whose boundary we are considering. Generally, if the Levi form signature has j plus signs and k minus signs at a point of $\mathcal{V}$, I shall say that this manifold has j degrees of pseudoconvexity and k degrees of pseudoconcavity at p.

An important result, effectively due to Hans Lewy (1956) (and Bochner (1943); cf. also Hörmander (1966)) states if f is a CR-function on $\mathcal{V}$—which means that

$$\bar{z}_1(f) = 0,\ldots,\bar{z}_n(f) = 0$$

—where we assume that $\mathcal{V}$ is realizable as a real hypersurface in a complex manifold $\mathcal{C}$, then f extends (locally) as a holomorphic function on any side of $\mathcal{V}$ having at least one degree of pseudoconvexity.

It should be remarked that any holomorphic function in $\mathcal{C}$ clearly restricts to a CR-function on $\mathcal{V}$. Moreover, any holomorphic function on one side of $\mathcal{V}$ which attains smooth values at $\mathcal{V}$ itself will also restrict to a CR-function on $\mathcal{V}$. The (Lewy) result just stated shows that a holomorphic function defined on a smoothly bounded region $\mathcal{D}$ of a complex manifold is always locally extendible beyond that boundary at points where $\mathcal{D}$ has at least one degree of pseudoconcavity. Thus smoothly bounded domains of holomorphy, on which locally inextendible holomorphic functions exist, have no points with any degree of pseudoconcavity—i.e. the Levi signature is always of the form $(+ \cdots + 0 \cdots 0)$.

Let us now consider the 5-dimensional real hypersurface $\mathbb{P}\mathbb{T}_0$ as a common boundary between $\mathbb{P}\mathbb{T}_-$ and $\mathbb{P}\mathbb{T}_+$. By simple direct calculation from (3.3), we find that the Levi form signature is $(+,-)$, while for the 7-dimensional real hypersurface $\mathbb{T}_0$ as a boundary between $\mathbb{T}_-$ and $\mathbb{T}_+$ the signature is $(+,-,0)$. (The zero arises from the holomorphic direction corresponding to complex rescalings $W \mapsto \lambda W, \lambda \in \mathbb{C} - \{0\}$.) From the results just stated we see that any CR-function defined in some open neighbourhood of a point in $\mathbb{P}\mathbb{T}_0$ [resp. $\mathbb{T}_0$] will extend to a holomorphic function in a neighbourhood in $\mathbb{P}\mathbb{T}$. [resp. $\mathbb{T}$.].

It turns out, however, that the most directly physical objects on these spaces are not holomorphic or CR-functions but objects of the next degree of abstraction, namely *first sheaf cohomology classes*. The simplest example of the physical interpretation of such a cohomology class can be illustrated as follows. Consider some suitable open neighbourhood $\mathcal{Q}$, in $\mathbb{C}M$, of a point $x \in \mathbb{C}M$. According to the discussion of §3, each point of $\mathcal{Q}$ will be represented by a projective line in $\mathbb{PT}$, so as the point varies throughout $\mathcal{Q}$ the corresponding line sweeps out an open region $\mathcal{L} \subset \mathbb{PT}$. We shall be concerned with an element

$$\Phi \in H^1(\mathcal{L}, \mathcal{O}(-2))$$

where, generally, $\mathcal{O}(r)$ denotes the sheaf of germs of holomorphic functions on $\mathbb{PT}$. "twisted by r"—i.e. represented in terms of holomorphic functions on $\mathbb{T}$. which are *homogeneous of degree r*. Under fairly liberal restrictions on the nature of $\mathcal{Q}$ (cf. Eastwood, Penrose and Wells (1981) for details) we find that Φ represents a solution ϕ of the wave equation

$$(4.2) \qquad \Box\phi \equiv \left(\frac{\partial^2}{\partial\tau^2} - \frac{\partial^2}{\partial\xi^2} - \frac{\partial^2}{\partial\eta^2} - \frac{\partial^2}{\partial\zeta^2} \right)\phi = 0$$

in $\mathcal{Q}$.

Suppose we have the simple situation in which Φ can be described by a representative Čech cocycle Φ_{12} $(= -\Phi_{21})$ for the 2-set cover $\mathcal{N}_1$, $\mathcal{N}_2$ of $\mathcal{L}$, so Φ_{12} is simply a holomorphic function (taken as homogeneous of degree -2 in $W_0, \ldots, W_3$) on $\mathcal{N}_{12} = \mathcal{N}_1 \cap \mathcal{N}_2$, where $\mathcal{L} \subset \mathcal{N}_1 \cup \mathcal{N}_2$. To obtain ϕ, we then perform the contour integral

$$(4.3) \qquad \phi(x) = \frac{1}{2\pi i}\oint_{\Xi} \Phi_{12}(W)\Delta W$$

with

$$(4.4) \qquad \Delta W = W_0 dW_1 - W_1 dW_0$$

where the 1-dimensional closed contour Ξ (in $\mathbb{PT}$.) lies in the intersection of $\mathcal{N}_{12}$ with the projective line $L_x \subset \mathbb{PT}$. representing $x = (\tau, \xi, \eta, \zeta)$. Note that since we are integrating over points of $\mathbb{PT}$. *incident* with x, we can use the explicit expression (3.2) to represent W_2 and W_3 in terms of W_0 and W_1. Moreover, because of the -2 homogeneity of Φ_{12} and the form of the expression (4.4), the exterior derivative of the integrand in (4.3) (for W incident with x) vanishes. Thus (4.3) is a genuine contour integral for fixed x, depending only on the homology class of the contour within $L_x \cap \mathcal{N}_{12}$. Simple direct calculation (using (3.2)) shows that (4.2) is indeed satisfied as required.

We could also envisage a more complicated cover $\mathcal{N}_1, \ldots, \mathcal{N}_r$ of $\mathcal{L}$, with Φ given a Čech description with respect to it by the family of holomorphic (homogeneous of degree -2) functions

$$(4.5) \qquad \Phi_{jk} = -\Phi_{kj} \quad \text{on} \quad \mathcal{N}_{jk} = \mathcal{N}_j \cap \mathcal{N}_k$$

subject to

$$(4.6) \qquad \Phi_{jk} + \Phi_{kl} + \Phi_{lj} = 0 \quad \text{on } \mathfrak{N}_{jkl} = \mathfrak{N}_j \cap \mathfrak{N}_k \cap \mathfrak{N}_l.$$

The expression for $\phi(x)$ is given, essentially as before, by (4.3), but now the contour Ξ is a branched one, consisting of various segments $\Xi_{jk} \subset \mathfrak{N}_{jk}$, over which Φ_{jk} is to be integrated, with endpoints $\mathbf{W}_{jkl} \in \mathfrak{N}_{jkl}$, and we sum over the contributions from these various Φ_{jk}. In fact (for appropriate $\mathfrak{D}$) the general solution of (4.2) can be obtained in this way (cf. Eastwood, Penrose and Wells (1981), Penrose and Ward (1980)).

Similarly Maxwell's free-space equations can be solved in terms of elements

$$(4.7) \qquad (\Theta, \Psi) \in H^1(\mathfrak{L}, \mathcal{O}(-4)) \oplus H^1(\mathfrak{L}, \mathcal{O}(0)).$$

The components of the Maxwell field tensor, in the (τ, ξ, η, ζ)-coordinate system, are given by

$$\begin{pmatrix} 0 & \theta_0 - \theta_2 + \psi_0 - \psi_2 & i(\theta_0 + \theta_2 - \psi_0 - \psi_2) & -2(\theta_1 + \psi_1) \\ \theta_2 - \theta_0 + \psi_2 - \psi_0 & 0 & 2i(\theta_1 - \psi_1) & -\theta_0 - \theta_2 - \psi_0 - \psi_2 \\ i(-\theta_0 - \theta_2 + \psi_0 + \psi_2) & 2i(-\theta_1 + \psi_1) & 0 & i(\theta_2 - \theta_0 - \psi_2 + \psi_0) \\ 2(\theta_1 + \psi_1) & \theta_0 + \theta_2 + \psi_0 + \psi_2 & i(\theta_0 - \theta_2 - \psi_0 + \psi_2) & 0 \end{pmatrix}$$

where

$$\begin{pmatrix} \theta_0 \\ \theta_1 \\ \theta_2 \end{pmatrix} = \frac{1}{2\pi i} \oint_{\Xi} \begin{pmatrix} W_0^2 \\ W_0 W_1 \\ W_1^2 \end{pmatrix} \Theta(W)\Delta W, \qquad \begin{pmatrix} \psi_0 \\ \psi_1 \\ \psi_2 \end{pmatrix} = \frac{1}{2\pi i} \oint_{\Xi} \begin{pmatrix} \partial^2/\partial W_2^2 \\ \partial^2/\partial W_2 \partial W_3 \\ \partial^2/\partial W_3^2 \end{pmatrix} \Psi(W)\Delta W$$

(cf. Penrose (1969), (1977)), the notation $\Theta(W)$, $\Psi(W)$ indicating the appropriate $\Theta_{jk}(W)$, $\Psi_{jk}(W)$ to be integrated over Ξ and summed as before. The quantities $\theta_0, \theta_1, \theta_2$ define the anti-self-dual part of the Maxwell field and ψ_0, ψ_1, ψ_2 define the self-dual part. For a *real* Maxwell field we require

$$\theta_0 = \bar{\psi}_0, \qquad \theta_1 = \bar{\psi}_1, \qquad \theta_2 = \bar{\psi}_2,$$

at real points x (and, at complex points, $\theta_0(x) = \bar{\psi}_0(\bar{x})$, etc.) so we can use *either* $H^1(\mathfrak{L}, \mathcal{O}(-4))$ *or* $H^1(\mathfrak{L}, \mathcal{O}(0))$ to specify such a field, employing complex conjugation in place of using the other.

Recall that the Levi form signature for a smoothly bounded domain of holomorphy—i.e. natural (smooth) boundary of the domain of some holomorphic *function*—is always of the form $(+ \cdots + 0 \cdots 0)$. But a function is a *zeroth* sheaf cohomology element (an H^0), whereas we are now more concerned with *first* cohomology elements (H^1's). There is an analogous concept to a domain of holomorphy for such objects, namely a smoothly bounded region $\mathfrak{D}$ on which is defined some holomorphic H^1 (i.e. an H^1 for a coherent analytic sheaf) which does not locally extend beyond the boundary $\partial\mathfrak{D}$. It follows from a theorem of Andreotti and Hill (1972) that, at each point of $\partial\mathfrak{D}$, $\mathfrak{D}$ either has precisely *one* degree of pseudoconcavity (i.e. Levi form signature of the form $(+ \cdots + 0 \cdots 0 -)$) or else is pseudoconvex with at least one zero in the Levi signature (i.e.

$(+ \cdots + 0 \cdots 00))$. Referring to the number of zeros in the Levi signature as the number, d, of *degrees of holomorphic flatness* we find, generally, that a smoothly bounded region on which is defined a locally inextendible holomorphic H^r has a number, q, of degrees of pseudoconcavity satisfying, at each point of its boundary,

$$(4.8) \qquad r \geqslant q \geqslant r - d.$$

We have seen that for ordinary functions (i.e. H^0's), we could refer to the concept of CR-function on a CR-manifold $\mathcal{V}$, realized as a hypersurface in a complex manifold $\mathcal{C}$, and ask whether holomorphic extendibility on one side or the other is locally possible. The appropriate concept for H^r's can be stated in terms of the $\bar{\partial}_b$-cohomology of Andreotti & Hill (1972) and Hill & McKichan (1977); cf. also Polking & Wells (1976), Hill (1973). We find that an appropriate concept of a "CR-H^r" is *not* taken as the restriction to $\mathcal{V}$ of a holomorphic H^r in $\mathcal{C}$ but may be defined, rather, in terms of the $\hat{H}^r$ provided by an exact sequence

$$(4.9) \qquad \cdots \to H^r(\mathcal{C}) \to \begin{matrix} H^r(\mathcal{C}^+) \\ \oplus \\ H^r(\mathcal{C}^-) \end{matrix} \to \hat{H}^r(\mathcal{V}) \to H^{r+1}(\mathcal{C}) \to \begin{matrix} H^{r+1}(\mathcal{C}^+) \\ \oplus \\ H^{r+1}(\mathcal{C}^-) \end{matrix} \to \cdots$$

where $\mathcal{V}$ separates $\mathcal{C}$ into the two pieces $\mathcal{C}^-$, $\mathcal{C}^+$ (so $\mathcal{C} = \mathcal{C}^- \cup \mathcal{C}^+ \cup \mathcal{V}$). Here $\hat{H}^r(\mathcal{V})$ denotes the required space of rth *hyperfunctional* $\bar{\partial}_b$-cohomology. (See Sato (1959–60), Kashiwara (1979).) The sheaf for each H is taken to be that of local cross-sections of some given holomorphic vector bundle over $\mathcal{C}$, and that for $\hat{H}$ to correspond in an appropriate sense to the restriction of this bundle to $\mathcal{V}$.

If we can arrange $H^{r+1}(\mathcal{C}) = 0$, which will be the case, for example, when $\mathcal{C}$ is *Stein* (i.e. a domain of holomorphy), then we obtain from (4.9)

$$(4.10) \qquad \hat{H}^r(\mathcal{V}) = \frac{H^r(\mathcal{C}^-) \oplus H^r(\mathcal{C}^+)}{H^r(\mathcal{C})}$$

—and if $\mathcal{C}$ is Stein we can ignore the $H^r(\mathcal{C})$ $(= 0)$ when $r > 0$. We can loosely interpret the elements of $\hat{H}^r(\mathcal{V})$, in this description, as the space of possible "jumps" in H^r as we cross from one side of $\mathcal{V}$ to the other. In the particular case $r = n = 0$ (where $\mathcal{C}$ has complex dimension $n + 1$), with $\mathcal{V}$ being (a portion of) the real axis in $\mathbb{C}$, (4.10) provides the standard definition of 1-dimensional hyperfunctions on $\mathbb{R}$ (Sato (1959–60), Kashiwara (1979)). Hyperfunctions are generalizations of distributions and so they include, as special cases, ordinary C^∞ functions. The C^∞ CR-functions on $\mathcal{V}$ can thus be taken to be *particular* elements of $\hat{H}^0(\mathcal{V})$.

The concept of local (Lewy) extendibility of C^∞ CR-functions on $\mathcal{V}$, into regions of $\mathcal{C}$ locally bounded by $\mathcal{V}$ wherever there is at least one degree of pseudoconcavity, applies also to CR-hyperfunctions (cf. Hill and MacKichan (1977)). Moreover, there is a corresponding result for elements of $\hat{H}^r(\mathcal{V})$ generally: local holomorphic extendibility into the complex manifold on one side of $\mathcal{V}$ can always be achieved, in the case of a nondegenerate Levi form, whenever the number, q, of degrees of pseudoconcavity is not exactly r. More generally, if there are d degrees of holomorphic flatness, then local extendibility can always be achieved wherever

$$(4.11) \qquad\qquad q > r \quad \text{or} \quad q + d < r,$$

which is the negation of (4.8).

Note that if both $\mathcal{C}^+$ and $\mathcal{C}^-$ have some pseudoconcavity everywhere along $\mathcal{V}$, as is the case when $\mathcal{V} = \mathbb{PT}_0$ or $\mathbb{T}_0$, then every CR-function (whether C^∞ or hyperfunctional) extends to both sides of $\mathcal{V}$ and is therefore always analytic. This shows that we cannot simply use CR-functions in a straightforward Čech approach to the CR-cohomology of such a $\mathcal{V}$, since nonanalytic elements of H^r will always occur for some $r > 0$. Indeed, when $\mathcal{V} = \mathbb{PT}_0$ or $\mathbb{T}_0$ we find that $\hat{H}^1(\mathcal{V})$ contains nonanalytic elements. In fact, solutions of the wave equation (4.2) or of Maxwell's equations on (regions in) *real* Minkowski space M can be represented by elements of an $\hat{H}^1(\mathcal{V})$ (respectively $\hat{H}^1(\mathbb{PT}_0, \mathcal{O}(-2))$ and $\hat{H}^1(\mathbb{PT}_0, \mathcal{O}(-4) \oplus \mathcal{O}(0)))$ and these fields certainly need not be analytic (cf. Wells (1981), Bailey, Ehrenpreis and Wells (1982)). Thus the direct Čech approach, with CR-functions on $\mathbb{PT}_0$, does not apply and a method such as using the sequence (4.9) or an explicit $\bar{\partial}_b$ approach is called for.[2]

This contrast between a direct Čech approach and such as the $\bar{\partial}_b$ method has its analogue in a related contrast between CR-structures which are realizable in terms of imbeddability in complex manifolds and those which are not. Let us consider two possible approaches to defining a complex manifold. In the first (method $\check{C}$), we consider the manifold to be built up from open patches on $\mathbb{C}^{n+1}$, a family of holomorphic transition functions F_{jk} being provided to "glue" these patches together, where the F_{jk} are subject to certain nonlinear relations closely analogous to (4.5) and (4.6). Likewise, the equivalence between two such complex manifolds so constructed is closely analogous to the condition for two Čech 1-cocycles Φ and Ψ to differ by a coboundary

$$\Phi_{jk} - \Psi_{jk} = \Lambda_k - \Lambda_j \quad \text{on } \mathfrak{N}_{jk}$$

where Λ_j is holomorphic on $\mathfrak{N}_j$. Here, the Λ_j are analogous to coordinate transformations on the individual patches. Finally, the concepts of taking refinements and direct limits are common both to Čech cohomology and to this method of complex-manifold building. Thus, method $\check{C}$ presents a complex manifold as a kind of nonlinear holomorphic H^1, presented according to a Čech prescription.

In the second method (method $\bar{\partial}$) the manifold is given first as a real manifold of dimension $2(n + 1)$. An operator J (with $J^2 = -1$) is defined to act in the tangent spaces and to satisfy an integrability condition like (3.6) with (3.5). When phrased suitably, method $\bar{\partial}$ resembles a nonlinear version of the Dolbeault ($\bar{\partial}$) approach to the definition of a holomorphic H^1.

---

[2] There are also other (essentially equivalent) methods of defining the concept of a (say C^∞) CR $- H^R$ on $\mathcal{V}$, which can provide the desired nonanalytic cohomology. One of these is to define the cohomology in terms of extensions of CR-vector-bundles over $\mathcal{V}$.

Because of the Newlander-Nirenberg theorem, these two methods of building a complex manifold are equivalent—and, likewise, these two approaches to constructing a holomorphic H^1 are equivalent. However, we have just seen that there is an inequivalence between the methods of defining the H^1 cohomology at the *boundaries* of complex manifolds. Such considerations led C. D. Hill to suggest, a number of years ago, that there ought to be a *difference between the $\check{C}$-method and $\bar{\partial}$-method of defining a complex manifold-with-boundary*, and that this difference should manifest itself when the manifold has one degree of pseudoconcavity at its boundary (or perhaps none, in cases where the Levi form is degenerate). The $\check{C}$-method would be as before, except that the patches would now have specified *boundary* portions (which would have to piece together suitably), given as (say) C^∞ real hypersurfaces in each coordinate patch. In the $\bar{\partial}$-method, the operator H would be required to attain C^∞-smooth boundary values at the boundary of the given $2(n + 1)$-real-dimensional C^∞ manifold-with-boundary.

Both methods would provide the boundary $\mathcal{V} = \partial\mathcal{D}$, of the complex manifold $\mathcal{D}$ with a CR-structure, but with the $\check{C}$-method, $\mathcal{D}$ would always be locally extendible to the other side of $\mathcal{V}$ (simply by a slight extension of the relevant coordinate patch) whereas with the $\bar{\partial}$-method we would have no guarantee that this can be done. Indeed, the existence of locally nonextendible holomorphic H^1's at (a realizable) $\mathcal{V}$, when there is just one degree of pseudoconcavity, suggests also the existence of locally nonextendible complex manifolds-with-boundary with just one degree of pseudoconcavity. Moreover, with a Levi form of, say, signature $(+\,-)$, like that of a perturbed version $\mathcal{V} = \mathbb{P}\mathcal{I}_0$ of $\mathbb{P}\mathbb{T}_0$, as described in §3, we may expect that the CR-manifold $\mathcal{V}$ can sometimes be realizable as a boundary *neither* on one side *nor* on the other.

One approach to the construction of such nonrealizable CR-structures is to produce $\mathbb{P}\mathcal{I}_0$, as in §3, where the space-time $\mathfrak{M}$ and hypersurface $\mathfrak{U}$ are suitably nonanalytic. However the mere nonanalyticity of the resulting CR-structure is not sufficient to ensure its nonrealizability. The reason is, of course, that the real hypersurface defining a realizable CR-manifold need not itself be analytic. Thus, LeBrun (1980) was forced into some more subtle (nonlocal) considerations in order to prove his nonrealizability result for this case.

The related but slightly different earlier suggestion due to Sparling was to use, instead, the Ward (1979) "twisted photon" construction. According to this construction (a special case of Ward's (1977) twistor method of generating self-dual Yang-Mills fields), self-dual Maxwell fields are represented by holomorphic line bundles over the appropriate regions of $\mathbb{P}\mathbb{T}$., rather than by elements $\Psi \in H^1(\mathcal{L}, \mathcal{O}(0))$, as in (4.7). The bundles are constructed using the multiplicative $\mathcal{O}_*$ (zeros excluded), rather than the additive $\mathcal{O}(0)$, the relation between the two being achieved via the familiar exact sequence

$$(4.12) \qquad\qquad 0 \to \mathbb{Z} \xrightarrow{\times 2\pi i} \mathcal{O}(0) \xrightarrow{\exp} \mathcal{O}_* \to 0.$$

Now, real Maxwell fields on (open) regions $\mathfrak{R}$ of real Minkowski space M are equivalent to self-dual Maxwell fields on $\mathfrak{R}$ (the relevant mutually inverse maps being "take the self-dual part" and "take twice the real part"). To express such fields we can thus either use elements of

$$(4.13) \qquad \hat{H}^1(\mathcal{K}, \mathcal{O}(0))$$

where $\mathcal{K}$ is the region of $\mathbb{PT}_0$ swept out by the projective lines representing the points of $\mathcal{R}$ (as in (4.9)), or else use the corresponding line bundles over $\mathcal{K}$. Such real Maxwell fields certainly need not be analytic, and may even be hyperfunctional. Let us take the field to be C^∞ but nowhere analytic. It may then be represented by a C^∞ CR-line bundle over $\mathcal{K}$ in accordance with the exponentiation procedure of (4.12) as applied to the element of (4.13) representing the field. This bundle is not extendible to a holomorphic line bundle over some open neighbourhood of $\mathcal{K}$ in $\mathbb{PT}.$, for if it were, the self-dual Maxwell field on the corresponding open region of $\mathbb{C}M$ that it represents would restrict down to (the self-dual part of) the nonanalytic field that we started with, which is a contradiction since Maxwell fields on open regions of $\mathbb{C}M$ are necessarily holomorphic (and therefore real-analytic). By choosing our original self-dual field to be either of positive or negative frequency, we can ensure *one*-sided extendibility of the resulting bundle into $\mathbb{PT}_-$ or $\mathbb{PT}_+$, respectively (cf. Wells (1981), Bailey, Ehrenpreis and Wells (1982)).

By a slight extension of Sparling's above line of argument, one can translate the above statements concerning inextendible or one-side-extendible CR-line bundles over $\mathcal{K}$ to nonrealizable or one-side-realizable CR-manifolds. The CR-line bundle over $\mathcal{K}$ is itself a CR-manifold with $(+0-)$ Levi signature (and the one direction of holomorphic flatness defines the fibre direction). Now the structure that a holomorphic line bundle $\mathcal{B}$ possesses, over and above that of its complex-manifold structure, can be represented in terms of the holomorphic vector field

$$(4.14) \qquad \lambda\frac{\partial}{\partial\lambda}$$

on $\mathcal{B}$, where λ is a fibre coordinate on each fibre. Likewise, the bundle structure of the above CR-line bundle over $\mathcal{K}$—which we now call $\mathcal{V}$—is also characterized by (4.14), which is now a CR-vector field on $\mathcal{V}$. Suppose that $\mathcal{V}$ forms a boundary to a complex manifold-with-boundary $\mathcal{D}$. It follows from the Lewy extension property (and the $(+0-)$ Levi signature) that the vector field (4.14) extends locally into $\mathcal{D}$ and defines $\mathcal{D}$ (locally near $\mathcal{V}$) as a holomorphic line bundle over some open neighbourhood in $\mathbb{PT}_+$ or $\mathbb{PT}_-$ of $\mathcal{K}$. (Though the Lewy extension property refers, in the first instance, only to CR-functions and not to vector fields, the extension to CR-vector fields is easy to achieve by various means, e.g. take components in local CR-coordinates.) It follows from the above arguments (essentially) that our inextendible, or one-side-extendible CR-line bundles are neither-side-realizable or one-side-realizable CR-manifolds, respectively. In the latter case, we have examples of Hill's complex manifolds-with-boundary of the $\bar{\partial}$-type which are not realizable by the $\check{C}$-method, since they are not extendible beyond their C^∞ smooth boundaries.

The above discussion has been given partly in outline. It depends to some extent upon the details of twistor geometry and the twistor constructions of physical fields. Not all of this is necessary, however, for the construction of neither-side-realizable and one-side-realizable CR-manifolds. A simplified example due to C. D. Hill, G. A. J. Sparling and the author can be given, where $\mathcal{V} = \partial\mathcal{D}$, with real dimension 5 and Levi signature $(0\ -\ )$, where the complex 3-manifold-with-boundary $\mathcal{D}$ cannot be extended as a complex manifold beyond its C^∞ boundary $\mathcal{V}$. This example has an advantage over what has been presented above in that the extendibility emerges as manifestly a *local* rather than a global obstruction. (The relation between cohomology classes on portions of $\mathbb{PT}$ and physical fields depends upon a nonlocal correspondence.) A fully detailed argument will be presented elsewhere, but the gist of the construction is as follows.

Let $\mathcal{S}$ be the unit 3-sphere in $\mathbb{C}^2$, and define $\mathcal{K}$ and $\mathcal{E}$ to be the intersections with $\mathcal{S}$ and with the closed exterior of $\mathcal{S}$, respectively, of the open ball of radius ε (< 2) centered about some point of $\mathcal{S}$. Then $\mathcal{E}$ is a complex 2-manifold-with-boundary, with $\mathcal{K} = \partial\mathcal{E}$ and Levi signature $(-)$. Choose an element $\Phi \in H^1(\mathcal{E} - \mathcal{K}, \mathcal{O})$ which has C^∞ boundary values on $\mathcal{K}$ but which is not locally extendible beyond $\mathcal{K}$. By the aforementioned standard exponentiation process, construct a holomorphic line bundle over $\mathcal{E} - \mathcal{K}$ which joins smoothly to a CR-line bundle over $\mathcal{K}$. These two pieces of line bundle together provide the required complex 3-manifold-with-boundary $\mathcal{D}$ (as its total space) and the portion over $\mathcal{K}$ provides the required C^∞ CR-5-manifold $\mathcal{V}$, of Levi signature $(0\ -\ )$, beyond which $\mathcal{D}$ is not locally extendible as a complex manifold. For the proof we appeal to the same argument as before, which uses Lewy extension of the vector field (4.14) to show that any extension of $\mathcal{D}$ across $\mathcal{V}$ as a complex manifold is also (locally) an extension of $\mathcal{D}$ as a holomorphic vector bundle. But such extension would imply that Φ extends across $\mathcal{K}$, which is a contradiction.

The Hill philosophy[3] would suggest that a corresponding example should exist in which $\mathcal{D}$ is a 2-complex-dimensional with Levi signature $(-)$ at its C^∞ boundary $\mathcal{V}$. Perhaps the original Nirenberg example bounds on one side and thus provides such a $\mathcal{V}$?

It is remarkable, and perhaps even somewhat ironic that such seemingly esoteric matters as nonrealizable CR-structures and the delicate dividing line between C^ω and C^∞ should have significant connections with physics. But it seems to be so. (The infinite-dimensionality of $H^1(\mathbb{PT}_+, \mathcal{O}(r))$—and hence of physical massless fields—can be attributed to Lewy nonextendibility across $\mathbb{PT}$.) What would Poincaré have made of all this?

I am grateful to T. N. Bailey, M. G. Eastwood, C. D. Hill, C. R. LeBrun and G. A. J. Sparling for many helpful suggestions. I am also grateful to NSF for financial assistance which enabled me to attend the Poincaré Symposium, and for

---

[3] Recently a large class of nonrealizable CR-manifolds has been presented by Jacobowitz and Trèves (1982). Also Kuranishi (1982) has shown that in dimension $2n + 1 > 7$ and for positive-definite Levi form all CR-manifolds are realizable. All these results are consistent with the Hill philosophy.

support under contract MCS 79-12938 while the author was at the Institute for Advanced Study, Princeton in March 1980, where some of this research was done.

REFERENCES

A. Andreotti and C. D. Hill (1972), *E. E. Levi convexity and the Hans Lewy problem. Part I: Reduction to vanishing theorems; Part II: Vanishing theorems*, Ann. Scuola Norm. Sup. Pisa Cl. Sci. (3) **26**, 325–363; (4) **26**, 747–806.

T. N. Bailey, L. Ehrenpreis and R. O. Wells, Jr. (1982), *Weak solutions of the massless field equations*, Proc. Roy. Soc. London Ser. A (to appear).

S. Bochner (1943), *Analytic and meromorphic continuation by means of Green's formula*, Ann. of Math. (2) **44**, 652–673.

H. Bondi (1965), *Relativity and common sense*, Heinemann, London.

É. Cartan (1923), Ann. Sci. École Norm. Sup. **40**, 325.

______ (1924), Ann. Sci. École Norm. Sup. **41**, 1.

______ (1932), *Sur la géométrie pseudo-conforme des hypersurfaces de deux variables complexes.* I, II, Ann. Math. Pura Appl. (4) **11**, 17–90; Ann. Scuola Norm. Sup. Pisa (2) **1**, 333–354.

S. S. Chern and J. K. Moser (1974), *Real hypersurfaces in complex manifolds*, Acta Math. **133**, 219–271.

M. G. Eastwood, R. Penrose and R. O. Wells, Jr. (1981), *Cohomology and massless fields*, Comm. Math. Phys. **78**, 305–351.

K. Friedrichs (1928), *Eine invariante Formulierung des Newtonschen Gravitationsgesetzes und des Grenzüberganges vom Einsteinschen zum Newtonschen Gesetz*, Math. Ann. **98**, 566.

R. P. Geroch (1970), *Domain of dependence*, J. Math. Phys. **11**, 437.

C. D. Hill (1973), *The Cauchy problem for $\bar{\partial}^n$*, Partial Differential Equations, Proc. Sympos. Pure Math., vol. 23, Amer. Math. Soc., Providence, R.I.

C. D. Hill and B. MacKichan (1977), *Hyperfunction cohomology classes and their boundary values*, Ann. Scuola Norm. Sup. Pisa Cl. Sci. Ser. IV **4**, 577–597.

L. Hörmander (1966), *An introduction to complex analysis in several variables*, Van Nostrand-Reinhold, New York.

H. Jacobowitz and F. Trèves (1982), *Non-realizable CR-structures* (to appear).

M. Kashiwara (1979), *Introduction to the theory of hyperfunctions*, Seminar on Micro-local Analysis, Ann. of Math. Studies, no. 93, Princeton Univ. Press, Princeton, N.J.

M. Kuranishi (1982), *Strongly pseudoconvex CR-structures over small balls. Part III: An embedding theorem* (to appear).

C. R. LeBrun (1980), *Spaces of complex geodesics and related structures*, Ph.D. thesis, Oxford Univ.

______ (1982), *𝓗-space with a cosmological constant*, Proc. Roy. Soc. London Ser. A **380**, 171–185.

J. Leray (1952), *Hyperbolic differential equations*, Institute for Advanced Study.

H. Lewy (1956), *On the local character of a solution of an atypical linear differential equation in three variables and a related theorem for regular functions of two complex variables*, Ann. of Math. (2) **64**, 514–522.

______ (1957), *An example of a smooth linear partial differential equation without solution*, Ann. of Math. (2) **66**, 155–158.

H. Minkowski (1908), *Space and time*, The Principle of Relativity (H. A. Lorentz, A. Einstein, H. Minkowski and H. Weyl, eds.), Dover, New York, 1923.

A. Newlander and L. Nirenberg (1957), *Complex analytic coordinates in almost complex manifolds*, Ann. of Math. (2) **65**, 391–404.

L. Nirenberg (1973), *Lectures on linear partial differential equations*, CBMS Regional Conf. Ser. in Math., no. 17, Amer. Math. Soc., Providence, R.I.

______ (1974), *On a question of Hans Lewy*, Russian Math. Surveys **29**, 251–262.

R. Penrose (1959), *The apparent shape of a relativistically moving sphere*, Proc. Cambridge Philos. Soc. **55**, 137–139.

______ (1968), *Structure of space-time*, Battelle Recontres, 1967 Lectures in Mathematics and Physics (C. M. DeWitt and J. A. Wheeler, eds.), Benjamin, New York.

______ (1969), *Solutions of the zero rest-mass equations*, J. Math. Phys. **10**, 38–39.

______ (1975), *Twistor theory, its aims and achievements*, Quantum Gravity, an Oxford Symposium (C. J. Isham, R. Penrose and D. W. Sciama (eds.)), Clarendon Press, Oxford.

______ (1977), *The twistor programme*, Rep. Mathematical Phys. **12**, 65–76.

R. Penrose and R. S. Ward (1980), *Twistors for flat and curved space-time*, General Relativity and Gravitation, Vol. II (A. Held (ed.)), Plenum Press, New York and London.

H. Poincaré (1905), *Science and hypothesis*, Walter Scott, London; reprinted, Dover, New York, 1952.

______ (1906), *Sur la dynamique de l'électron*, Rend. Circ. Mat. Palermo **21**, 129–176.

______ (1907), *Les functions analytique de deux variables et la représentation conforme*, Rend. Circ. Mat. Palermo **23**, 185–220.

______ (1954), *Oeuvres de Henri Poincaré*, vol. 9, Gauthier-Villars, Paris.

J. C. Polking and R. O. Wells, Jr. (1976), *Boundary values of Dolbeault cohomology classes and a generalized Bochner-Hartogs theorem*.

M. Sato (1959–60), *Theory of hyperfunctions*, J. Fac. Sci. Univ. Tokyo Sect. I **8**, 139–193, 387–436.

J. L. Synge (1960), *Relativity: the general theory*, North-Holland, Amsterdam.

N. Tanaka (1962), *On the pseudo-conformal geometry of hypersurfaces of the space of n complex variables*, J. Math. Soc. Japan **14**, 397–429.

J. Terrell (1959), Phys. Rev. **116**, 1041.

A. Trautman (1966), *Comparison of Newtonian and relativistic theories of space-time*, Perspectives in Geometry and Relativity (B. Hoffmann (ed.)), Indiana Univ. Press, Bloomington and London.

R. S. Ward (1977), *On self-dual gauge fields*, Phys. Lett. **61A**, 81.

______ (1979), *The twisted photon: massless fields as bundles*, Advances in Twistor Theory (L. P. Hughston and R. S. Ward (eds.)), Research Notes in Math. 37, Pitman Adv. Publ. Program, San Francisco, London, Melbourne.

R. O. Wells, Jr. (1979), *Complex manifolds and mathematical physics*, Bull. Amer. Math. Soc. (N.S.) **1**, 296–336.

______ (1981), *Hyperfunction solutions of the zero-rest-mass field equations*, Comm. Math. Phys. **78**, 567–600.

MATHEMATICAL INSTITUTE, OXFORD, GREAT BRITAIN

Proceedings of Symposia in Pure Mathematics
Volume 39 (1983), Part 1

The Cauchy-Riemann Equations
and Differential Geometry

R. O. WELLS, JR.

1. Introduction. In 1907 Poincaré wrote a seminal paper [35] on various topics in several complex variables. In this paper we shall discuss Poincaré's paper and its influence and relationship to certain directions of research in complex analysis and geometry over the past 70 years. Poincaré's paper discusses in some detail real 3-dimensional hypersurfaces in $\mathbf{C}^2$ and the relationship of the geometry of these hypersurfaces to the behavior of holomorphic functions and mappings defined near these hypersurfaces. The major topics treated by Poincaré include: (a) Hartogs' phenomenon, discovered only one year earlier by Hartogs [24]; (b) the Riemann mapping problem for domains in $\mathbf{C}^2$; (c) an analogous equivalence problem for smooth boundaries of such domains now known as the Poincaré equivalence problem; (d) automorphisms or self-equivalences of domains and boundaries of domains (as in (c) and (d)); and (e) the tangential Cauchy-Riemann equations for a real hypersurface in $\mathbf{C}^2$.

The paper of Hartogs [24] and the paper of E. E. Levi [29] stimulated the major developments in several complex variables in this century. The general theory of several complex variables was summarized first by Osgood [36] who treated the early developments. In 1932 Behnke and Thullen wrote their famous monograph [4] which summarized the state of the art and asked the major questions which occupied researchers for the next 40 years. This culminated in the general theory of Stein manifolds in the 1950's (described, for instance, in Gunning and Rossi [23], Grauert and Remmert [21]) and the later significant interaction of several complex variables with modern partial differential equations (as described by Hörmander [26]).

Poincaré's work on Hartogs' theorem relates to the general theory of several complex variables mentioned above. His question about the equivalence of domains was taken up by Bergman and Carathéodory who developed their invariant metrics, both very useful in modern developments (see [3 and 11]). His question about the equivalence of boundaries stimulated significant work by Segre and Cartan in 1931–32 which we discuss in more detail later. On the whole, however, the problems proposed by Poincaré have remained dormant for some time while the general theory of several complex variables was being developed. In some sense he was asking more difficult questions whose solutions required invariants of a considerably higher order

---

Reprinted from Bulletin Amer. Math. Soc. (N.S.) 6 (1982), 187–199.
1980 *Mathematics Subject Classification.* Primary 32-02, 53-01.

for their resolutions. In the past fifteen years these questions have been taken up again by a new generation of mathematicians who have a new perspective on these problems due to simultaneous developments in other areas of mathematics, such as differential topology, partial differential equations, algebraic topology, etc.

The study of the tangential Cauchy-Riemann equations on a real hyper-surface, for instance, is a more recent phenomenon. Manifolds equipped with tangential Cauchy-Riemann equations and their abstractions are called *CR-manifolds*. Poincaré discussed the tangential Cauchy-Riemann equations in his paper, as is mentioned in §2. These equations were not studied much at all until H. Lewy published his examples in 1956–60 [**30, 31**] which then stimulated a great deal of work. There were studies of the relation between function theory in the ambient space and the solutions of the tangential Cauchy-Riemann equations as well as investigations of the intrinsic questions concerning these equations (see the surveys by Folland and Kohn [**20**] and Wells [**46**] for further literature on this subject). Lewy's examples were also seminal for major developments in partial differential equations (see Hörmander [**25**]).

The question of the geometric *equivalence* of CR-manifolds (analogous to the equivalence of complex or Riemannian manifolds) is a question first raised by Poincaré in an extrinsic form in 1907. In this paper we want to give a summary of developments concerning this question. More generally we discuss the *geometry* of real hypersurfaces (considered as CR-manifolds), leaving the discussion of *analysis* on such manifolds to the works cited earlier.

In §2 we give a summary of Poincaré's original 1907 paper, including the topics discussed above. In §3 we discuss the solutions to the local equivalence problem for real hypersurfaces in $\mathbf{C}^2$ given by Segre and E. Cartan in 1931 and 1932 respectively. In §4 we discuss the developments of the past two decades, when these questions were again taken up by a new generation of mathematicians.

2. Poincaré's paper of 1907. We want to formulate Poincaré's equivalence problem in $\mathbf{C}^n$, even though Poincaré and other researchers of the early 20th century confined their research almost exclusively to $\mathbf{C}^2$. Following Poincaré [**35**] we have

A. *The local equivalence problem.* Let M and M' be two real-analytic real hypersurfaces defined in open sets U and U' in $\mathbf{C}^n$. Given $p \in M$ and $p' \in M'$, when does there exist a neighborhood V of p and V' of p' and a biholomorphic mapping $F: V \to V'$ so that $F(V \cap M) = V' \cap M'$? If such an F exists then M and M' are said to be *locally equivalent* at p and p', respectively. If such an F exists, then how unique is it?

B. *The global equivalence problem.* Suppose M and M' are two real-analytic compact hypersurfaces in $\mathbf{C}^n$ and $M = \partial D$, $M' = \partial D'$ where D and D' are bounded domains. When does there exist a neighborhood V of $\overline{D}$ and V' of $\overline{D}'$ and a biholomorphic mapping $F: V \to V'$ such that $F(D) = D'$ i.e., $F|_D$ is a biholomorphic mapping onto D'?

C. *The mixed equivalence problem.* Suppose M and M' are as in (B) above and suppose that for each point $p \in M$, M is locally equivalent at p to M' at some point $p' \in M'$. Then when is M globally equivalent to M' in the sense of (B)?

If $n = 1$, we see easily that two real-analytic arcs are always locally equivalent, since they are both locally equivalent to a segment of the real axis, by the very definition of real-analytic arc. The global equivalence problem is the Riemann mapping problem in the case $n = 1$, and the mixed problem clearly doesn't have a solution unless there are global geometric hypotheses on M and M' (e.g., for $n = 1$). The obstruction to uniqueness of the local and global equivalence problem is the group (or pseudogroup) of local and global self-equivalences, which in the case of $n = 1$, can be computed as a specific group of Möbius transformations (in the global case).

If $n > 1$, Poincaré observed that two local hypersurfaces are generically inequivalent, in contrast to the situation above. He based his observation on the fact that if M and $M' \subset \mathbf{C}^2$ were locally equivalent, then the three local parameters describing M' would have to satisfy 4 real partial differential equations with respect to the 3 local parameters describing M. These are nothing but the tangential Cauchy-Riemann equations, expressed in terms of local coordinates on M and M'. Such an overdetermined inhomogeneous system does not generically have a solution, since the compatibility conditions are generically not satisfied. To this author's knowledge this is the first time the tangential Cauchy-Riemann equations appear in the literature, if only in a peripheral manner. As mentioned in the introduction one can consult [20 and 46] for surveys of the analysis of these equations.

Poincaré discusses the solution of the equivalence problems above, and points out the need for a study of the automorphism groups of the structures involved. In particular, he singles out the distinction between an automorphism group or pseudogroup depending on a finite or an infinite number of parameters. If M is a real hypersurface, then let $\mathrm{Aut}_p(M)$ be the pseudogroup of local self-equivalences or automorphisms at the point p. Then it's obvious that M and M' being locally equivalent at p and p' respectively implies that $\mathrm{Aut}_p(M) \cong \mathrm{Aut}_{p'}(M')$. In other words the (pseudo-)group $\mathrm{Aut}_p(M)$ is an invariant of the structure and it can be used to derive other invariants. Let's consider the following two examples.

EXAMPLE 2.1. Let $z_j = x_j + iy_j, j = 1, 2$, be coordinates in $\mathbf{C}^2$, and let

$$M = \left\{ z \in \mathbf{C}^2 \colon y_2 = 0 \right\}$$

be a real hyperplane in $\mathbf{C}^2$. Consider the local mapping

$$T \colon \begin{cases} \zeta_1 = f(z_1, z_2), \\ \zeta_2 = \phi(z_2) \end{cases}$$

where f is holomorphic near $0 \in \mathbf{C}^2$, and ϕ is real-valued and real-analytic. Assume that T is locally invertible at 0. Then T is the restriction to M of a local biholomorphic mapping $\tilde{T}$ which maps M to M near $z = 0$. For instance, a special case includes T of the form

$$T: \begin{cases} \zeta_1 = z_1 + g(z_2), \\ \zeta_2 = x_2 \end{cases}$$

where g is an arbitrary holomorphic function of z_2 near $z_2 = 0$. There are an infinite number of independent parameters in the coefficients of the Taylor expansion of g, and thus $\dim \mathrm{Aut}_0(M) = \infty$, in this case.

EXAMPLE 2.2. Let

$$M = \left\{ z \in \mathbf{C}^2 : y_2 = |z_1|^2 \right\}$$

be a parabolic hypersurface in $\mathbf{C}^2$. Poincaré computed explicitly that $\mathrm{Aut}_p(M) = H$ where H is the group of transformations of $\mathbf{C}^2$ of the form

$$(2.1) \quad \begin{aligned} \zeta_1 &= \frac{\lambda(z_1 + az_2)}{1 - 2i\,\bar{a}\,z_1 - (r + i|a|^2)z_2}, \\[2ex] \zeta_2 &= \frac{|\lambda|^2 z_2}{1 - 2i\,\bar{a}\,z_1 - (r + i|a|^2)z_2} \end{aligned}$$

$\lambda \in \mathbf{C}^*$, $a \in \mathbf{C}$, $r \in \mathbf{R}$. This is a Lie group, which is, in fact a subgroup of $SU(2, 1)$ consisting of matrices of the form

$$\begin{bmatrix} \lambda & 0 & 0 \\ a & 1 & 0 \\ r + i|a|^2 & -2i\,\bar{a} & \bar{\lambda}^{-1} \end{bmatrix}$$

which induce mappings of $\mathbf{C}^2 \to \mathbf{C}^2$ which map M to M in terms of homogeneous coordinates considering $\mathbf{C}^2 \subset \mathbf{P}_2$ in a natural manner, and $y_2 = |z_1|^2$ being the restriction to $\mathbf{C}^2$ of a hyperquadric in $\mathbf{P}_2$. One main point of the computation of Poincaré is that *a priori* the self-equivalences at 0 were only *locally* defined, whereas, in fact, all such are globally defined rational mappings which are nonsingular on M.

From the group-theoretic point of view one sees that Examples 2.1 and 2.2 are not locally equivalent at $0 \in \mathbf{C}^2$. This follows more simply from the *Levi form*, first discovered 2 years later. The Levi form is the first fundamental invariant of a real hypersurface, although this was not singled out explicitly by Poincaré in his paper.

The Levi form is defined as follows: if M is a local hypersurface in $\mathbf{C}^n$ defined by

$$M = \{ z \in U : F(z) = 0 \}$$

and $dF \neq 0$ on M, where F is a real-valued C^2 function, then

$$(2.2) \quad L(M)(t) = \sum_{i,j=1}^{n} \frac{\partial^2 F}{\partial z_i \partial \bar{z}_j} t_i \bar{t}_j, \quad t \in \mathbf{C}^n,$$

where

$$(2.3) \quad \sum_{i=1}^{n} \frac{\partial F}{\partial z_i} t_i = 0.$$

This is an Hermitian form restricted to the holomorphic tangent space of M (defined by (2.3)). This was defined for $n = 2$ by Levi [29] and for $n > 2$ by Krzoska [28]. The hypersurface M is said to be *strongly pseudoconvex* if $L(M)$ is positive definite. The signature of $L(M)$ is a local biholomorphic invariant and an invariant of the local equivalence problem. In particular in Example 2.1 the Levi form has a zero eigenvalue, and in Example 2.2 the Levi form has a nonzero eigenvalue, so the signatures differ. The invariance of the signature is an easy fact to prove, and it is of critical importance in the geometric theory of functions of several complex variables.

In fact, Poincaré showed that on a real hypersurface $M \subset \mathbf{C}^2$ there is a natural exterior differential system on M (generated by the tangential holomorphic cotangent vectors and their complex conjugates) such that the complete integrability of this system was equivalent to the hypersurface being equivalent to Example 2.1. The *obstruction* to complete integrability of this exterior differential system is precisely the Levi form. This theorem was reformulated by Segre [36] and Cartan [12] sometime later in the form that a hypersurface in $\mathbf{C}^2$ is locally equivalent to a hyperplane (Example 2.1) if and only if the Levi form vanishes. This is also true in $\mathbf{C}^n$ (Sommer [38]).

Poincaré initiated a study of invariants of a real hypersurface by looking at relations between the Taylor series coefficients of a defining function $y_2 = F(z_1, x_1)$ and the Taylor series of a transformed equation $y_2' = F'(z_1', x_1')$. This he studied in terms of the coefficients of the expansion of the functions defining a local biholomorphic change of variables. By counting numbers of equations and unknowns he observed that the first nontrivial invariants (coefficients which couldn't be made to vanish by changes of coordinates) depended on the Taylor expansion up to order 9 in $\mathbf{C}^2$. This is not quite true, as the Levi form is of order 2, and the next higher order invariant in $\mathbf{C}^2$ turns out to be of order 6 (as was shown by E. Cartan in 1932). This process of finding invariants from the power series expansion point of view was carried out much later in a significant manner by Moser, as we see later in this paper. Basically, Poincaré recognized that at a point p of a hypersurface M there were an infinite number of nontrivial invariants $\{I_j\}$ which agrees with the corresponding invariants $\{I_j'\}$ at any locally equivalent hypersurface M' at a specific point p'. To determined what these invariants are and how to compute them has been the object of much of the research of the ensuing 70 years and will be discussed in the later sections of this paper.

In 1906, a year before Poincaré's paper, Hartogs gave the first examples of simultaneous analytic continuation of holomorphic functions of more than one variable, the study of which dominated research in several complex variables for the next 50 years as discussed earlier. The general Hartogs' theorem asserts that if D is a bounded domain in $\mathbf{C}^n$ with smooth boundary, such that $\mathbf{C}^n - D$ is connected, then any function f holomorphic near ∂D continues analytically in a single-valued manner to a holomorphic function on D. Poincaré gave a brief proof of the general result, and a much longer proof of the same result for the unit ball in $\mathbf{C}^n$ using spherical harmonics. The reason for the second proof is that there was no accepted rigorous proof of the general statement until Brown [7] and Bochner [5] gave their proofs thirty years later (it's now "elementary", cf. Hörmander [26]).

In the remainder of this wide-ranging paper Poincaré discussed among other things, the problem of when a perturbation of the sphere is still globally equivalent to the sphere. He sets up some nonlinear equations, and discusses, using spherical harmonics, a characterization of the linearization of the relevant equations, i.e., a first order solution to this particular equivalence problem in this special geometric setting. This was a first order approximation to a class of problems which have been resolved to any extent only quite recently (Webster [43], Burns, Shnider and Wells [10], Johnson [27], Greene and Krantz [22]) as is discussed in the next section.

3. Solutions of the Poincaré equivalence problem by B. Segre and E. Cartan. We now want to give a brief overview of two different types of solutions to the equivalence problem of Poincaré as formulated in the previous section. Let's restrict our attention to real-analytic hypersurfaces in $\mathbf{C}^n$. Many of the concepts and results are valid for more general smooth manifolds, but there are sometimes subtleties which we won't concern ourselves with here.

Let M be a (real-analytic) hypersurface in some open set in $\mathbf{C}^n$, i.e. for F a real-valued function in an open set $D \subset \mathbf{C}^n$,

$$M = \{ z \in D \colon F(z) = 0, \quad dF \neq 0 \}.$$

Consider the bundle $H(M) \subset T(M)$ defined by $H(M) = T(M) \cap JT(M)$ where $J \colon T(M) \to T(M)$ is the real-linear mapping corresponding to the multiplication of vectors by i (i.e., $J^2 = -I$). Thus

$$H(M) \subset T(M) \subset T(\mathbf{C}^n)|_M$$

and $H(M)_x$ is the maximal complex subspace of $T(\mathbf{C}^n)_x$ which is contained in $T(M)_x$ for $x \in M$. The bundle $H(M)$ is called the *holomorphic tangent bundle* to M. The pair $H(M) \subset T(M)$ is called the *CR-structure* (Cauchy-Riemann structure) of the hypersurface M (the name arises from the fact that sections of $\overline{H(M)}$ in the natural conjugations of vector fields on $\mathbf{C}^n$ are the tangential Cauchy-Riemann equations, alluded to above). We call M with its CR-structure a *CR-hypersurface*. The notion of CR-hypersurface can be abstracted in an appropriate manner, just as Riemannian manifolds are abstractions of submanifolds with the induced Riemannian metric. An abstract CR-hypersurface is a real $(2n - 1)$-manifold with a subbundle $H(M) \subset T(M)$ where $H(M)$ is a complex vector bundle of complex rank $(n - 1)$. The CR-structure is *integrable* if $[\Gamma(H(M)), \Gamma(H(M))] \subset \Gamma(H(M))$. This is automatic for hypersurfaces in $\mathbf{C}^n$ as is easily checked. An *integrable abstract CR-hypersurface* is the correct abstraction of a hypersurface with respect to the class of problems raised by Poincaré. For simplicity we will speak simply of "CR-hypersurfaces".

Just as with the isometric embedding problem one wants to know if all of the abstract ones can be embedded. This is a subject of current interest and is not completely resolved (see [6, 2] for some recent results). We mention here that for dim $M = 3$ the answer is no as is seen by an example of L. Nirenberg [33].

If M, M' are two CR-hypersurfaces and $f \colon M \to M'$ is a smooth mapping, then f is a CR-*mapping* if $df|_{H(M)}$ is a complex-linear mapping from $H(M) \to H(M')$. This is a natural generalization of the notion of holomorphic mapping

g of $\mathbf{C}^n$ to $\mathbf{C}^n$ where dg is required to be C-linear at each point of $\mathbf{C}^n$ for g to be holomorphic. A CR-*equivalence* is simply a CR-mapping $f\colon M \to M'$ such that f^{-1} exists and is also a CR-mapping.

The first fundamental result in the local Poincaré equivalence problem as posed in §2 is that M and M' are equivalent in the sense of Poincaré if and only if they are CR-equivalent. This latter is an intrinsic notion, and this result was proved by Segre in 1931 [37] in $\mathbf{C}^2$ and by Tanaka [39] for $n > 2$. It is an application of the Cauchy-Kowaleski theorem, or the equivalent Cartan-Kähler theorem. Thus the extrinsic equivalence of Poincaré can be reduced to an intrinsic study of the geometry of the manifold itself. The invariants which are described below are all invariants of the intrinsic CR-structure, although some of them can be expressed in terms of extrinsic structure (like Moser's normal coordinates). Segre, in the same paper, showed that if the Levi form is nondegenerate then the local automorphism group of CR-mappings $\mathrm{Aut}(M)$ is finite dimensional, and if the Levi form is identically zero then M is equivalent to the hyperplane with its infinite-dimensional automorphism group, as was mentioned earlier. The finite-dimensional result used a deep finiteness criterion for a transformation group leaving invariant solutions to certain second order ordinary differential equations due to Tresse [42]. An example of this is the differential equation $y'' = 0$, whose solutions have graphs which are straight lines in the plane, and the groups of diffeomorphism of the plane which leave the straight lines invariant is the projective group which is a finite-dimensional Lie group. Segre also constructed for each hypersurface $M \subset \mathbf{C}^2$, a family of analytic varieties in $\mathbf{C}^2$ parametrized by the points of M and such that CR-equivalence of the hypersurface was characterized by the biholomorphic equivalence of these varieties. These "algebro-geometric invariants" were utilized later in Webster's solution of the global equivalence problem for ellipsoids [43].

One year later E. Cartan gave a complete solution to the equivalence problem in terms of differential-geometric invariants. He constructed a suitable bundle of frames B adapted to the CR-structure of $M \subset \mathbf{C}^2$ and a natural (Cartan) connection ω and curvature Ω on this bundle $B \to M$ such that M and M' are CR-equivalent if and only if there is a bundle equivalence $B \xrightarrow{\simeq} B'$ which preserves the connection (and hence the curvature). This is completely analogous to the construction of a connection and curvature on the orthonormal frame bundle of a Riemannian manifold, where the curvature and its covariant derivatives are a complete set of invariants for Riemannian geometry. The major difference is that the bundle B for the CR-geometry is a 2nd order frame bundle, i.e., is a bundle of frames for the total space of a line bundle L determined by the (first order) CR-structure, namely L is the 1 real-dimensional subbundle of the cotangent bundle annihilated by the holomorphic tangent vectors. This was carried out in lengthy detail in two long papers [12] with many applications and examples worked out. He obtained the finiteness results mentioned earlier independent of the lengthy Tresse monograph. In these papers is also a complete classification of the hypersurfaces for which $\mathrm{Aut}(M)$ is transitive, which can be reduced to solvable algebraic problems. Moreover, the curvatures are defined on the frame bundle, but for a generic class of hypersurfaces (called *nonumbilic*, where the lowest order curvature is not zero) one can construct 9 *scalar* invariants, analogous to Gaussian curvature (or more generally scalar curvature) for Riemannian surfaces. These are functions defined on M so that M and M' are equivalent if and only if the scalar invariants are pointwise preserved. These scalar invariants are of 7th and 8th order, i.e., depend on 7th and 8th order derivatives of the defining function. On the other hand if the lowest order curvature terms vanish identically, then Cartan showed that the surface was locally equivalent to the "local model", the hyperquadric. In fact, the curvatures measure the deviation from the local model. The nondegeneracy of the Levi form insured that there was a good approximating hyperquadric, and the curvatures measure the deviation, just as in Riemannian geometry the curvature measures the deviation from a first order approximation by a plane. Cartan worked entirely intrinsically (often using extrinsic quantities to represent the intrinsic differential forms, however) and did not concern himself too much with the nature of the embedding of M in $\mathbf{C}^2$, except in his description of examples.

4. Higher-dimensional results in the past two decades. More recently there has been a lot of activity by various mathematicians extending the Segre and Cartan solutions to the Poincaré equivalence problems for hypersurfaces in $\mathbf{C}^2$ to hypersurfaces in $\mathbf{C}^n$ as well as applications of these results. In 1962 Tanaka [39] generalized Cartan's differential-geometric invariants to a special class of hypersurfaces in $\mathbf{C}^n$ (equipped with a 1-parameter family of automorphisms of the structure). In 1965 Tanaka announced the solution to the general local equivalence problem in $\mathbf{C}^n$ [40]. The details appeared in a long paper in 1976 in a very general setting of which the CR-hypersurface equivalence problem was a special case [41]. In the meantime Chern and Moser in 1974 [14] presented a dual version of the differential-geometric solution to the equivalence problem which included two main constructions: (a) the principal bundle with Cartan connection and curvatures on an appropriate frame bundle just as in Cartan for $\mathbf{C}^2$ and equivalent to Tanaka's work mentioned above (Tanaka's work is less explicit for real hypersurfaces), (b) a normal form called the *Moser normal form* (cf. [32]) which is a normalization of the Taylor series of the defining function for a real hypersurface with nondegenerate Levi form near a point p_0 up to the action of the finite-dimensional Lie group acting on $\mathbf{C}^n$ which leaves the best approximating hyperquadric to M at p_0 fixed. For instance in $\mathbf{C}^2$, Moser's normal form looks like the hypersurface given by

$$v = F(z, u) = z\bar{z} + C_{42}z^4\bar{z}^2 + C_{24}z^2\bar{z}^4 + \sum_{k+l>7} C_{kl}z^k\bar{z}^l$$

where $z = x + iy$, $w = u + iv$ are coordinates for $\mathbf{C}^2$, and F is a real-valued function of (z, u), and the coefficients C_{kl} are functions of u. The coefficients are unique up to the action of the group acting on $\mathbf{C}^2$ given in (2.1). The coefficients $\{C_{kl}\}$ are "curvatures". If they vanish then the hypersurface is equivalent to a hyperquadric (patently obvious in the expansion above). These normal form coefficients are equivalent to the curvatures and higher covariant derivatives of the curvature being evaluated for a specific choice of frame compatible with the CR-structure (just as for normal coordinates in Riemannian geometry).

In 1975 Chern [13] generalized the Segre invariants to $\mathbf{C}^n$ and studied these invariants in the language of exterior differential systems. Cartan had remarked in the introduction to his paper [12] that Segre's invariants were not a complete set of invariants (while Segre had claimed they were!). In fact this point was clarified only recently by Faran [16] who showed that, for instance in $\mathbf{C}^2$, there is an example of two distinct CR-structures on a specific 3-dimensional manifold (which the Cartan invariants distinguish), but for which the Segre invariants coincide. Moreover, this is the worst discrepancy possible for the two systems of invariants. Similarly, in higher dimensions the distinction between the invariants is a discrete phenomenon. Thus *local* variation in CR-structures can be described by either type of invariants.

We now turn to applications of these local invariants to various versions of the global equivalence problem. First, Webster used the Segre type invariants to classify the set of all real ellipsoids in $\mathbf{C}^n$ with respect to CR-equivalence [43]. Burns and Shnider [9] gave a general classification of real hypersurfaces M in $\mathbf{C}^n$ where $\mathrm{Aut}(M)$ is transitive, generalizing Cartan's classification in $\mathbf{C}^2$. The scalar invariants of Cartan were generalized to $\mathbf{C}^n$ for nonumbilic hypersurfaces (the lowest order curvature invariants were not zero in an appropriate sense; see Burns, Shnider and Wells [10] and Webster [45]). In particular this involved a refinement of Moser's normal form. By using Thom transversality and the scalar invariants one finds that "almost all" deformations of a given CR-hypersurface are inequivalent CR-hypersurfaces and whose automorphism group consists only of the identity transformation [10]. Fefferman proved an important extension theorem [17] for strongly pseudoconvex domains with smooth boundary. Namely a biholomorphic equivalence of such domains extends to a CR-equivalence of their boundaries.[1] Thus the invariants on the boundaries in such an equivalence must be preserved. This gives a way of distinguishing different complex structures on the interior of such domains. In Burns, Shnider and Wells [10] it was shown by using the scalar invariants, Thom transversality, and Fefferman's extension theory that there is an infinite-dimensional family of perturbations of the boundary of the open unit ball in $\mathbf{C}^n$ (or any strongly pseudoconvex domain) so that the deformed domains are biholomorphically inequivalent and have no automorphisms. Johnson wrote down explicit algebraic examples of such deformations in [27], basing his construction on earlier results of Webster who studied invariants in specific geometric settings [45]. For instance, one finds in [27], if

$$p(z) = z_1^5 + z_1^3 + z_1^2 + z_1 z_2 + z_2 z_3 + z_3 z_4$$

---

[1]This result was recently improved significantly by S. Bell, who showed that a biholomorphic mapping from a strongly pseudoconvex domain to a domain with a smooth boundary extends to the boundary. In particular the image domain must be strongly pseudoconvex, which was not hypothesized.

is a polynomial in $\mathbf{C}^4$, and if we set

$$D_t = \{ z \in \mathbf{C}^4 \colon |z|^2 + t^2 |p(z)|^2 < 1 \}$$

for $0 \leqslant |t| < \varepsilon$, for ε sufficiently small, then D_t is a 1-parameter family of strongly pseudoconvex domains in $\mathbf{C}^4$ with D_t not biholomorphic to $D_{t'}$ for $t \neq t'$, and with $\mathrm{Aut}(D_t) = \{\mathrm{id}\}$, $t \neq 0$. However, as is clear from the definition, D_0 is the unit ball in $\mathbf{C}^4$ with $\mathrm{Aut}(D_0) = SU(4, 1)/(\text{finite subgroup})$ of real dimension 24. All of the D_t have real-analytic boundaries and are diffeomorphic to the unit ball. This sheds some light on the subtlety of the original global Poincaré equivalence problem. Recently, Greene and Krantz [22] showed that the set of strongly pseudoconvex domains with no automorphisms is open in the set of such domains in general, while the set of domains equivalent to a given domain is a closed set. This complements the result in [10] that domains with no automorphisms are dense, as discussed above.

In the study of CR-structures on hypersurfaces the Cartan type invariants are described in terms of a principal bundle, a connection, and a curvature, just as in Riemannian geometry. In the CR-geometry there is a distinguished family of real curves, called *chains*, which are analogous to geodesics in a Riemannian manifold. The chains are invariants of the CR-structure and defined by second order differential equations depending on the connection in an appropriate manner (Cartan [12], Chern and Moser [14]). For the hyperquadric these chains are simply the intersection of the real hypersurface with complex lines in the ambient space, although this is not true in general. Fefferman [18] showed that if M is a strongly pseudoconvex manifold, then there is a naturally defined Lorentz metric on $M \times S^1$ such that the projection of the light rays (for the Lorentz metric) on $M \times S^1$ to M are precisely the chains of the CR-structure of M.

In a different direction, and related to the extension theorem mentioned earlier, Fefferman [17] found an asymptotic expansion of the Bergman kernel of a strongly pseudoconvex domain D in terms of the distance to the boundary $\rho = d(z, \partial D)$. This involved powers of ρ and also a logarithmic term. In [19] Fefferman undertook a deep and not yet well understood study of the relations between the coefficients of the expansion of the Bergman kernel and the invariants of the CR-structure on the boundary. This is a complex story, and the ramification and meaning of the relationships developed has yet to be explored.

Two additional topics are worth mentioning which relate to the variation in CR-structures. First, Dippolito [15] has taken an initial step in a description of the (infinite-dimensional) moduli space for generic classes of CR-structures on a CR-hypersurface. He obtains "coordinates" for the moduli structure by looking at transversal sections to orbits under the appropriate Lie group of the Moser normal coordinates in a suitable jet bundle. In a second direction there is a relationship of variation of CR-structure to descriptions of real spacetimes in relativity theory. Namely Penrose observed that if H is a spacelike hypersurface in a Lorentzian 4-dimensional real vacuum spacetime S (i.e., the Ricci curvature of the Lorentzian metric ds^2 is zero), then one can associate to each point of S the 2-dimensional cone of light rays emanating from that point (the cones of tangent vectors annihilated by ds^2). The totality of light rays for variable points of H can be given the structure of a 5-dimensional real manifold equipped with an integrable CR-structure with nondegenerate Levi form. Thus, to each H there is an M_H with a full set of invariants of either Segre or Cartan type. As H varies in S (is evolved by Einstein's equations), the M_H varies in the space of moduli of such structures (parametrized suitably by its CR-invariants). This has been described by LeBrun [28a] explicitly and by Bryant [8], who describes this in Cartan's language of moving frames. This interplay between real spacetime and CR-structures is a refinement of an earlier relationship between variations in complex structure on a 3-dimensional open complex manifold in $\mathbf{P}_3(\mathbf{C})$ and holomorphic solutions of Einstein's self-dual vacuum field equations (Penrose [34]; see Atiyah [1], Wells [47, 48] for expository descriptions of this recent relationship between complex geometry and mathematical physics).

Perhaps this relation between CR-structures and spacetimes will have some positive effects on our understanding of relativity theory and the difficult question of the relation between gravity and quantum mechanics (cf. Penrose [34]). This would be a pretty state of affairs to have Poincaré having asked questions in an area of pure mathematics which in the long run had applications to the description of the universe we live in as he was himself very involved in such problems throughout his career as a mathematician, physicist, and philosopher.

References

1. M. F. Atiyah, *Geometry of Yang-Mills fields*, Lezioni Fermione, Acad. Naz. dei Lincei, Scuola Normale Sup., Pisa, 1979.

2. M. S. Baouendi and F. Treves, *A property of the functions and distributions annihilated by a locally integrable system of complex vector fields*, Ann. of Math. (2) **113** (1981), 387–421.

3. S. Bergman, *The kernel function and conformal mapping*, Math. Surveys, no. 5, Amer. Math. Soc., Providence, R.I., 1950.

4. H. Behnke and P. Thullen, *Theorie der Funktionen mehrerer komplexer Veränderlichen*, 2nd ed., Ergebnisse der Math. und ihrer Grenzgebiete Band 51, Springer-Verlag, Berlin-New York, 1970.

5. S. Bochner, *Analytic and meromorphic continuation by means of Green's formula*, Ann. of Math. (2) **44** (1943), 652–673.

6. L. Boutet de Monvel, *Integration des equations de Cauchy-Riemann induites formelles*, Séminaire Goulaouic-Lions-Schwartz (1974-75), Exposé No. 9, Centre Math. Ecole Polytech., Paris, 1975.

7. A. Brown, *On certain analytic continuations and analytic homeomorphisms*, Duke Math. J. **2** (1936), 20–28.

8. R. Bryant, *Space-times and CR-manifolds*, Trans. Amer. Math. Soc. (to appear).

9. D. Burns and S. Shnider, *Real hypersurfaces in complex manifolds*, Proc. Sympos. Pure Math., vol. 30, Amer. Math. Soc., Providence, R. I., 1977, pp. 141–168.

10. D. Burns, S. Shnider and R. O. Wells, Jr., *Deformations of strongly pseudoconvex domains*, Invent. Math. **46** (1978), 237–253.

11. C. Carathéodory, *Über das Schwarze Lemma bei analytischen Funktionen von zwei komplexen Veränderlichen*, Math. Ann. **97** (1926), 76–98.

12. E. Cartan, *Sur la géométrie pseudo-conforme des hypersurfaces de deux variables complexes.* I, Ann. Math. Pura Appl. (4) **11** (1932), 17–90 (or *Oeuvres.* II, 2, 1231–1304); II, Ann. Scuola Norm. Sup. Pisa, (2) **1** (1932), 333–354 (or *Oeuvres.* III, 2, 1217–1238).

13. S. S. Chern, *On the projective structure of a real hypersurface in* C_{n+1}, Math. Scand. **36** (1975), 74–82.

14. S. S. Chern and J. Moser, *Real hypersurfaces in complex manifolds*, Acta Math. **133** (1974), 219–271.

15. P. Dippolito, *Universal bundles for deformations of asymetric structures*, Trans. Amer. Math. Soc. (to appear).

16. James Faran, *Segre families and hypersurfaces*, Invent. Math. **60** (1980), 135–172.

17. C. Fefferman, *The Bergman kernel and biholomorphic mappings of pseudoconvex domains*, Invent. Math. **26** (1974), 1–65.

18. ______, *Monge-Ampère equations, the Bergman kernel, and geometry of pseudoconvex domains*, Ann. of Math. (2) **103** (1976), 395–416.

19. ______, *Parabolic invariant theory in complex analysis*, Advances in Math. **31** (1979), 131–262.

20. G. B. Folland and J. J. Kohn, *The Neumann problem for the Cauchy-Riemann complex*, Ann. of Math. Studies, no. 75, Princeton Univ. Press, Princeton, N. J., 1972.

21. H. Granert and R. Remmert, *Analytische Stellenalgebren*, Springer, Berlin-Heidelberg-New York, 1971.

22. R. Greene and S. Krantz, *Deformation of complex structures, estimates for the $\bar{\partial}$-equation, and stability of the Bergman kernel*, Adv. in Math. (to appear).

23. R. C. Gunning and Hugo Rossi, *Analytic functions of several complex variables*, Prentice-Hall, Englewood Cliffs, N. J., 1965.

24. F. Hartogs, *Zur Theorie der analytischen Funktionen mehrerer unabhängiger Veränderlichen insbesondere über die Darstellung derselben durch Reihen, welche nach Potenzen einer Veränderlichen fortschreiten*, Math. Ann. **62** (1906), 1–80.

25. L. Hörmander, *Linear partial differential operators*, Springer-Verlag, Berlin-Göttingen-Heidelberg, 1963.

26. ______, *An introduction to complex analysis in several variables*, North-Holland, Amsterdam, 1973.

27. David S. Johnson, *Biholomorphic equivalence in a class of graph domains*, Indiana J. Math. **29** (1980), 341–348.

28. A. Krzoska, *Über die naturlichen Grenzen der analytischen Funktionen mehrerer Veränderlicher*, Dissertation, Greifswald, 1933.

28a. C. R. LeBrun, Jr., *Spaces of complex geodesics and related structures*, Thesis, Oxford University, 1980.

29. E. E. Levi, *Studii sui punti singolari essenziale delle funzioni analitiche di due o più variabili complesse*, Annali di Mat. **17** (1909), 61–87.

30. H. Lewy, *On the local character of the solutions of an atypical linear differential equation in three variables and a related theorem for regular functions of two complex variables*, Ann. of Math. (2) **64** (1956), 514–522.

31. ______, *On hulls of holomorphy*, Comm. Pure. Appl. Math. **13** (1960), 587–591.

32. J. Moser, *Holomorphic equivalence and normal forms of hypersurfaces*, Proc. Sympos. Pure Math., vol. 27, Part 2, Amer. Math. Soc., Providence, R. I., 1975, pp. 109–112.

33. L. Nirenberg, *Lectures on linear partial differential equations*, CBMS Reg. Conf. Ser. in Math., no. 17, Amer. Math. Soc., Providence, R. I., 1973.

34. R. Penrose, *Nonlinear gravitons and curved twistor theory*, Gen. Relativity and Gravitation **7** (1976), 31–52.

35. H. Poincaré, *Les functions analytique de deux variables et la représentation conforme*, Rend. Circ. Math. Palermo **23** (1907), 185–220 (or *Oeuvres*. IV, 244–289).

36. W. F. Osgood, *Lehrbuch der Funktionentheorie*, 2nd. ed., vol. 2, part I, Teubner, Leipzig, 1929.

37. B. Segre, *Questioni geometriche legate colla teoria delle funzioni di due variabili complesse*, Rend. Sem. Mat. Roma **7** (1931), 59–107.

Interno al problema di Poincaré della rappresentazione pseudoconforme, Rend. Acc. Lincei **13** (1931), pp. 676–683.

38. F. Sommer, *Komplex analytische Blätterung reeler Mannigfaltigkeiten im* C^n, Math. Ann. **136** (1958), 111–133.

39. N. Tanaka, *On the pseudo-conformal geometry of hypersurfaces of the space of n complex variables*, J. Math. Soc. Japan **14** (1962), 397–429.

40. ______, *Graded Lie algebras and geometric structures*, Proc. U.S.-Japan Seminar in Differential Geometry, (Kyoto, 1965), Nippon Hyoronsha, Tokyo, 1966, pp. 147–150.

41. ______, *On non-degenerate real hypersurfaces, graded Lie algebras, and Cartan connections*, Japan. J. Math. **2** (1976), 131–190.

42. M. A. Tresse, *Détermination des invariants ponctuels de l'equation différentielle ordinaire du second order $y'' = w(x, y, y')$*, S. Hirzel, Leipzig, 1896.

43. S. M. Webster, *On the mapping problem for algebraic real hypersurfaces*, Invent. Math. **43** (1977), 53–68.

44. ______, *On the Moser normal form at a non-umbilic point*, Math. Ann. **233** (1978), 97–102.

45. ______, *The rigidity of CR-hypersurfaces in a sphere*, Indiana Univ. Math. J. **28** (1979), 405–416.

46. R. O. Wells, Jr., *Function theory on differentiable submanifolds*, Contributions to Analysis, Academic Press, New York, 1974, pp. 407–441.

47. ______, *Complex manifolds and mathematical physics*, Bull. Amer. Math. Soc. (N. S.) **1** (1979), 296–336.

48. ______, *Complex geometry in mathematical physics*, Univ. of Montreal Press (to appear).

DEPARTMENT OF MATHEMATICS, RICE UNIVERSITY, HOUSTON, TEXAS 77001